T0340061

Janello Torriani and the Spanish Empire

Nuncius Series

Studies and Sources in the Material and Visual History of Science

Series Editors

Marco Beretta (*University of Bologna*)
Sven Dupré (*Utrecht University / University of Amsterdam*)

VOLUME 2

The titles published in this series are listed at *brill.com/nuns*

Janello Torriani
and the Spanish Empire

*A Vitruvian Artisan at the Dawn of the Scientific
Revolution*

By

Cristiano Zanetti

BRILL

LEIDEN | BOSTON

This book has been published with a financial subsidy from the European University Institute.

Cover illustration: View of Toledo with Janello's water-lifting Device called el Ingenio by Georg Braun, *Civitates Orbis Terrarum*, 1598. Courtesy of the Biblioteca de la Fundación Juanelo Turriano, Madrid. Picture editing by Greta Ferrari.

Virtual reconstruction of one of the lines of Janello Torriani's Toledo Device https://www.youtube.com/watch?v=YqAdX6cO860&t=150s

The Library of Congress Cataloging-in-Publication Data is available online at http://catalog.loc.gov
LC record available at http://lccn.loc.gov/2017022615

Typeface for the Latin, Greek, and Cyrillic scripts: "Brill". See and download: brill.com/brill-typeface.

ISSN 2405-5077
ISBN 978-90-04-32089-5 (hardback)
ISBN 978-90-04-32091-8 (e-book)

Contents

Acknowledgments

This book is a revision of my doctoral thesis called *Janello Torriani* (*Cremona 1500 ca.-Toledo 1585*): *a Social History of Invention between Renaissance and Scientific Revolution*, defended on October 27th 2012 at the European University Institute in Florence. I would like to thank the members of the examining board Prof Maria Antonietta Visceglia, from the Università di Roma La Sapienza, Prof Mario Biagioli, UC Davis School of Law, and my second reader Prof Bartolomé Yun-Casalilla, EUI and now Universidad Pablo de Olavide of Seville. Special thanks go to my patient supervisor Prof Antonella Romano, now director of the prestigious Centre Alexandre Koyré.

The research for my thesis and postdoctoral investigation was mainly carried out at the library of the EUI, at the Galileo Museum in Florence, the National Library of Florence, the Biblioteca di Stato di Cremona, the Library of the Fundación Juanelo Turriano in Madrid, the Saxo Institute of the University of Copenhagen, the Veneranda Biblioteca Ambrosiana of Milan, The Medici Archive Project in Florence, the Max-Planck-Institut für Wissenschaftsgeschichte in Berlin, and the Centre Alexandre Koyré in Paris. I have also visited different archives in Italy, Spain, Germany and France. I want to thank all the passionate librarians and archivists that have helped me in my study. I am especially indebted to Almudena Pérez de Tudela (Patrimonio Nacional de España) who shared with me important documents about Janello.

I would like to thank my friend Bjørn Okholm Skaarup, who first directed me to work on this topic and who gave me the very laptop with which I am writing this book, and Theo Pronk and Gojko Barjamovic for encouraging me. A special thank you goes also to Daniel Crespo Delgado of the Fundación Juanelo Turriano, for his constant and ready help and to the president Bernardo Revuelta Pol and to all the other friendly members of this institution. I have also to thank the Spanish committee that decided to award me with the 7th Garcia-Diego International Prise for the History of Technology. I need likewise to mention Prof Sven Dupré Director at the Max-Planck-Institut of the research group *Art and Knowledge in Premodern Europe*, Prof Luca Molà (EUI) and Dr Alessio Assonitis, Director of The Medici Archive Project, for their precious help. I cannot forget the walks and conversations held with the archaeologist and anthropologist Juanma Rojas and his lovely wife Teresa while exploring the ancient buildings of Toledo. I also want to thank Tommaso Munari, Joanna Milstein, Agnete Lassen, Marius Buning and many other colleagues who have provided me with their precious knowledge, passion and support. I appreciate the valuable help of Gary "Omobono" Kissick, for correcting my English and suggesting improvements to the draft, and the Department of History and Civilisation of the EUI, which has generously provided the necessary resources to fund his job. I also thank Mino Boiocchi, Andrea Bossola and Mario Liguigli and all the other photographers and institutions who allowed me to make use of the images present in this book.

From September 2016 to January 2017, the City of Cremona and Unomedia gave me the opportunity to create an exhibition about Janello. Beside the organisers and sponsors, I would like to thank every single one of the 21,768 visitors that came to see it. Without the help of Cinzia Galli, Alessandro Maianti and Hora (Associazione Italiana Cultori Orologeria Antica) and its generous president Antonio Lenner, it would have not been possible to create this exhibition. A special credit goes also to the scholars that have participated in the conference connected to the exhibit: Alberto Lualdi, Anthony Turner, Günther Oestmann, Dietrich Matthes, Rocío Sánchez-Barrios, Tiemen Cocquyt, Michael Korey, Marisa Addomine, Daniele Pons, Hanoch Ben-Yami, Federica Favino and Giancarlo Truffa.

I would like to thank the people who have been closer to me all these years: my mom Daria, my dad Annibale, my sister Greta, and my uncles Alberto and Gianni, my aunts Giuli and Nene, my lovely cousin Andrea, and my friends Marco, Anne, Alberto and Roberto. Finally, I owe a large debt of gratitude to my beloved Marta, marvellous partner and personal librarian.

I apologise for the people I forgot to mention and that are entitled to be listed here: please consider it just a *lapsus memoriae*. Not to mention, all the errors remain mine.

List of Illustrations

List of Abbreviations

AGP	Archivo General de Palacio, Madrid
AGR	Archives Générales du Royaume, Bruxelles
AGS	Archivo General de Simancas
AHN	Archivo Histórico Nacional, Madrid
AN	Archives nationales, Paris
AHPM	Archivo Histórico de Protocolos de Madrid
AHPT	Archivo Histórico Provincial de Toledo
ASDCr	Archivio Storico Diocesano di Cremona
ASCr	Archivio di Stato di Cremona
ASFi	Archivio di Stato di Firenze
ASMi	Archivio di Stato di Milano
ASMn	Archivio di Stato di Mantova
ASPr	Archivio di Stato di Parma
ASVe	Archivio di Stato di Venezia
ASV	Archivio Segreto Vaticano

Introduction

This book wishes to address an audience interested in the history of science and technology, attracted by the charming milieu of the *Cinquecento* and concerned with excellent patterns of creativity. It will show the central role that artisans formed in the Vitruvian tradition played in demonstrating through practical mathematics (also called mixed or applied) such as mechanics, perspective, practical astronomy, navigation, surveying, construction and the use of scientific instruments, etc., an increasing and positive control over Nature, a step rooted in humanist culture and foundational for the understanding of those historical processes known as the Scientific and the Industrial Revolutions. The educational and professional trajectories of these artisans, and their nobilitation thanks to humanist models, are key tools to reconsider the *vexata quaestio* introduced by Zilsel and Needham: why did "modern science" develop in Western Europe?[1] Moreover, it is an ambition of this work to unveil the contextual cultural reasons behind the development of certain technologies, setting them free from the anachronistic *clichés* that tend to explain the construction of human science and technology within a narrative of alleged natural evolution. On the contrary, this work will show how now irrelevant and unexpected factors of techno-scientific development, such as astrology, medieval medicine, aristocratic prestige and princely authority, could act as paramount driving forces in the progress of early modern technology.

The protagonist of this book is Janello Torriani, known in the Spanish-speaking world as Juanelo Turriano (Cremona, Italy ca. 1500 – Toledo, Spain 1585), the greatest among Renaissance inventors and constructors of machines. No one, until the end of his life, had ever designed and built more complex and bigger machines than him. From the four corners of the Habsburg Empire, and beyond, contemporary literates and mathematicians celebrated Janello Torriani and his creations in their writings. Those who have bestowed upon us

---

1   The Austrian Marxist philosopher of science Edgar Zilsel (1891-1944) saw in the growth of capitalism the necessary condition for the development of what he calls the "superior craftsman", a concept that Janello Torriani's story will help to reconsider. Zilsel built his thesis explaining why "Modern Science" originated in the West on the relation between economy, technological education and competition. The famous British sinologist and historian of science Joseph Needham's question owes much to Zilsel's thesis: Needham wondered why Modern Science developed in the West, and not in India or China, where scientific knowledge had been superior to that of the Western between the first and the fifteenth centuries of the Christian era. Joseph. Needham, *The Grand Titration : Science and Society in East and West.* (London: Routledge, 1969).

the memory of the greatest technological accomplishments of their time, and of their mysterious demiurge, range from the heroes of the Spanish *Siglo de Oro* – such as Miguel de Cervantes, Luis de Góngora, Juan de Herrera, Sebastián de Covarrubias, Francisco de Quevedo, Pedro Calderón de la Barca and Lope de Vega – to the Italians Gerolamo Cardano, Giovanni Paolo Lomazzo, Federico Zuccari, Federico Borromeo and Muzio Oddi, and from the Flemish Gulielmus Zenocarus, Gerhard Mercator and Jehan L'Hermite to the German Johannes Kepler and the English John Dee, to name just a few. Even the first book from China on European technology, published in the seventeenth century, was inspired by the amazement that news of the giant waterworks machine of Toledo by Torriani had provoked in that distant land.[2] It is striking how such fame turned into nearly complete oblivion, leaving only a few clues of a blurred and distorted memory dispersed here and there.

Oblivion and Rediscovery

Janello Torriani was very famous in his own time, but he is rather obscure today.[3] There are several reasons that explain why. First of all, one has to consider that Torriani was the court clockmaker, engineer, mathematician and inventor of Charles of Habsburg (fifth emperor, and first king of Spain of this name) and of Philip II of Spain. Political agendas originated in countries that since the sixteenth century desired to damage the image of Spain, and even political programs within Spain itself, especially in the last two centuries, have produced a harsh debate on the contribution of the Spanish monarchy to the development of modern science and technology. Janello, as the highest representative of the excellence of technology produced under Spanish royal patronage, was therefore a target for these biased and deceptive narratives. The so-called *Leyenda Negra*, or Black Legend, depicted Spain as a retrograde force of history, isolated and obsessed with religion, downplaying and ignoring its role in the development of science and technology during the early modern

---

2   Albert König, "Giulio Aleni sj (1582- 1649) and the Introduction of Western Water Supply Methods in 17th Century China," in *Proceedings of WWAC2016* (4th IWA International Symposium on Water and Wastewater Technologies in Ancient Civilisations, Coimbra, 2016).

3   "Giannello Torriano da Cremona – il più popolare orologiaio che si conosca [...] Tanto più strano e doloroso è constatare che una figura di primo piano come il Torriano, non abbia invogliato qualche storico del nostro tempo a intensificare le ricerche d'archivio": Enrico Morpurgo, ed., *Dizionario degli orologiai italiani: 1300-1800* (Milano: Nicola De Toma, 1974).Here the historian of clocks Enrico Morpurgo lamented the fact that during his time, despite Janello Torriani's fame, no historian had yet conducted intense archival research on him.

period. On the other hand, Spanish nationalists tried to represent an exaggerated portrait of the Spain of the *Siglo de Oro*, or Golden Age, as the precursor of all scientific and technological novelties.[4] In both cases, Janello Torriani was marginalised: the Black Legend could not take into account a man that demonstrated the technological superiority of the Spanish Crown, whereas Spanish nationalists could not accept that a foreigner came to Spain to bring knowledge that was not yet there.[5]

Moreover, the grand narratives in the history of science did not usually take craftsmen as Janello Torriani into account. Starting in the 1930s, Alexandre Koyré inaugurated a new stage in the historiography of science and was the first to coin the phrase "Scientific Revolution".[6] Koyré's view has been for a long time mainstream: it assumed that philosophers were responsible for the Scientific Revolution, and that they adopted the practical knowledge of the craftsmen, who were viewed merely as passive innovators. This historiographical tradition focused on those intellectual "contributors" who provided the bricks with which to build the idea of a progressive science based on a teleological and absolute truth. Ideological agendas tended to marginalise the role of artisans, considered instead by certain Marxist historians and philosophers of science not just as representative of the economic production that in Marx's

---

4   Víctor Navarro Brotóns and William Eamon, *Mas allá de la Leyenda Negra : España y la revolución científica = Beyond the Black Legend : Spain and the scientific revolution* (Valencia: Instituto de Historia de la Ciencia y Documentación López Piñero : Universitat de Valéncia : C.S.I.C., 2007); James Delbourgo and Nicholas. Dew, *Science and Empire in the Atlantic World* (New York, NY: Routledge, 2008); Daniela Bleichmar, ed., *Science in the Spanish and Portuguese Empires, 1500-1800* (Stanford, CA: Stanford University Press, 2009); María Luz López Terrada, José Pardo Tomás, and John Slater, eds., *Medical Cultures of the Early Modern Spanish Empire*, New Hispanisms : Cultural and Literary Studies (Farnham, Surrey: Ashgate, 2014). More specifically: Antonio Barrera-Osorio, *Experiencing Nature: The Spanish American Empire and the Early Scientific Revolution* (Austin, TX: University of Texas Press, 2006); Jorge Cañizares-Esguerra, *Puritan Conquistadors: Iberianizing the Atlantic, 1550-1700* (Stanford, CA: Stanford University Press, 2006); María M. Portuondo, *Secret Science: Spanish Cosmography and the New World* (Chicago: The University of Chicago Press, 2009); Bjørn Okholm Skaarup, *Anatomy and Anatomists in Early Modern Spain* (Farnham [u.a.: Ashgate, 2015).
5   See the works by Garcia-Tapia cited in this book and Jorge Cañizares-Esguerra, "Iberian Science in the Renaissance: Ignored How Much Longer?," *Perspectives on Science* 12, no. 1 (Spring 2004).
6   In 1939, the French philosopher Koyré introduced such a concept for the first time, soon followed by Butterfield and Hall: Edward Grant, *The Foundations of Modern Science in the Middle Ages: Their Religious, Institutional, and Intellectual Contexts* (Cambridge: Cambridge University Press, 1996), xii; A. Rupert Hall, *The Scientific Revolution, 1500-1800 : The Formation of the Modern Scientific Attitude* (London: Longmans, Green, 1954).

view drove history, but also as subaltern heroes who mirrored the twentieth century working class in its struggle for emancipation. Already at the beginning of the 1940s the Jewish Austrian philosopher of science Edgar Zilsel had singled out a certain type of Renaissance artisan, whom he labelled as "artist-engineer" or "superior craftsman", whom he considered a precursor of the Scientific Revolution:

> Beneath both the university scholars and the humanist literati there were some groups of superior craftsmen who needed more knowledge for their work than their colleagues did. The most important of them may be called artist-engineers, for not only did they paint their pictures, cast their statues, and build their cathedrals, but [they] also constructed lifting-gears, earthworks, canals and sluices, guns and fortresses, found new pigments, detected the geometrical laws of perspective, and invented new measuring tools for engineering and gunnery ... They are the immediate predecessors of science. The two components of scientific method were still separated: methodical training of intellect was preserved for upper-class learned people, for university-scholars and humanist literati; experiment and observation were left, more or less, to plebeian workers. Real science is born when, with the progress of technology, the experimental method of the craftsmen overcomes the prejudice against manual work and is adopted by rationally trained university-scholars. This is accomplished with Galileo (1564-1642).[7]

Within the scholar-craftsman debate generated by Zilsel, several historians have tried to demonstrate the importance of artisans in the processes that led to the Scientific Revolution. Thanks to empirical and scholarly observation, the appreciation of technology and its practitioners is now well established in historiography.[8] We shall soon go back to this problem, reconsidering Zilsel's categories, and anticipating his chronology.

---

7   Edgar Zilsel and Joseph Needham, *The Social Origins of Modern Science*, ed. Diederick Raven, Wolfgang Krohn, and R. S Cohen, 1942 as The Sociological Roots of Science (Dordrecht; Boston: Kluwer Academic Publishers, 2000), 4-5.

8   Besides Marxist figures such as Boris Hessen, Henryk Grossmann, and other scholars endorsed the idea that there had been an artisanal influence in the making of the Scientific Revolution: among them we can remember Leonardo Olschki and Robert Merton. For an overview on this issue, see chapter I of Pamela Long's *Artisan/practitioners and the Rise of the New Sciences, 1400-1600* (Corvallis, OR: Oregon State University Press, 2011); and the introduction by Antonella Romano of the catalogue of the 2016-17 exhibition on Torriani held in Cremona: *Janello Torriani, a Renaissance genius*. Cristiano Zanetti, ed., *Janello Torriani, a Renaissance Genius*

The marginalization of Torriani is also based upon a lack of physical evidence of what once made him universally famous. It has been said of Brunelleschi that "the contemplation of his creations alone says enough about him".[9] The opposite may be true of Janello Torriani: nothing of his great creations is left, despite the fact that during his lifetime, as seen before, in and beyond Europe the machines he built for the Emperor and his son the King of Spain, the first two "global rulers", were beyond comparison. As in an archaeological layer, only the negative imprint left by the great volume these lost technological wonders once occupied points to the role Torriani's inventions have in the history of technology. Under Habsburg patronage, Torriani had created a number of technological devices that were hailed by his contemporaries as mechanical marvels. These included the first known machine-tool to cut gears, something he used to create the most complex and compact machine ever seen up to his days: the Microcosm. This was a planetary automaton, enriched by almost two thousand mechanical components; moreover, he built the Crystalline, another planetary clock enclosed in a rock-crystal case so that the eye could see the intricate clockwork in motion; and, above all, he created the first giant machine of history, the Toledo Device, a 300 meters-long complex structure that could elevate water over a slope of about 100 meters. Besides these amazing achievements, Torriani also participated in the Gregorian reform of the calendar, contributing a tract and mathematical instruments for calculus, and he served the Crown of Spain, and other minor patrons, with waterworks surveying, bell designing and casting, celestial observations, and the production of automata and other curious creations, such as a combination lock, ingenious applications of gimbals, hydraulic pumps, portable mills, and scientific instruments.

Janello Torriani's story is at the same time exceptional and exemplary. On the one hand, it represents the distinctive path of the most impressive constructor of machines of the Renaissance. On the other, we can also consider Torriani as a paradigmatic figure: he perfectly represents the traditional categories of the

---

(Cremona: Comune di Cremona, 2016), 10-12. See also A. Rupert Hall, "The Scholar and the Craftsman in the Scientific Revolution," in *Critical Problems in the History of Science*, ed. Institute for the History of Science and Clagett (Madison: University of Wisconsin Press, 1969); and Paolo Rossi, *I filosofi e le macchine, 1400-1700*, [1962] (Milano: Feltrinelli, 2007); Eugenio Garin, *Umanisti, artisti, scienziati: studi sul rinascimento italiano* (Roma: Riuniti, 1989); and Pamela H Smith, *The Body of the Artisan: Art and Experience in the Scientific Revolution* (Chicago: University of Chicago Press, 2004).

9   Paolo Galluzzi, "Dall'artigiano all'artista-ingegnere: Filippo Brunelleschi uomo di confine," in *Arti Fiorentine. La grande storia dell'artigianato*, ed. Franco Cardini and Riccardo Spinelli, vol. 1, Il Medioevo, 1998, 285.

"Renaissance genius" and "polymath", the "superior craftsman" and the "court artisan", all problematic definitions that call for a close analysis. This book wants to retrieve from oblivion one of the most important and forgotten actors of Renaissance technology, and at the same time, by following his trajectory, this inquiry attempts to partially deconstruct the above-mentioned categories, rethinking them, and to explore how innovative knowledge was created, and how technological invention and innovation were practiced and how they circulated in sixteenth-century Europe.

My research is informed by the work of that group of historians who, since the time of Edgar Zilsel, and even more intensively in the last decades, have attempted to address the concerns raised by social and cultural issues in the history of science and technology. This book aspires to be a contextualised biography, in which unfold the education, social rise, transnational mobility, and multifaceted activities of an amazingly skilled artisan, located between theoretical knowledge and practice, between urban space – where he had his guild-controlled workshop, and where he was invested with mathematical public offices – and the court.[10] In this latter context, since the appearance

---

10   In so doing, I have been particularly inspired by the new biographical genre that aims
     to use an individual life to describe and analyse social and cultural changes. In recent
     years scholars have shown an increasing interest in the problem of biography as a his-
     toriographical genre. Although French and German sociologists and social historians
     traditionally attributed scientific value only to macro-analysis of structures through
     quantitative examination, and rejected biography as anecdotal and event-based, the rise
     of microhistory in Italy during the 1970s showed a way of integrating the level of daily
     life into social history. Some historians have even talked of a "biographical turn". See for
     example Jacques Revel, *Giochi di scala: la microstoria alla prova dell'esperienza* (Roma:
     Viella, 2006); Hans Erich Bödeker, ed., *Biographie schreiben* (Göttingen: Wallstein, 2003);
     Volker R Berghahn and Simone Lässig, *Biography between Structure and Agency: Central
     European Lives in International Historiography* (New York: Berghahn Books, 2008); and
     recent monographs such as: Nigel Hamilton, *Biography: A Brief History* (Cambridge, Mass.:
     Harvard University Press, 2007); Barbara Caine, *Biography and History* (Basingstoke; New
     York: Palgrave Macmillan, 2010); Michael Rustin, "Reflections on the Biographical Turn in
     Social Science," in *The Turn to Biographical Methods in Social Science: Comparative Issues
     and Examples*, ed. Prue Chamberlayne, Joanna Bornat, and Tom Wengraf (London; New
     York: Routledge, 2000), 33-52. Mary Terrall, "Biography as Cultural History of Science," *Isis*
     27 (2006): 306-13; Lorraine Daston and H. Otto Sibum, "Introduction: Scientific Personae
     and Their Histories," *Science in Context* 16, no. 1 (2003): 1-8. See also my recent contribution
     Cristiano Zanetti, "Videmus nunc per speculum in aenigmate : ¿y si además miramos con
     una lupa? La biografía en la Historia de la Ciencia y de la Tecnología," in *La historia biográ-
     fica en Europa: nuevas perspectivas*, ed. Isabel Burdiel and Roy Foster, Historia global 7
     (Zaragoza: Institución Fernando el Católico, 2015), 119-44.

of Norbert Elias' studies in historical sociology,[11] some scholars, such as Bruce T. Moran,[12] have shifted the focus of their analysis in the history of science and technology to courtly structures. Within this perspective, the problem of "patronage" has become a central issue and a necessary field of analysis for those who desire to understand *ancien regime* society, its economy, its dynamics and its rules; in a word, its culture.[13] This book aims to demonstrate how the career of a successful Renaissance engineer offers an extremely effective means of analysing these patronage structures, and to understand how princely patronage influenced technological evolution: the problem of technological innovation and of the professions involved with it, especially within the fields of micro- and macro-mechanics, has been explored in relation to "power". Another structural analytical tool that informs my inquiry is "network" and the

---

11   Norbert Elias, *The Court Society* (New York: Pantheon Books, 1983).

12   Bruce T Moran, "German Prince-Practitioners : Aspects in the Development of Courtly Science, Technology, and Procedures in the Renaissance," *Technology and Culture* 22 (1981): 253-74; and "Princes, Machines and the Valuation of Precision in the 16th Century," *Sudhoffs Archiv* 61 (1977): 209-28.

13   Ronald G Asch and Adolf M Birke, *Princes, Patronage, and the Nobility: The Court at the Beginning of the Modern Age, ca. 1450-1650* ([London, England]; Oxford [England]; New York: German Historical Institute London ; Oxford University Press, 1991). Anthony Molho, in an essay on Cosimo de Medici the Elder, has given a forceful example of such a structure in the fifteenth-century Italian context, highlighting the complexity and the brokerage function of patronage: Anthony Molho, "Cosimo de Medici: 'Pater patriae' or 'Padrino?,'" *Stanford Italian Review* I/I (1979). In defining "patronage", I borrow the definition R. Weissman attempted to standardise in 1987 for the Mediterranean world. Regarding Weissman's geographical distinction, I could not find a diversified patronage system in the Habsburg Empire stretching from the Netherlands down to Sicily passing through Burgundy, Germany, Italy and Spain. This Mediterranean geographical definition may have more to do with Weissman's perception of continuity between Ciceronian ancient Roman models of networks of power and a more recent view of the Mediterranean as being more corrupt than central and northern European countries. According to Weisseman, Mediterranean patron-client relationships share the following attributes: first, there exists an inequality of power or resources between patron and client; second, patronage is a long-term relationship, with a moral or social rather than a legal basis; third, a patronage relationship is not restricted to a single kind of transaction, it is multi-standard and multipurpose; fourth, patronage is a relationship in which the patron provides more than simple protection. He provides brokerage, mediation, favours, and access to networks of friends of friends; and, fifth, Mediterranean patronage has a distinctive ethos, standing outside the officially proclaimed social morality. Ronald F. E Weissman, "Taking Patronage Seriously: Mediterranean Values and Renaissance Society.," in *Patronage, Art, and Society in Renaissance Italy*, ed. Francis W. Kent and Patricia Simons (Canberra: Huanities Research Centre/Clarendon Press, 1987), 25.

related process of "circulation". These concepts allowed me to investigate how and why this man from the periphery of the empire could become one of the closest companions of Emperor Charles V.

Other processes highly relevant for the narrative of this book are "education" and "ennoblement". The success of people like Janello Torriani, with his late but remarkable professional blossoming, did not spring from mere chance, but from educated talent. This book will illustrate how Torriani's education was strictly dependent on humanist culture. Humanist pedagogy, the rise of mathematical public offices in the medieval Italian city republics and in nu Renaissance courts, allowed artisans to draw from different disciplines in order to perform invention – or, to put it better, innovation

Paradoxically, despite this book seeking to avoid adopting the panegyric agenda of celebrative biography, the myth of Janello Torriani was the starting point of my investigation: following the clockmaker's entrance into the Habsburg sphere, a series of acclamatory materials, both visual and literary, was created, giving the historian a generous quantity of evidence to analyse. These abundant documentary traces – administrative, narrative, poetic and visual – have enabled me to raise questions about sixteenth-century technological innovation, its practitioners and their relation to power. I have looked at sculptures, medals, poems, paintings and literature celebrating Janello not simply as tokens of success, as they were "natural" reactions to his outstanding creations; my view is that they were part of a language spoken mainly at court, a language which was articulated for specific reasons: as such, those signs should be read according to a syntax of power in which the artisan negotiated a higher social status and his patrons exploited his ingenuity to increase their prestige, or glory, as Jacob Burckhardt and Edgar Zilsel would have it.[14] Janello Torriani and his powerful brokers and patrons made use of his successful creation of mechanical devices, as they were "instruments of credit"[15] to build and strengthen, together with the craftsman's reputation as a unique creator, "*the prince among the architects of clocks*" – as Charles V called him – their prestigious humanist role as protectors of a new Archimedes.

As previously seen, Janello Torriani's marvelous creations had a relevant impact on early modern Spanish literature, and one can find still today a revival

---

14   Paul Zanker, *The Power of Images in the Age of Augustus* (Ann Arbor: University of Michigan Press, 2002); and David Freedberg, *The Power of Images: Studies in the History and Theory of Response* (Chicago: The University of Chicago Press, 1989).

15   Mario Biagioli, *Galileo's Instruments of Credit: Telescopes, Images, Secrecy* (Chicago: University of Chicago Press, 2006).

of a production of novels inspired by the clockmaker and his myth.[16] Besides poems and novels, there are even folk sayings connected to Janello surviving in different parts of Spain and Latin America. Indeed, in Spain, until some decades ago, one could still hear the phrase *"Eres un Juanelo!"* meaning *"How smart you are!"*,[17] while in Central and Southern America people continue to use the saying *"el huevo de Juanelo"*, first recorded in the seventeenth century by Calderón de la Barca.[18] This is an adaptation of the story relating to Brunelleschi's egg that in Italy later became Columbus's egg.[19] It is not rare to

---

16   See the contribution Daniel Crespo Delgado, "Juanelo Turriano: Janello Torriani in Spanish Literature," in *Janello Torriani, a Renaissance Genius*, ed. Cristiano Zanetti (Cremona: Comune di Cremona, 2016); of the same author see: "Juanelo Turriano: Genius and Fame," in *Renaissance Engineers*, ed. Alicia Cámara Muñoz and Bernardo Revuelta Pol, English edition 2016, Juanelo Turriano Lectures in the History of Engineering (Madrid: Fundación Juanelo Turriano, 2016), 9-24. See also the anthology edited by Luis Moreno Nieto and Ángel Moreno Santiago, *Juanelo y su artificio: antología* (Toledo: D.B. ediciones, 2006); José Cristóbal Sánchez Mayendía, "El artificio de Juanelo en la literatura española," *Cuadernos Hispanoamericanos* nº 103 (1958): 73-93; and the novels Klaus E Erichson, *Juanelo, der Schmied von Toledo Roman* (Norderstedt: Books on Demand, 2011); Jesús Ferrero, *Juanelo O El Hombre Nuevo*, Punto de Lectura (Madrid: Alfaguara, 2000); Baltasar Magro, *El Círculo de Juanelo* (Madrid: Brand Editorial, 2000); Joaquín Valverde Sepúlveda, *Juanelo Turriano: el relojero del emperador* (Madrid: Rubiños 1860, 2001); Antonio Lázaro, *Memorias de un hombre de palo* (Madrid: Santillana, 2009); Gilimón Gaetano Blancalana, *Disertaciones Y Opúsculos Sobre Toledo* (Toledo: Celya, 2011); José Antonio Ramírez Lozano, *El relojero de Yuste: los últimos días de Carlos V* (La Coruña: Ediciones del Viento, 2015).

17   I thank Daniel Crespo Delgado of the Fundación Juanelo Turriano for this information. Daniel informed me that when his mother used to go to school, smart people were addressed in this way.

18   Ana María Carabias Torres, ed., *Las Relaciones entre Portugal y Castilla en la época de los descubrimientos y la expansión colonial* (Salamanca, España: Ediciones Universidad de Salamanca, Sociedad V Centenario del Tratado de Tordesillas, 1994), 173; Milton A Buchanan, "Short Stories and Anecdotes in Spanish Plays," *The Modern Language Review* 4, no. 2 (1909): 183.

19   The story which inspired the latter saying was that of Filippo Brunelleschi and his refusal to show his model for the dome of Santa Maria del Fiore in Florence; fearing the idea might be stolen, the architect devised a stratagem to challenge his competitors: "he who could make an egg stand firmly on the smooth marble, would by doing show his skill and construct the dome. And an egg being brought, all the masters tried to make it stand upright, but none found the way. When the egg was given to Filippo, he picked it up gracefully and hitting it on its base upon the marble floor ... made it stand upright. Rumbling the architects that similarly they would have been able to do that, Filippo answered them, laughing, that would they have still been able to curve the dome, seeing previously his model and drawing. And so it was resolved that he should be charged with conducting

travel across Castile and Extremadura and find people attributing without basis hydraulic systems and sundials to Janello. Even the monumental monolithic columns put up in Franco's time at the sides of the street leading to the controversial monument of *El Valle de los Caídos*, columns that had been lying in the countryside for centuries and whose origin remains a mystery, are commonly believed to have been creations of Janello, and are indeed called "*los Juanelos*".

This image of Janello Torriani as an ingenious inventor, sometimes distorted into a necromancer, is especially rooted in Toledo, where his waterworks with their imposing physical presence in the metropolitan space revealed Torriani's rare skills to the admiration of a wide section of society. From the time he built these machines, and in the decades that followed his death, the Toledo Device became what today we would define as a "tourist attraction". Literary production targeting an audience that went far beyond the court repeatedly mentioned this machine, inviting visitors to Toledo to admire it. Once it had disappeared, the memory of this incredible machine was turned into a phantasmal legend of a wooden automaton that used to collect food at the Archbishop's palace for his old and poor master Janello.[20] In our own days, landscapes in Madrid, Toledo and Cremona carry the blurred and mysterious memory of Janello Torriani,[21] and his appearance in local historiographies (Spanish and Lombard) as a genius worthy of celebration, reflecting the excellence of the *genius loci*, bringing honor and prestige to both his ancestral and adoptive countries, calls out for a trans-national investigation, which points to two different and largely independent traditions. For the first time, the two historiographies and already published and unpublished archival documents from both Italy and Spain are integrated into an organic narrative, which represents one of the most interesting technical and scientific careers of the early modern period.

---

this work". Giorgio Vasari, *Le opere di Giorgio Vasari, pittore e architetto aretino*, vol. Parte 1, Volume 1 (Firenze: David Passigli e Soci, 1832), 259.

20   Zanetti, *Janello Torriani, a Renaissance Genius*.

21   In Madrid *calle de Juanelo* was named after Torriani even during his lifetime; in fact, he had a house there. In Toledo and Cremona political administrations chose to name two streets and two high schools after him. In Madrid, a foundation supporting research into the history of technology was named after him. The founder of this cultural institution, the philanthropic engineer José García-Diego, wrote a biography on Torriani in the 1980s that was of great inspiration and assistance for my own work. The Fundación Juanelo Turriano has published a copious number of Spanish documents related to Janello Torriani, of which I have made abundant use.

The Vitruvian Artisan between Renaissance and Scientific Revolution[22]

Janello Torriani's story is nestled in the interwoven chronological strands of the Renaissance and the Scientific Revolution. These two historiographical ideas, that Westerners traditionally consider as crucial *momenti* in the making of our present world, have a common conceptual ancestry: more than simple chronological markers, these concepts are tools that historians had used to highlight problems of origins, continuity and ruptures in knowledge, often having searched for the mirage of the dawn of Modernity. The very concept of Modernity is a projection cast upon something in the past whose quality or shape resembles structures or institutions that we may recognise as specifically belonging to us. Perhaps paraphrasing Benedetto Croce, who stated that history is always contemporary, in his famous book on *The Scientific Revolution*, Steven Shapin observed: "the people, the thoughts, and the practices we tell stories about as 'ancestors,' or as the beginning of our lineage, always reflect some present-day interest".[23]

In 1940 Ferguson referred to the Renaissance as "the most intractable problem child of historiography".[24] The child has now grown old, but shows no sign of recovery. The concept of the Renaissance has changed over time in terms of its function within the intellectual *milieu*, bias and agendas of historians, giving birth to one of the most problematic fields of historiography. Torriani embodies the elements that Burckhardt observed and described as the foundations

---

22   With the concept of "Vitruvian artisan" I try here to identify a most distinctive group within what Edgar Zilsel had called "superior craftsmen" and "artist-engineers". See the last paragraphs of this introduction.

23   Steven Shapin, *The Scientific Revolution* (Chicago, IL: University of Chicago Press, 1996), 7.

24   Wallace K Ferguson, *The Renaissance* (New York: H. Holt and Co., 1940), 7. Ferguson's comment was a play on Burckhardt's earlier formulation: "In the character of these States, whether republics or despotisms, lies, not the only, but the chief reason for the early development of the Italian. To this it is due that he was the firstborn among the sons of modern Europe". See Jacob Burckhardt, "The Development of the Individual," in *The Civilization of the Renaissance in Italy* (Mineola, NY: Dover Publlications, 2010), 82-103.In recent years, there has been a vigorous debate as to whether to consider the Renaissance as a period, as a cultural movement, or as a style. Each of these three conceptualizations of the Renaissance has complex chronological limits: historians have adopted various different boundaries according to their analytical foci. Recently, Marguerite L. King has suggested a lattice in order to embrace all different conceptions of the Renaissance. She has gathered all different mainstream positions in three categories: small, medium, and large-sized Renaissance: Margaret L. King, *The Renaissance in Europe* (London: Laurence King Publishing, 2003), viii-xii..

of what he theorised as modern man: the manifestation of the individual, his desire for glory, the power of will, and the expression of a remarkable wit. The contemporary array of representations of Janello Torriani as an individual with unique skills and personality overlaps with the problem of Renaissance individualism or selfdom.[25] Beside the written material in which the clockmaker's personality emerges, Torriani appears to be the first artisan coming from the populace to be celebrated during his lifetime – and therefore to have been bestowed upon us – with the full humanist linguistic armory: medals, paintings, sculpture, assimilation to Classical icons, eulogy and poetry. As we shall see, another focal point in Torriani's story that belongs with the problem of the Renaissance is the role played by humanist culture in creating the necessary conditions for a constant specular game played with the myth of Classical Antiquity. It is a statement of this book to consider as an essential step to the transition to a new science the role played by artisans like Janello, who were formed in humanist classical culture.

However, the concept of Humanism, for me so essential to understand Janello Torriani's education and career, has also been the object of intense debate in historiography. Some historians see in Humanism a philosophical movement (for instance Eugenio Garin and Hans Baron), while others, such as Paul Oskar Kristeller, consider it a rhetorical style. However, Kristeller's authoritative view seems not to be able to confute completely the philosophical essence of what Hans Baron has conceptualised as "Civic Humanism", i.e., a moral philosophy fashioned in Florentine society (and especially in the chancery of Coluccio Salutati and Leonardo Bruni) consequent to the dramatic siege of the city by the despot of Milan at the beginning of the fifteenth century. I here consider Humanism as a moral philosophy that goes beyond Florence, rooted in medieval Italian city republics' administration and in Italian medieval universities, and that aims to readapt ancient Roman values (real, rhetorical or mythical) to Christian modernity. The power of Humanism was, in my eyes, the capability of empowering any discourse (even politically opposite ones, such as republicanism and despotism) with the prestige of Antiquity. The power of its rhetoric was able to influence a society highly receptive to ancient Roman

---

25   The concept of individualism has been widely criticised and reformulated by scholarship. Hans Baron, "The Limits of the Notion of 'Renaissance Individualism': Burckhardt after a Century," in *In Search of Florentine Civic Humanism: Essays on the Transition from Medieval to Modern Thought*, [1960, then modified and enlarged in 1973], vol. II (Princeton, N.J.: Princeton University Press, 1988), 155-81; John Jeffries Martin, "The Myth of Renaissance Individualism," in *A Companion to the Worlds of the Renaissance*, ed. Guido Ruggiero (Malden, MA: Blackwell Publishers, 2002), 209-24; Stephen Greenblatt, *Renaissance Self-Fashioning: From More to Shakespeare* (Chicago: University of Chicago Press, 1980).

glory.[26] Models coming from Vitruvius, Archimedes, Aristotle and many other popular writers from ancient Rome and Greece provided a powerful tool for change: through the stages of philological reconstruction of a text, its assimilation through testing the knowledge it conveyed, its consequent amendment and eventual implementation or even conscious surpassing of it.

Humanist values are not just here considered as causes of Janello's formation, but also as rhetorical filters we need to be aware of: this confrontation with classical culture emerges constantly in Janello's life, from his education to his professional objectives and to the way he self-fashioned his public image. This book aims to move beyond the myth of the categories of "Renaissance genius" and "Renaissance polymath", deconstructing them.[27] Katharine Park and Lorraine Daston have written that: "The multifaceted 'Renaissance man' is to some extent a trick of historical perspective, which creates polymathesis out of what was simply a different classification of knowledge and a different professional division of labor." For this reason, "early modern career trajectories can often appear to modern eyes at once dazzlingly diverse and oddly

---

26   Hans Baron, *The Crisis of the Early Italian Renaissance : Civic Humanism and Republican Liberty in an Age of Classicism and Tyranny*, 2 vols. (Princeton, N.J.: Princeton University Press, 1955); Eugenio Garin, *Medioevo e rinascimento: studi e ricerche*, 2a ed, Biblioteca di cultura moderna 506 (Bari: Laterza, 1961).

27   On the 29th of January 2017 the exhibition *Janello Torriani, genio del Rinascimento* (i.e. in English: *Janello Torriani, a Renaissance Genius*) closed its doors with the unexpected success of almost 22,000 sold tickets. Together with Professor Cinzia Galli, I had curated this exhibition, basing it on my PhD thesis defended in 2012 at the European University Institute. My idea was to name the exhibition *"Genius ex Machina"*, a play on words on the popular Latin expression *Deus ex machina*. In my eyes, this title was able to convey the idea of somebody capable as Torriani to solve mechanical problems that nobody else was able to overcome succesfully, and at the same time to transmit an idea of artificiality to the role played by Janello as a genius. My PhD thesis, like the named exhibition and the present book, aims to reject the idea of "genius" as an inborn phenomenon and tries instead to unfold the educational processes that shaped his skills. However, the press office of the exhibition decided to change the title with the excuse that "Janello Torriani, a Renaissance Genius" was going to be better understood by a general audience. Despite my useless protests that such a title had an assertive tone, proposing a reading of the term that was opposite to my intent, this story demonstrates how the idea of the inborn genius is still strongly embedded in society to explain successful careers. The controversial use of the term "genius" has recently attracted new attention in academia: for instance, since 2014, at CRASSH – University of Cambridge – Dr Alexander Marr is directing an ERC-funded research project named *Genius Before Romanticism: Ingenuity in Early Modern Art and Science*.

circumscribed".[28] In order to accomplish this task, I have focused on the problem of the education received by an individual such as Torriani whom a general audience may acknowledge as a typical "Renaissance genius". However, superior craftsmen, or more appropriately "Vitruvian artisans", such as Torriani overcame the epistemic boundaries of their time, embodying to a certain extent a contextualised case of polymathesis. In our case, Classical models provided Janello with an innovative curriculum and a mythical aura.

Mario Biagioli's book on Galileo's instruments of credit opens an innovative analysis of "the aura of greatness, genius, and perhaps even sacredness" of scientific characters. Biagioli argues that the historian can consider this "aura as a mappable effect of negotiations ... The aura, therefore, is not just a result (the a posteriori recognition of one's work), but a resource for producing that work in the first place, as well as for securing its acceptance from patrons and fellow-practitioners (the way the financial backing of a project is a necessary step toward its possible, but by no means necessary, success)."[29] In this perspective, I will consider Janello Torriani's aura of a new Archimedes not just as the result of "a posteriori recognition", but also as a possible effect of a strategy to obtain resources and credit to realise innovative technologies and win a better social and economic position. For instance, nearly nothing is known about the first thirty years of his life, and such an obscure background has enabled not just later panegyrists, but also Torriani's coeval ones to manipulate it according to humanist models, representing him as a natural born genius. For this reason, I have attempted to shed light on the "prehistory of the Renaissance genius" – or as we may call it more appropriately today, of the Vitruvian artisan – when Janello had put together the toolkit that made him desired and employable at the Mount Olympus of the talented: the imperial court.

Before proceeding on the concept of Vitruvian artisan, we should introduce another of the problem-siblings of the Renaissance, the "Scientific Revolution".[30]

---

28   Lorraine Daston and Katharine Park, "Introduction: The Age of the New," in *The Cambridge History of Science*, ed. Lorraine Daston and Katharine Park, vol. 3. The Early Modern Science (Cambridge: Cambridge University Press, 2006), 5-6.

29   Biagioli, *Galileo's Instruments of Credit.*, See chapter 1: "Financing the Aura: Distance and the Construction of Scientific Authority".

30   In this book Janello's career will enable us to take a brief look at another "problem child" of historiography: the Reformation. It seems that Emperor Charles v and his son Philip ii were to use Toriani's inventions to boast of a Catholic scientific superiority over the Protestant princes. We cannot at present say whether this was the result of a conscious policy. But, for certain, the flaunting of Janello Torriani's mechanical marvels, whose inner mysteries were kept secret, triggered a vibrant competition among the Protestant courts as well, whose princes started to patronise similar projects.

The Scientific Revolution is usually described as pertaining to the moment around 1600 when in Europe a new knowledge about the natural world and the growth of technological crafts surpassed the ones of the ancient Greco-Roman world, setting out an allegedly new scientific method and confuting the previous authorities. These changes were wrought mainly thanks to new discoveries in the fields of technology, geography, astronomy and anatomy. Historians and philosophers of Science have interpreted this moment as a radical and dramatic break from the medieval past, coining the phrase *Scientific Revolution* to define it. Though this presumed revolution has no clear chronology, the importance attributed by Galileo and Kepler to Copernicus' *De Revolutionibus Orbium Coelestium* (1543) and Vesalius' contemporary printing of the *De Humani Corporis Fabrica* made these two works an ideal point of reference for the narrative of the Scientific Revolution. Like the Renaissance, the chronology attributed to the Scientific Revolution depends much on the focus of individual scholars, and reflects a long-standing problem relating to continuity and ruptures in the conquest of "modernity". It has been observed that the narrative of the Scientific Revolution "will last as long as the myth of modernity, of which it is part and parcel".[31] As Shapin also wrote: "The past is not transformed into the 'modern world' at any single moment: we should never be surprised to find that seventeenth-century scientific practitioners often had about them as much of the ancient as the modern".[32] Western historians consider the dynamics reflected by the definitions of Renaissance and Scientific Revolution (together with the Reformation and the Industrial Revolution) to be critical in the history of the world. Koyré approached the problem of science from a philosophical angle, as an intellectual history of ideas. He identified the works of those philosophers who reshaped the perception of the structure of the Cosmos as the most significant representatives of this drastic turn. In his seminal book *From the Closed World to the Infinite Universe*, published in 1957, he claimed that the so-called seventeenth century Scientific Revolution reflected a radical shift in the very basis of European philosophy. Refusing simplified contrapositions between old spiritual approaches and modern pragmatic ones (for instance, the substitution of an alleged medieval *vita contemplativa* with a modern *vita activa*), Koyré described this turn, or crisis of European consciousness, as a dramatic change in the very foundations of Western philosophy, namely, in its very vision of the Whole. This was the reason why he considered this change as the most important moment in the history of science, which he saw as the birth of the "Modern World". Thomas Kuhn's analysis of the

---

31   Daston and Park, "Introduction: The Age of the New," 16.
32   Shapin, *Scientific Revolution*, 7.

structure of scientific revolutions helps us today to view this shift in European epistemology as a long and complex phenomenon, and not as a sudden revolution.[33]

While the idea of a Scientific Revolution has been challenged and deconstructed, it still stands because of the magnitude of changes that it describes: changes that occurred in European culture during that period and gave birth to a different science based on a new epistemology built upon the experimental method and mathematics. It has been noted that since the sixteenth century, mathematicians and philosophers have demonstrated an awareness of the innovative character of their contributions to the understanding of Nature. Writers associated with the natural sciences included with increasing boldness the adjective "new" in the titles of their works. And so, we see this adjective used in the *Nova Scientia* by Nicolò Tartaglia (1537), in the *New Attractive* by Robert Norman (1581), in the *Machinae Novae* by Fausto Veranzio (1595), in the *Nova horologiorum inventio* by Cherubino Sandolino (1599), in the *Nuovo Teatro di Machine et Edificii* by Vittorio Zonca (1607), in the *Astronomia Nova* by Johannes Kepler (1609), in the *Novum Organon* (1620) and in the *New Atlantis* (1627) by Francis Bacon, in the *Discorsi e dimostrazioni matematiche intorno a due nuove scienze* by Galileo Galilei (1638) and in the *De Mundo Nostro Sublunari Philosophia Nova* by William Gilbert (1651), and this is citing but a few of only the most famous titles. These thinkers were aware that their work stood in opposition to something, which they perceived, or wanted to represent, as traditional. Such claims to novelty have challenged the minds of philosophers and historians, who have speculated about the existence and consistency of such a turn.

Recent historiography has named this period, roughly covering the *Cinquecento*, "The Age of the New".[34] Though even Kuhn endorsed Koyré and Hall's mainstream idea that it was the *literati* and not the craftsmen who were the agents of this shift of paradigm towards "modern science", the increasing relevance technology has gained in our world, plus the crisis of ideologies and the social changes in Western society – now less dismissive of manual labour – has stimulated new investigation of the technical people during the Scientific Revolution, and once marginalised theses such as Zilsel's are now attracting new interest. From geographical discoveries to the invention of new machines and techniques that proliferated in sixteenth century Latin Christendom, we can see that it was the field of technology and practical mathematics that was

---

33   Thomas S Kuhn, *The Structure of Scientific Revolutions* (Chicago: University of Chicago Press, 1970).

34   Daston and Park, "Introduction: The Age of the New," 1-5.

first giving strength to this mental revolution in consciousness for western Europeans, who believed for the first time that they could surpass the knowledge of their ancient models, acquiring a new self-confidence to advance *ars* and *scientia*. Thanks to the mechanical clock, to Filippo Brunelleshi's revolutionary mathematical techniques in dome construction and in mathematical perspective, to Aristotle of Bologna's incredible engineering performances, to Leonardo's visionary projects, to naval architecture, to new geographical and natural discoveries, and to Torriani's unbelievable mechanical creations, the intellectuals of the Renaissance understood that their society had overcome the giants of Antiquity.[35] And this had happened through sensible experience and practical mathematics. The famous quote attributed to Bernard of Chartres during the twelfth century – "*sumus nanos super homeros gigantium*" – that is to say, if we see far, it is not of our own merit, but because we are like dwarves standing upon the shoulders of giants, turned now towards a paritarian relationship with classical Greco-Roman science. The present book wants to contribute to the scholar-craftsman debate showing how unexpected technical achievemnets could influence the consciousness of Renaissance society, and demonstrating how the application of mathematics to practical problems could efficiently empower humanity over Nature, overcoming limits previously considered insurmountable.

As late as 1991 Jim Bennet wrote: "There were many different sorts of mathematics in the early modern world ... and the branch most neglected by historians – practical mathematics – in many ways proved the most progressive, thanks to urgent new contexts of use, such as navigation". Practical or applied mathematics, also described in the Aristotleian tradition as *scientiae mediae* or μηχανη, are here to be intended as mechanics, perspective, and speculative astronomy.[36] A couple of decades after Bennet, we can say that the role of technology and of its practitioners in the scientific development of theories has occupied a paramount position in historiography. Indeed, it has been argued that the "new" scientific knowledge of the Early Modern Period was rooted in the practical one of the sixteenth century. With the end of the Cold War, which often led to the marginalization of Marxist theories such as

---

35   Alex Keller, "A Byzantine Admirer of 'Western' Progress: Cardinal Bessarion," *Cambridge Historical Journal* 11, no. 3 (1955): 343-48. Of the same author: "A Renaissance Humanist Looks at 'New' Inventions: The article 'Horologium' in Giovanni Tortelli's 'De Orthographia,'" *Technology and Culture* 11, no. 3 (1970): 345-65. See also A.C. Keller, "Zilsel, the Artisans, and the Idea of Progress in the Renaissance," *Journal of the History of Ideas* 11, no. 2 (1950): 235-40.

36   Aristotele, *Meccanica*, ed. Maria Fernanda Ferrini (Milano: Bompiani, 2010), 164-65.

Zilsel's, historians looked at these practitioners of applied mathematics with a more balanced eye.[37] The American scholar Pamela Smith has emphasised the artisan's workshop as the place where early modern science was first "disseminated and inculcated." However, she has also pointed out an ongoing resistance in the Anglo-Saxon academic world towards narratives that consider artisans as actors of the Scientific Revolution. Smith argues for full epistemological dignity for the artisanal world in its reading of Nature through the senses of its practitioners. Nonetheless, it seems that she, like Zilsel more than half a century earlier, accepts a dichotomy between a "high culture" and a technical one as a faithful representation of Renaissance epistemology.[38] In order to emphasise the scientific dignity of craftsmen's sensibly justified understanding of Nature, Smith coined the phrase "artisanal epistemology". This book calls for a less sharply separated representation of Renaissance artisanal and university knowledge, as for a less Manichean contraposition between humanists and

---

37   In the last decades there have been many seminal works published in this field: Steven
      Shapin and Simon Schaffer, *Leviathan and the Air-Pump : Hobbes, Boyle, and the Experi-
      mental Life* (Princeton, N.J.: Princeton University Press, 1985); Antonio Sánchez, "La voz de
      los artesanos en el Renacimiento científico: cosmógrafos y cartógrafos en el preludio de la
      'nueva filosofía natural,'" *Arbor* 186 (2010): 449-60; Harold J. Cook, "The Cutting Edge of a
      Revolution?: Medicine and Natural History near the Shores of the North Sea," in *Renais-
      sance and Revolution: Humanists, Scholars, Craftsmen and Natural Philosophers in Early
      Modern Europe*, ed. J.V. Field and Frank A.J.L. James (Cambridge: Cambridge University
      Press, 1993), 45-62; Jürgen Renn and Matteo Valleriani, "Galileo and the Challenge of the
      Arsenal," *Preprint of the Max Planck Institute for the History of Science* 179 (2001): 1-32. This
      very year has seen the publication of a book in the Springer series "Studies in History and
      Philosophy of Science": Lesley B. Cormack, Steven A. Walton, and John Andrew Schuster,
      eds., *Mathematical Practitioners and the Transformation of Natural Knowledge in Early
      Modern Europe*, (Cham: Springer, 2017). Unfortunately, I was not able to use this new work
      that testifies for a revival of this kind of interest in the philosophy of science. I was par-
      ticularly inspired by the work of the American scholars Pamela O. Long and Pamela
      Smith, who have shaken the debate by taking more cogent material to testify for an arti-
      sanal influence in the Scientific Revolution. Long, *Artisan/practitioners*; Smith, *Body of
      the Artisan*.

38   Zilsel's view of the importance of technical but learned men in the shift towards what we
      call "modern science" was still biased by his belief in a strict dichotomy between scholars
      and craftsmen in the Renaissance, a dichotomy that, according to him, came to an end in
      the seventeenth century, when scientists endorsed the knowledge of the artist-engineer.
      Despite the contributions of Keller and Rossi, who pointed to the presence of the belief in
      scientific and technological progress in both university and artisanal contexts, even
      recent historiography has tended to endorse a narrative of two parallel dimensions of
      knowledge: the province of learned higher education and the mechanical and illiterate
      sphere of the workshop.

Aristotelians:[39] the case study of Janello Torriani, together with some other well known examples, shows how this artisanal *scientia* was to be found only in certain workshops – those of Zilsel's "superior craftsmen" which hosted *literati* who transferred the categories of Latin and Greek theoretical knowledge to these curious and talented artisans. The latter, in turn, brought the power of their fresh imagination, their curiosity, and their refined practical experience – the long-trained skills of their hands – to interpret scientific texts. Pamela O. Long, another influential American scholar, has shown how artisans interacted with scholars during the Renaissance in specific "trading zones", to be understood as common areas of interest where both scholars and craftsmen could cooperate, such as antiquarianism, water-supply projects, and the text by Vitruvius itself.[40] In our story the "trading zones" that allowed Janello to overcome epistemological boundaries by interacting with scholars are also medicine, urban mathematical public offices, and courtly-related practical mathematical projects.

The necessary role played by these "learned practitioners" in the construction of a new consciousness of scientific dignity calls now for a more precise definition which can describe the specific European cultural context that produced them. Zilsel's concepts of "superior craftsman" and "artist engineer" are here substituted by the one of the "Vitruvian artisan". This figure developed during the Quattrocento, when Vitruvius was most celebrated and influential, and his famous statement about knowledge being the offspring of theory and practice was held in great esteem, something that Pamela O. Long has shown in detail in the chapter *Artisans, Humanists and De Architectura of Vitruvius*.[41]

---

39 Many influential scholars of the history of philosophy and of the Renaissance have noticed that it is impossible to speak of a single Aristotelianism in the Middle Ages and the Renaissance, in light of the different interpretations (often contradictory) that were in circulation, or even by Aristotelianism seen as a school, given the mobility of professors in the universities of the era. Kristeller has also shown that the contraposition between humanism-scholasticism should be treated with great caution: in the Renaissance this dichotomy was not absolute, but it was more a part of a discourse made up by some influential humanists. For an introduction to the theme, well accompanied by examples, see the essay: Eugenio Garin, "Aristotelismo veneto e scienza moderna," in *Umanisti, artisti, scienziati: studi sul rinascimento italiano* (Roma: Riuniti, 1989), 205-28; and Paul Oskar Kristeller, "Humanism and Scholasticism in the Italian Renaissance," in *Humanism and Scholasticism in the Italian Renaissance*, ed. Benjamin G. Kohl and Alison Andrews Smith, Major Problems in European History Series (Lexington, Mass.: D.C. Heath and Co., 1995), 285-96.

40 Pamela O. Long, "Hydraulic Engineering and the Study of Antiquity: Rome, 1557-70," *Renaissance Quarterly*, 2008, 1098-1138; Long, *Artisan/practitioners*.

41 See Long, *Artisan/practitioners*.

Though Vitruvius's *De Architectura* had been known during the Middle Ages,[42] being mainly studied in monasteries and Cathedral schools, it was in a booming urban society obsessed with Classical culture and shaken by a thriving social mobility, like the one of the late Middle Ages, that such books could meet the interests of ambitious craftsmen. The pervasion of urban culture with humanist ideology transformed Vitruvius and other classical literature into a "prestige-maker": the craftsman could use it as a powerful tool for social climbing, whereas the wealthy patron (both as an individual or as an institution) interested in civil or military constructions could dress himself with a Vitruvian garment woven by the scholar and the superior craftsman, adopting the charisma of a new Augustus. I consider the phrase "superior craftsman" too general a definition that does not acknowledge the specific humanist background of careers such as the ones of Lorenzo Ghiberti, Filippo Brunelleschi, Aristotle of Bologna, Filarete, Bramante, Leonardo da Vinci, Michelangelo and Janello Torriani, to name a few. Their educational background and social aspirations were strictly dependent on Vitruvius' idea of "architectura". The concept of "artist-engineer" is instead too specific, and has to do with the Vasarian idea of fine arts (painting, sculpture and architecture intended mainly as building design), excluding many practitioners such as Torriani who did not paint, sculpt or draw the beautiful facades, but who were able to design machines and construct them inspired by classical texts and aspiring to an intellectual dignity.

---

42   There are around 80 extant medieval copies of Vitruvius. *Ibid.*, 64. For instance we know
      that Aegil, carolingian abbot of Fulda, predecessor of Rabanus Maurus, studied Vitruvius:
      see Fabrizio Crivello, "Tuotilo: l'artista in età carolingia," in *Artifex bonus: il mondo
      dell'artista medievale*, ed. Enrico Castelnuovo (Roma: Laterza, 2004), 28.

PART 1

A Centre of Education for the Vitruvian Artisan at the Periphery of the Empire

∵

CHAPTER 1

Janello Torriani's First Education

Cremona, the Italian Wars and the Desire for a Better Life

Our story begins around the year 1500, at the very centre of the Po valley where, on the north bank of the greatest Italian river, lay Cremona (Fig. 1), Janello Torriani's *patria*. The historical origins of Cremona go back to the time Hannibal was crossing the Alps with his elephants: it was founded in Celtic land by the Romans in 218 B.C. together with the twin colony of Placentia (the modern Piacenza, around 30 km away on the southern bank of the river) in order to control Cisalpine Gaul. The city grew rich and prosperous. Despite two major destructions, suffered in AD 69 and at the time of the Germanic Langobardic invasion at the beginning of the seventh century, the city flourished once again in the Middle Ages, becoming one of the most important City Republics or Communes of Northern Italy. By the fourteenth century the city had lost its ancient freedom and had become a part of the Visconti dominion that had its capital in Milan, some 80 km away. By 1535, at the time of Janello Torriani's adulthood, the duchy of Milan had lost its last autochthonous ducal family, the Sforza, and the state was now ruled by imperial governors under the direct control of Charles V, and later by Spanish ones. Milan was the metropolis of

FIGURE 1   South view of Cremona. COURTESY OF MINO BOIOCCHI.

© KONINKLIJKE BRILL NV, LEIDEN, 2017 | DOI 10.1163/9789004320918_003

the duchy, perhaps the largest Italian city at the end of the fifteenth century, with more than 100,000 inhabitants. Its wealth came from different factors, the most important being probably its grasp on the commercial routes that ran along the lakes of Como and Maggiore to the Swiss passes that led to Northern Europe. Cremona, on the other hand, had some 40,000 residents at its peak during the sixteenth century, and provided a 250,000 *scudi* revenue during the second half of the century, 1/4 of the taxes of the whole duchy and 1/2 of what Milan paid. Cremona, beside agriculture, was mainly involved in textile manufacture: it has been calculated that by 1580, 92% of its export consisted of fustian – a cloth with a cotton weft and a linen warp.[1] Cremona stood at the centre of a neuralgic network of cities, which belonged to different states: on the West one could easily reach Lodi (49 km) and Pavia (72 km), both part of the same duchy; on the East stood Mantua (65 km) and the small universe of tiny states ruled by the many branches of the Gonzaga family, some of them just a few kilometers outside of the city-walls; southwards lay Parma (54 km), together with Piacenza once part of the duchy of Milan and latter both dominion of the dusky dynasty of the Farnese; and on the northern side were Brescia (50 km), Bergamo (74 km) and Crema (40 km), all of them parts of the dominions of the Republic of Venice. Still today, approaching Cremona from each of these directions, the cityscape appears dominated by the Torrazzo, the medieval brickwork tower where Janello Torriani appears in 1529 for the first time mentioned in a document while he was taking care of the public clock, a tower that at that time was the oldest among the three tallest of the world, and the pride of the community (Figs 24-25). Just aside emerges the spire-crowned bulk of the Romanesque Cathedral.

Upon entering the Cathedral of Cremona, despite the Romanesque and Gothic style of the structure, the observer's attention is captured by the high quality cycle of frescos running all around the principal nave and painted between 1514 and 1521. This long sequence of religious scenes depicts the stories of the Virgin Mary and Christ, and it is famous among art historians for its encyclopedic range of Renaissance styles influencing the five great masters who were charged with this endeavor, drawing on local, Northern, Roman, Florentine, Venetian and Emilian elements. The frescoes painted on the last two spans on the southern side and on the counter-façade leave the quiet and equilibrate composition of the mature Renaissance style of the older part of the painted register for a grandiose and expressive representation of Jesus' last hours. Livid stormy colors dominate the background and a crowding of

1   Giovanni Vigo, "Il volto economico della città," in *Storia di Cremona*, ed. Giorgio Politi, vol. 4, L'età degli Asburgo di Spagna: 1535-1707 (Bergamo: Bolis, 2006), 220-23.

men-at-arms and infantry covered with metal plates, shouting, pushing, pulling and moving with their tense musculature, reveals the encounter between
Michelangelo's Tosco-Roman powerful anatomy, Venetian atmosphere-mak-
ing, transalpine *horror vacui*, and sharp wood-cut detailed expressionism,
giving an astonishing, dramatic atmosphere to the *Via Crucis* and to the
Golgotha's final sacrifice that expands over a wide painted superfice of 9.20 for
12 meters (Figs 2, 4, 5). These characteristics of Pordenone's frescos, painted
around 1520-1521, can be seen as exemplary visual articulations of their own
time. The first decades of the sixteenth century in Lombardy[2] were not just a
time of cultural influences from the four cardinal points, but equally a time
when real armed tides were smashing the country from all sides. The soldiers
populating Pordenone's frescos are dressed like the very French, Swiss, German,
Venetian, North Italian, Papal, Tuscan, Neapolitan and Spanish contemporary
soldiers that were constantly ravaging the country, dreadful, powerful and violent, bringing diseases to the body and, in the authorities' eyes, even heretic
poison to souls.

Such was the image of war that faced the people of these districts, and this
was the traumatic political context in which Janello Torriani came to life, grew
up, got educated, learned a profession, got married and had his children. Local
chronicles remind us that just before the beginning of the new century, alarm-
ing signs were observed in the sky preluding, for the astrologers, sorrowful
times to come: a 60 years-long conflict, known later as the Italian Wars, was
going to bring traumatic changes to the history of Italy.[3] Torriani was born
under Venetian dominion (1499- 1509). In 1509, as a consequence of the
Cambrai League and the initial victory of the latter over the *Serenissima*,
Cremona was returned to the duchy of Milan, now the possession of the Valois
King of France Luis XII. A few years later, in 1512, the Swiss Confederation drove
out the French from the State of Milan and put Massimiliano Sforza, first son
of Ludovico il Moro, on the ducal throne. The Venetians, now allied with the
French, managed to retake the city for a couple of months (1513), before the
Swiss and Sforza were back again. The Battle of the Giants, at Marignano (1515),
returned once more the reins of government of the duchy to another French
monarch: Francis I, who kept it until 1522 when he was defeated at the Battle of
Biccoca. Prospero Colonna, general captain of the Pontifical Army, seized the
city in the name of the last Sforza, Francesco II, the younger son of Ludovico il
Moro. Having lost the duchy, King Francis lost even his personal freedom at the

---

2   Lombardy should not be mistaken with the present administrative region, but it has to be
     understood as the historical area that roughly covers the whole of Northern Italy.
3   "Cronaca di Cremona dall'anno 1494 al 1525," *Bibliotheca historica Italica*, 1876, 189-276.

FIGURE 2   *Giovanni Antonio de Sacchis, called Pordenone,* Golgotha, *fresco,*
1520-1521, Cathedral of Cremona. COURTESY OF THE DIOCESE OF
CREMONA.

battle of Pavia in 1525. But once again the fragile political balance was soon
tipped, and in the same year Emperor Charles V fell into bad terms with his
former ally Duke Francesco II, and Spanish soldiers together with the German
imperial mercenaries took Cremona away from the Sforza. Only in the following year, after a siege, did the Sforza, now members of the League of Cognac,
recapture the city and eventually, at the end of the decade, the Emperor even
recognised him as the legitimate Duke of Milan bequeathing upon him as a

bride his niece Christina, Princess of Denmark. However, the star of the House of Sforza was destined to an irrevocable decline: the last Sforza died in 1535 without an heir. From this moment Emperor Charles V became *de facto* ruler of the duchy, conveying it later into the hands of his son Philip. Janello thus witnessed 11 changes of lordship in Cremona during his first 35 years of life, which he lived against the background of an almost constant state of war.[4]

These were not easy times. The numerous clashes between Italian, Swiss, Spanish, German and French armies were not restricted to the battlefield: the country was constantly ravaged, towns sacked, people abused, agriculture ruined and trade and production weakened. The numbers of these armies were increasing: combat was constantly breaking out and the evolution of artillery meant that a few minutes' bombardment could cause casualties by the hundreds as the battles of Ravenna (1512), Novara (1513), Marignano (1515), Bicocca (1522) and Pavia (1525) had shown, signaling an increase in the massive use of efficient cannons and portable artillery. This was one of the reasons why the number of soldiers in the ranks of the armies was rising to unprecedented levels. This had a tremendous impact on the populations and regions that had to feed, shelter and host undisciplined hoards of professional assassins, together with their horses, trains of carriages, beasts and accompanying villains, adventurers, suppliers and prostitutes following behind them. Violence and plague were all too familiar to Janello's countrymen, and personal security was more often than not a short-lived illusion.

Since the time of Archimedes, it was well-known that war, while a deadly threat, was also an opportunity for the practitioners of practical mathematics to display their skills. Leonardo da Vinci, who had worked for about 20 years for the Sforza in Milan, fashioning himself as a new Archimedes able to invent new dreadful war-machines, decided to leave the duchy as soon as fortune turned her back to his Sforza lord Ludovico il Moro, and the French had taken over, showing a quite destructive attention towards his works, demolishing his equestrian clay model for a cyclopean bronze sculpture. Leonardo was leaving Lombardy when a new generation of local Renaissance mathematicians that would be indebted to his legacy -such as Girolamo Cardano, Niccoló Tartaglia

---

4   Antonio Campi, *Cremona, fedelissima città et nobilissima colonia deRomani: rappresentata in disegno col suo contato, et illustrata d'una breve historia delle cose più notabili appartenenti ad essa, et dei ritratti naturali deduchi et duchesse di Milano, e compendio delle lor vite* (Milano: Bidelli, 1645), 137-50; Letizia Arcangeli, "La città nelle guerre d'Italia (1494-1535)," in *Storia di Cremona*, ed. Giorgio Chittolini, vol. 6, Il Quattrocento. Cremona del Ducato di Milano: 1395-1535 (Bergamo: Bolis, 2008), 42; Carlo Bonetti, "L'assedio di Cremona (Agosto-Settembre 1526)," *Rivista Militare Italiana*, 1916, 8-28.

RAGIONAMENTI DE NICOLO
TARTAGLIA SOPRA LA SVA TRAVAGLIATA
INVENTIONE.

*Nelli quali ſe dechiàra uolgarmente quel libro di Archimede Siracuſano Intitolato.
De inſidentibus aquæ, Con altre ſpeculatiue pratiche da lui ritrouate ſopra le
materie, che ſtano, ꝛ chi non ſtano ſopra lacqua, Vltimamente ſe aſſe-
gna la ragione, et cauſa naturale di tutte le ſottile, et oſcure
particularità dette, et dechiarate nella detta ſua
trauagliata inuẽtione cõ molte altre
da quelle dependenti.*

Apreſſo di Lautore.
Cõ gratia, et priuilegio del Illuſtriſſ. Senato Veneto che niun poſſa ſtãpare ne far ſtãpa
re la preſente operina ne parte di quella, uẽder ne far uẽdere in Venetia, ne in alcun al
tro loco, o terra del dominio Veneto per anni dieci ſenza conſentimento del Autore
ſotto pena de ducati. 300. ꝛ perder le opere, Come che nel preuilegio appare.

FIGURE 3　　Portrait of Niccolò Fontana, called Tartaglia. *From his:* Ragionamenti de Nicolo
Tartaglia sopra la sua Trauagliata inuentione. Nelli quali se de chiara uolgarmente
quel libro di Archimede Siracusano intitolato. De insidentibus aquae, con altre
speculatiue pratiche da lui ritrouate sopra le materie, che stano, & chi non stano
sopra l'acqua ..., *Venice, 1551.*

and Janello Torriani- came to life in the region, destined to suffer the effect of the same war the Tuscan engineer was escaping. Leonardo would come back to Milan in the following years before moving eventually to France and in Milan he would join a cultural milieu which remained a fertile land for practical mathematics.

The Italian Wars had caused certain changes that were influential in Janello's formation. The constant warfare provoked the reduction and for long periods the shutting of many universities such as the *Studium* of Pavia, where one Giorgio Fondulo of Cremona, a physician, had been studying and later teaching. According to Antonio Campi's account of Janello's youth, contained in his *Cremona fedelissima* (1585), Fondulo was charmed by the brilliant intellect of Janello Torriani when the latter was still a child, and he taught him what he knew in the field of mathematics, with special regard to astrology.[5] We do not know whether Giorgio Fondulo returned to his native Cremona to practice his medical profession of his own accord or because war pushed him to do so during the long-periods of extraordinary closure and reduced activities of the University of Pavia. What we do know is that Fondulo provided youngsters with special tutorials in Cremona. Besides Giorgio Fondulo, the military operations had brought to Cremona people of different origins, with their technological culture and their networks. In Cremona Janello could witness the revolution of the fortification-system of his town (especially in the years 1516-1526) that transformed the huge castle and the proud medieval tall city-walls coroneted by castellated battlements into a modern system of low ramparts protected by a wide fosse and bastions. Practitioners of applied mathematics were employed in the defensive tasks, from the casting of cannons to the construction of long bridges over the river, to the strengthening of fortifications, and sometimes their capabilities provoked common admiration, as was recorded during the siege of 1526 when Spanish engineers led the general applause.[6] A young man like Janello, with his interest in mechanics, had certainly plenty of opportunities to observe, and perhaps even cooperate with, military engineers. After all, the first records about his employment in Milan in the central administration of the duchy, as we shall see, describe him as an engineer. It is possible to imagine that, in the year 1526, right after Spanish and German troops left the city, when Duke Francesco II Sforza had to reside in Cremona for several months, Milan being still in the hands of the enemy, Janello Torriani took the chance to make himself known by the political and

---

5   Campi, *Cremona fedelissima città*, 137-50.
6   "Cronaca di Cremona dall'anno 1494 1525"; Bonetti, "L'assedio di Cremona (Agosto-Settembre 1526).".

military hierarchies of the duchy, something that may have had important consequences in the development of his future career.

However, the suffering and constant danger provoked by war must had a unforgettable impact on Janello, though not as evident as in the case of the unfortunate Niccolò Tartaglia, another paramount protagonist of the development of Renaissance mathematics (Fig. 3). The latter was born in the neighboring town of Brescia, and he is in many senses an exemplary victim of these ruthless conflicts. Since Brescia, like Cremona, was a possession of the Venetian Republic, after the attack of the Cambrai Legue, it fell into French hands. However, in 1512, the population of Brescia participated in a riot and managed to take control of the city for two weeks until fresh French troops arrived to reinforce the expelled ones, crushing the rebels without mercy. Brescia was sacked and its population was brutally slaughtered, violated and robbed. Tartaglia was a 12 year-old boy at the time. He would later describe how he escaped together with his mother and his sister into the Cathedral when enemy troops broke into the city-walls, believing this holy building to be a safe shelter for them. It was not: a French soldier struck the young boy's head with a sword at least fives times in front of his mother. One of these violent blows penetrated Niccolò's jaw and palate, breaking through his teeth. He survived only because, covered by the blood and by the arms of his mother, and after so many blows in the head, he was believed to be dead. Niccolò could neither talk nor eat for a long time, and after his recovery, he would suffer a lifetime's speech impediment. He became a stutterer, and for this reason he was given the nickname "Tartaglia", meaning in Italian exactly this disability.[7] Niccolò chose to keep this name even when he became a renowned mathematician, in order to remember this traumatic event.

In the neighboring city of Cremona the situation was slightly better but still fraught with danger. In the same year of the sacking of Brescia, Cremona avoided by the skin of its teeth the same destiny that Ravenna, Novara, Prato, Pavia, Pordenone, Treviso, Rome itself, and many other villages and cities had suffered during the Italian Wars. The community managed to escape being sacked by paying a huge ransom, and providing food and lodging for the Swiss forces of the Holy League. Similar situations occurred in 1521 and 1526, when the city came really close to being stormed by enemy armies. Even if Torriani's

---

7 Niccolò Tartaglia, *Quesiti et inuentioni diuerse de Nicolo Tartaglia: di nouo restampati con una gionta al sesto libro, nella quale si mostra duoi modi di redur una città inespugnabile. La diuisione et continentia di tutta l'opra nel seguente foglio si trouara notata.*, [1546] (In Venetia: Appresso de l'auttore : per Nicolo de Bascarini : ad instantia et requisitione, et a proprie spese de Nicolo Tartaglia autore, 1554), 69v.

FIGURE 4   *Giovanni Antonio de Sacchis, called Pordenone,* Christ Nailed to the Cross, *fresco, 1520, Cathedral of Cremona.* COURTESY OF THE DIOCESE OF CREMONA.

hometown was not plundered, violence, extortion, humiliation and death were ever present. On two occasions Cremona had been captured by the Swiss and Sforza side (1512) and by the Holy League and the Sforza (1522). On both occasions, for more than a year, French troops continued to hold the castle of S. Croce, within the very city. The effects of the French artillery's bombardment are still visible today on the northern walls of the Cathedral and on the Torrazzo. Besides the blasts falling on the city from the castle, and the desperate forays by the hungry troops inside to pillage anything edible, Cremona had to pay for half of the costs of maintenance of the troops constantly sieging the castle in order to prevent these very attacks. Moreover, waves of dreadful diseases cyclically hit the town: since 1495 syphilis, known as the French disease, appeared in Cremona frequently, killing many and horribly disfiguring the unfortunate survivors. Since the same time contagions of petechial fever and plague killed large numbers of people in 1503, 1504, 1505, 1511, 1512, 1513 and 1524. In the year 1523 the local administration considered it to be necessary to build a *Lazzareto* (a hospital for infective disease) outside the city-walls to host the numerous victims of the plague. The incessant work of destruction of the armies in the countryside, together with unfortunate climatic conjunctions, brought terrible famines that raged in Cremona, especially in the years 1501, 1505, and 1518. During the famine of 1501 artisans armed themselves and tried to sack the deposits of food in the palaces of the gentry, but they were captured and hanged. In 1505 some people even died of hunger, and in the winter of 1511, because of the great cold, the river froze, blocking the mills and leaving the city without bread for several days. Occupying armies had to be fed by the locals– whom they treated with disdain–, and it happened that even noble citizens were killed.

It was not rare to see wealthy members of the local gentry kidnapped on the specious accusation of high treason and tortured in order to extort their golden ducats, and during the siege of 1526 the local nobility was even obliged to work manually together with people from the lower classes in the reinforcement of the town's defenses. When the foreign troops were not responsible for atrocities, it was the bitter grudge of the local factions that precipitated despair: Guelphs and Ghibellines, allied with one or the other Transalpine power, were as ruthless as devils in oppressing the peasants and in taking the most cruel vendettas on each other. Janello had to witness murders and terrible executions during his youth, as when in 1521 the French authorities slowly burned alive two Ghibellines and quartered another two, after having tormented them with red-hot pincers. No surprise that the populace, exhausted by the unbearable situation, began to run after Franciscan preachers calling for peace as if they were the messiah himself, or to denounce several apparitions of the Virgin Mary, On two occasions, even the Church itself urged caution because these apparitions were to be read as misinterpretations of signs produced by Nature itself and not by God.[8]

Looking closer to Janello we know that even his father, one Gherardo, suffered economic consequences from this condition of constant warfare: having rented a mill from the "lord" Cornelio Meli, "magnificent knight" of Cremona (*magnificus eques dominus*), Gherardo was sued by Meli for failing to pay his rent. At law court Janello's father advanced the reasonable excuse that he could not use the mill for the reason that the canal in which it was situated had run dry because of the war between French and Imperial forces. Despite this, he was ordered to pay the rent![9]

Even if it was the entire population that suffered the effects of this neverending war, a commoner like Janello's father still had to endure the unjust

---

8   "Cronaca di Cremona dall'anno 1494 al 1525."

9   Rita Barbisotti, "Janello 'Torresani', alcuni documenti cremonesi e il 'baptismum' del Battistero," *Bollettino Storico Cremonese*, Nuova Serie, 7 (2001): 255-68.

FIGURE 5   *Giovanni Antonio de Sacchis, called Pordenone,* Pilate judging Christ, *fresco, 1520, Cathedral of Cremona.* COURTESY OF THE DIOCESE OF CREMONA.

oppression of the gentry which, in accordance with a central government in desperate need of taxes to sustain the army, managed to regain authority, depriving the merchants of the power they had slowly acquired in the last 300 years.[10] With such a background, it is not striking that when Janello entered the entourage of the Habsburgs, he seemed to avoid working on military engi-

---

10   Giorgio Politi, *Aristocrazia e potere politico nella Cremona di Filippo II* (Milano: SugarCo, 1976), 20, 33-47; Marco Bellabarba, *Seriolanti e arzenisti: governo delle acque e agricoltura a Cremona fra Cinque e Seicento* (Cremona: Biblioteca statale e libreria civica di Cremona : Distribuzione, Libreria del convegno, 1986), 57.

neering, and it is even less surprising to observe how hard he worked in order to keep his post at court, despite leaving reluctantly forever his unfortunate patria for an unknown but safe world. To do so ensured a secure existence for himself and for his household.

Fashioning the Aura of the Genius

The first writer to introduce Janello to fame was Girolamo Cardano in 1544, in his first edition of his *De Libris Propriis*, where he attributed to the Cremonese the reconstruction of a planetary instrument.[11] Six years later (1550) Cardano added in the first edition of his bestseller *De Subtilitate* that Janello, to whom he attributed several inventions, was "a man of great ingenuity in anything that concerns machines".[12] In the same year, Marco Girolamo Vida (1480-1566), bishop of Alba and the most famous Cremonese humanist of the century (Fig. 6), definitively consecrated Janello's aura of a genius when the clock-maker accomplished the planetary automaton for Charles v. Besides this great accomplishment it is important to note that in these 1550 writings, both Cardano and Vida mentioned that Janello had also constructed a *machina ctesibica*, a bronze pump with two alternating pistons which slide in two cylinders with valves below, a device described by Vitruvius in his *De Architectura*, a piece of news we shall discuss later.

The portrait that Vida made of Janello illustrates the rhetorical taste for contrapositions of the time. The contrast between repulsiveness to behold and attraction for genius fits very well with the mannerist taste for the *monstrum*, i.e., something perturbing that has to be seen:

> If somebody was to observe this man, nothing would appear less obvious than the light of his genius, much of his aspect and figure and appearance are coarse, unkempt, almost wild and even indecorous, an aspect that suggests lack of ability, of talent, and even of the hope that he could be capable of accomplishing something. This increases from his lack of decorum in the seeing him always with his face, hair, and beard blackened,

---

11   Girolamo Cardano and Ian Maclean, *De libris propriis: the editions of 1544, 1550, 1557, 1562, with supplementary material* (Milano: FrancoAngeli, 2004), 71, 148.

12   "*His igitur demonstrates tanquam principiis, ratio consurgit machinae Ctesibicae, quae sic constat, ut etiam Ianellus Turrianus Cremonensis, vir magni ingenii in omnibus quae ad machinas pertinet, opere ipso expressit*": Girolamo Cardano, *Hieronymi Cardani Medici Mediolanensis De svbtilitate Libri xxi* (Norimbergae: Apud Ioh. Petreium, 1550), 7.

and covered with ash and raw soot, with his hands and his big, enormous fingers, always full of rust, badly roughened, and inappropriately dressed, such that you would believe him to be a Brontes, or a Steropes, or some other [cyclops] servant of Vulcan, for he moulds everything he makes at the anvil with his own hands, being a born worker at the forge. However, lest anyone should imagine that some excellent master in mathematics has prepared for him the calculations of the orbits, of the motions or of the stars and has solved it all for him before because he understands nothing about these things, but is only skilled in craftsmanship, let him know that he invents them all and fabricates them by himself with no help of any kind, using his own talent, his own research, his own fancy, as they say. He is both inventor and executor at once, uniquely outstanding amongst all those fellow-citizens of ours–to whom I alluded before–in what concerns astrology, though his calculations were never culturally prepared or worked out under any master. Not only does he penetrate in this way into the phenomena of the firmament itself but besides with his intelligence he reaches the very causes by his calculations and has a very good knowledge of them. Moreover, often, with assurance and wisdom he contradicts recognised authorities on Astrology, authors of books on it. On not a few occasions he convinces them with incontrovertible arguments, obliging them to admit their errors and blunders.[13]

Vida's hyperbolic speech, which was supposed to be read in front of the Senate of Milan (indeed, there was a harsh dispute for precedence between Cremona and Pavia within the hierarchy of the duchy), never took place, but the text circulated. It is remarkable that in Vida's eyes, among the several portraits of great men and women that brought glory to Cremona demonstrating its superiority over Pavia, the one of a craftsman deserved the larger space. Perhaps, the emperor's enthusiasm for Janello Torriani's skills made of the clockmaker a powerful "superconductor" to transmit the imperial favour upon Cremona, especially when the judge of the Cremona-Pavia dispute was supposed to be the Imperial Governor of the State Ferrante Gonzaga, who happened to be the employer of Janello as well. Vida indeed makes it clear:

---

13   Marco Girolamo Vida, *Cremonensium Orationes III adversus Papienses in Controversia Principatus* (Cremonæ: Giovanni Muzio e Bernardino Locheta, 1550); I have used here – sometimes readapting it – the translation in English by Charles David Ley made after the Spanish one by by García-Diego: José Antonio García-Diego, *Juanelo Turriano, Charles V's Clockmaker: The Man and His Legend* (Madrid: Castalia, 1986), 54.

I believe Senators, that there is not one of you who has not seen this admirable, extraordinary and, in a certain way, portentous work already finished and concluded in all its parts and numbers ... You will not lack information, senators, as to how, some time ago in this same royal city of yours, by the initiative of our friend Gonzaga, a certain citizen of Cremona, renowned in the mechanical art for all his excellent qualities, was ordered to construct for the Emperor a clock of admirable, unusual and incredible skilfulness ...[14]

In the fragment from Vida's speech what was meant to strike the audience was the contrasting double nature of a craftsman like Torriani, Cyclops-like, big, sooty and a "born worker at the forge," yet at the same time in possession of a sharp mind that could handle the most precise theoretical knowledge. To the features depending on a craft cultivated since youth, Vida adds a scientific

---

14   Vida, *Cremonensium Orationes III*, 53-57.

knowledge that allowed the vile and illiterate worker (i. e., ignorant in the Latin language) to design a mechanical marvel and to discuss with (and even correct) learned people well taught in the science of the stars. Torriani seemed to be especially confident in one mathematic discipline: we shall see that once in Spain he told Ambrosio de Morales, royal historiographer to Philip II, that he had never met anybody with a better command than his in arithmetic.[15] It has been noted that since its birth in the late thirteenth century, clockmaking was one of these "trading zones" in which practical and theoretical knowledge were merging.[16]

Interestingly, in Vida's encomium, something seems to draw upon a literary topos: it is the passage where Janello is said to have been correcting authorities and books' authors in the field of astrology. This is a discourse that must have been well established in the duchy of Milan: indeed, some decades earlier, Leonardo da Vinci had used similar arguments for the introduction to his manuscript treaty on the art of painting:

> I know well that, not being a man of letters, it will appear to some presumptuous people that they can reasonably belabour me with the allegation that I am a man without learning. Foolish people! They do not know that I might reply as Marius did to the Roman patricians by saying that they who adorn themselves with the labours of others do not wish to concede to me my own; they will say that since I do not have literary learning I cannot possibly express the things I wish to treat, but they do not grasp that my concerns are better handled through experience rather than bookishness. Though I may not know, like them, how to cite from the authors, I will cite something far more worthy, quoting experience, mistress of their masters.[17]

Historians of Science have emphasised the role played by certain works that from the second half of the sixteenth century had expressed the idea of tech-

---

15   Ambrosio de Morales, *Las antiguedades de las ciudades de España que van nombradas en la Coronica, con la aueriguacion de sus sitios, y nóbres antiguos* (En Alcala de Henares: en casa de Iuan Iñiguez de Lequeríca, 1575), fols. 91-94.

16   Gerard Dohrn-Van Rossum has observed that until the beginning of the sixteen century, clockmaking attracted individuals with a combination of sophisticated theoretical knowledge and highly-skilled craftsmanship: Gerhard Dohrn-van Rossum, *History of the Hour: Clocks and Modern Temporal Orders* (Chicago: University of Chicago Press, 1996), 185.

17   Leonardo, da Vinci, *Leonardo on Painting: An Anthology of Writings*, ed. Martin Kemp and Margaret Walker (New Haven: Yale University Press, 1989), 9.

nical arts and experimentation as dignified sources of knowledge. Giuseppe Ceredi's *Tre discorsi sul modo di alzare le acque* (1567), Guidobaldo del Monte's *Mechanicorum libri* (1577), Bernard Palissy's *Dans l'Art de terre* (1580) and Robert Norman's *The Newe Attractive* (1584) are among the most representative works in this "new scientific trend".[18] The craftsman William Gilbert, who in his *De Magnete, Magneticisque Corporibus, et de Magno Magnete Tellure Pysiologia Nova* (1600) wrote he was dedicating his treatise to "true philosophers, ingenious minds, who not only in books but in things themselves look for knowledge" called this approach "a new style of philosophizing",[19] but this empirical method of investigation had already been at the centre of discussion for more than a century, as the passage by Leonardo clearly illustrates.

Italy was not the only place where this discourse on empiricism was taking place. For example, Philippus Aureolus Theophrastus Bombastus von Hoenheim (1493-1541), known as Paracelsus, is famous for his writings in favour of empiricism against bookishness. In the case of Paracelsus the attack was not just on the milieu of literacy, but even against ancient knowledge.[20] In 1547, in Nuremberg, the mathematician, writer and calligrapher Johann Neudörffer the Elder wrote that a locksmith with no letters (he emphasised), a certain Hanns Bulman,[21] was able to build a mechanical *Theoretica Planetarum*, something similar to Janello's first great creation. In the same years also Giorgio Vasari's

---

18   Giuseppe Ceredi, *Tre discorsi sopra il modo d'alzar acque da' luoghi bassi* [...], ed. M. Favia Del Core (Parma: Appresso Seth Viotti, 1567); Guidobaldo Dal Monte, *Guidiubaldi e marchionibus Montis Mechanicorum liber* (Pisauri: apud Hieronymum Concordiam, 1577); Rossi, *I filosofi e le macchine, 1400-1700*, chap. 1. If Shapin finds so many mathematicians from the Seventeenth century endorsing this discourse it is probably because it was already part of the scientific paradigm of their time. The increasing employment from the Fifteenth century by cities and by courts embedded in humanist rhetoric of experts in mathematics who were empirically experimenting with new methods to find innovative approaches in different fields of applied mathematics, enriching their patrons and themselves, created new forms of self-promotion and self-fashioning. This process emerged from the tensions between the mainstream method of claiming scientific authority and this new strategy.

19   Shapin, *Scientific Revolution*, 68.

20   Ibid.

21   This manuscript was used in the seventeenth and eighteenth centuries; but it was left unpublished until the nineteenth century. The editor of the manuscript writes wrongly Hanns instead of Jacob. The surname is also recordes as Büllmann or Püllmann: Johann Neudörffer, *Des Johann Neudörfer, Schreib- und Rechenmeisters zu Nürnberg, Nachrichten von Künstlern und Werkleuten daselbst aus dem Jahre 1547*, ed. Andreas Gulden and Georg Wolfgang Karl Lochner (Osnabrück: Zeller, 1970), 65-66; Jeffrey Chipps Smith, "Nuremberg and the Topographies of Expectation," *Journal of the Northern Renaissance*, Novem-

account of Brunelleschi's friendship with the physician and mathematician Paolo dal Pozzo Toscanelli shows the same literary topos:

> One evening Messer dal Pozzo Toscanelli returned from his work [*da studio* in Vasari's text, i.e., Padua university] and happened to be in his garden having supper with some of his friends, and he invited Filippo to join them. Having listened to him discussing the mathematical sciences, Filippo struck up such a friendship with Paolo that he learned geometry from him. And although Filippo was not a learned man [*non aveva lettere*, i. e., he did not read Latin], he was able to argue everything so well from his own practice and experience that he confounded Paolo on many occasions.[22]

What is here important in terms of Torriani's educational trajectory is the existence of a cultural milieu in which theory and practice were converging in full dignity into a discourse that supported the blooming of practical mathematics. The contrast of two different systems of knowledge, one theoretical taught at the university and one practical taught at the workshop, was indeed mainstream but not absolute. First of all, one could argue that technological knowledge at the workshop, to a certain extent, may be considered as scientific.[23] But even if we refuse this idea, we must admit that there were certain artisans – Zilsel's *superior craftsmen* – that mastered theoretical knowledge, even if, as in the case of Leonardo, Brunelleschi, Bulman and Janello, they were illiterate, that is to say that they were not educated in Latin, the lingua franca of European high culture. Such artisans were nevertheless discussing a series of mathematical problems more commonly associated with university culture. It was in their workshops that both artisans and academicians found and spoke the same language. Within urban space and at court (and not anymore at the monasteries, as it was chiefly happening in previous centuries), the best practical skills were interacting with the most refined theoretical knowledge. If at the university the scientific curriculum was theoretical, just outside of it, in the

---

ber 5, 2009 <http://www.northernrenaissance.org/nuremberg-and-the-topographies-of-expectation>. Last access: 29/07/2016.

22   Giorgio Vasari, *The Lives of the Artists*, ed. Julia Conaway Bondanella and Peter E Bondanella (Oxford: Oxford University Press, 1991), 114; *Le vite de piú eccellenti architetti, pittori, et scultori italiani: da Cimabue insino a' tempi nostri : nell edizione per i tipi di Lorenzo Torrentino, Firenze 1550*, ed. Luciano Bellosi and Aldo Rossi (Torino: G. Einaudi, 1986), 296.

23   As the economic hisorian Joel Mokyr put it: "Technology is knowledge, even if not all knowledge is technological": Joel Mokyr, *The Gifts of Athena: Historical Origins of the Knowledge Economy* (Princeton, [N.J.]: Princeton University Press, 2002), 2.

workshop, a literate man had the chance to interact with practice, thanks to common interests with ingenious artisans and because of the impulse of communal regimes' administrations that at least since the twelfth century had increasingly served to mould urban space and the countryside, creating offices for functions connected with engineering. In the years around 1550 in Florence, Nuremberg and Cremona, Vasari, Neuddörffer and Vida demonstrated how the idea that a craftsman could be learned in mathematics as much as, if not even more than, an academician had become widely accepted. The profitable role of empirical experience in mathematical learning, exemplarily proved by Brunelleschi in linear perspective and engineering, and so dramatically expressed by Leonardo in the above-mentioned words, had been the subject of debate in Christendom for a long period. The English Franciscan Roger Bacon (1214-1294), in his *Opus Majus'* famous chapter on *Scientia Experimetalis*, had emphasised that "*sine experientia ... nihil scire potest*", i. e., "without experience nothing can be known".[24] Of course, the semantic meaning of Roger Bacon's "Experimental Science" is not the same as -today;[25] however, this is one of the medieval roots of the New Science. Renaissance urban societies, embedded with humanist culture, expressed this conception in different ways: at the beginning of the fifteenth century, in the *De ingenuis moribus et liberalibus adolescentiae studiis*, the humanist Pier Paolo Vergerio (1370-1444), discussing liberal education, underlined the importance of action in the process of learning: "For we are not able to give evidence that we know a thing unless we can reproduce it."[26]

Vergerio was not talking about mathematics, which he considered purely theoretical, yet his statement stigmatises the fictional possession of a knowledge provided by passive learning. More than 150 years later, Giuseppe Ceredi (1520-1570), a physician from Piacenza – twin city of Cremona – whose book on the Archimedean screw (1567) contributed to the field of hydraulic engineering, provides us with an authoritative justification of the scientific practice of demonstrating with experience applied mathematical knowledge; indeed, Ceredi states that both Aristotle and Galen supported the idea that:

---

24   Roger Bacon, *The "Opus Majus" of Roger Bacon: Ed., with Introduction and Analytical Table Volume 2*, ed. Henry Bridges (Nabu Press, 2010), 167.

25   Peter Dear, "The Meanings of Experience," in *The Cambridge History of Science*, ed. Lorraine Daston and Katharine Park, vol. 3 (Cambridge: Cambridge University Press, 2006), 106-31.

26   Pietro Paolo Vergerio, "Pier Paolo Vergerio Defines Liberal Learning," in *Vittorino Da Feltre and Others Humanist Educators*, ed. William Harrison Woodward (Cambridge: Cambridge University Press, 1897), 102-9.

> No science nor art, whose last aim is placed into practical operation can
> be perfectly mastered; if one, who has absorbed the rules, does not con-
> firm them with many and different successful experiments.[27]

It seems that since the late Middle Ages, practice based on mathematics
enjoyed as *scientia media* a high status. *Scientia media* was a concept used by
Aristotle in his *Posterior Analytics* (I, 41) where he states that there are pure
mathematical sciences (arithmetic and geometry) and "mixed", "subaltern" or
"middle" ones, such as optics and harmonics. By the end of the sixteenth century there would be an intense academic debate on this issue. However, the
concept of *"scientiae mediae"* was not a Renaissance rediscovery: among medieval commentators of the *Organon*, Thomas Aquina had already discussed this
passage of the *Posterior Analytics*.[28] By the end of the fourteenth century
Milanese engineers had used Aristotle to claim a necessary role of practical
skills in the making of scientific knowledge. By 1399, the first Duke of Milan,
Gian Galeazzo Visconti – owner of the first planetary automaton as well –
summoned south of the Alps the Parisian architect Jean Mignot to advise local
engineers on the construction of the new enormous gothic Cathedral of his
capital city: some static problems were emerging because of the unusual
dimension of the building that had to surpass any other previous temple. There
soon arose a dispute between local masters and the French architect, who
had concluded his criticism with the cutting sentence *"Ars sine Scientia nihil
est"*, insinuating that the local masters were ignorant in geometry. The local
masters, on the other hand, quoted Aristotle – showing therefore that their

---

27 "*Nessuna scienza, od arte, il cui ultimo fine sia posto nell'operatione, si può perfettamente
    possedere; se chi ha appresso i precetti di lei, non conferma lor poi con varie esperienze molte
    volte, & sicuramente riuscite*": Ceredi, *Tre discorsi*, 7.

28 See: Mario Helbing, "La scienza della meccanica nel Cinquecento," in *Il Rinacimento ital-
    iano e l'Europa*, ed. Antonio Clericuzio and Germana Ernst, vol. 5, Le scienze (Costabis-
    sara (Vicenza): Angelo Colla Editore, 2008), 574; for what concern the meaning of
    experience and the debite on the scientificity of mixed mathematical sciences in the
    Early Modern Period, see: Dear, "The Meanings of Experience," 119-22;; About Thomas
    Aquina commenting on the *Posterior Analytics*: Alberto Strumia, *Introduzione alla filoso-
    fia delle scienze* (Bologna: Edizioni Studio domenicano, 1992), 26; for what concerns Regi-
    omontanus' claim of astronomy as a *scientia media* even superior to pure mathematical
    sciences, see: H. Darrell Rutkin, "L'astrologia da Alberto Magno a Giovanni Pico della
    Mirandola," 2008, 53; Aristotele, *Analitici secondi*, ed. Mario Mignucci, vol. Organon IV
    (Roma: Laterza, 2007).

reasoning was backed by authority- and concluded that "*Scientia sine Arte nihil est*".[29]

Another influential book attributed to Aristotle that gave dignity to another *scientia media* was the so-called *Mechanics* or *Quaestiones Mechanicae*: brought into Italy by the Greek cardinal Bessarion in the fifteenth century, and given as a present with the rest of his great library to the Republic of Venice. It circulated in manuscript form and was first printed by Aldo Manuzio in 1497. This work stated that mechanical problems were related to the fields of both mathematics and natural science, putting the method (mathematics) into the place of practice (Nature). The *Quaestiones Mechanicae* had the effect of reinforcing Aristotle's notion that mathematics could be applied to the reading of nature, which also dealt with the problems of motion. The properties of the scale, the lever, the pulley, the winch and the wedge were geometrically analysed and tested in the construction of machinery. These operations took place despite a prejudice held by some toward scientific experimentation at that time.[30] Though this text is not anymore attributed to Aristotle, during the fifteenth and sixteenth centuries it was received as such, and today it is still incorporated in the *Corpus Aristotelicum*.[31] In this context, the workshops of Vitruvian craftsmen such as Brunelleschi or Torriani presented an alternative but complementary way to the reading of Nature, comprised of a joyful interaction between people with varied education, but unified in the desire for a more complete and empirical knowledge. Vida and Fondulo, like Paolo dal Pozzo Toscanelli, Leon Battista Alberti or Luca Pacioli, constituted a sector of humanist university-trained culture that enthusiastically interacted with artisans. With this association they saluted the encounter of theory and practice, which, according to Vitruvius, were the parents of knowledge. The habit of prizing artisans was not confined to Italy: other highly urbanised places as, for instance,

---

29   James S. Ackerman, " 'Ars Sine Scientia Nihil Est' Gothic Theory of Architecture at the Cathedral of Milan," *The Art Bulletin* 31, no. 2 (1949): 84-111.

30   As mentioned in the introduction, many great scholars of the history of philosophy and of the Renaissance have noticed that it is impossible to speak of a single Aristotelianism in the Middle Ages and the Renaissance, in light of the different interpretations (often contradictory) that were in circulation, or even by aristotelismi seen as a school, given the mobility of professors in the universities of the era. For an introduction to the theme, well accompanied by examples, see the essay: Garin, "Aristotelismo veneto e scienza moderna."

31   Aristotele, *Meccanica*; Paul Lawrence Rose and Stillman Drake, "The Pseudo-aristotelian 'Questions of Mechanics' in Renaissance Culture," *Studies in the Renaissance* 18 (1971): 65-104; Helbing, "La scienza della meccanica nel Cinquecento"; Rossi, *I filosofi e le macchine, 1400-1700*, 36-37.

the Low Countries, some parts of France, and Franconia, as in the above-mentioned case of Nuremberg, offered examples of humanist apologetic discourses
of artisanal knowledge.[32] Pamela Smith, in the book where she presented her
theory of an "artisanal epistemology", has collected some of these learned
claims produced around a northern European university context. For instance,
she has noted that in 1531 the Valencian humanist Luis Vives remarked that:

> Scholars should not 'be ashamed to enter into shops and factories, and to
> ask questions from craftsmen, and to get to know about the details of
> their work'. When living in the southern Netherlands, Vives also advised
> his students to imitate the example of the fifteenth-century Louvain
> scholar Carolus Virulus, who sought out the fathers of his students in
> order to learn from them about their trades.[33]

Perhaps in 1568 the French philosopher Petrus Ramus (Pierre de la Ramée,
1515-1572) had Vives' words in mind when he visited a workshop in Nuremberg
for four days that he wished (he claimed) had been four years. Nuremburg
(together with Augsburg) was the German capital of sixteenth century
mechanical production. In such a fervid urban context, a respected teacher of
mathematics and calligraphy such as ohann Neuddörffer the Elder – who
remained outside university disputes and wrote in the vernacular – praised in
his 1547 *Nachrichten von Künstlern und Werkleuten* the artisans of Nuremberg
with 79 biographies of artists and artisans, claiming his city to be blessed by
God with a number of skilled craftsmen and artists greater than any other
town. The most famous artisan of Nurenberg was, of course, the painter
Albrecht Dürer, who, as Leonardo in Italy, aspired to the dignity of a scholar
writing treatises. In 1538 Vives had used Albrecht Dürer's figure in one fictional
dialogue between the painter from Nuremberg and two scholars. Vives' goal
was to stigmatise the stupidity of learned pedantry versus smart practical
knowledge. But Vives, Ramus, and Virulus were scholars too, though very much
open to a certain typology of Aristotelism and enthusiasts of interacting with
skilled artisans. It seems that their polemic attacks had to do more with
controversies internal to the world of the university, such as Erasmus' or
Montaingne's attacks against pedantry. The emerging enthusiasm for empiricism might have provoked turmoil in the traditional hierarchical structure of
scientific authority, resulting in a radicalized position of some conservative
scholastic academicians against empiricism. We should also not forget that

---

32   Smith, *Body of the Artisan*, 66.
33   Ibid., 66-7.

FIGURE 7   *Francesco Casella,* Aristotiles, *1513, Studiolo Landriani, Cloister of Sant'Abbondio,*
             Cremona. P[er]hypateticus (Aristotle) holding his book on Metaphysics, and on
             whose pages one can read in Latin: "All men, by nature, desire knowledge".
             COURTESY OF MINO BOIOCCHI AND OF THE DIOCESE OF CREMONA.

humanists were far from considering manual work the equal of intellectual. In
the eyes of educators such as Pier Paolo Vergerio, it was the less gifted child
who had to be directed to the study of crafts instead of the liberal arts.[34] Only
the Vitruvian craftsman, able to bridge practice and theory, could aspire to be
considered differently. With these illiterate, but talented and eager to learn
craftsmen, the scholar spoke a common language: mathematics.

Mathematics can be defined as the discipline of measurement and proportions, a skill that artisans with great command in precision-manufacture, such
as clockmakers, engineers and painters, could undoubtedly master. When
Pope Gregory XIII called for scientific contributions from all Catholic countries, the most powerful monarch of the time, the King of Spain Philip II,
commissioned the task to those whom he considered to be the three most
authoritative institutions of his realm in one of the traditional four branches of
mathematics: astronomy. These institutions were the universities of Salamanca
and Alcalá, and none other than his court clockmaker Janello Torriani! In 1579,
working therefore at this project, an elderly Janello Torriani performing one of
his last duties for the Spanish Crown, wrote to Philip II:

---

34   Vergerio, "Pier Paolo Vergerio Defines Liberal Learning."

I do not recognise myself capable of any rhetorical skill [*facilità di dire*], which is necessary to explain problems of such a difficulty. For this reason I took as an expedient Mathematical tables that are the instruments for the operations. Thus, I propose to Your Majesty not just the way to operate thez reduction [of the Calendar], but the tables that also explain it, and these are three: two circular ... and one squared.[35]

An illiterate craftsman such as Torriani was able to design complex and accurate mathematical instruments. Literate academicians trained in the *quadrivium* could read them. Both groups thus spoke a common language. Janello's trajectory shows that by the end of the sixteenth century, craftsmen were officially welcomed into mathematical discussions at the university level, and courts played a central role in this process. In 1552, when Philip II's father granted Janello a life-long pension, he called him "*mathematicus*," a shocking recognition for a guild-craftsman.[36] Janello accepted this investiture and he made use of it, as it emerges in several documents confirming his aspirations towards an intellectual status that went far beyond the practical dimensions of a craftsman, clockmaker or engineer – e. g., the transcription of a petition sent to the Doge of Venice: "Serene Prince, Giannello Turriano, Cremonese Horologist, and Mathematician of His Catholic Majesty ...".[37] However, it seems that the aspirations of the clockmaker to the dignity of mathematician had been hindered: In 1584, Juan de Herrera, in a letter to the Spanish ambassador in Venice, wrote that:

35　From Janello's letter and treatise on the reduction of the calendar sent to the King by the 19th of June 1579: "*io che nè in me conosco quella facilità di dire, che seria necessaria per dichiarare difficoltà tanto importante, ho pigliato per espediente con una superficie delle Matematiche, che sono gli istrumenti delle operationi, di proponere a V.M.tà non solo il modo della riduttione, ma le Tavole, che la dimostrano ancora et queste sarano tre; Due in forma circolare ... ed una quadra. Le quali s'io havessi l'oro et l'argento de gli Alessandrini et Babilonici, d'oro et di argento le haverei scolpite a V.M.tà ... Ma forse sarà più accertato che io le habbia poste in carta, acciocchè no riusciendo al proposito di quello si desidera, si possano con manco pensiero et lacerare, et ponere in pezzi*" (my translation); Juanelo Turriano, *Breve discurso a su majestad el Rey Católico en torno a la reducción del año y reforma del calendario: con la explicación de los instrumentos inventados para enseñar su uso en la prática*, ed. José A García-Diego and José María González Aboin ([Spain]: Fundación Juanelo Turriano : Castalia, 1990), 73-74.

36　Imperial Privilege for a pension granted to Janello Torriani by Charles V, ASCr, Comune di Cremona, Miscellanea Jurium, vol. 10, f. 149.

37　ASVe, Senato, Terra, filza 50, doc. 2.

The method to find the longitude has been improved by those who so far have not had and still today do not enjoy any credit in mathematics: they are the Marquis de los Velez (God has him in glory!), Janello Torriani and those who have sailed to India from Portugal ...[38]

Beside the establishment at court of great libraries and academies where both theoretical knowledge and practical use of mathematical instruments were involved,[39] Renaissance princes built and patronised court workshops.[40] In this framework craftsmen and literati interacted and courtiers could be educated in theoretical and material knowledge. We will also later see in detail how even princes sometimes entered the workshop.[41] The prince replicated at court the successful social model of the urban "superior craftsman's" workshop. Janello in Spain would run a court workshop with several servants and with both practical and theoretical duties.[42] In his court-workshop he built clocks, instruments and components for the Toledo Device. On a more intellectual level, Janello and his court-workshop had to keep his creations going, write technical treatises for the Crown, partake in astronomical observations with

---

38   "... el medio para hallar las longitudines que ha seido aprobado de los que por acá han tenido
      y tienen algun crédito en lo de las matematicas, que fue el marqués de los Vélez, que sea en
      gloria, y de Juanelo Turriano y de los que han navigado a la India de Portugal" (my transla-
      tion); María Isabel Vicente Maroto, "Juan de Herrera: un hombre de ciencia," in Actas del
      Simposio Juan de Herrera y su influencia (Camargo : 14/17 de julio, 1992), ed. Miguel Angel
      Aramburu-Zabala and Javier Gómez Martínez (Santander: Fundación Obra Pía Juan de
      Herrera, 1993), 81.

39   For court Spanish academies see: Mariano Esteban Piñeiro and Maria Isabel Vicente
      Maroto, "La Casa de la Contratación y la Academia Real Matemática," in Historia de la
      ciencia y de la técnica en la Corona de Castilla, ed. José María Lopez Piñero, vol. III, Siglos
      XVI y XVII (Valladolid: Junta de Castilla y León, Consejería de Educación y Cultura, 2002),
      35-52; Victor Navarro Brotóns, "El Colegio Imperial de Madrid: el colegio de San Telmo de
      Sevilla," in Historia de la ciencia y de la técnica en la Corona de Castilla, ed. José María
      Lopez Piñero, III, Siglos XVI y XVII (Valladolid: Junta de Castilla y León, Consejería de
      Educación y Cultura, 2002), 53-72; Nicolás Garcia Tapia and Maria Isabel Vicente Maroto,
      "Las escuelas de artillería y otras instituciones técnicas," in Historia de la ciencia y de la
      técnica en la Corona de Castilla, ed. José María Lopez Piñero, III, Siglos XVI y XVII (León:
      Junta de Castilla y León, Consejería de Educación y Cultura, 2002), 73-82.

40   Luis Cervera Vera, Documentos biográficos de Juanelo Turriani (Madrid: Fundación
      Juanelo Turriano, 1996), doc. 126. From this point, all translations of this book are mine.

41   See the section: Virtus vera Nobilitas est.

42   See the section: Janello Entrepreneur.

academicians, survey other people's work, and draft projects.[43] Enthusiastic literati, like Ambrosio de Morales and Esteban de Garibay, influential officials like the architect Juan de Herrera, and nobles of the court visited Janello's working space.

Powerful *exempla* from Antiquity, both from books and ruins, accelerated the enhancement of the "Vitruvian artisans" of the Renaissance. A discourse based on a classical tradition, of central importance for European elites, brought more rapid and definitive changes than any revolution: great shifts were made in the name of the restoration of ancient virtues. Beside the use of powerful figures such as Archimedes and Daedalus, even scholastic and humanist discourses entered together with scholars in the workshop: in the workshop where literati and craftsmen co-operated, Aristotelian and Neo-Platonist (or perhaps it is more correct to call them neo-Pythagorean) approaches to the reading of the world, though among them different, partook in an interesting fusion. If the senses were the necessary gates for experience, the idealistic geometrical and proportional obsession of Pythagorean philosophy enforced the use of a strict mathematical common language, consecrated with a hallowed aura by the successful results obtained. The Vasarian supremacy of drawing should probably be understood as a degree of capability in geometrical designing. The *Academia del Disegno* (1563), which trained nobles and craftsmen, considered drawing the basic and necessary form of knowledge. The warranty provided by this practice was in the process of mastering the World. Following the creation of this Medici academy, many other such institutions were established all around Europe. The platonic name of these associations, in its Pythagorean declination and fertilised by the Archimedean

---

43   When Janello died, the King asked for *"los ynstrumentos y otras cossas del dicho Juanello que se avran de tomar para nuestro servicio"*: these consisted in six chests full of papers, books and iron instruments that were brought to San Lorenzo's monastery. They may have been lost during the fire of the El Escorial. On March 1575 Juanelo was ordered to finish within a year the book he had already begun for the functioning and management of the two clocks he had created for the Emperor. In 1577 Juan López de Velasco, royal cosmographer, organised a systematic scientific observation of a an eclipse of the Moon: Juan López de Velasco and Andrés García de Céspedes observed the phenomenon from Madrid, Janello Torriani and a certain Dr. Sobrino from Toledo, Rodrigo Zamorano from Sevilla, and Jaime Juan from Mexico: Cervera Vera, *Documentos biográficos*, doc. 24, 50, 125, 126 and 132; Carlos Enrique Esteve Secall, "Aspectos histórico – gráficos de una observación a escala intercontinental: Las Instrucciones del Cosmógrafo Lopez de Velasco" (XVI congreso internacional de ingeniería gráfica, Zaragoza, 2004) <http://www.egrafica. unizar.es/ingegraf/pdf/Comunicacion17110.pdf>; María M. Portuondo, "Lunar Eclipses, Longitude and the New World," *Journal for the History of Astronomy* 40 (2009).

and Aristotelian mechanical tradition, created by royals and nobles, celebrated the triumph of practical mathematics and their utility, and what we now call technoscience. European elites came to recognise that well-taught technicians could augment and strengthen their power. One need only consider the crucial role of engineers in the most expensive and basic activity of the period: war. The institution of the academy was moulded on the successful experience provided by the workshop of the superior-craftsman. It was of vital interest to the state to replicate such experts. The court had a great demand for those who were in possession of this mixed knowledge. This was the context in which Torriani's rise must be understood.

In order to better zoom in on this context as a function of Janello's individual educational experience, we have to start from a couple of accounts written at the time of his death. A certain Camillo Capilupi (1531-1603), a Mantuan nobleman connected to the patronage-network of Torriani that I will later illustrate, wrote a couple of manuscripts collecting contemporary anecdotes. Two of these are related to Janello Torriani. In the second one, put on paper between 1579 and 1592, describing the precocious intelligence of the clockmaker, Capilupi claimed that:

> Maestro Janello Cremonese was born in the countryside, and by his pure virtue and strength of his genius alone, whilst he looked at the herd in the countryside, contemplated the movement of the stars and the sky so that he became in a short space of time so very intelligent in that art that he both began to craft clocks with his hands and to devote himself to the mathematics, accomplishing in both professions marvellous things ...[44]

There is no evidence of Torriani's early career as a shepherd. Similar stories were circulating about astonishing devices and their creators. There is greater

---

44   *"Quesito elegantísimo di Maestro Gianello a Carlo v Imperatore. Maestro Gianello Cremonese nato in contado, et per propria virtù / et forza del suo ingegno solito mentre guardava gli animali / alla Campagna colla contemplatione al moto delle stelle / et di cielo ch'egli vedeva divenne in brevísimo spatio di tempo / intelligentissimo di quell'arti [co]si tanto che datosi colle mani a fabri= / car orologii, et alle matematiche fece nell'un et nell'altra / professione così maravigliose ..."* (my translation); Camillo Capilupi, ms. Vittorio Emanuele 1062, Biblioteca Nazionale di Roma. There is a record written by Camillo Capilupi in a previous manuscript (ms. Vittorio Emanuele 1062, ca. 33v, Biblioteca Nazionale di Roma) that presents the same title, but it is shorter and does not contain this preface. I browsed the pages of Garcia-Diego's draft manuscript on Janello Torriani's biography to find some clues about this documents. I thank the Fundación Juanelo Turriano, for allowing me to check these papers.

amusement in the narrative of a wonder than in a story of rigorous study and constant commitment: the clock of the Cathedral of Cambrai was believed to have been made by a shepherd, whose eyes, after the construction, were put out to prevent him from replicating his marvel.[45] The same story was told about Felice di Salvatore of Fossato, who in 1552 had put a sophisticated astronomical clock in a tower at the Piazza Grande of Arezzo.[46] Clocks with astronomical dials and automata were most impressive things for a simple mind to behold and they were apt to provoke marvel. However, what is more interesting to observe is the unfolding here of the typical Renaissance *topos* of ingenuity that depends on a rhetoric of *sprezzatura*, fashioning *a posteriori* the origin of a brilliant career, in which the element of skill is considered to be an inborn feature.[47] The artificious *sprezzatura* tended to credit skills obtained by hard work and exercise to the influence of the heavens: in the scientific paradigm of the time, it was commonly held that one's ingenuity was the result of sidereal influences, and sometimes even of a supernatural process involving divine Providence. In Vida's pamphlet, though Torriani's physical appearance

---

45   David S. Landes, *Revolution in Time: Clocks and the Making of the Modern World* (Cambridge, Mass.: Belknap Press of Harvard University Press, 1983), 198.

46   Enrico Morpurgo, ed., *Dizionario degli orologiai italiani: 1300-1800* (Roma: La clessidra, 1950), 76.

47   In the early sixteenth century, a humanist of the Gonzaga family called Baldassare Castiglione, author of the famous *The Book of the Courtier*, defined a method for acquiring a certain social elegance, a courtesy. Describing the reasons behind the power of a superiorly refined manner he introduces the concept of *sprezzatura*: this consisted of a system which masked the achievements born of hard work in favor of those granted by "grace", which etymologically derives from the Latin *gratia*, meaning "gift". This word well defines the objective of such comportment that shrewdly conveys and manipulates the idea of an individual being superior through no effort of their own, admirable because he or she was gifted with intrinsic superior abilities and worth. Castiglione wrote: "But having before now often considered whence this grace springs, laying aside those men who have it by nature, I find one universal rule concerning it, which seems to me worth more in this matter than any other in all things human that are done or said: and that is to avoid affectation to the uttermost and as it were a very sharp and dangerous rock; and, to use possibly a new word, to practise in everything a certain nonchalance [*sprezzatura*] that shall conceal design and show what is done and said without effort and almost without thought. From this I believe grace is in large measure derived, because everyone knows the difficulty of those things that are rare and well done, and therefore facility in them excites the highest admiration; while on the other hand, to strive and as the saying is to drag by the hair, is extremely ungraceful, and makes us esteem everything slightly, however great it be". Baldassare Castiglione, *Il Libro Del Cortigiano*, ed. Giulio Preti (Torino: Einaudi, 1965), bk. I, chapter XXVI.

lacked dignity, with his huge hands and his beard and bizarre clothes all covered with soot, the clockmaker's mind was described as divine:

> I believe Senators, that there is not one of you who has not seen this admirable, extraordinary and, in a certain way, portentous work already finished and concluded in all its parts and numbers, in which its egregious artificer, with his eminent talent and goaded eagerness for investigation, has emulated the divine, indescribable, never sufficiently remembered and, until this moment, inimitable activity of God Himself ... This is, in fact, transferring what is divine to men and like contending with Nature herself and – if it is lawful to say so – emulating the Eternal Craftsman.[48]

The literary topos of the "*divine genius*" was well rooted in the humanist milieu.[49] For instance, when a hundred years before Torriani, Carlo Marsuppini (1399-1453) dictated the tomb's epitaph for the famous painter, architect, sculptor, clockmaker and engineer Filippo Brunelleschi, he had used the same rhetoric: "*plures machinae divino ingenio ab eo adinventae*", which means that many machines were invented by his divine ingenuity.[50] Vasari as well represents this belief introducing the character of Leonardo da Vinci: "The greatest gifts often rain down upon human bodies through celestial influences as a natural process, and sometimes in a supernatural fashion a single body is lavishly supplied with such beauty, grace, and ability that wherever the individual turns, each of his actions is so divine that he leaves behind all other men and clearly makes himself known as a genius endowed by God (which he is)

---

48   Vida, *Cremonensium Orationes III*, 53-57.

49   When in 1436, Leon Battista Alberti, in his *Della Pittura* presented his idea of the artist as an "*alter Deus*", it was not clear, as Rudolf and Margot Wittkower suggested, whether he had only drawn upon the Medieval tradition of the *deus artifex*, or upon Platonic philosophy too: Rudolf Wittkower and Margot Wittkower, *Nati sotto Saturno: la figura dell'artista dall'antichità alla Rivoluzione francese* (Torino: Einaudi, 1996), 112.

50   "*Non fu purtroppo attuato il progetto originario di decorare la sua tomba con lastre marmoree nelle quali sareebbero state delineate anche le machine da lui inventate ... un segno eloquente del radicale mutamento del clima culturale e, insieme, della rapida trasformazione dell'artefice medievale nell'artista-ingegnere del Rinascimento. La fortuna di Brunelleschi marca infatti emblematicamente l'inizio del distacco delle funzioni del tecnico tradizionale, semplice operatore, privo di aspirazioni letterarie e quasi sempre condannato all'anonimato. Il suo inaudito successo aprì la via all'affermazione sociale e alla definizione di una nuova e ben più ambiziosa identità professionale, che caratterizzerà gli artisti-ingegneri delle generazioni successive ...*": Galluzzi, "Dall'artigiano all'artista-ingegnere," 292.

rather than created by human artifice".[51] Also Janello's amazing skill was per-ceived as a divine gift, and therefore admired and respected even by people from a higher status.

This belief downplayed his hard work and study, leaving nearly no trace in the contemporary records and giving us few clues to reconstruct in detail the educational path and daily practice of the foremost among Renaissance con-structors of machines. When he had the chance, the clockmaker made no secret of his hard work to Ambrosio de Morales, court historiographer of Philip II, telling him that in twenty years of assiduous study for the arduous project of the first great planetary clock he felt twice so ill that he had been close to die.[52] In Capilupi's narrative, Janello Torriani's ingenuity is still more striking because presented in a contrasting picture, where economic poverty exalts a natural intellectual richness. Such a model was widely popular in the Renaissance as one can see in Vitruvius' narration of Ctesibius's origins (Fig. 8) or in Giotto's youth narrated by Vasari, and it would still be popular in the seventeenth century when Vincenzo Viviani, describing Galileo Galilei's youth, would em-phasise the contrast between his inborn genius that led him still as a child to recreate models of machines he had the chance to observe, and the economic difficulties his father's financial condition imposed on his education.[53]

To grasp something more about Janello Torriani's youth and education, we have to look at the only contemporary printed account that Antonio Campi (1524-1587), a fellow countryman of Torriani, published in 1585, the very same year of the clockmaker's death (Fig. 9). Antonio came from a famous family of painters. He was a painter too, and a sculptor, a map-maker and a historio-grapher. Beside his vast and highly praised pictorial activity, Antonio Campi's most famous work was the *Cremona fedelissima città, et nobilissima colonia dei*

---

51 Vasari, *The Lives of the Artists*, 285.

52 Morales, *Antiguedades de las ciudades de España*, fol. 90v-94r. I wonder if we are not fac-ing here another Renaissance literary topos: Leon Battista Alberti as well, in his autobiog-raphy, wrote that because of hard study, he twice got so sick as to almost lose his life. Leon Battista Alberti, *Autobiografia e altre opere latine*, ed. Loredana Chines and Andrea Severi (Milano: Rizzoli, 2012), 69.

53 Vitruvius Pollio, *De Architectura libri X*, English translation from: Vitruvius Pollio, *The Ten Books on Architecture*, trans. Morris H. Morgan, De Architectura (Cambridge, etc.,etc.: Harvard university press, 1919), 325; Vasari, *Vite de più eccellenti architetti, pittori, et scultori italiani*, 299-300. Twentieth century Torriani's biographer Garcia Diego, instead, took this feature as a matter of fact, and not as a topos. Vincenzio Viviani, "Racconto istorico della vita del sig. Galileo Galilei, nobil fiorentino," in *Fasti consolari dell'Accademia Fiorentina*, ed. Salvino Salvini (Firenze: nella stamperia di S.A.R., per Gio. Gaetano Tartini, e Santi Franchi, 1717), 397-431.

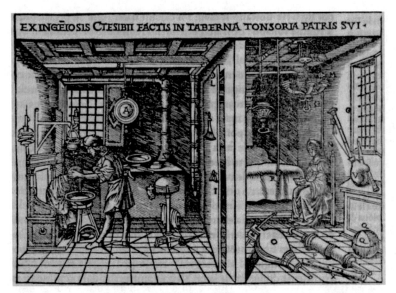

EX INGĒIOSIS CTESIBII FACTIS IN TABERNA TONSORIA PATRIS SVI ·

FIGURE 8   *Cesare Cesariano,* Ctesibius as a child inventing mechanical instru-
            ments (on the right), while his father in the next room exercises the
            humble profession of barber (on the left), *from the first illustrated
            vulgarised edition of Vitruvius, 1521.*

Romani, a historical work printed in 1585 and offered to Philip II. The elegant
volume, rich in fine engravings, executed by Agostino Carracci on Antonio's
original drawings, was made to accompany the first topographic map of the
city of Cremona, and the first one of its county, an enterprise that in Cremona
had won Antonio the title of knight and the exemption from taxation (year
1571). The *Cremona fedelissima,* dedicated to King Philip II, who was Duke of
Milan as well, is a well-written text, clearly testifying to a humanist-influenced
education, though vernacular. Antonio Campi's library suggests his strong cultural commitment: he owned a remarkable collection of 2,000 titles divided
into 6,000 volumes. An evaluation of every single book was made in 1587 and
takes up 73 pages in folio.[54] Antonio Campi had to be well acquainted with the
kinds of mathematical instruments necessary both for drawing a correct fore-
shortening of volumes and for reproducing a topographical map. He therefore
must have been sensitive to mathematics and well informed on such a famous
mathematician as Janello. In his *Cremona fedelissma* he wrote of Torriani:

---

54   Carlo Bonetti, "La libreria dello storico e pittore Antonio Campi," *Cremona* IV, no. I (1932):
      5-11.

Of all craftsmen our city had, no-one had given to it more honour than Lionello [sic] Torriano, a man of low origins, but given by God such a sublime ingenuity that he astonished the world, and everybody reckoned him a miracle of Nature, because, even though he has always been illiterate, he was able to talk about astrology and about the other mathematical arts so deeply and with such a fundament, that he seemed to have always attended nothing but that. He learned astrology before reading, his teacher being Giorgio Fondulo, doctor in medicine, philosopher and excellent mathematician, who very deeply loved Torriani, reckoning his genius supernatural.[55]

Since he was born around 1524 and worked in Milan in the middle part of the century, Antonio Campi may have known Janello personally. However, it sounds strange that he would deform Torriani's first name into *"Lionello"*. This could demonstrate an indirect knowledge of Torriani, or could be more likely the result of an attempt to ennoble a name considered inelegant: Lionello, meaning "little lion", was a name in use of the Italian nobility. As we shall soon see, Janello had his name and family name changed several times during his rise to success.[56] Even if Antonio did not know Janello Torriani personally, he

---

55   *"Ma di quanti artefici ha havuto la nostra Città, niuno più l'ha illustrata di Lionello Torriano, huomo nato bassamente , ma dotato da Iddio di così sublime ingegno, che ha fatto stupire il mondo, & e stato riputato da ogn'uno un miracolo di Natura, poiché non havendo giamai imparato lettere, parlaua dell'Astrologia, & dell'altre arti Matematiche tanto profondamente, & con tanto fondamento, che pareva non haver giamai atteso ad altro studio , haveva egli imparato Astrologia ancora che non sapesse pur leggere, insegnandoli Giorgio Fondulo Dottore di Medicina, & Filofofo, e Matematico preclarissimo, che molto l'amava, conoscendolo d'ingegno sopranaturale"*: Campi, *Cremona fedelissima città*, LV.

56   It may even have been that Antonio confused the identities of two different members of the same family. There are two documents from 1544 and 1550 mentioning Torriani as Leonello: an engineer called *Leonello Torriano* was paid at the time of Governor del Vasto (year 1544). In 1550 the guild of the blacksmiths of Milan elected a certain *Leonello Torriani q(uondam) Gerardo* as their abbot. The homonymy of the father and the identical place of residence (Porta Romana, parish of S. Nazzaro in Brolo) suggests an erudite transformation of the rude Janello into a more elegant Leonello. Janello himself could have been the author of such a change: after all this was not the only time he altered his name. If this Leonello does not coincide with the very Janello Torriani, he must have been a brother of his: we shall later encounter Janello's nephew called Bernardino, father to Leonardo, future engineer to Philip II. Another contemporary author, Angelo Baronio, had used a similar variant of the name in order to describe Torriani: *"Insignis virtute Leo"*, but he was probably influenced by Antonio Campi, as was the case for certain other local writers. For the archivial references of these documents see: Silvio Leydi, "Un cremonese

had many acquaintances that could have informed him about the clockmaker, among them the very Giorgio Fondulo, father of the painter Giovanni Paolo, an apprentice of his, who, as Torriani before him, had entered the service of the governors of Milan.[57] Therefore, we should take Campi's words as a precious testimony to spread light on the very obscure education of Janello. Indeed, the relationship between the goldsmith Brunelleschi and the university-trained physician Paolo dal Pozzo Toscanelli is one of the most significant examples in the process of interaction between talented artisans and curious university-trained scholars, and we can argue that Vasari's model for understanding these two men is the same we can recognise embodied in the relationship between Janello Torriani and his mentor Giorgio Fondulo, as we shall later see in detail.

This interaction made possible the implementation of the Hellenistic ideal of mixed knowledge embodied by great ancient inventors of machines such as Archimedes or Ctesibius, and conceptualised by Vitruvius, architect of the first Roman emperor: as we shall see, Janello considered himself an "architect of clocks", adopting a term that has a very different meaning from our modern one. Vitruvius, in the first book of his *De Architectura*, wrote:

> The knowledge of the architect is adorned with many disciplines and various skills: from which judgement he can evaluate any work of art.

---

del Cinquecento 'aspectu informis sed ingenio clarus': qualche precisazione per Giannello Torriani a Milano (con una nota sui suoi ritratti)," *Bollettino Storico Cremonese*, Nuova Serie, 1997, no. 4 (1998): 133-38.

57   Antonio knew Senator Danese Filodoni, who in 1587, two years after Torriani's death, and in the very same year of Antonio Campi's demise, sent to Cremona the portrait of the clockmaker together with a model of his Toledo Device. Moreover, Alessandro Lamo (1555-1612) mentioned that the painter Bernardino Campi, whose niece married Antonio Campi's son Claudio, had given to Torriani one of his paintings to take with him to the Habsburg court, most probably to promote the artist there. Antonio Campi had access to a special witness: Janello Torriani's mentor, Giorgio Fondulo, had a son named Giovanni Paolo, who had been apprentice to Antonio Campi, as the painter himself declares in his *Cremona fedelissima*. Ibid.; Silla Zamboni, "Campi, Bernardino," *Dizionario biografico degli Italiani* (Roma: Istituto della Enciclopedia italiana, 1974); Alessandro Lamo, *Discorso di Alessandro Lamo intorno alla scoltura, e pittura, doue ragiona della vita, ed opere in molti luoghi, ed a diuersi principi, e personaggi fatte dall'eccellentissimo, e nobile pittore cremonese* (Cremona: Nella Stamperia del Ricchini, 1774), 52; Campi, *Cremona fedelissima città*, LV; Francesco Arisi, *Cremona literata*, vol. II (Cremona: typis A. Pazzoni & P. Montii, 1706), 198; Francesca Campagna Cicala, "Fondulo, Giovan Paolo," *Dizionario biografico degli Italiani* (Roma: Istituto della Enciclopedia italiana, 1997); Lidia Azzolini, *Palazzi Del Quattrocento a Cremona* (Cremona: Editrice Turris, 1994), 134. The piece of news that Gio. Paolo was son of Giorgio (unknown to Campagna Cicala) comes from a notary act of the State Archive of Cremona: ASCr Not. Rog. G.F. Ghisolfi, fil. 908, year 1566, 26th of March.

This science is the offspring of practice and of theory. Practice is the fruit of continuous and incessant contemplation of the way of executing any work made with one's hands for the best transformation of materials according to each project. Theory is the fruit of reasoning through demonstrations and explication. And because of that those architects that are without letters, despite their manual capability, could not reach any authority in their work. But those architects who have just theory also fail, grasping in the shadows in this art and miss the true goal. But those that have one and the other are better equipped to achieve the authority of a project and turn it into an execution.[58]

The Vitruvian context in which Torriani moved is very clear, owning a copy of *De Architectura* and gaining fame reconstructing two of the machines described in it, as we shall see in detail in the second part of this book. The cultural milieu in which Torriani was educated, was well acquainted with figures of ancient inventors, such as Hero of Alexandria, who had consciously adopted this idea of construction based upon mathematical theory and practice. As we read in Pappus of Alexandria, a mathematician's of the late Roman Empire:

> Now, the mechanicians of Hero's school tell us that the science of mechanics consists of a theoretical and a practical part. The theoretical part includes geometry, arithmetic, astronomy, and physics, while the practical part consists of metalworking, masonry, carpentry, painting, and the manual activities connected with these arts. One who has had instruction from boyhood in the aforesaid theoretical branches, and has attained skill in the practical arts mentioned, and possesses a quick intelligence, will be, they say, the ablest inventor of mechanical devices and the most competent master-builder. But since it is not generally possible for a person to master so many mathematical branches and at the same time to learn all the aforesaid arts, they advise a person who is desirous of engaging in mechanical work to make use of those special arts which he has mastered for the particular ends for which they are useful.[59]

---

58   Cesare Cesariano and Vitruvius Pollio, *Di Lucio Vitruuio Pollione De architectura libri dece traducti de Latino in vulgare affigurati: commentati & con mirando ordine insigniti* (Impressa nel amoena & delecteuole citate de Como: P[er] Magistro Gotardo da Po[n]te, 1521).

59   Pappus of Alexandria, *Mathematical Collection*, trans. D. Jackson, vol. Book 8, 1970 <http:// archimedes.mpiwg-berlin.mpg.de/cgi-bin/toc/toc.cgi?page=2;dir=pappu_coll8_095_ en_1970;step=textonly>. Last access: 07.09.16.

FIGURE 9   *Agostino Carracci (engraving)*, Self-portrait of Antonio Campi, *Cremona fedelissima*, Cremona 1585.

Family, Social Status, Education

One of most disorienting features of Torriani is the changeability of his name – Giannello, Lionello, Janello, Giano, Juanelo – and of his surname – de Torrexanis, Torresani, Torresan, Torriani, della Torre, Torriano and Turriano. The clockmaker was originally called Janello Torresani, and he was the son of a certain *Gherardus*. From one of the chirographic notary deeds it emerges that Gherardo Torresani[60] was in turn the son of a certain *Ianellus*; therefore, Janello was not a nickname, but it was in fact a family legacy.[61] In 1572, only Alessandro Lamo from Cremona will remember correctly the original name and surname of the clockmaker, calling him *"Gianello 'l Torresan"*.[62] As we shall see, probably different strategies of ennoblement actuated by Janello and his admirers reshaped his name for a series of variants more suitable to his fortunate career.

The first player in this lucky professional trajectory was Gherardo, Janello's father. Janello's alleged poverty was but a myth, and his father had sufficient economic resources to pay for his son's expensive education in the workshop.[63] At the time of the Ancien Régime fathers often chose the educational path of their offspring. Indeed, traditionally education and the expenses of a practical training in a workshop were paid and accepted, if not directly dictated, by fathers. The revival of classical authors on the theme of education created a humanist literature on pedagogy on which we shall draw in order to represent the educational context of Renaissance North Italian urban space.[64] With two

---

60   See: Barbisotti, "Janello Torresani"; and Leydi, "Un cremonese del Cinquecento.". I found
      members of a Torresani kin living during the fifteenth century in Asola and Verona. Asola
      is a town located in the present-day region of Mantua. A famous member of this family
      was Andrea Torresani who had joined Aldo Manuzio in Venice in 1508. Other Torresani
      included Bartolomeo and Lorenzo, painters from Verona. Asola is 30 km from Cremona,
      25 from Mantua and 60 from Verona. Further investigations in the archives might reveal
      relations between these Torresani. During the fifteenth century, Asola and Verona
      belonged to the Venetian Republic, and since 1499, they were joined with Cremona. We
      know that clockmakers from the Venetian Republic entered Cremona during this period.

61   A large part of the historiography that deals with Torriani, especially the Spanish one,
      accepts the theory that Juanelo was a nickname with origins in the name Giovanni.
      Thanks to a recent article, we can be reasonably sure about the evolution of Torriani's
      name and surname. Barbisotti, "Janello Torresani."

62   *Gianello [e]l Torresan*, meaning "Gianello the Torresan". Alessandro Lamo, *Sogno non
      meno piacevole, che morale* (Cremona: Appresso Cristoforo Draconi, 1572), 53.

63   For the documents here quoted on Gherardo de Torresani see: Barbisotti, "Janello Torre-
      sani."

64   On education in humanist culture and on the different roles of mothers and fathers as
      pedagogical actors see: Paul F. Grendler, *Schooling in Renaissance Italy: Literacy and*

a   b

FIGURE 10a-b   Bronze seal matrix (right) and seal (left) bearing the inscription S. IANELI +
                TVRIANI, *in the style of the fourteenth century.* COURTESY OF THE MUSEO ALA
                PONZONE, PINACOTECA CIVICA, CREMONA.

notary deeds dated 28th June 1520, Gherardo Torresani, inhabitant of the parish of S. Silvestro in Cremona, bought in a joint-ownership a floating mill on
the river Po for 155 *libras*. Gherardo Torresani appears in another couple of
documents dated 1523 that deal with a deed of sale of 11 perches of land with a
small house for the total value of 198 *libras*. The same notary drafted two further documents (dated 1524 and 1529) involving Gherardo. The first one deals
with the rent of the above-mentioned mill by the Ciria canal. Gherardo could
not pay the rent because of the effects of war, which had incapacitated him to

---

Learning, 1300-1600 (Baltimore: Johns Hopkins University Press, 1989); Christopher Carlsmith, *A Renaissance Education: Schooling in Bergamo and the Venetian Republic, 1500-1650*
(Toronto: University of Toronto Press, 2010); Margaret L. King, "The School of Infancy: The
Emergence of Mother as Teacher in the Early Modern Times," in *The Renaissance in the
Streets, Schools, and Studies: Essays in Honour of Paul F. Grendler*, ed. Konrad Eisenbichler
and Nicholas Terpstra (Toronto: Centre for Reformation and Renaissance Studies, 2008),
41-86; "Cronaca di Cremona dall'anno 1494 al 1525."

use the mill. Following this default, the knight Cornelio Meli, owner of the mill, summoned Gherardo to appear in court, in front of the *podestà* of Cremona. The lawsuit ended with a mutual agreement to extend the period of the payment. The second document reveals that Gherardo eventually managed to pay his debt in full.

At this stage, Gherardo was living in the parish of Santa Lucia. In the same year he bought "*sex lapidum*", perhaps six pieces of marble, we do not know what for. Since Gherardo's work was connected with mills, I am tempted to interpret these *lapides* as having related to a millstone. Furthermore, the Cremonella canal, which cut the city from North to South, left the city-walls in this very area and the banks of the river Po, where there were many floating water-mills, were very close to the new house of Gherardo. Documents dating from some decades later show that a great part of the population of this parish was employed in textile manufacturing, which often involved mills.[65] In 1530, when Janello married, Gherardo was now living at a different address from his son, but in the same *vicinia* of Sant'Agata, not far from his first documented residence. Before 1536, Gherardo died.[66] Janello Torriani's wife was a certain Antonia de Sigella, daughter of *quondam dominus* Bernardino de Sigella. Two documents drafted on the 14th of March 1530 outline Antonia's considerable dowry: she brought to Janello 50 *libras* cash plus another 100 in goods. The status of *dominus* (usually an honour connected with the equestrian order) of Antonia's father is underlined, while Janello is not even referred to as master. This may suggest that the marriage signalled a rise of social rank for the clockmaker and a more secure economical life for the de Sigella family (which now lacked a *pater familias*). Janello's profession probably gave him the chance to penetrate a higher social class.

However, a further hypothesis is plausible, based on two clues, which relate to the patronymic of the deceased father-in-law, and the customary ways master craftsmen might come to possess a workshop. So far, I could not find any family bearing the surname "de Sigella". This surname might be related to the profession of a respected man such as Bernardino. Sigella could be translated

---

65   Giorgio Politi, "Ultimi anni d'attività di Gianfrancesco Amidani, mercante-banchiere cremonese (1569-1579)," *Archivio Storico Lombardo: Giornale della società storica lombarda* XI, no. I (dic 1984): 45.

66   Barbisotti published a document dated 26th of July 1536, where it is stated that *Magister Ianello*, living in the *vicinia* of San Prospero, and son of *quondam* (that means Janello's father was dead) Gherardo, takes as apprentice Giovan Francesco Botti, son of quondam Marsilio from the *vicinia* of Maioris Mercatelli (just behind the Cathedral). Barbisotti, "Janello Torresani."

as "seals", *sigella* being the plural form of the Medieval Latin *sigellum* (Classical Latin: *sigillum*). Was his father-in-law involved with the production of seals? This was a profession usually connected to metalworkers and mint officials. Until the reign of Francesco II Sforza, who died in 1535, Cremona had its own active mint.[67] This could also explain how Janello was able to obtain possession of a workshop. Indeed, we know that Janello Torriani became a master clockmaker and during the 1530s had a workshop in Cremona. At that time there were only three possible ways to become a workshop-owner: to inherit it from one's father, to buy it, or to marry the daughter or the widow of a master.[68] Most probably, Gherardo was not a master, even though Janello, in a document drafted in Milan when the father was already dead, had once claimed to be son of master *Gherardus* to whom he even attributes the status of *dominus*.[69] This seems to be an attempt by Janello to ennoble his own past: after all, the metamorphosis of his very surname points in the same direction. In fact, the Torriani household was one of the oldest and most noble in Milan, and its assonance with Torresani probably appealed to the ear of an ambitious Janello, eager to elevate his status. His successful employment in the capital of the duchy likely

---

67   Germano Fenti, *La zecca di Cremona e le sue monete: dalle origini nel 1155 fino al termine dell'attività* ([Cremona]: Linograf, 2001).

68   The Mazzoleni family of Padua offers here a term of comparison. Alberto Mazzoleni was a blacksmith. He had two sons: Giorgio, who walked in his father's footsteps becoming blacksmith and Giovanni Francesco who became clockmaker. In 1509 at the age of 18, Both Giorgio and Giovanni Francesco signed a marriage contract with two orphan sisters who, it seems, gave them access to the ownership of the workshop of their deceased father, a certain Leonardo Falso della Brentella. Elda Martellozzo Forin, *La bottega dei fratelli Mazzoleni, orologiai in Padova, 1569 : la sorprendente attività dell'artigianato padovano nella età di Galileo svelata da inedita documentazione archivistica*, Quaderni dell'artigianato padovano (Saonara Pd: Il prato, 2005, 2005), 23.

69   García-Diego, *Juanelo Turriano, Charles v's Clockmaker*, 148, n. 19: *Notario Dionigi Allegranza Seniori, di Milano, que le concerne. Di quel giorno ed un anno sono i patti di Magister Janellus de Torrianis fil. Q. m. domini Girardi, abit in Milano, a Porta Nuova, nella parrochia de S. Benedetto, con i quali promette di accettare in sua casa e instruire Sigismondo de Bacilieri di Ferrara ad adiscendum artem, et exertitium conficiendi orologios, et ad laborandum in apotheca dicti domini Ianelli*. There are other cases of successful clockmakers, sons of humble people who had nothing to do with this refined art: For instance see the brothers Dionisio and Pier Domenico di Cecco from Viterbo, moderators of the public clock of Siena from 1469 to 1475. They were famous for their clocks with complex carillons. The brothers brought their marvellous clock around Italy, from Naples to Florence and to Rome. Dionisio and Pier Domenico made also some hydraulic works for the Republic of Venice in 1481. They seemed to be sons of a simple master mason called Cecco. E. MORPURGO, *Dizionario biografico degli orologiai Italiani*, Roma 1950, pp. 58-59.

gave the Cremonese clockmaker the chance to fabricate a better pedigree. Although we cannot systematically exclude the possibility that Gherardo was a master (In fact, even Janello did not call himself "master" in the wedding documents, though he boasted of such a title the year before when regulating the civic tower-clock),[70] we know he was involved with some business that provided a young Janello with the necessary financial coverage to become an apprentice. In this case he could have helped his talented son to buy a workshop. The third hypothesis implies that Bernardino de Sigella, Janello's deceased father-in-law, was a master metalworker himself and that Janello inherited his workshop by marrying his daughter Antonia. Further archival research may spread more light on this obscure point.

From the marriage with Antonia de Sigella, Janello had at least two sons, one of whom died in an indeterminate year between 1541 and 1556 (the sculptor Leone Leoni is said to have lent Janello the money necessary to his burial),[71] and a daughter, Barbara Medea, who would accompany her father to Spain and later became his universal heir. In a letter of Barbara Medea to King Philip III dated 1601, Torriani's daughter attributed to herself the age of 70.[72] She was in all probability the oldest born from the clockmaker and Antonia, just after their wedding in 1530.

What emerges from these documents is that Campi's reference to Janello's low origin, "*bassamente nato*," does not mean he was from an indigent family, as commonly accepted by Spanish historiography, but merely a non-noble.[73] Janello Torriani's future remarkable accomplishments in hydraulic engineering suggest an early predisposition of the young boy towards this kind of mechanics. Had Gherardo's involvement with mills and agriculture likely influenced Janello's predisposition towards mechanics, his father must have noticed and encouraged such an inclination. It seems that Italian urban culture stimulated similar behaviours. If not a direct influence of humanists' pedagogical ideas such as Pier Paolo Vergerio's or Leon Battista Alberti's on Gherardo

---

70   ASDCr, Fabbrica del Duomo, Libri Provisionum, I, ca. 159r. Record of a payment of 15 *libras* (i.e. pounds) dated to the 6th July 1529, to be paid to "*magister Ianellus de Torresanis ... adaptandi seu reformandi horolia existentia super Toratio*".

71   Information coming from a letter of Leone Leoni written in 1556: Leydi, "Un cremonese del Cinquecento," 132.

72   García-Diego, *Juanelo Turriano, Charles v's Clockmaker*, 42.

73   Another engineer from Cremona, a contemporary of Torriani, who was employed by two kings of France and by the Most Serene Republic of Venice as Governor of Candia, was taken as an example of fame achieved through personal virtue by the same Antonio Campi, especially because he was also "*nato bassamente*". Campi, *Cremona fedelissima città*, 169.

Torresani, sixteenth-century Italy had its cultural roots in this kind of educational tradition. What might have been relatively extraordinary in his education was the mentorship provided by Giorgio Fondulo. I say relatively, because Fondulo's mentorship is to be considered as a practice consistent with humanist pedagogy.

Humanist Pedagogy in Cremona

You teach children, you perform a task for the state
PETRARCH[74]

Janello had to be educated in vernacular grammar and mathematics: in fact, he could write in good Italian and he knew arithmetic, geometry, astronomy and some music, the basic four subjects of the quadrivium, namely mathematics. A young boy was usually required to know how to read and write before entering into a workshop as an apprentice. From pedagogical treatises (based on classical models), we know that children began to attend school when they were around seven years of age. The first stage of education related to alphabetization. This involved the use of *tabula*, or *carta*, the *salterio* and the *pseudo-Donatus*, otherwise known as *Ianua* (from the first word of this textbook). Some inventories written in Cremona during the fifteenth century show the use of such textbooks.[75] In the neighbouring city of Brescia, Nicolò Tartaglia was sent to school when he was 5 or 6 years old. His father died shortly after and Nicolò had to quit. According to him, when he was 14, he decided autonomously to attend school for only 15 days learning the letters until "k". However, since he could not afford the payments, he was forced to leave. Tartaglia claimed proudly he learned the rest he knew as an autodidact.[76] It is likely that Torriani also attended such a school as Tartaglia.

The other basic knowledge provided by elementary education in Renaissance Italy was the so-called *abacus*. The students of *abacus*-schools were mainly

---

74   From a letter by Petrarch dated 1352 and collected in the *Rerum familiarum*; English translation from: Grendler, *Schooling in Renaissance Italy*, 3.

75   King, "School of Infancy," 46-48; Giuseppe Mainardi, "Due biblioteche private cremonesi del secolo xv," *Italia medievale e umanistica* 2 (1959): 449-51; Mariarosa Cortesi, "Libri memoria e cultura a Cremona nell'età dell'Umanesimo," in *Storia di Cremona*, ed. Giorgio Chittolini, vol. 6, Il Quattrocento. Cremona del Ducato di Milano: 1395-1535 (Bergamo: Bolis, 2009), 212; Nicola Ircas Jacopetti, "Il censimento annonario cremonese nel 1576," *Bollettino Storico Cremonese* XXII (1964 1961): 147, n. 22.

76   Tartaglia, *Quesiti et inuentioni diuerse de Nicolo Tartaglia*, 69v-70r.

sons of merchants and artisans whose interests were primarily practical. The teachers at *abacus*-schools were well taught in the *quadrivium* (arithmetic, geometry, astronomy and music), but their main educational activity concerned arithmetic and geometry and other practical subjects strictly related to them, such as perspective and topography (important for civil and military engineering, architecture, painting, etc.). Some vulgarizations of the *"Liber Abaci"* by Fibonacci (1202) were already circulating at the end of the thirteenth century. Geometry was taught mainly through Euclid's VI book of the Elements (which addressed the proportionality of similar triangles). The adoption of Hindu-Arabic numbers, which began in the time of Fibonacci and was already established long before Torriani's birth, had permitted easier calculations and the use of equations in order to find the value of an unknown quantity. Pupils were asked to resolve realistic arithmetical and geometrical problems; for example, they were asked to calculate the price of goods, the change among different currencies, the surface of a field, the capacity of a barrel or the elevation of a tower. Numerous *abaco* textbooks circulated among Italian cities, both as manuscripts and printed. For instance, one can list the *Ludi matematici* by Leon Battista Alberti (1452 ca.), the *Trattato d'abaco* by Piero della Francesca, or the abaco book par excellence by Piero's fellow countryman and admirer: Luca Pacioli. Pacioli wrote the famous *Summa de arithmetica, geometria, proportioni et proportionalità* (1494), which enjoyed great diffusion.[77] In Milan, Leonardo da Vinci illustrated Pacioli's text.

It has been observed that Italian cities and towns from the Centre and the North of the peninsula witnessed the rise of a pedagogical shift starting already in the thirteenth and exploding during the fourteenth century. Italian communal administrations needed a large number of officials, notaries, secretaries and especially merchants able to write, read and count (and sometimes their wives too) to carry out private trade and public administration. City councils paid a salary to public teachers under the justification of the "common good". The Ciceronian *topos* of the *rei publicae utilitas* (or simply *publica utilitas*) was plainly the legitimisation used in order to hire public teachers paid with the community's money. Indeed, by the *Trecento*, in the majority of the Italian communal and seignorial regimes, public teachers, together with the *medico condotto* (the physician) and the surgeon, became institutional figures. Public teachers were paid by the tax-office, and they could also ask for additional fees from their students. In the *Statuta* of the Commune of Cremona approved in

---

77   Filippo Camerota, "L'eredità di Euclide: la tradizione dell'abaco," in *I Medici e le scienze: strumenti e macchine nelle collezioni granducali*, ed. Filippo Camerota and Mara Miniati (Firenze: Giunti, 2008), 23 and following; Grendler, *Schooling in Renaissance Italy*, 13.

the years 1339, 1349 and 1356, teachers of the public schools, though probably schools of higher education and not elementary schools, were to be paid partly by the community and partly by their students.[78]

In the Italian mercantile environment, fathers were aware of the fact that, beside its practical use for book-keeping, contracts, etc., education could also elevate their scions' social status, especially in an urban cultural context in which the upper class considered learning and literary consumption as a sign of distinction. Basic education in grammar and arithmetic could easily give to anybody the possibility to cultivate one's personal interest. Economic efforts were made by humble parents to push their children up the social ladder. According to the tradition, the *tribune of the people* Cola di Rienzo, whose father was an innkeeper, studied to become a notary, and Donatello, whose father was a *ciompo* (a simple worker in the wool industry), could attend a goldsmith's workshop. In Rome by the first half of the sixteenth century, many workers and craftsmen could write. It seems that in mercantile cities like Florence and Venice, by the beginning of the sixteenth century, around a third of the male population was able to read and write.[79] Since the thirteenth century, Cremona also had a strong mercantile tradition, and this figure could easily apply to it as well.[80]

Even if Janello's father could not have afforded to pay for the elementary education of his son, free education of indigent talented children was a feature of humanist culture as the theoretical and practical works of pedagogues such as Vergerio, Guarino da Verona and Vittorino da Feltre show. As Cardinal Silvio Antoniano said, even though a poor boy should follow his father's vile profession, he:

> might have a natural disposition toward something more noble. He went happily to school, learned more quickly than others, loved books, and

78   Emilio Giazzi, "Letteratura specialistica e biblioteche professionali a Cremona tra Medio-
      evo ed età Moderna," in *I professionisti a Cremona: eventi e figure di una storia centenaria*,
      ed. Valeria Leoni and Matteo Morandi (Cremona: Libreria del Convegno, 2011), 15.
79   Giuseppe Micheli, *I fatti di Cola di Rienzo* (Roma: Sovera Edizioni, 2002), 77-79; Horst W.
      Janson, "Bardi, Donato, detto Donatello," *Dizionario biografico degli Italiani* (Roma: Isti-
      tuto della Enciclopedia italiana, 1964); Grendler, *Schooling in Renaissance Italy*, 108.
      Najemy, who considers Villani's numbers on Florentine pupils' education alike (68% of
      boys and girls in the 1330s!), offers an even more remarkable picture for fifteenth century
      Florence, where 80% of households' heads were able to write with their own hands fiscal
      declarations: John M. Najemy, *A History of Florence 1200-1575* (Malden, MA: Blackwell,
      2006), 45.
80   Vigo, "Il volto economico della città," 222-23.

disliked the 'humble crafts'. If the teacher confirmed that the boy possessed superior intelligence, a penniless father need not despair.[81]

This phenomenon was later institutionalised by the Tridentine legislation.[82] The rise of a moral commitment to promote the cultivation of children's natural predispositions appeared already in the writings of the above-mentioned Pier Paolo Vergerio. Born under Venetian rule, he lived in Bologna and Florence, where he also taught. At the very beginning of the fifteenth century, he wrote *De ingenuis moribus et liberalibus adolescentiae studiis*, the first treatise of humanist pedagogy, in which he stated that one should teach a child according to the pupil's natural predispositions.[83] The famous humanist Leon Battista Alberti argued that education was to be understood as the improvement of natural capacities.[84] Humanists bestowed upon their societies some of the opinions of the Ancients on education. Works such as Plato's *Republic*, translated by Pier Candido Decembrio of Pavia (1439), supported the idea of teaching according to one's talent, despite social class and even gender. Probably the most famous pedagogical places of Renaissance Italy to support teaching in accordance with "natural inclinations" were the schools of Guarino Veronese and the *Ca' Zoiosa* – meaning "the Joyful House" – of Mantua created by Vittorino da Feltre under the Marquis of Mantua's patronage. This institution had up to 70 students: among them, the sons of the marquis, the sons of noble families and some talented poor boys. In fact, Vittorino believed in the idea of identical cultural opportunities for high intellectual capacities, in spite of social origin.[85] Even if one were not destined to an eminent position, one had duty towards society and therefore merited an education. The core of such a civil ethic was based on Christian tradition, Classical knowledge, and the discipline of the knightly courtesan.

Cremona was geographically close to Mantua (a few kilometres from the border with this state) and to the Venetian Republic. In fact, for the period

---

81   Grendler, *Schooling in Renaissance Italy*, 9, 88, 102-8; Pietro Verri, *Storia di Milano*, vol. II (Milano: Soc. Tipografica deClassici Italiani, 1835), 371.

82   Grendler, *Schooling in Renaissance Italy*, 107.

83   Pietro Paolo Vergerio, "De Ingenuis Moribus et Liberalibus Adolescentiae Studiis," in *Vittorino Da Feltre and Other Humanist Educators: Essays and Versions : An Introduction to the History of Classical Education*, ed. William Harrison Woodward (Cambridge: Cambridge University Press, 1897), 102-9.

84   Leon Battista Alberti, *I libri della famiglia*, ed. Ruggiero Romano and Alberto Tenenti (Torino: G. Einaudi, 1994); William Harrison Woodward, *La pedagogia del Rinascimento, 1400-1600* (Firenze: Vallecchi, 1923), 58.

85   Cortesi, "Libri, memoria e cultura."

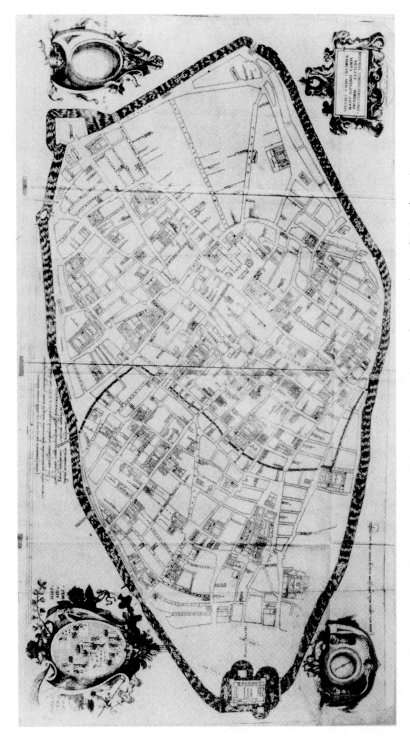

FIGURE 11   Topographic map of Cremona, designed by Antonio Campi and engraved on copper by David De Laude in 1582.

1499-1509, the Lion of Saint Mark ruled Cremona. Beside the local tradition based since the thirteenth century on the classical readings of notaries and teachers of the local *Studium*,[86] and without forgetting about the enormous relevance of Tuscan humanism, the Northeast of Italy can be considered an outstanding humanist pedagogical workshop. The fertile urban soil provided by the area in which Venetian influence was stronger (i.e., the Most Serene Republic itself, the Marquisate of Mantua and the Este territories of Ferrara, Modena and Reggio) was sown by the cultural models yielded in the Greek dominions of the Republic and in its privileged relation with the East, in the reception of the Byzantine diasporas, in the library of Cardinal Bessarion,[87] in the legacy of Petrarch, and in the ancient humanist milieu at Padua University. The result was a conscious revival of the Classical ideal of a harmonious balance between the cultivation of both intellect and body. This had a great influence on the schools of these humanists: grammar, arithmetic, geometry, and rhetoric were accompanied by games providing physical exercise, military drilling, dance, music and singing. Of course there were differences among the pedagogues. Vittorino's complex curriculum required the study of Latin and Greek authors, music and mathematics, composition in prose, rhetoric, grammar, poetry, history and physical exercise. Bartolomeo Sacchi, nicknamed Platina, recalls how Vittorino used to praise the Greek concept of *enkuklopaideia*, i. e. a 360° education.[88]

Also people from Cremona attended the humanist schools led by Guarino da Verona and Vittorino da Feltre, among them Iacobus Cremonensis of San Cassiano and Bartolomeo Platina. As concerns mathematical studies, Jacobus was an important figure in the long and meticulous process of philological research and emendation of classical authorship. Jacobus was a pupil of Vittotino da Feltre, and likely the latter's most beloved protégé. Indeed, at the death of his master, Jacobus inherited Vittorino's library and was appointed by the Marquess of Mantua at the direction of the *Ca' Zoiosa*. After three years, Jacobus moved to Rome where Pope Nicolas V ordered him to translate Archimedes. It seems that this translation, because of the complexity and the difficulty of interpretation, had little circulation. Nevertheless Regiomontanus received a copy of it, which he used in order to prepare his own posthumous

---

86   Giazzi, "Letteratura specialistica," 15.

87   Gianni Micheli, "L'assimilazione della scienza greca," in *Scienza e tecnica nella cultura e nella società dal Rinascimento a oggi*, vol. 3, Storia d'Italia. Annali (Torino: Einaudi, 1980), 199-257.

88   Woodward, *La pedagogia del Rinascimento, 1400-1600*, 36-44.

edition.[89] Bartolomeo Sacchi was born in Piadena, a village in the ecclesiastic diocese of Cremona, in 1421. He was called Platina after the Latin name of this place. He died in Rome in 1481. Platina is a paradigmatic example of the pedagogy of the Mantuan school: after attending Vittorino's *Ca' Zoiosa*, he served as a man-at-arm under the command of the captains Francesco Sforza and Niccolò Piccinino, and he taught the sons of Lodovico Gonzaga, Marquis of Mantua. He moved to Florence to learn Greek, attending Giovanni Argiropulos' classes and worked at Virgil's opera under the patronage of the ruler of Mantua. From 1461 he was in Rome, where Paul II imprisoned him twice and even tortured him under the accusation of paganism. Platina belonged to the circle of the academics of Pomponio Leto. The next pope, Sistus IV, was instead extremely friendly to Platina, appointing him curator of the Vatican Library. Author of many works, Platina was committed to the idea of civic education as one can infer from moral works such as *De Optimo Cive*, *De Principe Institutio* – dedicated to Federico Gonzaga and inspired by Cicero's *De Officis*. Platina is also known to be the biographer of Vittorino da Feltre and the author of the lives of the popes, but he is mostly famous for the fresco by Melozzo da Forlì, where he is depicted while kneeling in front of Sistus IV. Interest in Cremona about this North-eastern Italian humanist pedagogy is also documented by the edition of a textbook by Guarino da Verona: by 1494 or 1495, the *Carmina differentialia* was published in Cremona.[90]

There were a number of other relevant educators from Torriani's city who had a significant role in the pedagogy of the time. Among them we can remember Bartolomeo Petronio, who taught grammar and rhetoric to the sons of Francesco Sforza and Bianca Maria Visconti, Dukes of Milan and Lords of Cremona.[91] Beside this magnificent court, the duchy of Milan had in its capital

---

89   Marshall Clagett, *Archimedes in the Middle Ages*, vol. 1, The Latin tradition (Madison: The University of Wisconsin Press, 1964); *Archimedes in the Middle Ages*, vol. 2, The translations from the Greek by William of Moerbeke (Philadelphia: The American philosophical society, 1976); *Archimedes in the Middle Ages*, vol. 3, The fate of the medieval Archimedes (Philadelphia: The American philosophical society, 1978); *Archimedes in the Middle Ages*, vol. 4, A supplement on the medieval Latin traditions of conic sections (Philadelphia: The American philosophical society, 1980); *Archimedes in the Middle Ages*, vol. 5, Quasi-Archimedean geometry in the thirteenth century (Philadelphia: The American philosophical society, 1984).

90   Guarinus Veronensis, *Carmina differentialia* (Cremona: Rafainus Ungaronus & Caesar Parmensis, 1494).

91   Rita Barbisotti, "Gli inizi della stampa a Cremona (1473-1500)," in *Storia di Cremona*, ed. Giorgio Chittolini, vol. 6, Il Quattrocento. Cremona del Ducato di Milano: 1395-1535 (Bergamo: Bolis, 2009), 240.

city important humanist schools. Many leading Renaissance intellectuals had been trained at the schools based around the Sforza and, after Louis XII's invasion, the Valois-Orléans. For instance, the humanist *Scuole Piattine* (founded by Tommaso Piatti) opened in 1503. It was here that Fazio Cardano and his son Girolamo taught mathematics. Cremona was subject to the same cultural trends as Milan and other neighbouring cities: at the beginning of the sixteenth century Cremona had ten public lectors of *humanitates*.[92] As we shall see, in these very years pupils from other cities were coming to study Latin letters in Cremona.

Another member of Giorgio Fondulo's family, Girolamo Fondulo (Cremona, ca. 1490-Paris 1540), was a respected humanist and the preceptor to the king of France's second son, the future Henry II. In 1529 Girolamo Fondulo purchased for King Francis I some 50 volumes, forming the core of his humanist library.[93] Marco Girolamo Vida (Cremona ca. 1480-Alba 1566) was probably the foremost among local humanists. He was the author of the vivid portrait of Janello written in 1550, which presented an enthusiastic eulogy to the craftsman and his planetary clock, baptizing it with the Greek name of *Microcosm*.[94]

---

92   Giovanni Da Pozzo, *Storia letteraria d'Italia*, vol. 1, Il Cinquecento (Padova: Piccin Nuova Libraria, 2007), 149-51; Paolo C. Pissavino, "Le forme della conservazione politica: ragion di stato e utopia," in *Le filosofie del Rinascimento*, ed. Cesare Vasoli (Milano: Bruno Mondadori, 2002), 457; Domenico Bordigallo, *Urbis Cremonae syti designum*, ed. Emanuela Zanesi (Cremona: Associazione ex Alunni del Liceo-ginnasio "Daniele Manin," 2011), 89; Arcangeli, "La città nelle guerre d'Italia (1494-1535)," 50.

93   Francesca Piovan, "Fondulo, Girolamo," *Dizionario biografico degli Italiani* (Roma: Istituto della Enciclopedia italiana, 1997); Francesco Arisi, *Cremona literata*, vol. 1 (Parmae: Pauli Montii, 1706), 139-40; Campi, *Cremona fedelissima città*, xxvi.

94   Vida started his classical education in Cremona, before moving to Mantua, Padua, Bologna and then to Rome. He was especially well-connected with the popes of the Medici family. Leo x, after Vida become popular because of two Latin poems on chess, "*Scacchia ludus*" and the silkworm "*De bombyce*", commissioned him to produce an epic work on the life of Christ: the *Christias*, exampled on Virgil's Aeneid. This text was wildly successful and seems to have led Clement VII to appoint him bishop of Alba in 1533. Vida fashioned himself as a new Christian Virgil. The ancient poet once wrote "*Mantua vae miserae nimium vicina Cremonae*" (Mantua, alas, too near unfortunate Cremona! Virgil, *The Works of Virgil*, ed. Malcolm Campbell (New-York: Printed for E. Duyckinck, 1803), 45) and indeed the city of the Gonzaga was very close to Vida's home. What could be more suitable for such a Virgilian humanist than to study at Mantua as Virgil had studied at Cremona? Marco Antonio Vida (as he was called before taking monastic vows), after his elementary education moved to study at Mantua, most probably at the convent of the Lateran Canons of San Marco. He became one of them and thanks to their connections he moved to Leo's x court in Rome. Later, again because of their support, he received the priory of Santa Pelagia in Cremona, a church that he would order Giulio Campi to rebuild and splendidly decorate in the 1540s. Vida was made bishop of Alba in the period that

Even though the careers of many Cremonese humanists were based in Milan, Paris, Padua, Mantua, Florence, Bologna or Rome, it is not improbable that their prestige and their relations in Cremona exerted a certain influence on the pedagogical systems of Torriani's city. A small community always enjoys the external successes of its scions: Torriani's many encomia in Cremona offer a good example of this. Moreover, the links with the fatherland were never really cut: many of these characters, especially nobility, managed to come back and live for certain periods in Cremona, influencing the old *patria*.[95]

In conclusion, humanist culture created the conditions for a wide penetration of educational practices within urban space outside universities, and made the promotion of individual talent independent of social background socially acceptable. This may explain why in Northern Italy, Germany, and in the Low Countries, cities with great social mobility connected to knowledge (both practical and theoretical) produced and exported large numbers of innovative technicians.

Federico Gonzaga became Marquis of Monferrato (of which Alba is a part). Vida was championed by not only the two Medici popes, but also by the two that favoured Torriani: Pius IV and Cardinal Alessandrino, the future Pius V. Vida was also believed to have exercised an influence on Carlo Borromeo, especially as a result of his *Constitutiones synodales*, written in 1562: Vincenzo Lancetti, *Della vita e degli scritti di Marco Girolamo Vida Cremonese* (Milano: G. Crespi, 1831), 36-37.

[95]   A good example of these elastic roots that once in a while brought back to Cremona local humanists with international careers appears in one of the most significant series of images of 16th century Lombardy: in 1535, outside the city-walls of Cremona, a new and imposing cycle of frescos was begun at the church of S. Sigismondo. The local noble, and ducal architect and engineer Bartolomeo Gadio (1414-1482) had rebuilt the church some decades before on the spot where in 1441 Francesco Sforza had married Bianca Maria Visconti. The lavish decoration that was supposed to celebrate the union between the Visconti and the Sforza was curiously started in 1535 when the dynasty was beginning to die out. Guazzoni suggests a possible link between the beginning of the decoration of this building and the meeting in Cremona of Girolamo Fondulo with three fellow humanists: Benedetto Lamprido, Bernardo Regazzola and Bishop Vida: Valerio Guazzoni, "Pittura Come Poesia: Il Grande Secolo Dell'arte Cremonese," in *Storia Di Cremona*, ed. Giorgio Politi, vol. 4, L'età degli Asburgo di Spagna: 1535-1707 (Bergamo: Bolis, 2006), 350-415.

CHAPTER 2

The Theoretical Clock

The Science of the Stars

Despite its wealth, Cremona was in Italy a second-rank city concerning high culture: it neither had a long lasting university,[1] nor a notable production of books,[2]

---

1   During the fifteenth century Cremona had for some decades its university, a *Studium Generale*, acknowledged with a bull from Emperor Sigismund in 1413. However, since 1292, and during almost all of the fourteenth century, a *Universitas Scholarium* was open in Cremona for students of law, logic, Latin grammar and (at least from the year 1387) medicine. The faculties that Sigismund granted to the *Studium* were: Holy Theology, Civil and Canonical Law, Medicine, Natural Philosophy, Moral Philosophy and Liberal Arts. It seems that the divisions within the Visconti family kept the local *Studum* alive, despite Galeazzo's attempt in 1361 to centralise (and therefore control) university education in Pavia. Most probably Gian Galeazzo's successive elimination of his uncle Bernabò (lord of Cremona), and the unification of the Visconti dominions in 1385, contributed to the decline of the local institution. This *Studium Generale* was granted the same privileges as Paris, Bologna, Orleans and Montpellier, and, as such, it had the power to deploy several chairs, among them one in mathematics and astrology. The curriculum in mathematics and astrology consisted of the teaching of *De imaginum coelestium significatione* and *De horis planetariis*. Grendler utterly neglected Cremona in his work on Renaissance Italian universities, and even some Italian scholars consider this *Univarsitas* more a gymnasium propaedeutic to university education than a proper university: see in Valeria Leoni and Matteo Morandi, eds., *I professionisti a Cremona: eventi e figure di una storia centenaria* (Cremona: Libreria del Convegno, 2011); Gabriella Zuccolin, "Le figure sanitarie: secoli XIV-XVII," in *I professionisti a Cremona: eventi e figure di una storia centenaria*, ed. Valeria Leoni and Matteo Morandi (Cremona: Libreria del Convegno, 2011), 96-97; and Giazzi, "Letteratura specialistica," 15-16.

2   The number of incunabula printed in Cremona was small: only 32 known printed editions between 1473 and 1500 or 60 if one adds the 28 incunabula printed in Soncino, a community belonging to the county of Cremona. Printers in 15th century Cremona included foreigners and locals and they were often involved in companies with the same business in other towns. One of them, Rafaino Ungaroni, or Ongaroni, was an engineer and he came from a family practicing the same job since the first half of the 15th century. An Anna Ungaroni got married in the 16th century with the painter Bernardino Campi. The French Pierre Maufer (de Maliferis) was also active in Cremona. In contrast, a significant number of printers from Cremona were working in Venice. Book production in Cremona saw a sharp rise after 1492. In this last decade of the century, 31 incunabula were edited. The majority of these texts related to Petrarch's philosophy and classics: among them the prestigious *Castigationes Plinianae* by Ermolao Barbaro (1495): Bonetti, "La libreria dello storico e pittore Antonio Campi," 7;

FIGURE 12
*Giovan Pietro da Rho
(?)*, Funerary
monument erected in
1501 in memory of the
famous physician and
astrologer Giovanni
Battista Plasio, *Church
of St. Augustine,
Cremona.* The right
hand of the doctor
holds an armillary
sphere. COURTESY OF
MINO BOIOCCHI AND
OF THE DIOCESE OF
CREMONA.

nor an especially rich public library.[3] Pavia, the third centre of the state, though

Barbisotti, "Gli inizi della stampa a Cremona (1473-1500)," 238. In other Italian centres, beside
the main printing capitals (Venice 3,747, Rome 2,074, Milan 1,119, Florence 881) Brescia had a
production of 293 printed incunabula, Mantua 55 (Sabbioneta 1), Pavia 336, Ferrara 120,
Bergamo 1, Piacenza 4, Parma 80, Modena 87, Reggio Emilia 29, Verona 43, Padua 185, Vicenza
120, Treviso 109, Bologna 572, Genova 4 or 5, Turin 59, Siena 83, Pisa 18, Urbino 3, Perugia 64,
Naples 325, and Palermo 1. For a general perspective on the European production of incunabula: England 408, France and francophone Switzerland 5266, Germany and other German-
speaking countries 9908, Italy 10426, Duchy of Burgundy (Flanders included) 2335, Spain 892.
Outside of Italy, the main printing capitals for the fifteenth century were: Lyons 1422, Basel
847, Nuremberg 1059, Paris 3171, Augsburg 1256, Cologne 1616; this numbers are related to the
census taken so far by the ISTC (here uploaded to April 2009).

3   By 1487, when the University had probably definitively sacrificed to the ducal one of Pavia,
the *Massari* of the Cathedral (administrators of the Cathedral's buildings and possessions,
elected by the civil authorities of the city) created "*ad utilitatem publicam*" (for public utility)
a public library connected to the scriptorium of the bishopric, which for many centuries had
been a place of schooling. At the opening of the Cinquecento the patrimony of the public
library of Cremona consisted of 460 titles. This institution was structurally a typical humanist
library (like those which exist to this day at Cesena and Florence), equipped with lines of
lecterns and benches to which the books were chained and a large window. Anyone who had
a desire for learning could enter it. The library had a curator who had to live in a room next

smaller than Cremona, had the most important university of the duchy, and it had an incunabula production ten times bigger than Janello's city. However, it is interesting for our story, as concerns astronomical knowledge connected with the construction of planetary clocks, that Cremona was home to a Dalmatian printer, a certain Dobric Dobrida from Ragusa (Boninus de Boninis) who had escaped from Lyon. Before the French period, Boninus had previously worked in Brescia, printing in 1485 the *Statuta* for the City Council of Cremona and for its mayor tradesmen's guild: the *Universitas Merchatorum*. Owing to his sudden flight from Lyon, in Cremona he immediately reprinted the astronomical book just edited in the French city in the year 1494: Willelm Gilliszoon de Wissekerke's (Gulliermus Aegidii Zelandinus) *Liber desideratus super celestium motuum indagatione sine calculo*. As we shall see, books such as Zelandinus' were at the core of the interests of Torriani's mentor, Giorgio Fondulo, and formed the base of Janello Torriani's astronomical knowledge.

At this time, Latin astronomical knowledge was mainly based on Ptolemy's *Almagest*, translated by Gerard of Cremona in Toledo during the twelfth century, This book was circulating in all of Latin Europe. Thus, the fact that Gerard of Cremona sent home from Toledo copies of his translations had no great impact on the scientific environment of Cremona, except for the reverence that local scholars felt for their famous countryman. Other important works circulating at the time included the hugely successful *Sphaera Mundi* by

---

to it. By contract, he could also give classes in an adjacent room. According to the sixteenth century notary and chronicler Domenico Bordigallo, close to the library there were two schools instituted by Bishop Girolamo Trevisano (1507-1523), a Venetian elected during the dominion of the *Serenissima* (1499-1509). These two schools, one for grammar and the other for music, were also open to the indigent children of the city. In the early 1520s, the bishop also had a "deputed grammar master" on his payroll. As already mentioned, at least ten public lectors in *humanitates* completed the picture. Bonetti supported the idea that this library already existed at the beginning of the fifteenth century: Bonetti, "La libreria dello storico e pittore Antonio Campi," 10; Emilio Giazzi, "Fragmenta codicum: la biblioteca e lo 'scriptorium'presso la Cattedrale di Cremona: sulle tracce di una biblioteca dispersa," in *Cremona: una cattedrale, una città : la cattedrale di Cremona al centro della vita culturale, politica ed economica, dal Medio Evo all'Età Moderna*, ed. Giancarlo Andenna (Milano: Silvana, 2007), 92-93; Antonio Manfredi, "Gli umanisti e le biblioteche tra l'Italia e l'Europa," in *Il Rinascimento italiano e l'Europa*, ed. Annalisa Belloni and Riccardo Drusi, vol. 2, Umanesimo ed educazione (Treviso: A. Colla, 2007), 267-84; Damiana Vecchia, "Nuove ricerche sulla Biblioteca Capitolare di Cremona (secc. IX-XVI)" (Università degli studi di Parma- Facoltà di Lettere e Filosofia, relatore Prof. A. Belloni, 1997), 102-54; Emilio Giazzi, "Frammenti di codice a Cremona: testimonianze per una storia della cultura cittadina," in *Cremona: una cattedrale, una città : la cattedrale di Cremona al centro della vita culturale, politica ed economica, dal Medio Evo all'Età Moderna*, ed. Giancarlo Andenna (Milano: Silvana, 2007), 42-43.

Johannes de Sacrobosco, the most simple of all astronomical text-books, Ptolemy's *De Planisphaerio*, the *Theorica Planetarum*, often but wrongly attributed to Gerard of Cremona (or to a thirteenth-century homonymous Gerard from Cremona, an astrologer who came from the village of Sabbioneta in the diocese of Cremona). This was the only circulating Latin version of Ptolemy's *Theoretica Planetarum* before Georg von Peuerbach's posthumous publication of his *Theoricae novae planetarum*. The German mathematician translated Ptolemy's *work* directly from ancient Greek, at Cardinal Bessarion's request: indeed, the Greek cardinal had called for a more philological edition of ancient knowledge, highly corrupted by Arab and Latin translators. Another thirteenth-century *Theoretica Planetarum*, probably by Campanus of Novara, was than focussing on the geometrical representation of the Ptolemaic system, the so-called *equatorium*.[4]

The *equatorium* is a device constructed upon a geometrical projection of the motions of the heavenly bodies: their movement is expressed in circular shape, like a clock's dial with its pointer indicating here the position of a planet on the Zodiac. The Christian medieval tradition was based on the Ptolemaic model, filtered by Arabic and Persian commentaries. The most important Islamic models for *equatoria* were produced between the eleventh and the early twelfth century El-Andalus: Abulcasim, Azarchel and Abu-l-Salt. After Campanus' *Theorica Planetarum*, there had been other authors attempting to improve the geometrical system, such as Richard of Wallingford's *Albion*, Giovanni de Dondi's *Tractatus de Astrarii* or Zelandinus's Parisian manuscript. Sixteenth-century planetary horology would be based on this mathematical tradition: indeed, from a theoretical perspective, the science behind Janello Torriani's planetary clock was no different than that accessible to any other

---

4   The *Sphaera Mundi* had a huge success and was printed in more than 30 editions before the sixteenth century. Francis R Johnson, "Astronomical Text-Books in the Sixteenth Century," in *Science, Medicine, and History : Essays on the Evolution of Scientific Thought and Medical Practice Written in Honour of Charles Singer*, ed. E. Ashworth Underwood, vol. 1, 2 vols. (Oxford: Oxford University Press, 1953), 285-302. Ptolemy's Planispherium was translated to Latin from the Arabic in 1144, it was printed for the first time in 1507 by the Celestinian monk Marco Beneventano in appendix to Ptolemy's *Geography* and dedicated to Pope Giulius II. A second edition was printed in 1536 in Basel by Valdero and a third one in 1558 by Federico Commandino for the types of Aldo Manuzio: Claudi Ptolemeu, *Il planisferio di Tolomeo*, ed. Rocco Sinisgalli and Salvatore Vastola (Firenze: Cadmo, 1992), 37; Georg von Peurbach, *Theoricae novae planetarum* (Norimbergae: Johannes Müller Regiomontanus, 1472); Michael H. Shank, "L'astronomia nel Quattrocento tra corti e università," in *Le scienze*, ed. Antonio Clericuzio and Germana Ernst, Il Rinacimento Italiano e L'Europa 5 (Treviso: Angelo Colla Editore, 2008), 3-20.

part of Latin Christendom and was based, except for minor improvements, on Campanus da Novara's Ptolemaic *equatorium*.[5]

Astronomical knowledge was very important for the medical profession. Between the second half of the fifteenth century and the beginning of the sixteenth, there were in Cremona different physicians well read in astrology. Torriani's mentor, Giorgio Fondulo, a member of the local gentry, was one of them. Beside him, numbered other nobleman such as Giovan Battista Plasio (1410-1492), remembered by Bishop Vida and later by Bernardino Baldi (1553-1617) in his work on the lives of mathematicians, and Leonardo Mainardi (ca. 1410-1480 ca.).[6] Plasio's sculpted tomb is still visible in Cremona, in the church of Sant Agostino (Fig. 12), and it reflects iconographically the importance in Renaissance medicine of an epistemological field, the results of which would now seem completely irrelevant – and even absurd – to our idea of medical knowledge. The simulacrum shows a frowning man, sitting with an armillary sphere in his right hand and an open book in the left. On the pages one can read *"SUPERATA TELLUS SIDERA DONAT,"* a sentence from Boethius' *Consolatio Philosophiae*.[7] The meaning is that the virtuous man has not to fear death, because like Hercules, after his earthly fatigues he will be granted the heavens: "if you overcome the Earth, your reward will be the stars". This physician was celebrated after his death in a funeral sermon, as was common use for respected people. From this source we learn also that the ducal family of the Sforza employed Plasio in Milan. Later the physician found employment at Ferrara too. Domenico Bordigallo, in his already mentioned manuscript *Urbis Cremonae syti designum* (1515-1527 ca.), reported that this famous astronomer used to

---

5   Some manuscript and printed versions of the *Theorica Planetarum* were circulating together with paper versions of the volvelles. These paper displays were probably inspired by texts on the astrolabe. We have examples of some of these Renaissance mathematical instruments made out of paper: for instance, Zelandinus' and Johannes Schöner's works on the *equatorium*, Petrus Apianus' *Astronomicon Caesareum* and the paper-instruments in the shape of volvelles that Torriani sent to Gregory XIII for the reform of the calendar at the beginning of the 1580s. Emmanuel Poulle, "L'equatoire de Guillaume Gilliszoon de Wissekerke," *Physis* 3, no. 3 (1961): 223-51; Vicente de Cadenas y Vicent, *Hacienda de Carlos V al fallecer en Yuste* (Madrid: Hidalguia, 1985), 22; Turriano, *Breve discurso*.

6   Francesco Novati, "Due matematici cremonesi del secolo XV: frà Leonardo Antonii e Leonardo Mainardi," *Archivio Storico Lombardo*, IV – anno XXXII, 7 (1905): 218-25; Bernardino Baldi, *Cronica de matematici, overo Epitome dell'istoria delle vite loro* (Urbino: Angelo Ant. Monticelli, 1707), 104-5. Marika Leino and Charles Burnett, "Myth and Astronomy in the Frescoes at Sant'Abbondio in Cremona," *Journal of the Warburg and Courtauld Institutes*, 2004, nn. 40-43.

7   Anicius Manlius Torquatus Severinus Boethius, *La consolazione della filosofia*, ed. and trans. Ovidio Dallera (Milano: Rizzoli, 1977), 336.

write prognostics from the top of a tower where he observed the movement of stars.[8] This was probably a common activity for learned men interested in astrology. Indeed, since the thirteenth century, tower-astrological observations had been performed in Northern Italy. In the city of Forlì, one of the most famous Italian astrologers, Guido Bonatto, was said to have performed similar astrological observations:

> When constellations which augured victory appeared, Bonatto ascended with his book and astrolabe to the tower of San Mercuriale above the Piazza, and when the right moment came, gave the signal for the great bell to be rung.[9]

Observing the debates on astronomical knowledge in the Renaissance, and in the specific in Cremona, it is interesting to note that a common geographical identity could push a scholar to endorse one scientific tradition and to exclude another. After all, it was up to urban elites to decide which activities deserved to be sponsored on the basis of which were consistent with the celebration of their *ethos*, and with it, of themselves. Vida's celebration of Torriani (referred to above) appeared in a pamphlet that tried to demonstrate the right of precedence of Cremona over Pavia in terms of etiquette. Etiquette is a mirror that reflects the grade of dignity in a hierarchy. Such urban elites promoted their offspring's education. With this agenda, they sponsored institutions of knowledge, such as universities, schools, academies, public or private libraries and collections. Amidst this practice of competition among cities, it seems that even scientific authority could be used as a battlefield to demonstrate the superior dignity of one group over another.[10] For instance, Regiomontanus, thanks to philological arguments, had written a *Disputationes contra Cremonensia*

---

8   "*Adest etiam ibi prope Turris nuncupata illorum de Plasiis, ubi astronomus dominus Baptista Plasius preclarus sua pronostica componebat ex stellarum fluxu vagantium*": Bordigallo, *Urbis Cremonae syti designum*, 128.

9   Silvio A. Bedini, "Falconi, Renaissance Astrologer and Astronomical Clock and Instrument Maker," *Nuncius* 19, no. 1 (2004): fasc. 1, pp. 35-36; Cesare Vasoli, "Bonatti, Guido," *Dizionario biografico degli Italiani* (Roma: Istituto della Enciclopedia italiana, 1969).

10   Regarding such attitudes during the Early Modern period see Giorgio Vasari's discourse about the superiority of Florentine artists, and in the field of mathematics Muzio Oddi's celebration of the School of Urbino: Alexander Marr, *Between Raphael and Galileo: Mutio Oddi and the Mathematical Culture of Late Renaissance Italy* (Chicago: The University of Chicago Press, 2011), chapter 1. Marr elaborating on the sixteenth century Italian concept of *patria* (fatherland), endorses Denys Hay's definition: "For most Italians, *patria* meant, not the entire peninsula, but those narrower localities with which they had immediate

deliramenta, attacking popular translations made by scholars from Cremona of ancient texts. Indeed, the Greek scholar George of Trebizond noted that Cardinal Bessarion's Byzantine version of the Almagest was different from Gerard's. George decided to make a new translation and sought papal patronage. Cardinal Bessarion insisted on using Theon of Alexandria's comment as a guide for the translation. After only nine months, in 1451, George finished the translation with a 600 pages comment. The papal entourage refused this version. Therefore, Cardinal Bessarion, irritated by the attacks against Theon, sought an alternative introduction to Ptolemy. During a trip to Vienna (1460-1461) he convinced Peuerbach to write a new epitome to the Almagest. However, Peuerbach died shortly after, and his pupil Regiomontanus picked up the baton. He followed Bessarion to Italy where he worked on the Cardinal's Greek text. He terminated the task in 1462 but the work was only printed in Venice in 1496.[11]

Bernardino Baldi, in his *Cronica dei matematici*, remembered that Plasio once wrote a lost treaty defending, against Regiomontanus, his thirteenth century fellowcountryman, the astrologer and astronomer Gerard of Sabbioneta. Even Vida, in the middle of the sixteenth century, in the same pamphlet in which he had praised Torriani, stated:

> Look at the eminence in philosophy of Gerard of Sabbionetta [sic]. He was easily the leader of his age among philosophers, famous as much for his knowledge of many arts, and the best of them, as for his eloquence! Since he excelled in languages and tongues of many nations, he put the books of Avicenna the Arab and Almansor [Rhazes' Liber ad Almansorem] into Latin.[12]

The same Giorgio Fondulo, Torriani's mentor, in a letter sent on the 16th of March 1507 to a friend of his, ardently defended the authority of another mathematician and astrologer from Cremona, a certain Leonardo, with a peculiar argument: "I do not want to listen to what You are saying because Leonardo was a most ingenious man. To confirm this, I want to tell You that You should never doubt Leonardo to have made any mistake, but firmly believe that he

---

      sentimental and political ties", p. 34: We can observe that such concept of *patria* is still mainstream in twentyfirst century Italy.

11   According to Shank this is to our days the best introduction to Ptolemy. Shank, "L'astronomia nel Quattrocento tra corti e università."

12   Leino and Burnett, "Myth and Astronomy in the Frescoes at Sant'Abbondio in Cremona," 287-28, nn. 40-43.

FIGURE 13 *Francesco Casella, Frescoed vault at the Cloister of Sant'Abbondio, 1513, Studiolo Landriana, Cremona.* COURTESY OF MINO BOIOCCHI AND OF THE DIOCESE OF CREMONA.

wrote what he wrote using the most sure demonstrations. Thus, I dare to say, not for the glory of a fellow-countryman, but for the sake of truth, that since Ptolemy, there has not been any man of a deeper knowledge in the mathematical matters than Leonardo Cremonese".[13]

---

13 My translation, from: *ms Latin 7192*, From the Mazzarino collection (n. 5437), Bibliothèque Nationale de France Paris:"*qual cosa perho non auditer Voleti dire per esser stato Leonardo homo ingeniosissimo. Ad confirmatione Di questo ve voglio dire che nullo modo doveti dubitare Leonardo haver errato: ma firmiter credere esso haver dicto cio ha scritto*

FIGURE 14   *Francesco Casella,* Alpetragius (the twelfth century Moroccan-Spanish
             astronomer of Cordoba al-Bitruji, who was able to harmonize the physics of
             Aristotle with Ptolemy's Almagest), *1513, Studiolo Landriani, Cloister of Sant'Ab-
             bondio, Cremona.* COURTESY OF MINO BOIOCCHI AND OF THE DIOCESE OF
             CREMONA.

Fondulo's need to stress his flamboyant defence not being a matter of local
pride reveals the very possibility that it actually was. However, personal belief
in a home-made scientific tradition might also be the unconscious result of the
educational trauma set on stage by a scientifically stubborn, self-confident and
self-referential tradition. Confuting such a tradition without a clear demonstration of its errors must have been tantamount to patricide! One can find
other examples of this means of sacralising a local tradition of knowledge: for
instance, in the context of Cremona, Domenico Bordigallo, writing a chronicle
of his city, remembered all the professors of law to have flourished in the city
along many centuries. In order to give more credit to this tradition, Bordigallo

cum cortissime demonstratione: *Per tanto io non per gloria de uno compatriota: ma per la
veritade oiso dire da Ptholomeo et in qua non esser stato homo de più profunda scientia nele
cose mathematice che Leonardo Cremonese;"* from the chronological indications present in
the manuscript, we know that this was written within the first decade of the sixteenth
century by a certain Bernardino Alieri, a notary, poet and public accountant of the commune of Cremona. The existence of this manuscript was given in: FNovati, "Due matematici cremonesi del secolo XV: frà Leonardo Antonii e Leonardo Mainardi." About
Bernardino Alieri see: Bordigallo, *Urbis Cremonae syti designum.* About Bernardino Alieri
see: Bordigallo, *Urbis Cremonae syti designum,* 43.

reported that the local (medieval) school was the same that Virgil was supposed to have attended in the first century B.C.[14] Any kind of knowledge,
especially the kind difficult to demonstrate, depends very much on credit, and
credit stands on different legs, one of them being tradition.

In the very context of Cremona, in connection with this spirit of pride
for local scholarship, we can turn to a problematic but very interesting document: it is a fresco-painting representing Ptolemaic astrological scholarship in
a room of the cloister of the church of Sant'Abbondio in Cremona (Fig. 13).
The commissioner was a powerful man, a certain Gerolamo Landriani, provost
of Sant'Abondio and the general of the Umiliati order (1485-1525).[15] He
belonged to a most influential banker family. He was brother of the ducal treasurer and was appointed several times to relevant ducal offices. Among them,
he was made Curator of the State at the time of Duke Massimiliano Sforza
(1512-1515), when the fresco was executed. Landriani hired a painter, probably
Francesco Casella, to depict this complex scholarly system of figures of astronomers all connected with the *Almagest* and therefore positioned within the
translating *opus* of the double-faced Gerard (the above-mentioned and often
confused Gerard of Cremona and Gerard of Sabbioneta). In this fresco, which
is still visible in the archivolt and on the upper part of the walls, a series of ten
philosophers and scholars are depicted with their names, and with scrolls and
books quoting lines connected with the *Almagest*, with some of them pictured
holding mathematical instruments. Aside, the coat of arms of the Landriani
family is visible. On the external wall of the room, engraved on a corner stone,
one can see the initials of the name and title of General Gerolamo Landriani
together with his motto *"Gott weiss"* with the date 25th of June 1511.

Did the vibrancy of the debate against *"Cremonensium deliramenta"* have
any influence on Torriani's path of education and on his interests directed to
clockmaking and astronomy? Perhaps. What is sure is the relevance of these
arguments in early sixteenth century Cremona. This atmosphere provided
Janello with a favourable cultural environment interested in celebrating its
mathematical tradition, with a special focus on astronomy. My comments
on this localistic defence of a scientific tradition may contradict what French
historian of technology Emmanuel Poulle has stressed about scientific instruments. Poulle stated that medieval scientific instruments were basically the
same all over Europe, since they were based on the same "universal" astrological
technical Latin literature. Poulle probably meant that the geocentric Ptolemaic

---

14   Ibid.
15   The Umiliati order was disbanded in 1571, after one of its members made an attempt on
      Cardinal Borromeo's life, unsuccessfully shooting at him with a harquebus.

FIGURE 15   *Francesco Casella,* Timocharis of Alexandria (third century B.C.). *1513, Studiolo Landriani, Cloister of Sant'Abbondio, Cremona.* A source cited in the Almagest, is here represented stargazing with the help of a quadrant. COURTESY OF MINO BOIOCCHI AND OF THE DIOCESE OF CREMONA.

system was by far the mainstream theory.[16] In this perspective Peuerbach's and Regiomontanus' reformulation of the *Theorica* can be considered as a nuanced version of the same theory. However, political issues and their geographical impact are also relevant: for instance, as we shall see, once the Republic of Venice occupied Cremona, two clockmakers from its territory and loyal to Saint Mark were appointed in succession as moderators of the public clock of Cremona. This meant that craftsmen with both a theoretical and practical knowledge shaped in the territory of Venice were appointed with authoritative offices in Cremona, transferring there within a mathematical institution their science and craft.

The paramount role of astrology in Renaissance society, and the increasing number of university students during the fifteenth century, produced an increase in the production of technical books and mathematical instruments. Both items had a central role in the dissemination of knowledge during Torriani's lifetime. Moveable types technology increased the circulation of identical and carefully edited copies at a cheaper price. Moreover, it

---

16   E Emmanuel Poulle, "La produzione di strumenti scientifici," in *Produzione e tecniche,* ed. Philippe Braunstein and Luca Molà, vol. 3, Il Rinascimento italiano e l'Europa (Costabissara Vicenza: Fondazione Cassamarca : Angelo Colla Editore, 2007), 345-47.

has been noted that the market of mathematical instruments experienced a price-differentiation: the prices of the instruments could change considerably. For instance some astrolabes were produced on a smaller scale together with minimal decoration, a factor that testifies to accessibility to such instruments for a non-elitist market.[17] We shall return to this point when considering urban mathematical workshops as places of knowledge.

Renaissance Scientific Instruments

As previously seen, in the fourteenth century Cremona fell under the influence of nearby Milan, the place where the first known public clock,[18] the first complete planetary one,[19] and the first known watches[20] were created. It seems that the art of clockmaking was likely already well established in Cremona by this century when a certain Mondino da Cremona moved to Venice: he worked all his life on a complex clock that later, in 1334, he sold to the King of Ciprus for 400 ducats.[21] Another fellow-countryman clockmaker, Antonio da Cremona,

---

17   Gerard L'E. Turner, "Two Early Renaissance Astrolabes by Falcono of Bergamo," ed. Marco Beretta, Paolo Galluzzi, and Carlo Triarico, *Musa Musaei : Studies on Scientific Instruments and Collections in Honour of Mara Miniati*, 2003, 53-62.

18   Dohrn-van Rossum, *History of the Hour*, 130.

19   Giovanni de Dondi's Astrarium, purchased or perhaps even commissioned by the first duke of Milan Gian Galeazzo Visconti, was built in the second half of the fourteenth century, presumably between 1365 and 1384, according to the most reliable dating based on astronomical references cited by Dondi for the construction of the whole mechanism.

20   For the first time, in 1490, small, portable watches, in the shape of chiming timekeepers attached to ball costumes (*horologini piccoli et portativi*), were mentioned at the court of Ludovico Sforza. Enrico Morpurgo, *L'origine dell'orologio tascabile* (Roma: La clessidra, 1954); *L'orologeria italiana dalle origini al Quattrocento* (Roma: La Clessidra, 1986); Jürgen Abeler, *In Sachen Peter Henlein* (Wuppertal: Selbstverlag, 1980). Günther Östmann, quoting Giuseppe Brusa, has justly underlined that the dispute over the invention of the watch by a single person has always had more to do with national pride than with historical sources: Giuseppe Brusa, "Early Mechanical Horology in Italy," *Antiquarian Horology* 18, no. 5 (1990): 510.). Giuseppe Brusa, "I primi orologi da persona in Italia: nuovi indizi e nuove eccellenti testimonianze," *Voce di Hora* 3 (1997): 3-20. For a well-documented and balanced research, see: Thomas Eser, *Die älteste Taschenuhr der Welt?: Der Henlein-Uhrenstreit* (Nürnberg: Germanischen Nationalmuseums, 2014); "Die Henlein-Ausstellung Im Germanischen Nationalmuseum: Rückblick, Ausblick, Neue Funde," *Deutsche Gesellschaft Für Chronometrie: Jahresschrift* 54 (2015): 23-34; and Günther Östmann on early watches in Germany in "The Origins and Diffusion of Watches in the Renaissance: Germany, France, and Italy," in *Janello Torriani: A Renaissance Genius*, ed. Cristiano Zanetti (Cremona: Comune di Cremona, 2016), 141-43.

21   Morpurgo, *Dizionario degli orologiai italiani*, 1950, 130.

FIGURE 16   *Francesco Casella,* Ptolemy, *1513, Studiolo Landriani, Cloister of Sant'Abbondio,*
             Cremona. Ptolemy is holding in his left hand a compass and a book with
             astronomical drawings, and a set square in his right hand. The author of the
             Almagest is here erroneously represented as a king of the Macedonian-Egyptian
             royal family. COURTESY OF MINO BOIOCCHI AND OF THE DIOCESE OF
             CREMONA.

is said to have accompanied a Venetian embassy to Delhi (year 1335). According
to this tradition, the Venetian ambassador carried a clock to that city, which
had perhaps been designed by Jacopo de Dondi and made by Antonio himself.[22] This is still a very blurred issue: we know nothing about this clock,
or about Antonio. For sure we have local clockmakers working in Cremona
during the fifteenth century, at an earlier date than scholarship so far has
acknowledged, as we shall later see.

    Cremona was at the centre of the crossroads at which Renaissance Italian
clockmaking had its most productive and innovative experiences: Milan,
Mantua, Parma, Reggio and the university-cities of Padua, Bologna and Pavia.
Clocks can be included in that set of production embracing scientific instruments: indeed, measurement was the main task of such instruments. A
scientific instrument in Torriani's time was a point of contact between craftsmanship and the current theoretical mathematical knowledge. Among clocks,

---

22   Raimondo Morozzo della Rocca, "Sulle orme di Polo," *L'Italia che scrive* XXXVII, no. 10,
      Numero speciale dedicato a Marco Polo (1954): 121-22.

the most complex ones were the astronomic and the planetary clocks that beside the hours of the day had also cosmographic functions: in fact they were displaying the position of the heavenly bodies in the Zodiac.

At the dawn of the sixteenth century, there were more than a dozen mathematical instruments dealing with cosmography: some represented the universe, while others could be used to ascertain time, date and, more roughly, geographical position.[23] Among such scientific astronomical instruments featured the circle of 360°, the quadrant (Fig. 15), the meridian ring, the sextant, the armillary sphere (Fig. 28), the astrolabe (Fig. 18), the sundial (Fig. 17), the torquetum, the triquetrum (Fig. 19), Peuerbach's regula and geometrical square, the so-called Jacob's stick, the nocturnal, the equatorium and Ptolemy's slide-ruler. Each of these European apparatuses was portable, usually held in hands, therefore moveable and not particularly accurate for precise collection of data through observation, more instruments of representation and explanation than investigation.

In the year 1505, when Janello Torriani was a young boy, at the time of the Venetian dominion of Cremona, it seems that a certain Falconi from Bergamo (alias Falchone, Falco, or Falconus) "who had great fame in astrology"[24] became moderator of the public clock. The public clock was set in the tallest tower of the city, a suitable location for stargazing. In these years of residence in Cremona, Falconi produced some of his signed or attributed scientific instruments, mainly very fine astrolabes, nocturnals (Fig. 26) and horary-quadrants.

Also clocks were scientific instruments: it is assumed that the weight-driven mechanical clock with an escapement system appeared by the end of the thirteenth century somewhere in Europe, perhaps in present day Northern Italy or England.[25] Scholarship has divided clocks into different categories, considering their utilization: the first set contains timepieces (chronometers and alarm-clocks) dividing the time in days of 24 equal hours. Then we have astronomical clocks (in Latin: *astraria*). These added to the hours the position of Sun and Moon in the Zodiac. There are different types of astronomical clocks depending on the geographical area where they were produced. In Italy,

23   Poulle, "Produzione di strumenti scientifici," 360; Giorgio Strano, "The in-Existent Instruments," in *Musa Musaei: Studies on Scientific Instruments and Collections in Honour of Mara Miniati*, ed. Marco Beretta, Paolo Galluzzi, and Carlo Triarico (Firenze: L.S. Olschki – Istituto e museo di storia della scienza, 2003); Jim Bennett, "Gli strumenti astronomici prima del Seicento," in *Galileo : immagini dell'universo dall'antichità al telescopio*, ed. Paolo Galluzzi (Firenze: Giunti, 2009), 219-25.

24   "*che godeva assai lustro nell'astroloxia*": Turner, "Two Early Renaissance Astrolabes by Falcono of Bergamo." 53-62.

25   See: Marisa Addomine, "Cenni di storia dell'orologeria da torre," in *Orologi da torre: MAT, Museo arte tempo di Clusone*, ed. Marisa Addomine and Daniele Pons (Milano: Skira, 2008).

by the end of the fifteenth century, a number of different cities had a public astronomical clock: Padua, Brescia, Cremona, Mantua, Ferrara, Reggio and Bologna.[26] Finally we have the most complex clock, the planetary one, which shows the hours of the day and the movements in the zodiac of the seven heavenly bodies: the Moon, Mercury, Venus, the Sun, Mars, Jupiter and Saturn. The planetary clock derives its forms from Campanus of Novara's *equatorium* to whom a mechanical engine was applied.[27] Campanus did not invent the *equatorium*. In fact, we know of the existence in classical times of several *planetaria*: the Antikythera machine, perhaps based on Archimedes' planetary devices, was lost in a shipwreck in the 1st century B.C. and found 2000 years later. It has been interpreted as a planetarium showing the motions of the seven heavenly bodies and is considered the oldest surviving geared device.[28] In the second century A.D. Ptolemy standardised the geometrical technique behind any later *equatorium* and Campanus popularised it. The great difficulty about planetary machines was to represent for every planet the regular movement on the epicycle, the irregular motion of the centre of the epicycle on the deferent, and the irregular (not for Ptolemy, but for his commentator Thabit ben Qurra, whose theory was widely accepted in Christendom)[29] rotation of the centre of

---

26   In this city there was an astronomical public tower-clock, which, according to some histo-
      rians, depended on Pythagorean cosmology, and especially on Philolaos of Taranto's the-
      ory, therefore representing a non-geocentric system. It has been observed that Nicholaus
      Copernicus, who studied in Bologna, had chance to observe it. However, already in 1973,
      Antonio Simoni confuted such theory with some consistent observations. The problem is
      nevertheless still open, because despite Simoni's arguments being strong, the dial and the
      mechanism of that clock were lost: Carlo Maria Cipolla, *Le macchine del tempo: l'orologio
      e la società* (Bologna: Il mulino, 1981), 19-20; Enrico Morpurgo, "L'orologeria italiana:
      l'orologio di Bologna e il cardinale Bessarione," *La Clessidra* anno 30, no. 12 (1974): 32-33;
      Antonio Simoni, "L'orologio pubblico di Bologna del 1451 e la sua sfera," *Culta Bononia.* 5
      (1973): 3-19.

27   *Equare* is the Medieval Latin verb used to define the action of determining the position of
      the sun in the sky, i.e., to see in which part of the zodiac it appears.

28   Derek John de Solla Price, "An Ancient Greek Computer," *Scientific American* 201, no. June
      (1959): 60-67. Michael T. Wright has made a marvellous reconstruction of the device and
      has supported the original theory as expounded by Price (though he later changed his
      mind). An illuminating virtual reconstruction by Mogi Vicentini, made after Wright's
      model, can be seen at the following website: <http://www.mogi-vice.com/Antikythera/
      A-W-M.zip>; the only problem remains the extremely fragile state of the extant parts of
      the device.

29   The Dominican friar Gasparo Bugato, an eyewitness, reported that Janello's clock was
      organised according to this theory: Gasparo Bugati, *Historia uniuersale* (In Vinetia:
      Appresso Gabriel Giolito de Ferrarii, 1571), 1025-26.

FIGURE 17   *Francesco Casella,* Tebithcore (Thābit Ibn Qurra, ninth century), *1513, Studiolo*
             Landriani, Cloister of Sant'Abbondio, Cremona. Tebithcore is holding an altitude
             sundial. This astronomer is believed to have introduced the theory of trepida-
             tion of the equinoxes. Janello's only extant armillary sphere and his lost
             planetary clocks reproduced mechanically this theory. COURTESY OF MINO
             BOIOCCHI AND OF THE DIOCESE OF CREMONA.

the deferent around the centre of the Earth.[30] The first known completed
medieval planetary clock was Giovanni de Dondi's Astrarium, a property of the
dukes of Milan (Fig. 20).[31] Janello Torriani reached popularity studying and
then reconstructing a more compact and more complex version of the
Astrarium, as we shall later see.

   Whereas the first type of clock is the ancestor of our chronometers, the
other two categories died out. The reason was that their applications were relevant for astrology: Medieval and Renaissance astrology was very significant
for casting horoscopes and for other human trades, among them medicine. It
has been observed that in the two previous centuries of the existence of the

30   Poulle, "L'equatoire de Guillaume Gilliszoon de Wissekerke," 229.

31   There are at least 11 manuscripts about the construction of Dondi's Astrarium: Silvio A.
      Bedini and Francis Maddison, *Mechanical Universe: The Astrarium of Giovanni de Dondi*
      (Philadelphia: American Philosophical Society, 1966), 40; Poulle, "Produzione di stru-
      menti scientifici," 363-64.

mechanical clock *"there was no such thing as a typical clockmaker"*[32] or a well-defined profession of clockmaking. In fact, while for medieval water-clocks and early mechanical ones their constructors were mainly monks, it has been argued that prominent fourteenth century builders of mechanical astronomical and planetary clocks were usually physicians. They were the designers who then employed metalworkers to carry out their projects. Except for the abbot Robert of Wellingford, all the other recorded scholars involved with the theory and construction of planetary clocks during the Middle Ages were physicians. The above-mentioned Johannes Campanus of Novara, famous for the *Theoretica Planetarum*, was familiar with medicine, and was involved in the papal court where, though not officially papal archiater, Nicolaus III appointed him *"phisicus et capellanus papae"*. Henry Bates of Malines, author of the medical-astrological treatise *De diebus criticis*, declared to have made around the year 1274 a new type of astrolabe with his own hands. One of two Jews in charge of writing the *Alfonsine Tables* for the King of Castile, a certain Judah ben Moses ha-Cohen was a physician. At the beginning of the fourteenth century, a Danish astrologer and physician, a certain Petrus Philomena (canon at Roskilde and teacher of mathematics and astronomy at Bologna and Paris), made or directed the construction of an instrument for the calculation of the ecliptic longitude of the heavenly bodies. Jacopo and Giovanni de Dondi were father and son, and both astronomers, astrologers and physicians. A friend of Giovanni de Dondi, a certain Philippe de Mézières even claimed that this physician had made the Astrarium with his own hands. Guido da Vigevano's *Texaurus regis Francie* (1335), and Conrad Kyeser's *Bellifortis* (left unfinished in 1405) are two treatises written by two court medical-astrologers dealing with mechanics, specifically related to the art of sieging: *poliercetica*. In France, Jean Fusoris was both a physician and an astrologer and he is said to have built with his hands numerous complex clocks for kings, popes and other wealthy princes. A student of his, the medical astrologer Henry Arnault of Zwolle, was court physician to the Duke of Burgundy, for whom it seems he made a complex planetary clock. He also worked for the King of France. During the fourteenth century, the Jewish cultural community around the Crown of Aragon, famous for its contribution to cartography and active in translating for the Christian kings, provided the court with some medical-astrologers who constructed clocks. In Venice Giovanni Fontana (ca. 1395- ca. 1455), who had studied medicine at Padua, was familiar with the construction of organs, weaponry,

---

32   Dohrn-van Rossum, *History of the Hour*, 175.

fountains and clocks.[33] As we shall later see, at the time of Torriani, physicians like Cardano or Ramusius were deeply involved with planetary clocks and astrology. This happened because at the beginning of its history, the design of astronomical and planetary mechanical clocks was not reliant on the mastery of a specific craft, but was instead based on a well defined science: mathematics, to be understood in its speciality devoted to celestial bodies: astrology/astronomy.

Ancient astrology attempted to bridle the geocentric cosmos into a geometric system of alignments and superimpositions of planets and the Zodiac. This system was based on the belief that the celestial bodies had active influences on planet Earth. Name-related analogies made think astrologers that, for instance, planet Mars was influencing war or Venus love. The representation of the Ptolemaic cosmos was the result of an optical illusion: the misunderstanding of the existing proportions and real distances of the heavenly bodies from each other, from the fixed stars and from planet Earth. Indeed, the main idea behind this science was the interaction between the seven heavenly bodies and their position in the ecliptic, the belt within which, from planet Earth, we can see inscribed the movements of these seven errant stars. This is possible because all the orbits of the solar system move around the equator of the Sun. The zodiac, meaning in ancient Greek "the route of the living creatures", was divided into twelve constellations: to each group of stars was attributed the name of a real or mythical creature, with magical qualities. All of the zodiac's stars were believed to be fixed in one single sphere: the eighth. The science of stars – astrology – was a multifaceted discipline whose practice included direct observation, reduction to geometrical models through mathematical calculations, and magical theories based on influence-theories. Not even the new Copernican theory of the structure of the cosmos, when adopted, destroyed the notion of the Zodiac as the basis of ancient astrology.

Astrology has often been divided into two main branches: *astrologia quadrivialis* and *astrologia judiciaria*. The first comprised of the study of the motions of the heavenly bodies and their sequence of appearance over the horizon and their sidereal distances.[34] The second focused instead on the mag-

---

33   Lynn Townsend White, "Medical Astrologers and Late Medieval Technology," *Viator*, 1975, 295-308; John David North, *God's Clockmaker Richard of Wallingford and the Invention of Time* (London; New York: Hambledon Continuum, 2006), 303; Agostino Paravicini Bagliani, "Campano da Novara," *Dizionario biografico degli italiani* (Roma: Istituto della Enciclopedia italiana, 1974).

34   The epistemological division of Astrology in sub-branches, did not always follow this model. For instance, Campanus of Novara, thirteenth century theorist of the *equatorium*, put the *judiciaria* within the category of *astrologia quadrivialis* together with *astrologia*

ical qualities of the heavenly bodies, and on their influences, claiming to be able to provide precise forecasting. This side of astrology regarding influences (and mainly *judiciaria* in its claims of foretelling the future) was from a philosophical and theological perspective, the thorniest. This was indeed a major problem for the Christian Catholic doctrine, which feared, with this science of correspondences, a weakening of the basic idea for man to be able to reach salvation throughout the choice of a pious life or sincere repentance. Judiciary astrology implied not just that the influence of the stars could give a man certain characteristics but even a fixed destiny. Free will was at stake.[35] Not just ecclesiastical environments revealed their doubts on the reliability of this science. Astrology was already in the sixteenth century a controversial discipline. Eugenio Garin made of this debate the central issue of a book, highlighting that the so-called "modern science" was not originated as a bolt from the blue or as a revolution.[36] Open scepticism was sometimes strong even among the very patrons of clockmakers, and their entourages: "If one has to believe astrologers" (*Se agli Astrologi si dee credere*) wrote a doubtful Gosellini, secretary and biographer to Ferrante Gonzaga, describing the astrological features that were supposed to have determined the elective affinities between Charles v and the Mantuan governor of Milan.[37] In 1477 Ferrante's great-grandfather, the Marquis of Mantua Ludovico III Gonzaga, patron of the marvellous public astronomical clock of that city, wrote to the son of the clockmaker who built it: "Let's say that we neither trust you nor trust other astrologers" (*dicemo che a ti né ad altri astrologi credemo*).[38] However, divinatory practices were still popular in the seventeenth century. The example of Kepler or Galileo casting horoscopes is

---

demonstrantis, interested just in the geometry of the celestial motions. This last one was then divided by Campanus into theoretical and practical: Campano da Novara, *Equatorium planetarum*, ed. William R. Shea and Tiziana Bascelli, trans. A. Bullo (Padova: Conselve, 2007), 8.

35   The life of the Florentine theologian and judicial astrologer Francesco Giuntini offers a practical example see: Germana Ernst, "Giuntini, Francesco," *Dizionario biografico degli Italiani* (Roma: Istituto della Enciclopedia italiana, 2001); Lynn Thorndike, *A History of Magic and Experimental Science*, vol. VI (New York: The Macmillan Company, 1941), 129-33.

36   Eugenio Garin, *Lo zodiaco della vita: la polemica sull'astrologia dal Trecento al Cinquecento* (Roma; Bari: Laterza, 1976).

37   Giuliano Goselini, *Vita di don Ferrando Gonzaga principe di Molfetta*, Collezione di ottimi scrittori italiani in supplemento ai classici milanesi 16 (Pisa: Presso N. Capurro co'caratteri di F. Didot, 1821), 6.

38   Alberto Gorla and Rodolfo Signorini, *L'orologio astronomico astrologico di Mantova: le ore medie e solari, lo zodiaco, le fasi lunari, le ore dei pianeti, i caratteri umani, le attività giornaliere e le previsioni astrali*, ed. Rosa Manara Gorla (Mantova: A. Gorla, 1992), 36.

probably the most famous. Burckhardt had already focussed on the practice of astrology in the Renaissance, noticing its importance in Italy, from the time of Frederic II up to the sixteenth century and contemptuously overemphasizing its picturesque and superstitious side. Burckhardt used this argument to construct his theory on the birth of modernity during the Renaissance: he reckoned the diffusion of astrological superstition as one of the reasons behind the rise of a new scepticism:

> With these superstitions, as with ancient modes of thought generally, the decline in the belief of immortality stands in the closest connection. This question has the widest and deepest relations with the whole development of the modern spirit.[39]

Historians of Science, in their attempt to search for the origin of Modern Science, have perhaps overemphasised the division between these two branches, underestimating and rejecting entirely *judiciaria*. However, as Garin has noted, Renaissance astrology was considered as the science of the Whole. One could argue that the science of invisible correspondences is not foolish, but a consequential phase of the observation of Nature. The Moon governs tides and human, animal and vegetal cycles of fertility; magnetism attracts iron; and above all, the Sun rules seasons and life on Earth. Still today, as it was with astrology, contemporary Science addresses the problems of invisible forces. As recently observed of divination, the modern eye is often unable to put astrology in its context without an attempt at rationalisation or even to adopt a position of smugness, considering it as external and less relevant than other scientific fields.[40]

Even *astrologia quadrivialis* accepted the theory of influences, but called for a major methodological rigidity in opposition to the hermetical and mysterious claims of *astrologia judiciaria*. In particular, it denied the possibility of precise previsions. Humanists, in their attempt to recover the "true" wisdom of the Ancients, found in Ptolemy's *Tetrabiblos* an authoritative and early negative critique of the less rigorous and pro-predestination approach of *judiciaria*

---

39   Jacob Burckhardt, *The Civilisation of the Renaissance in Italy* (Project Gutenberg, 2000), 202-15; Paola Zambelli, *White Magic, Black Magic in the European Renaissance* (Leiden; Boston: Brill, 2007); Mario Biagioli, "The Social Status of Italian Mathematicians, 1450-1600," *History of Science* 27 (1989): 41-95; White, "Medical Astrologers," 309-10.

40   Addomine, "Cenni di storia dell'orologeria da torre," 20; Guido Giglioni, "La divinazione: motivi filosofici e aspetti sociali," in *Le Scienze*, ed. Antonio Clericuzio and Germana Ernst, Il Rinascimento italiano e l'Europa 5 (Costabissara (Vicenza): Angelo Colla Editore, 2008), 247-59.

FIGURE 18   *Francesco Casella,* Massahalla (the Jewish Masha'allah, eighth-ninth century).
1513, Studiolo Landriani, Cloister of Sant'Abbondio, Cremona. The medieval Latin
tradition erroneously attributed a famous treaty on the astrolabe to Massahalla,
the instrument he is here holding. COURTESY OF MINO BOIOCCHI AND OF THE
DIOCESE OF CREMONA.

Ptolemy did not reject the fact that physical qualities of the planets and Zodiac
could interact with the Earth, its meteorology and life; but this interaction had
to be read according to the rules of geometry and without the intervention of
demoniac and supernatural forces. Ptolemy did not accept the possibility of
making precise previsions about the future, but he believed in the possibility of
foretelling negative or positive conditions. The only field he found to be of any
use for this practice was medicine: indeed the prevision of a certain physical
condition could be used to change the physical condition of illness of the
patient through the prescription of a certain medication. In this way Ptolemy
contradicted the belief of predestination, sustaining the possibility of interacting with the conditions created by the position of the stars.[41] This is a very
important point as regards the problem of astronomical and planetary clocks
as diagnostic and therapeutic instruments, as we shall see. Despite this scepticism towards judicial astrology, the discipline was widely practiced, even

---

41   Ornella Pompeo Farancovi, "La riforma dell'astrologia," in *Le Scienze,* ed. Antonio Clericuzio and Germana Ernst, Il Rinascimento italiano e l'Europa 5 (Costabissara (Vicenza):
Angelo Colla Editore, 2008), 59-61.

within the Church.[42] In fact, since intellectuals were mainly clergymen, the institutions of the Church mirrored every single component of the intellectual panorama of that time: as some opposed judiciary astrology, others supported it.

Thus, in the Cinquecento, it was a widely accepted proposition that in a positive or ill-omened way the stars ruled human activities. War, voyages, business, sexual activity, etc. were supposed to be undertaken only under certain auspices and astrologers were responsible for forecasting the different planetary alignments. Furthermore, astrology was also used for divination and for the *genethliologia*, the art of casting horoscopes and calculating nativities: according to the position of the stars in the zodiac at the moment of birth, astrologers claimed they could predict an individual's fortune, qualities and death. This belief was not just diffused in the Renaissance or limited to Christendom. All the Mediterranean civilisations used the same symbols and shared an ancient Hellenistic tradition that in turn was indebted to Egyptian and Babylonian astrology.[43] The Eastern Roman empire and Islam cultivated

---

42   Before the actuation of the Trent Council the year of birth was often not recorded. On the contrary, the day, with its astrological influence and the patron saint, was remembered, especially among well-taught figures. A suitable example is provided by the archbishop of Valencia, former professor of philosophy at the University of Alcalá: in 1566, on his deathbed, he did not know exactly his age, but he remembered the day of his birth when the power of the stars, in that moment of the year, had ill influences on Earth. This astrological conjunction caused, he believed, his misfortunes. Leon Battista Alberti in his *I Libri Della Famiglia* (1434) invites fathers to take careful note of the hour, date and place of one's child's birth. Leon Battista Alberti did not explain the use for that though he says: "These records should be kept with our dearest treasures. There are many reasons for doing this, but, all else aside, it shows the consciousness of a father." Dohrn van Rossum, interested in demonstrating the rise of a new rational perception of time connected with mercantile society such as "conflicts over inheritance", does not take in account what I consider the most common use for such a record in this time: astrological speculation. As a visual example of the relevance of astrology in these years in Florence are the painted vaults of Sacrestia Vecchia at San Lorenzo and Cappella de'Pazzi at Santa Croce. This diffused belief in the power of stars did not necessarily take to an idea of fixed destiny. Cardano observed that with an ethical effort one could contrast the imperfections given to him by the stars. David C. Goodman, *Power and Penury: Government, Technology, and Science in Philip II's Spain* (Cambridge: Cambridge University Press, 1988), 1; the translation of Alberti's passage is made by Dohrn-van Rossum in *History of the Hour*, 228; Giglioni, "La divinazione: motivi filosofici e aspetti sociali," 255.

43   Even Augustus' iconographic program of power that accompanied the creation of the empire was based on the belief in his extraordinary horoscope: it is for this reason that one can find the representation of the Capricorn associated with the first emperor. In reality Augustus' star-sign was Libra. However the adopted star-sign was believed to

the great bulk of ancient astronomical and astrological knowledge. Even in distant cultures, such as the Chinese one, star signs (though different from the Mediterranean ones) were believed to have influences on human life.[44] Jean-Claude Schmitt, in his essay on the invention of the birthday, observes that since Late Antiquity Christianity had rejected this celebration as pagan. From the late Middle Ages kings began to record the day of their birth. It was only from the sixteenth century that this practice began to regain some popularity among the urban population of Europe. This was largely the result of the fortune of horoscope casting.[45]

Unlike rulers and soldiers, common people had little interest in the forecasting of victories and military deeds. Their concerns about horoscopes were more mundane: trade, travel, agriculture and health. Health is a dominant issue in human life: fear of death and pain push it to this central position. The Milanese engineer Cesare Cesariano, famous for his pioneering vernacularization of Vitruvius, commenting on his IX book, wrote that the Roman architect called also for his profession for a basic knowledge of astrology. But, he adds, it was the physicians "*i medici*" who "*maxime*" i.e., in the highest degree, had interest in the movement of the planets in the Zodiac, because of the power these heavenly bodies exercise on humans.[46] Medicine is therefore a most important discipline, and in this scientific *milieu* based on Ptolemaic geo-cen-

---

portend special skills in political and religious leadership. Perhaps, being born under the sign of Libra – which meant that Octavian had been conceived nine months earlier under the sign of Capricorn – helped his program. Stars were powerful symbols even in the process of divinization of his adoptive father: Caesar. Iconography emphasised that the new god was saluted by the appearance in the sky of a comet: the *Sidus Iulium*. Both celestial symbols were extremely important in Christian narrative: the comet announced the birth of Christ and the event was finally set in the calendar in December under the sign of the Capricorn: Zanker, *The Power of Images in the Age of Augustus*, 34-48, 168, 220, 231.

44   Marco Polo witnessed that in the city of Quinsai in China (present-day Hang-Zhou) and in some parts of India parents carefully recorded the hour of the birth of their offspring, in order to determine, according to the position of the stars, their future deeds and qualities. Garin, *Lo zodiaco della vita*, 3-49; Rodolfo Signorini, ed., *Fortuna dell'astrologia a Mantova: arte, letteratura, carte d'archivio* (Mantova: Sometti, 2007), 20-21; Marco Polo, *Il Milione di Marco Polo*, ed. Giovanni Battista Baldelli Boni (Firenze: Da'Torchi di Giuseppe Pagani, 1827), 333, 419.

45   Jean-Claude Schmitt, *L'invenzione del compleanno* (Roma; Bari: Editori Laterza, 2012).

46   Cesare Cesariano, *Volgarizzamento dei libri IX (capitoli 7 e 8) e X di Vitruvio, De architectura, secondo il manoscritto 9-2790 Seccion de Cortes della Real Academia de la Historia Madrid*, ed. Barbara Agosti (Pisa: Scuola normale superiore, 1996), 32.

trism, astrology dictates diagnosis, cure, time of medication and prognosis.[47] Renaissance physicians believed that if two equal human bodies, generated under the influence of the very same stars, and suffering from the same illness, were cured with the same medication, administered in equal quantity but at different times, because of astrological influences, they would respond in different ways. This was the theory of *critical days (dies critici)*.[48] Furthermore, according to Medieval and Renaissance medicine, each human body's single limb had strict links with one of the star signs. Each of the constellations of the zodiac was then believed to have special qualities: for instance Aries, the first sign, was said to be oriental, warm, igneous, choleric and masculine.[49] These characteristics helped the physician to find the correct cure according to Galenic medical tradition, which saw disease as a lack of balance among the four humours and their related qualities. Willelm Gilliszoon de Wissekerke's (Gulliermus Aegidii Zelandinus) *Liber desideratus super celestium motuum indagatione sine calculo*, printed in the last decade of the fifteenth century in Lyon and Cremona, offers a good example of the close relation between astronomy and medicine, an example which was likely known to Giorgio Fondulo and to his pupil Janello: the book describes the cosmos in dimension, motion and mathematical relations, and then gives information about the practical use of this knowledge, described to be first of all medical, and then judicial. Hence, practitioners of the *ars medica* had to be able to cope with astrology and mathematical calculations concerning the position of the planets in the

---

47   A good example of such a practice in medical astrology can be found in the incunabulum printed in 1473 in Mantua by Pietro Adamo de Micheli and reproduced in: Gorla and Signorini, *Orologio astronomico astrologico di Mantova*; on astrological influences on the practice of prognosis in the Sforza court of Milan, see the chapter "*A Web of Correspondence: Gian Galeazzo's Illness and Renaissance Medical Astrology*", in: Monica Azzolini, *The Duke and the Stars: Astrology and Politics in Renaissance Milan* (Cambridge, Mass.: Harvard University Press, 2013).

48   "*Quante medicine totalmente uguali, in un medesimo corpo parimente disposto, per essere date dagli ignari medici in diversi tempi et hore, diverse fanno l'operationi!*": from Pietro Adamo de Micheli in: Gorla and Signorini, *Orologio astronomico astrologico di Mantova*, 91; Giuseppe Dell'Anna, *Dies critici: la teoria della ciclicità delle patologie nel 14. secolo*, 2 vols. (Galatina (Lecce): M. Congedo, 1999).

49   These relations were ichnographically expressed with the *homo anatomicus* or *signorum* (anatomical man or star-signs man) a human figure whose head was accompanied by the star-sign of Aries, throat and neck and shoulders by Taurus, arms by Gemini and so on, following the anatomic and zodiacal orders, until the feet connected to the sign of Pisces: from Pietro Adamo de Micheli in: Gorla and Signorini, *Orologio astronomico astrologico di Mantova*, 92 and following.

zodiac.[50] For this reason, in 1571 Philip II, King of Spain, responded to the *Cortes* which attributed the failures of physicians to their ignorance of planetary motions, forbidding universities to allow physicians to graduate "without a degree of bachelor in astrology".[51] This was a trend that started in Italian universities, where since the fifteenth century astronomy/astrology had become an academic discipline, taught by physicians, with courses covering the span of four years.[52] In her book on the Sforza and astrology, Monica Azzolini suggests that it was the fortune enjoyed at court by certain astrologer-physicians, such as Ambrogio Varesi da Rosate, a character that we shall soon meet, that "may have led other physicians to embrace astrological medicine more boldly than in the former period".[53]

Medical science, already since the twelfth and thirteenth centuries, had increasingly absorbed the knowledge transmitted by the translation of classical texts and their theories about the influences of the heavenly bodies on the world, and therefore strictly devoted to mathematics, since Marcianus Capella (fourth-fifth centuries), organised in the so-called *quadrivium*: arithmetic, geometry, astrology and music. In the academic *curriculum*, arithmetic and geometry were taught during the first year, and they were propaedeutic to astrology (which, as we have shown, was not a monolithic discipline), whose teaching was added to the following years. The mathematical curriculum in Italian universities at the time of Fondulo was based on Boethius' *De Arithmetica*, Euclidean geometry, Ptolemaic astronomy, Arabic astrological scholarship, the thirteenth century Sacrobosco's *De Sphera Mundi*, and the *Alfonsine Tables*. Practical exercises provided by the *Theorica Planetarum*, giving some information about the use of astronomical instruments such as the astrolabe, by the *De quadrante* (also by Campanus of Novara), and by the *Legatur liber de urina non visa*, taught students to trace the relations between celestial movements, illness and human urine, an important exercise for medicine at that time. Professors of medicine were also asked to produce predictions for the current year to be held by the beadle so that anybody could consult them before the end of the year. The previsions were both of judicial character (foretelling the future of singular human beings or groups of persons) and of a natural astrological nature (predicting natural events). The professor was also

---

50   Guillermus Aegidii, *Liber desideratus super celestium motuum indagatione sine calculo*, ed. Bonino Bonini (Cremona: Carolus de Darleriis, 1494). See also on Gallica's website the digital copy of the 1511 edition of this book. Garin, *Lo zodiaco della vita*, 36.

51   Goodman, *Power and Penury*, 7-9.

52   Shank, "L'astronomia nel Quattrocento tra corti e università," 5-7.

53   Azzolini, *Duke and the Stars*, 88.

FIGURE 19   *Francesco Casella,* Abrachius (medieval Latin version of the Arabic for Hippar-
              cus, the Ancient Greek astronomer of the second century B.C.), *1513, Studiolo*
              Landriani, Cloister of Sant'Abbondio, Cremona. Hipparcus is here holding a
              Triquetrum or Parallatic Rulers. COURTESY OF MINO BOIOCCHI AND OF THE
              DIOCESE OF CREMONA.

required to write an almanac containing a calendar with the positions of the
planets. At Bologna University professors of medicine had also to produce a
yearly forecast for the local Commune, and to provide their colleagues at the
University with free astrological consulting, when required. By the end of the
fifteenth century, at the University of Pavia, as in many other institutes of learning, the title of the teaching of *astrology* shifted to *mathematics*.[54] As a result of

54   Paul F. Grendler, *The Universities of the Italian Renaissance* (Baltimore: Johns Hopkins
     University Press, 2002), 408-17; Shank, "L'astronomia nel Quattrocento tra corti e univer-
     sità," 6. As concerns the practice of Renaissance medicine, in order to treat patients in
     relation to their horoscopes and to the planetary time, physicians needed efficient instru-
     ments for astronomical calculations: tables of planetary positions, and calculators such as
     astrolabes, armillary spheres and eventually planetary clocks. Astrolabes and astronomi-
     cal tables had however already been criticised by Robert Grossateste († 1253), the bishop
     of Lincoln and an authoritative scholar, who considered them too inaccurate. The enthu-
     siasm for a science that could measure Nature and mathematically forecast physical phe-
     nomena gave great impetus to the diffusion of the discipline. In Lynn White Jr.'s eyes, the
     necessity for less crude astronomical instruments in the practice of medicine pushed
     physicians to improve them and to draw up more accurate astronomical tables. Some of
     the harshest criticism towards Arabic astrology, in favour of a less hermetic and more

his university training at Pavia the physician Giorgio Fondulo must have had a broad astrological knowledge, which he would later impart to Janello Torriani.

We have seen how at Janello Torriani's times, there was a particular connection between medicine, astrology and planetary clockmaking. Since this knowledge was contained in texts mainly written in Latin and since Janello was illiterate, we shall now focus on the person who undertook the task of levelling these linguistic barriers for Torriani's sake.

A Physician as a Mentor: Giorgio Fondulo (Cremona 1473-1545)

We have seen that Antonio Campi wrote that when still a child, Janello Torriani had as mentor a fellow citizen physician, philosopher, famous mathematician, and former lecturer at Pavia University. His name was Giorgio Fondulo and, as Francesco Arisi noted,[55] he had written four books dealing with medicine: *De Podagra lib. 3*, *De Modo componendi Theriacam*, *De Morbo Gallico*, and *De Arborum, & Herbarum natura*, confirming his status of prolific scholar. Unfortunately, none of them seems to have survived. Campi reports that Giorgio Fondulo loved Janello for his genius and predisposition towards astrology and mathematics, deciding to teach him about these things even before he had learned to read. It seems that this humanist impulse to encourage precocious inclinations to academic education was received in various parts of Renaissance Europe: for instance Regiomontanus, *enfant prodige* of astronomy, enrolled at a university when he was only 11 years old, and in Scotland John Napier could do the same when he was slightly older.[56]

Giorgio Fondulo is a rather obscure character; however it is possible to build up a contextualised portrait of the man, setting him in the cultural environ-

---

mathematical method conforming with Ptolemy's concerns expressed in the *Tetrabiblos*, came from physicians such as Agostino Nifo (1473-1540s). For a practical demonstration about the use of an astrolabe, check the web site of the Museo Galileo of Florence. Here one can find a well-made virtual guide to the use of such an instrument. Pompeo Farancovi, "La riforma dell'astrologia," 60-61.

55   Arisi, *Cremona literata*, 1706, II:186.
56   Grendler refers to university professors teaching children for free within the framework of communal schools. However, *"whether they did so spontaneously or in conformity with communal regualtion is unknown"*. As we have previously seen, Tartaglia's experience undermines any notion of a general institutionalisation of such a care at the beginning of the sixteenth century: Grendler, *Schooling in Renaissance Italy*, 105; Christopher Walker and Elena Joli, *L'astronomia prima del telescopio* (Bari: Dedalo, 1997), 265; James Stuart Tanton, *Encyclopedia of Mathematics* (New York: Facts on File, 2005), 345.

ment of his times.[57] Giorgio Fondulo most probably studied and certainly taught at Pavia University. According to a list (based on the records from the University's *Rotoli Antichi* and Parodi's manuscripts) published in the second half of the nineteenth century, one Giorgio Fondulo from Cremona is said to have taught in the faculty of Arts. He was registered as professor *"ad lecturam Philosophiae moralis, in festis"*. The list usually records the first year of teaching. In Fondulo's case it was 1497.[58] Moral philosophy was a minor lectureship in Italy, and in Pavia it was taught in holyday-classes by one or two professors (sometimes friars) and with a low salary. It was comprised of the study of humanist studies, natural philosophy and "Christian Aristotle". We could find a parallel with the same teaching in Padua, where new graduates usually held these lectureships as a temporary position. Giorgio Fondulo was born in 1473, therefore he should have been 24 years old when appointed to this lectureship. Giorgio Fondulo might even have followed his medical studies while teaching in the holydays. The fact that all Cremonese sources describe him as a physician reinforces the hypothesis that the appointment of the lectureship in moral philosophy was just a post of a young graduate student. When did Giorgio Fondulo teach Torriani at Cremona? If Janello followed the general trend of education in Renaissance Italy, we can assume his encounter with Giorgio Fondulo took place between 1505 and 1507.[59]

---

57   What is known so far about Giorgio Fondulo comes from the already quoted Campi, *Cremona fedelissima città*, LV; Arisi, *Cremona literata*, 1706, II:186, and from other countrymen: Lodovico Cavitelli, *Annales: quibus res ubique gestas memorabiles a patriae suae origine usque ad annum salutis 1583 breviter ille complenus est* (Cremonae: Draconius, 1588), ca. 320r; Lamo, *Sogno non meno piacevole, che morale*, canto II, 35. New material from a Parisian manuscript, from the *rotuli* of the University of Pavia and from the State Archive of Cremona will be here presented to draw a better protrait of this character.

58   Not at Padua University as Spanish historiography keeps saying; Giorgio Fondulo could have taught as well in other periods, but unfortunately the rolls for the following academic years are missing: 1501-1504; 1506-1509; 1511; 1513-1514; 1516-1519; 1522-1529; 1531-1532: *Memorie e documenti per la storia dell'Università di Pavia e degli uomini più illustri che v'insegnarono*, vol. 1, Serie dei rettori e priofessori con annotazioni (Pavia: Bizzoni, 1877), 168; Grendler, *The Universities of the Italian Renaissance*, 400.

59   Most scholars agree that Janello was born between 1500 and 1505. We have seen that the trend described by humanists and historians about children's education in Renaissance Italy has estimated the starting point for schooling around the age of seven. According to Campi, Fondulo is said to have taught Torriani in astrology even before he learned to read. If this is true, we could suppose that the tutorship started when Janello was under seven. In this case the limits of a hypothetical chronological spread (considering the age of five a reasonable limit *post-quem*) for the beginning of this process goes from 1505 to 1512. However, I prefer to exclude all those theories that place Janello's year of birth after 1500.

While we have ascertained the time at which Fondulo began his academic career (1497), the point at which it ceased is still unknown. The University of Pavia had students attending the courses until 1512: then, because of war, from the June of this year until the first part of 1516, the university was shut down.[60] However, Giorgio Fondulo had left his position much earlier. We do not know exactly when, but the first French-Venetian campaign against the duchy of Milan must have brought some troubles to Pavia and its university already in 1499. However, Fondulo might have left his position for reasons other than warfare. Already by the years 1506 and 1507 Fondulo was practicing medicine in Cremona. We know this from information provided by the only known manuscript from the Bibliothèque Nationale de France which contains a transcription of a text by Giorgio Fondulo: this is a copy of an epistolary exchange between the physician and a professor teaching medicine and astrology and philosophy at Pavia University named Paolo Trizio, or da Trezzo, referring to a village not far from Milan.[61]

Indeed, the only contemporary witness to report Torriani's age was the royal historiographer Esteban de Garibay: attending Torriani's funeral at Toledo in 1585, he said that the clockmaker had died at the age of 85. For Esteban de Garibay's text see: Eugenio Llaguno y Amirola, *Noticias de los arquitectos y arquitectura de España desde su restauración, por –, ilustradas y acrecentadas con notas, adiciones y documentos por Juan Agustín Ceán Bermúdez*, vol. II (Madrid: Imprenta Real, 1829), 250.

60   By the end of the fifteenth century, the University of Pavia, the most important of the Duchy, could boast an average of 60 professors and an enrolment of nearly 700 students, a quarter of whom were Germans. Since Lombardy was the centre of the Italian Wars, the University of Pavia, as well as the very city, suffered several misfortunes: Pavia's population decreased from approximately 16,000 souls at the beginning of the century to just under 5.000 in 1536, as a consequence of being sacked twice. Grendler, *The Universities of the Italian Renaissance*, 85.

61   In the letters of the Parisian manuscript Lat. 7192 (Bibliothèque Nationale de France) Paolo signs both in Italian (Paulo di Trezo), and in Latin: (Paulus Tricius). Therefore not "*Paolo Frizo*" as recorded by the inventory of the Bibliothèque Nationale de France and by Favaro who follows Benedetti (Antonio Favaro, "Nuove ricerche sul matematico Leonardo cremonese," *Bibliotheca matematica*, III, 1905, 326-41). Moreover, in the transcription of the names inscribed in the *Rotuli* of the University of Pavia is recorded a certain Paolo Trizio from Milan, who we may identify with Fondulo's friend: this Paolo Trizio was teaching in 1510: *Ad Lect. Medicinae et Astrologiae vel Philosophiae*. He is said to have been writing a book about the construction of astrolabes and inscription of circles (*Memorie e documenti per la storia dell'Università di Pavia e degli uomini più illustri che v'insegnarono*, 1, Serie dei rettori e priofessori con annotazioni:122). According to Bedini, the Ambrosiana Library of Milan holds a manuscript on the Astrarium attributed to Paolo Tricio: I have browsed through the manuscript which describes the construction and the use of the astrolabe, not the Astrarium. It is relevant to report that two member of the da Trezzo

Considering what we know about Torriani's age, it is possible that Giorgio returned to Cremona and taught him in this very period. The first letter by Paolo reveals that war prevented him from writing to his dear friend Giorgio Fondulo.[62] Because of constant warfare, it is difficult to determine which particular military operation he was referring to. In fact, although by 1500 the duchy of Milan had fallen within French control, the war continued in Romagna and Naples until 1503 and in Southern Lombardy until 1506, when Pope Giulio II took Bologna. Cremona, once included with Pavia as part of the duchy of Milan, had subsequently been lost to the Venetians (1499-1509).

This amicable correspondence provides a number of interesting insights. We have previously discussed the astrological curriculum for physicians at Pavia University. Both Giorgio and Paolo were educated in astrology and developed a strong interest in the discipline. The first striking feature of their correspondence is that Paolo (from the university town of Pavia), asks Giorgio for new books in astrology and the works of the mathematician Leonardo Cremonese. Fondulo answered from Cremona that "it is not that easy to add water to the sea", a reference to the fact that new books were expected to come to Pavia first. However, Giorgio sent to Paolo di Trezo a list of books he had found in Cremona: among them, featured astrological medieval works by Michael Scot, Abram Avenemre, and the *Ispalense*.[63] Giorgio added he had also found the book Paolo was searching for: the *Pratica Artis Metricae* by Leonardo Cremonese, a kind of manual dealing with geometrical figures and Euclid's and Petro de Curte's calculation of their measures, proportions, areas and volumes. The treaty contains also a *Cosmography* by the same author and

---

family had previously had to do with medicine and with the Astrarium. Jehan de Trizio was a physician at the time the Astrarium was given to Gian Galeazzo Visconti. A certain Antonio de Trizio was instead the Milanese ambassador who asked master Gulielmus (probably Gulielmus Zelandinus) who was employed at the French royal court, to come to Milan to restore the Astrarium for the second time: Bedini and Maddison, *Mechanical Universe*, 28-29.

62   "*Le varie et diverse occupationi et disturbi temporali, si de guerre infirmitade, como anchora daltri assai variabili casi orrorosi, e stato mio pigro calamo somnolento in fare quello che fra benivoli si sole usare*": manuscript Lat. 7192, letter 1: *Datum Papie die 23 Septembri 1506*, Bibliothèque Nationale de France.

63   Michael Scot, a well known twelfth-thirteenth century translator, astrologer and courtier to Emperor Frederic II. He is considered to be one of the most important promoters of Aristotelian texts in Latin Christendom. Abram Avenemre is perhaps the author of the astrological book *In Judiciis*. With the name Ispaliense Giorgio perhaps refers to Johannes Hispalensis (*i.e.*, of Seville), a twelfth century Jew converted to Christianity, who translated books of astrology and astronomy.

refers to an obscure instrument "in shape of a galley for sailing around the world".[64] This might be the clock described in a tract within the manuscript entitled: *Ars instrumenti horologici pro tempore sereno editum per reverendum magistrum Leonardum Cremonensem*. This tract, most probably dedicated during the fifteenth century to a duke of Milan, is accompanied by eighteen illustrations depicting an instrument with components similar to an astrolabe, an armillary sphere and a clock. In this letter, Giorgio Fondulo manifested his passion about astronomy and astrology, lamenting he had little time to look into astrological books, because he had to devote most of his time to the practice of medicine. However, if asked, he would be very glad to challenge himself with mathematical demonstration and with some research browsing "dusty old books". Meanwhile he asked Paolo to find him some books on astrological medicine and some horoscopes of famous people. Paolo would later refer him to some books about astrology and alchemy.[65]

Paolo was unable to complete all the demonstrations related to Leonardo's manual, which he had copied previously in Pavia. He wrote to Giorgio Fondulo that he had been reading the whole of Euclid, but that he could not find how to complete some demonstrations. Giorgio helped him, confuting Paolo's fears that Leonardo might have made some mistakes. As we have already seen, Giorgio Fondulo claimed Leonardo to have been the most skilled mathematician of his times. And here comes an interesting passage of Paolo's letters that

---

64   "*uno instromento in forma /de galea col quale se po navigare per tutto el mondo*": Manuscript Lat. 7192, Bibliothèque Nationale de France.

65   Ibid., "*Ben che male al mar aqua se pote agiungere Cum sit che Pavia sia fior de tutti li studii et dove ogni zorno qualche cosa nova in ogni scientia doveria sorgere, niente di mancho avisovi ma haver visto a Cremona Michael Scotto in Astrologia; Abram Avenemre in Iudiciis: Lo Inspalense: de Leonardo Cremonise una certa pratica quale credo sia quella de che scriveti de modo mensurandi; la qual comincia in questo modo: Artem metricam sive mensurativam occasione quodam prospiciens ut ulterius uno tractato de Cosmographia; Et insuper uno instromento in forma di galea col quale se po navigare per tutto el mondo. E vero che io poca opera dago ad Astrologia perché el me bisogna pur sulicitar la pratica de li infermi; Tamen se qualche bella dificultade in Astrologia alla fiata quive avanza tempo me scrivereti; Sforceromi voltar qualche volume libro pulverulento per farvi cosa grata et iocunda … Ulterius scrivetime se qualche cosa appare stampata qua de novo circa le cose Astronomice et Medecinale; Et quando lo accade qualche Iudicio de homo notabile o valente in Atrologia non ve rincresca mandarmilo: perché consimilmente faremo verso de vui*": letter 2 "*Datum Cremone die 27 Septembris 1506*", and "*Ars instrumenti horologici pro tempore sereno editum per reverendum magistrum Leonardum Cremonensem*". And Paolo answered thae: "*Vero è sono stampate tutte le opere de Arnaldo da Villanova et in Alchimia et in Astrologia et de Somnis, et altre qualcose sono bellissime: piacendovi cosa che qua si ritrova prego me scrivati*": letter 3 "*Datum Papie die 18 februarii 1507*".

opens new considerations about Janello Torriani's later career as an automata-maker. Paolo referred his friend to a number of experiments he had made after a corrupted version of *De Spiritalibus* by Hero of Alexandria. Should he be interested Paolo promised to send Giorgio a description of the machine he made.[66] Giorgio did not know the work by Hero; therefore he asked for the sketches, which clearly had an explanatory function.[67] Giorgio Fondulo's personal interests may have put Janello in touch with classical knowledge on automata already at this early stage. Of a great interest is the report of physicians involved with mechanical experiments inspired by Hellenistic treaties: Lynn White Jr. has noted that physicians were the first to produce tracts on engineering and mechanics, and also that nearly all of those medical-astronomers involved with mechanics were court physicians. White argued that this position put them in privileged contact with warfare and stimulated their interest in military machines.[68] It seems to be the strong link between Renaissance medicine and mathematics in the form of astronomy/astrology, based on the geometrical study of rotary motion that provides the best explanation for such interests. After all, among university superior faculties – Theology, Law, and Medicine – the latter was most firmly connected to mathematics. In this respect, it is significant that Renaissance physicians were among the most interested in ancient mathematical texts, in whose bulk mechanical treaties were to be found.

Another very important point emerging in Fondulo's epistolary document is that both ancient and medieval mechanical knowledge was disseminated through the circulation of manuscripts. I have often used dates of printed editions as evidence of a text's circulation. However, this should not be strictly interpreted as a *terminus non ante quem*. This epistolary document shows that Hero's opus circulated long before the first printed editions of his *Spiritalium*

---

66   Ibid., *"Item me accaduto uno altro libro de Herono philosopho de Spirabilibus intitulato, è vero che è tutto falsificato: niente di mancho ne ho cavato una praticha, et ho fato fare uno bochale da un sol tubo del qual uscisse Aqua pura: vino pur et vino lymphato ad plantum del operatore quando a vuy piacerà mandarovi el disegno":* letter 1: *"Datum Papie die 23 Septembri 1506".*

67   Ibid., *"Del tractato dicesi de spirabilibus de Herono philosopho io non intendo la materia circha la quale lui debia tractare: Tamen molto grato mi sara veder et intender quella pratica dicesi del bochale":* letter 2 *"Datum Cremone die 27 Septembris 1506".* And: *"Dele cose medicinale sel accadera altro me avisireti. Le opere de Arnaldo già sono apresso di me 18 mesi fa quale pur hebe a Pavia. Non altro se non che aspecto ogni zorno qualche vostra litera con quelle gentilezze de Herono philosopho et insuper quelle vostre demostratione facte circha le pratiche de Leonardo":* letter 4 *"Cremone die 16 marcii 1507 a Nativitate".*

68   White, "Medical Astrologers."

liber curated in the late sixteenth century by Federico Commandino, Bernardino Baldi and Giovan Battista Aleotti.[69] The example given by Giorgio Fondulo and Paolo di Trezzo is not isolated: in the territories around Pavia and Cremona the circulation in manuscript versions of Hero's works, together with those of Pappus and others, was also noted by Giuseppe Ceredi in 1567 (whom we shall meet again), a physician from Piacenza who acquired some manuscripts of Hero and other ancient authors dealing with automata and mechanics from the library of deceased fellow countryman and humanist Giorgio Valla († 1500), who had previously taught at Pavia, Milan and Venice.[70] Also the Milanese self-declared professor of architecture Cesare Cesariano (1475-1543), first author of a vernacular printed and illustrated version of Vitruvius, was acquainted with Hero's and Pappus' manuscripts.[71] This mode of knowledge dissemination was especially connected with university practices, where professional copyists provided textbooks for students and professors.[72]

---

69   Hero of Alexandria, *Heronis mechanici Liber de machinis bellicis necnon Liber de geodae-sia*, ed. Francesco Barozzi and Francesco de Franceschi (Venetiis: Apud Franciscum Franciscium Senensem, 1572); *Heronis Alexandrini Spiritalium liber*, ed. Federico Commandino (Paris: Apud Ægidium Gorbinum, 1583); *Di Herone Alessandrino De gli automati, ouero, Machine se mouenti, libri due*, trans. Bernardino Baldi (In Venetia: Appresso Girolamo Porro, 1589); *Gli artifitiosi et curiosi moti spiritali di Herrone. Aggiontoui dal medesimo Quattro Theoremi ... Et il modo con che si fà artificiosamte salir vn canale d'acqua viua, ò morta, in cima d'ogn'alta torre*, trans. Bernardino Baldi (Ferrara: per Vittorio Baldini Stampator ducale, 1589); *Hero Alexandrinus Spiritali di Herone Alessandrino ridotti in lingua volgare da Alessandro Giorgi da Vrbino (Versi da G.B. Fatio).*, trans. Alessandro Giorgi (Urbino: Appresso B. e S. Ragusij fratelli, 1592). A partial printed edition of Hero was already published in 1501: Elio Nenci, "Mechanica e machinatio nel De subtilitate" (Cardano e la tradizione dei saperi, Milano: F. Angeli, 2003), 67-82.

70   " ... *avenga che quasi a sorte mi fur venduti da chi lor non conosceva, certi scritti di Herone, di Pappo, et di Dionisidoro tolti dalla libraria, che fu gia del dottissimo Giorgio Valla nostro Piacentino, il quale per gli meriti suoi inalzato dalla liberalità dell'Illustrissimo Signor Giovan Giacomo Trivulzi, che allhora governava lo stato di Milano*", Ceredi, *Tre discorsi*, 6; for what concerns Giorgio Valla, look: Gianna Gardenal, Patrizia Landucci Ruffo, and Cesare Vasoli, *Giorgio Valla tra scienza e sapienza* (Firenze: L.S. Olschki, 1981).

71   Alessandro Rovetta and Maria Luisa Gatti Peter, eds., *Cesare Cesariano e il classicismo di primo Cinquecento* (Milano: Vita e pensiero, 1996), fig. 20, p. 42.

72   In spite of the late printing in the advanced sixteenth century, already by the end of the fifteenth century the great mathematical works of ancient Greece had been recovered and circulated. These included Archimedes' corpus, Diophantus' *Arithmetica*, Pappus of Alexandria's mathematical collection, Hero of Alexandria's mechanical books, Apollonius' *Conica*, and the pseudo-Aristotle's *Mechanica* (Already printed in a Greek version by Aldo Manuzio in 1497). Influential medieval mathematicians were also rediscovered during the same period. Grendler underlines that the entire bulk of Leonardo Fibonacci's

Another significant clue about this cultural *milieu* is given by Giorgio Fondulo as he lamented he was a busy physician with little time to dedicate to his passion, the study of mathematics and especially astrology. Indeed, in his letters he confirmed a great joy in discussing, demonstrating and searching for mathematical issues. This "love for knowledge," or "*lamor* [sic] *de la scientia*" as Paolo calls it,[73] was expressed not only by epistolary inquiries or by the exchange of books. Giorgio Fondulo was also committed to supporting young students, a factor that became crucial in terms of shaping Torriani's trajectory. The civic commitment to education of such Renaissance characters emerges in this correspondence: indeed, Paolo recommends to his friend in Cremona a 13 year-old boy, son of a deceased professor of Pavia University. The boy, Aurelio Grasso, was indeed going to study Latin grammar in Cremona.[74]

---

corpus had been ignored throughout the Middle Ages by academics. It was thanks to Luca Pacioli that the works of the Pisan were brought to the attention of the public: Grendler, *The Universities of the Italian Renaissance*, 414; concerning the issue of manuscript copies circulating at Pavia University: Luciano Gargan, "«Extimatus per bidellum generalem studii Papiensis». Per una storia del libro universitario a Pavia nel Tre e Quattrocento," in *Per Cesare Bozzetti: studi di letteratura e filologia italiana*, ed. Simone Albonico (Milano: Fondazione Arnoldo e Alberto Mondadori, 1996), 19-36; for other universities and the problem in a more general context: Stefano Zamponi, "'Exemplaria', manoscritti con indicazioni di pecia e liste di tassazione di opere giuridiche," in *La Production du livre universitaire au Moyen Âge: exemplar et pecia : actes du symposium de Grottaferrata, mai 1983*, ed. Louis-Jacques Bataillon, Bertrand Georges Guyot, and Richard H. Rouse (Paris: Éditions du CNRS, 1988), 125-32; Stefano Zamponi, ed., "Manoscritti con indicazioni di pecia nell'Archivio Capitolare di Pistoia," in *Università e società nei secoli XII-XVI: nono Convegno internazionale : Pistoia, 20-25 settembre 1979* (Pistoia: Centro italiano di studi di storia e d'arte, 1982), tab. 1-12; Jean Destrez, *La Pecia dans les manuscrits universitaires du XIIIe et du XIVe siècle* (Paris: Éd. Vautrain, 1935); Giulio Battelli, "Il libro universitario," in *Civiltà comunale: libro, scrittura, documento : atti del Convegno, Genova, 8-11 novembre 1988*, vol. 2, Atti della Società ligure di storia patria, XXIX (Genova: Società ligure di storia patria – Associazione italiana dei paleografi e diplomatisti – Istituto di civiltà classica cristiana medievale: Università di Genova, 1989), 281-313; Luciano Gargan, "Le note conduxit: libri di maestri e studenti nelle Università italiane del Tre e Quattrocento," in *Manuels, programmes de cours et techniques d' enseignement dans les universités médiévales : actes du Colloque international de Louvain-la-Neuve, 9-11 septembre 1993*, ed. Jacqueline Hamesse (Louvain-La-Neuve: Institut d'Études Médiévales de l'Université Catholique de Louvain, 1994), 385-400.

73   manuscript Lat. 7192, letter 3 *Datum papie die 18 february 1507*, Bibliothèque Nationale de France.

74   "*Post scripta venie li uno putto chiamato Aurelio figliolo del quondam Magistro M(agnifico?) Petro Grasso legente in studio qui rason Canonica il qual starà li in dozena per imparare le Gramaticale littere. Prego vogliati cercarlo et offerirli quello posseti per lui per amor mio, et*

Fondulo answers that he planned to offer to him intellectual and material support. Giorgio Fondulo's dedication to the young protégé remains a constant through all the letters.[75] Even Giorgio's testaments testify to the social pedagogical commitment of the physician, who left enough resources to maintain, besides four physicians and surgeons to take care of the indigents, an academy dealing with any subject involving Latin or vernacular letters. If the academy was not to be established, the bequest had to be doubled in order to maintain eight poor youths in their studies, providing them with money to purchase books and to pay teachers. Giorgio's brother, Cristoforo Fondulo, was said to have imitated his sibling.[76]

As we have seen, the books involved with astronomical knowledge relevant for planetary clockmaking were used at universities and therefore were written in Latin. Thanks to the initiative and commitment of a literate man like Fondulo, "higher knowledge" could be handed over to illiterate but talented pupils like Janello. The academic background and connections of Fondulo afforded Janello Torriani the chance to access the knowledge of the library

---

velo racomando accadendo: è di età circha 13 anni bono filiolo et accostumato": manuscript Lat. 7192, letter 1: *Datum Papie die 23 Septembri 1506*, Bibliothèque Nationale de France. Pietro Grassi's family was a noble Milanese one. A homonymous uncle of his had been bishop of Pavia († 1426). Pietro Grassi taught at Pavia University until 1505, when he died. He started teaching at Pavia University in 1472 where he lectured Institutiones, and was ordinary lector of canon law, civil law: *Memorie e documenti per la storia dell'Università di Pavia*, 1, Serie dei rettori e priofessori con annotazioni:59; Gigliola Di Renzo Villata, "Grassi, Pietro," *Dizionario Biografico Degli Italiani* (Roma: Istituto della Enciclopedia italiana, 2002).

75 "*... perché pigliati qualche ardire in recomandarme uno vostro amico et filiolo de homo singulare, quale per amor vostro ho ritrovato et offerto lopera mia le facultate, et ogni poter mio in qualunche occurrentia sua sempre esser paratissimo et simile a vui ve aricordo de non poter mancho de mi disponere che de vui medesimo. Ad Aurelio fareti et intendere che non mi refuta ne habia alcun rispecto in ogni suo bisogno ricorrere da me como da patre proprio: per che veramente faroli intendere le littere vostre non esser stati de poco momento apresso di me ...*": manuscript Lat. 7192, letter 2 *Datum Cremone die 27 Septembris 1506*, Bibliothèque Nationale de France. It is interesting here to observe the practice of early sixteenth century education, scattered among different cities: from Aurelio Grasso's story we learn of a geographical distribution of schools providing different levels of education: a university town like Pavia was probably not considered as suitable for pre-academic education as Cremona. Unfortunately it is not possible to be sure of all the possible reasons Grasso may have arrived at Cremona; indeed, this could well have been influenced by some family strategy unknown to us.

76 Francesco Robolotti, "Dei medici cremonesi," in *Effemeridi delle scienze mediche, compilate da Giovambattista Fantonetti*, ed. Giambattista Fantonetti, vol. IX (Milano: Paolo-Andrea Molina, 1839), 312.

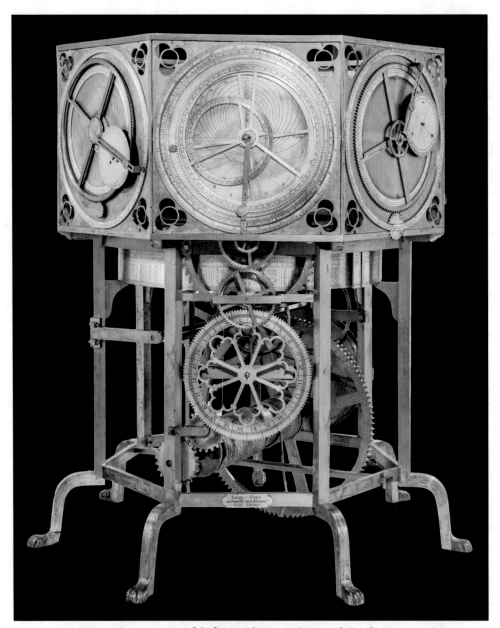

FIGURE 20   *Reconstruction of the fourteenth century Giovanni de Dondi's Astrarium, Luigi Pippa, 1961-1963, cm 100 × 80.* COURTESY OF THE MUSEO NAZIONALE DELLA SCIENZA E TECNOLOGIA "LEONARDO DA VINCI" OF MILAN.

patrimony enjoyed by academics: the books circulating, printed and kept in Cremona were no longer the sole context for Janello's education. Fondulo had access to printed books and manuscripts as well in Cremona, and through his friends in Pavia he was able to draw on another cultural background, more vast than the one kept within the city-walls of his little *patria*. Giorgio Fondulo was accustomed to keeping himself updated with new publications, eagerly requesting them, if he could not get any in Cremona. Another remarkable point here relates to the fact that Janello Torriani was the craftsman who would rebuild Dondi's Astrarium: considering Giorgio Fondulo's personal interests and the fact that he taught at the University of Pavia, it is possible that the physician may have linked Janello to the planetary clock, which was kept in the castle of Pavia, and was considered the most astonishing astronomical machine of Christendom.[77]

In Pavia, Giorgio Fondulo also had some other old common friends that he asks Paolo to greet: Rosati, Bobio, Rustico, Balbo, de Laquila.[78] From the *rotuli* of Pavia University it seems they were all involved with the teaching of medicine, and therefore also with astrology. Some of them also seem to have been employed at the ducal court. Rosati can be identified with two different characters. The first was Giovanni Rosate, called Monferrato, from Rosate, who was appointed in 1467 "*ad lect. Chirurgiae*", then made administrator of the hospital of S. Matteo, and physician of the count of Pitigliano (chef-general of the duke of Milan), and latterly from 1500 onwards, because of his political connections with the Sforza, he was exonerated from teaching, eventually dying in 1515. The second hypothetical identification, more alike than the first one, has to do with Ambrogio Rosate, called Varesio, astrologer of Duke Ludovico il Moro: in the year 1461, this famous astrologer was appointed "*ad lecturam Fisicae naturalis*", and in 1486, and 1500 "*ad lecturam Almansoris*". He was physician and astrologer of the Duke Ludovico il Moro, and he was given the fief of Rosate from which it took the name. By 1495, Duke Lodovico had Dondi's Astrarium transported from Pavia to Rosate castle, where it remained for a year, where the phisician could study it. Friar Luca Pacioli praised him as an "expert investigator of the celestial bodies and interpreter of future events".[79] Bobio can be

---

77 For the circulation of astrological books between the University of Pavia and the Sforza court in the fifteenth century see chapter 1 in: Azzolini, *Duke and the Stars*. For the piece of news regarding university professor borrowing books from the ducal library see: p. 51, note 119.

78 "*Che me ricomandati alli nostri amici vechi: principalmente accadendo ali nostri Rosati: al Bobio: al Rustico: al Balbo: de Laquila*": Memorie e documenti per la storia dell'Università di Pavia.

79 Bedini and Maddison, *Mechanical Universe*, 26. Azzolini, *Duke and the Stars*, 23.

identified with Francesco Bobbio, appointed "*ad lect. Logicae ordin. De mane et Sophistariae*" in 1480, and "*ad Medicinae ordin. de mane*" in 1483, and 1511.[80] Rustico may be identified with Pietro Antonio Rustico, perhaps from Piacenza, who was appointed in 1486 to lecture "*philosophia*", and in 1487 "*ad lect. Medicinae ord.*". Balbi should instead be recognised with that Agostino Balbi of Pavia who in 1483 and 1499 was in charge of lecturing "*Praticae Medicinae extraord. de nonis*"; he died in 1512. Finally, de Laquila must be that Sebastiano Aquila, perhaps from Pavia, who in 1500 was appointed "*ad lect. Medicinae*".[81] It is difficult to imagine that Giorgio Fondulo, with this array of friends, and especially with Paolo da Trezzo and Ambrogio Rosate (if our identification is correct), did not have access to the Astrarium. At least Dondi's machine must have been at the core of many discussions Giorgio had with his friends regarding planetary clockmaking. Moreover, as mentioned above, the professors of Pavia had access to the ducal library in the Castle, where astrological books and the Astrarium were kept.

A further link between the Astrarium and Giorgio Fondulo may be found in the public administration of Cremona: since the Venetians had taken possession of Cremona in 1499, two famous clockmakers were appointed to the management of the public clock: Zanino of Clusone (a village near Bergamo) and Falconi of Bergamo. Zanino could have been the same old clockmaker of the Sforza who in 1473 had tried unsuccessfully to restore the Astrarium; or perhaps a homonymous. Furthermore, it seems that Falconi was made keeper of the public clock of Cremona in the very same years in which Fondulo acted as Janello's mentor. Although Falconi was probably not in any direct contact with the Astrarium, he was an expert in astronomy, and therefore a most interesting character in Fondulo and Janello's eyes. We will soon come back to this scientific instruments maker.

---

80   The rotuli mention another professor named Bobbio, but I do not think this other homonymous could fit into this consistent group of physicians: Uberto Bobbio was professor of law between 1510 and 1525: *Memorie e documenti per la storia dell'Università di Pavia*.

81   Ibid.

CHAPTER 3

The Practical Clock

The Guild

In 1586, a year after Janello's death, and following Juan de Herrera's inspection of the water supply machineries made in Toledo, the famous Spanish architect royal wrote to Juan de Ibarra, royal secretary, who as a bureaucrat was very worried about the efficiency and maintenance cost of these structures. Juan de Herrera wrote that the two devices alternated were providing more than the amount of water Janello was obliged to make available in the signed contract and, in a polemic way (it seems) he added: "everything has been carefully evaluated except for the workmanship that is the work and the industry of the master".[1] It was precisely the elements *"trabajo e industria"* that the Spanish architect found to be the basic attributes of a master. Juan de Herrera's laconic sentence points to a society in which people involved with practical but ingenious and industrious enterprises enjoyed a full self-conscious dignity. Mario Otto Helbing has observed that the *Didascalicon* of Hugh of Saint Victor (1096-1141) was greatly appreciated during the Middle Ages. The scholar argues that the *Didascalicon* defined *artes mechanicae* as all those activities involving manual work, therefore inferior to *artes liberales*, specifically intellectual and consequently nobler. However, the distinction between the two categories is not always as clear as sometimes expected. For example, among mechanical arts, thanks to the impact of Aristotle's *Mechanics* and to his definition of *scientia media*, the *ars mechanica* dealing with transmission of movement had already gained an intellectual status in fifteenth century Italy. Moreover, despite the strictly hierarchical society of Medieval Europe, Christian religion had also offered some openings to a more favorable perception of manual work: St. Joseph was a carpenter, St. Peter and other apostles humble fishermen, and both Bazalel from *Exodus* and St. Luke were believed to have been inspired by God in their manual enterprises, namely the construction of the Tabernacle and the portrait of Mary.[2]

---

1   6th of March 1586, Madrid: "... *la maestría que es el trabajo e industria del maestro que esto no se ha tasado*": Cervera Vera, *Documentos biográficos*, doc. 129.

2   Helbing, "La scienza della meccanica nel Cinquecento." For a broader discussion of the high intellectual status that some Medieval artists had already enjoyed, see: Enrico Castelnuovo, ed., *Artifex bonus: il mondo dell'artista medievale* (Roma: Laterza, 2004).

© KONINKLIJKE BRILL NV, LEIDEN, 2017 | DOI 10.1163/9789004320918_005

The title defining the worker's excellence in a craft was "master". The first known document mentioning Torriani, as already seen, dates back to 1529, when the clockmaker was already called a master and he was in charge of the modification and reform of the civic clock of Cremona. What did it mean to be a master in the sixteenth century? The Latin *"magister"*, in the late Middle Ages, had a broad semantic sphere. The term could designate a magistrate (i.e., an officer in lay, military or ecclesiastic institutions), a teacher, a graduate student in a university context, or an artisan with a specific knowledge. Torriani belonged to this last category. Covarrubias, the author of the most important Early Modern Castilian dictionary (published posthumously in 1611), can help us to better frame the role of a *maestro*:

> Master is called everybody that, well-taught in any branch of knowledge, discipline or art, teaches to others explaining it; if does he not educate other people, he had usurped the name of master.[3]

Here is, in fact, a most characteristic feature of this title: a master craftsman was supposed to teach an apprentice. This occurred within the guild institutional framework and was regulated by private contract. The guilds were corporate structures that organised the work of certain categories of practitioners around a religious ritual and based on a charter.[4] These regulations varied

---

3 *"El que es docto en qualquiera facultad de sciencia, disciplina o arte, y la enseña a otros dando razón della, se llama maestro; porqué si en esto falta, ha usurpato el nombre de maestro."* : Sebastián de Covarrubias Orozco, *Tesoro de la lengua castellana o española según la impresión de 1611, con las adiciones de Benito Remigio Noydens publicadas en la de 1674* (Barcelona: S.A. Horta, 1943).

4 The artisanal professions were since the Late Antiquity traditionally hereditary. Since the end of the thirteenth century these restrictions had been abolished in many Italian towns: the guilds were indeed important institutions in the economic and social reform of the period, when the feudal restrictions tying the peasants to the land were abolished, permitting a flow of cheap labour into urban space. In Justinian's *Corpus Iuris Civilis*, the guild is called: *schola, corpus, universitas* and *officium*; Caterina Santoro, *Collegi professionali e corporazioni d'arti e mestieri della vecchia Milano: catalogo della mostra* (Milano: Edizioni dell'Ente manifestazioni milanesi – Archivio storico civico di Milano, 1955), 10: where is mentioned for the Middle Ages as well the synonyms of *collegium* or *paraticum*; see also: Marina Gazzini, "Confraternite/ Corporazioni: i volti molteplici della schola medievale," in *Corpi, "fraternità", mestieri nella storia della società europea*, ed. Danilo Zardin et al. (Roma: Bulzoni, 1998). For a general history of the guilds: Antony Black, *Guilds and Civil Society in European Political Thought from the Twelfth Century to the Present* (London; New York: Methuen, 1984); Alberto Guenzi, Paola Massa, and Fausto Piola Caselli, eds., *Guilds, Markets, and Work Regulations in Italy, 16th-19th Centuries* (Aldershot, Hampshire, Great Britain; Brookfield, Vt., USA: Ashgate, 1998); Stephan

from town to town, even though the core was essentially the same. In order to earn the title of master, a candidate had to stay at a master's workshop for a long period (at least two years) and he should be in receipt of a specific contract, which endowed him with a status higher than that of a simple worker. After this period the apprentice usually became a journeyman and could work at another master's workshop, sometimes in a different city. The master, after having received a conspicuous fee from the apprentice's father, rewarded his apprentice with all the secrets of his art and with clothing, board and lodging. The apprentice, who was educated by the master, at the same time had to produce objects for him, and to sell them on his behalf. The master was obliged to teach his art to an apprentice, but took on only one such pupil at a time (or more than one after the first apprentice had already stayed at the workshop for a couple of years), whereas he could hire as many helpers as he needed. The status of apprentice implied much more than a simple teacher-learner relationship; indeed, it went far beyond this. The master became a tutor. The apprentice entered the master's household, and usually lived permanently with the master and with his family. He was educated both as a craftsman and as a citizen.

Medieval and modern craft-guilds had several functions: artisan-manufacturing control (quality, quantity and methods of production), protection of the affiliated, mutual aid (and sometimes assistance to the poor of the town), religious common ritual and fiscal charge.[5] In Cremona the taxation was paid collectively to the state by the officials of the guilds, who had previously collected the money among their members according to the rules of the guild written in the book of the statute.[6] The major power (emperor, pope, king, duke, republican government, etc.) recognised those *"universitates"* or corporations, and transferred authority to them with a bill. Scholarship has often

---

R. Epstein and Maarten Roy Prak, *Guilds, Innovation, and the European Economy, 1400-1800* (Cambridge; New York: Cambridge University Press, 2008).

5   The problem of guilds in the social system of the *Ancien Régime* relates to the larger issue of associations. For instance membership to a major guild did not prevent the affiliated from creating smaller associations. Already Muratori's attention was attracted by the history of the *"confraternitae"* which he analysed in his *"Antiquitates"*: Ludovico Antonio Muratori, ed., *Antiquitates Italicae medii aevi* (Mediolani: ex typographia Societatis Palatinae, 1742); The overlapping of religious and social practice and the sometimes difficult borderline between guilds and confraternities is analysed by Gazzini, "Confraternite/Corporazioni: i volti molteplici della schola medievale," 56 and 309.

6   Carla Almansi Sabbioneta, ed., *L Università dei Mercanti e le corporazioni d'arte a Cremona dal medioevo all'età moderna: mostra iconografica e documentaria; catalogo; sala contrattazioni 3-15 giugno 1982* (Cremona: Linograf, 1982), 11-14.

emphasised the double-faced action of these corporate bodies: on the one hand, the long-established holders of a craft were able to maintain a monopoly in their field of production, preventing newcomers from entering into hard competition with them, while on the other hand, the craftsmen of a special art had to maintain a certain degree of quality in their products. The guild was generally responsible for determining the conditions for the apprenticeship, the rules in the production and trade of the artefacts, and it imposed the obligation of inscription within this institution to everybody that wanted to manufacture and sell a certain type of product.[7] For all these reasons, traditional historiography has considered guilds conservative social bodies. This view was opposed by a scholarship that saw guilds as "associations of equals embodying the principles of urban freedom".[8] In both cases, the guilds played a paramount role in European urban life and in its material production and related knowledge. Sometimes, in republics like Florence, they even monopolised the whole political power where, since the end of the thirteenth century, they substituted the former feudal political system based on blood with a new one exclusively expressed through guild-membership and based on wealth, trade and production.[9]

---

7   Dominique Fléchon, *L'orologiaio: mestiere d'arte* (Milano: Il saggiatore, 1999), 61.

8   Black, *Guilds and Civil Society in European Political Thought from the Twelfth Century to the Present*, 8-9; A remarkable example of guilds' conservative attitude has been presented in the case of Hanns Spiachl. Friedrich Klemm, *Technik: eine Geschichte ihrer Probleme* (Freiburg: Alber, 1954). However, in 2008 Epstein and Prak published a volume that provides an in-depth analysis of the relation between guilds and innovation. Their work tries to downplay the significance of the interpretations of influential historians such as Cipolla, Landes and Mokyr: (Landes, *Revolution in Time*, 210; Mokyr, *The Gifts of Athena*, 31). These essays criticised the idea of guilds as chiefly conservative bodies. According to this analysis, the representation of guilds as responsible for technological, economic and social stagnation was originated by a prejudice rooted into the theories of so-called Enlightenment and Smithian liberalism (Adam Smith, *Wealth of Nations* (Raleigh, NC: Hayes Barton Press, 2005), bk. 1, chap. 1, part 2). Epstein and Prak, *Guilds, Innovation, and the European Economy, 1400-1800*; Guenzi, Massa, and Piola Caselli, *Guilds, Markets, and Work Regulations in Italy, 16th-19th Centuries*; Stephan R. Epstein, "Trasferimento di conoscenza tecnologica e innovazione in Europa (1200-1800)," *Studi storici. Rivista trimestrale dell'Istituto Gramsci* 50, no. 3 (2009): 717-46; Bartolomé Yun Casalilla, "Misurazioni e decisioni: la storia economica dell'Europa preindustriale oggi," *Studi storici. Rivista trimestrale dell'Istituto Gramsci* 50, no. 3 (2009): 581-605; John David North, "Cultura e storia economica: il mercato dell'arte europeo," *Studi storici. Rivista trimestrale dell'Istituto Gramsci* anno 50, no. Luglio-Settembre 2009 (2009).

9   Najemy, *A History of Florence 1200-1575*, 81-87.

In terms of Janello's experience, the clockmaker picked up his practical knowledge about metal and clockworks within the guild system, and until he was 50 years old, he participated in its social and economic system as a master. However, Janello could not have been educated within the framework of a guild of clockmakers: the first one we know about was established in Paris in 1544.[10] In what guild did Torriani receive the title of master?

Janello Torriani the Blacksmith

At least from the year 1529 *"Janello de Toresanis"* was a master: he was in charge of the reformation of the Torrazzo's clock, the main civic tower in Cremona. How did he come to possess such a title, and to which guild did he belong? From a later document of the early 1550s, when Torriani had already lived in Milan for about a decade, it emerges that he was affiliated to a local guild of blacksmiths.[11] After all, blacksmiths, locksmiths, crossbow-makers, cannon-makers, bell-smelters and brass and tin workers were often unified under the same guild: the blacksmiths.[12] Sometimes clockmakers could also be included in the guild of the goldsmiths. The production of clockworks required the use of different metals, precision and also craftsmanship on a micro-scale similar to that used in jewellery. As we shall see, portable clocks were often worn as jewels. Brunelleschi is probably one of the best-known examples of a goldsmith who created clocks.[13]

---

10   Steven Epstein, *Wage Labor and Guilds in Medieval Europe* (Chapel Hill: University of North Carolina Press, 1991), 230.

11   Leydi, "Un cremonese del Cinquecento," 138.

12   For instance, three clockmakers from fifteenth century Reggio Emilia bore the surname Clavario, which refers to the profession of locksmithing (Lat *clavis* = key). Another example of this typical professional syncretism in the years of Janello's formation is provided by masters Andrea and Marco from the neighbouring town of Crema: these master clockmakers, active at Ferrara from 1458-1506, were also crossbow-makers. Morpurgo, *Dizionario degli orologiai italiani*, 1974; and Gian Carlo Del Vecchio and Enrico Morpurgo, *Addenda al Dizionario degli orologiai italiani edizione 1974 di Enrico Morpurgo* (Milano: Tipografia Nava, 1989); Cipolla, *Le macchine del tempo*, 24; García- Diego reports that the first record of a clock-maker entering a blacksmith guild was in 1541: García-Diego, *Juanelo Turriano, Charles v's Clockmaker*, 32. About clockmakers in blacksmiths' guilds North of the Alps, see Eva Groiss, "The Augsburg Clockmakers' Guild," in *The Clockwork Universe: German Clocks and Automata, 1550-1650*, ed. Klaus Maurice and Otto Mayr (Washington; New York: Smithsonian Institution ; N. Watson Academic Publications : National Museum of History and Technology : Bayerisches Nationalmuseum, 1980), 57-86.

13   Mario Fondelli, *Gli "Oriuoli mechanici" di Filippo di ser Brunellesco Lippi: documenti e*

Thus, in Milan, on the 22nd of January 1550, *"Leonello* [sic] *Torriani q. Gerardo"* was elected abbot of the blacksmiths' guild. We know that some years earlier in Cremona Torriani worked as a locksmith: he had a contract for the gilded copper doors of the monumental stoup of the Baptistery (Figures 12a-12b).[14] Moreover, Cardano, in the second and third editions of his *De Subtilitate libri XXI* (1554 and 1560), mentioned a special combination lock whose invention he attributed to Janello.[15] For these reasons it is difficult to imagine Torriani in Cremona as a member of anything other than the guild of the blacksmiths.

In 1474, the Duke of Milan Galeazzo Maria Sforza had approved the oldest preserved statute of the blacksmiths' guild of Cremona. This manuscript statute was valid during Janello's lifetime, covering the period that goes from 1474 until 1592, when a new one in vernacular was printed and presented to Philip II for approbation. Except for a few chapters, the two statutes have the same structure. Unfortunately, some of the pages of the fifteenth century manuscript statute are missing. The statute, entitled *Satuta Paratici et Artis Ferrariorum Civitatis et Districtus Cremonae* (i.e. Statute of the Guild and Art of the Blacksmits of the City and District of Cremona),[16] was written in Latin and contains 37 chapters or *rubricae*. It comprises two ratifications by the third Sforza duke and by the Republic of Venice, and some additions of the following century, made under Spanish rule (Figures 22-23).[17]

---

    notizie inedite sull'arte dell'orologeria a Firenze: l'orologo dipinto da Paolo Uccello nel Duomo fiorentino : nuovi studi e precisazioni per la sua lettura, ed. Umberto Baldini, Quaderni della critica d'arte (Firenze: Le lettere, 2000); Arnaldo Bruschi, *Filippo Brunelleschi* (Milano: Electa, 2006), 15.

14   There are still four small decorated gilded-copper doors with iron locks closing the *batismum* (a marble structure containing the Holy Water) of the Baptistery of Cremona. Barbisotti recently attributed them to Janello Torriani: Barbisotti, "Janello Torresani."

15   Girolamo Cardano, *Hieronymi Cardani De svbtilitate libri XXI: nvnc demum recogniti atq perfecti* (Basileæ: per Lvdovicvm Lvcivm, 1554), 472.

16   *Paraticum* is one of the most common Lombard terms for a guild. The most used synonyms in this context are *universitas*, *ars* and sometimes *collegium*, although the latter was usually used to define the upper guilds of the notaries, physicians, engineers and lawyers.

17   The cover reports the year MDXXVIIII in Roman ciphers (1529), the very year Janello, already a master, was mentioned for the first time in a document. The coloured drawings on the two covers show on one side the Virgin assumed (to whom the Cathedral of Cremona is dedicated, and in whose fest all the *paratica* had to go in procession to pay their tribute), and on the other Saint Eligius and Saint Anthony (also said "of the fire"), two saints connected with metalwork. In Cremona, all the guilds used to participate in the grand festival of the Assumption when, after gathering in squadrons armed with sticks in

The paragraphs 12-13-14-15 of the vernacular version (1592) of the statute, corresponding to the lost *rubricae* 13-14-15-16 – except for the n. 16 – in the 1474 manuscript,[18] stated that all persons working iron, steel, brass, bronze and tin or trading these materials within the city and the district of Cremona, had to enter this guild by the means of an inscription fee and the insertion of the name and the fabric-mark in the book of the art.[19] Unfortunately, the lists or *matriculae* of this guild's members are lost. Nevertheless, one can reasonably

---

order to bring the offerings to the Cathedral, a bull was released in the square and the procedure culminated in a brawl. Cardinal Borromeo forbade this profane side of the festival in 1575. The guild gathered in the church of Saint Antony of the fire (destroyed in 1702 by the French troops after the so-called *Surprise of Cremona*). The election of the officers of the *paraticum* took place during the festival of Saint Laurentius, another patron saint connected with fire. Carla Bertinelli Spotti, Maria Teresa Mantovani, and Giovanna Ferrara Bondioni, *Cremona: momenti di storia cittadina* (Cremona: Turris editrice, 1996), 166-164. According to the *Statuta et Ordinamenta Comunis Cremonae* (the Statutes of the Commune of Cremona in force since fourteenth century, when they were rewritten after the Visconti regime was extended over this city), no corporate body could have any statute and rules without the supervision of a specialist elected by the College of the judges of Cremona: *Statuta et ordinamenta comunis Cremonae facta et compilata currente anno Domini MCCCXXXIX.* (Milano: A. Giuffrè, 1952), 261, "*rubrica XXXVI: Rubrica de statutes colegiorum, vilarum et paraticorum*". The statute was ratified in 1485 by Duke Gian Galeazzo Maria Sforza (fol. 15), and in 1499 by the Venetian Republic (fol. 16). Some notes were added between 1553 and 1595. One of the major differences between these two versions relates to the entrance of blade-smiths into the *Paraticum Ferrariorum* at an unspecified moment between 1474 and 1592.

18   Here are listed the rubricae or paragraphs partially or entirely lost in the 1474 statute. In brackets are the correspondent numbers of the paragraphs in the vernacular version of 1592: *12-(11) De pena tenentis seraturam novam falsam gariboldellum vel clavem ("ad formam cere"* in the text) *13-(12) De libro paratici fiendo super quo describantur omnes de dicto paraticoet quantum solvere debeant volentes; exercere artem ferrarizie et describi in paratico; 14-(13) Quantum solvere debeat fabricantes campanas; 15-(14) De personis vendentibus ferrum vel azale et quantum solvere debeant; 30-(28) De pena laborantium vel facientes laborare in certis festis; 31-(28) De pena recusantium dare pignu;s 32-(29) Quod condenationes debeant exigi per consules infra decem dies et de pena consulum eas non exigentium; 37- De ducalibus litteris inpetrandis pro confirmatione statutarum; Statuta paratici ferrariorum* 1474-1590 (cover painted in the year MDXXVIIII), Biblioteca Statale di Cremona, Deposito Libreria Civica, B.B. 1. 7. / 17.

19   Here are the paragraphs of the statute of the guild of the blacksmiths' from 1592, which imply the compulsory inscription of anybody working with the above-mentioned metals into the very art: *12- Del libro del Paratico, e del pagamento per intrare in esso; 13- Delli campanari e lavoranti d'ottone e stagno; 14- Delli venditori di ferro o acciajo nuovo o vecchio, lavorato o non lavorato; 15- Delli venditori suddetti fuori della città e suoi borghi;* Carla Almansi Sabbioneta, "L'arte ferrarecia a Cremona tra i secoli XV e XVIII," in *Il Ferro nell'arte:*

a

FIGURE 21a   *Cistern of the Baptistery of the Cathedral of Cremona.* COURTESY OF MINO
               BOIOCCHI AND OF THE DIOCESE OF CREMONA.

b

FIGURE 21b   *Detail of the cistern with one of the four decorated doors by Janello Torriani, 1543*

infer that a clockmaker and a locksmith like Torriani (because of the metals he worked with) likely had no option other than to enter this guild.[20]

Furthermore, *rubrica* 13 of the statute deals with bell-smelters:[21] it is known that during his Spanish period Janello's knowledge in this art was put to use by Philip II in the 1570s for the casting of El Escorial's concert of bells. To serve Philip II properly, Janello wanted to evaluate and choose personally the metal (the bronze for bells was more expensive than that used for casting cannons) and he also stressed his desire to be present during the casting of the bells, "and His Majesty was very happy about that". Moreover, the officer writing the relation affirms that the designated bell-maker was ill, so another one in Toledo, at the service of His Majesty, had been chosen for the casting, a master whom Janello appreciated very much. The royal officer also added that: "it is not necessary to talk about the metals and the quantity of each thing involved in the process, because it is Janello's business", assessing with these words the authority Janello enjoyed in that craft. Another letter by Janello Torriani, probably written in the same year 1578, contains a description of the quantity of metal for the different types of bells. One clear feature of this letter is Janello's tendency to economise. Indeed, Torriani demonstrated how to reduce the scheduled twelve bells down to nine (which according to Janello would have served in the same way), saving no less than 57 *quintales* of metal (2.565 kilograms) for the total considerable value (making a calculation after the prices indicated by the a previous letter) of 216,600 *maravedis*, i.e., 577.6 *ducados*, almost as much as Torriani's yearly wage plus his imperial pension at the peak of his career as royal clockmaker.[22] It is interesting to note that in a letter dated 1462 written by the Commune of Cremona to the Duchess of Milan, one

---

       documenti e immagini (Cremona: Camera di commercio, industria, artigianato e agricoltura, 1989).

20   In brackets the number of the correspondent paragraph in the later statute of 1592 13-(12) *De libro paratici fiendo super quo describantur omnes de dicto paraticoet quantum solvere debeant volentes exercere artem ferrarizie et describi in paratico; 14-(13) Quantum solvere debeat fabricantes campanas; 15-(14) De personis vendentibus ferrum vel azale et quantum solvere debeant; 16-(15) De vendentibus ferrum vel azale in districtu cremonae; Statuta paratici ferrariorum* 1474-1590, Biblioteca Statale di Cremona, Deposito libreria civica Deposito libreria civica, B.B. 1. 7. / 17.

21   This corresponds to *Rubrica* n.12 in the 1592 vernacular version.

22   Cervera Vera, *Documentos biográficos*, doc. 23, 13th of November 1570, and doc. 79, year 1578 (?): relation by a royal officer about the issue of the bells for El Escorial. The bronze for the bells is here said to be more expensive than that which was used for the artillery: a quintal (a *quintal* consists of 45 kilograms: 100 pounds) of the first cost 3,800 *maravedis*, whereas the second cost only 3,000.

reads that clockmaker Antonio Tezano was promising to cast the bells of the public clock of Cremona. This shows that clockmaking and bell-casting were very tightly linked competencies in Cremona, at least in connection to the office of public-clockmaker: indeed, Antonio Tezano, like Janello 70 years later, was keeper of the clock of the civic tower of that city.[23]

It has been calculated that around 60% of medieval and Renaissance clock-makers were master blacksmiths and locksmiths.[24] The locksmiths of Cremona were included in the guild of the blacksmiths: the *Paraticum's* statute made clear that this profession was a competence of the blacksmiths' guild.[25] Being ranked within the guild of the blacksmiths was not to be considered a shame for a refined craft such as Janello's, as we may anachronistically think today. There were indeed some practical reasons that pushed clockmakers to enter these guilds in different parts of Europe.[26] This is evident in the statute of the blacksmiths of Cremona: beside the fact they worked similar materials, all these metalworkers had a common interest in having a regulated supply of coal guaranteed. Four paragraphs, or *rubricae*, deal with the coal supply and are meant to protect the associates of the guild from any fraud perpetrated by the coal-sellers. The procurement of black-market coal was punished: all coal had to be measured with approved gauges.[27] Looking at these details, it seems that guild-structures depended as much as on the production process as on the

---

23   Archivio di Stato di Milano, Fondo Autografi : Artisti diversi (1447-1842), pezzo 93: Orolo-
        giai e Orologi: Letter of the 27th of September 1462 from Cremona to the Duchess of
        Milan: "... *come optimo mag[ist]ro se obliga grati[s] qua[n]to p[er] manufactura a questa
        co[mun]itade / fare le campane necessarie suso lo turazio et cosi quelle che i[n] futuro ca[s]
        u aliq[u]o / fosseno rotte ...* ".

24   Dohrn-van Rossum, *History of the Hour*, 193.

25   In statute one can read that the guild forbade locksmiths to make more than a single key
        for a lock or to make wax copies of it, or to make picklocks. This rule was clearly an
        attempt to prevent the sale to miscreant third persons of already sold locks' keys or pick-
        locks to thieves: *Statuta paratici ferrariorum* 1474-1590 (cover painted in the year MDXX-
        VIIII), Biblioteca Statale di Cremona, Deposito Libreria Civica, B.B. 1. 7. / 17: see especially
        chapter 12- *De pena tenentis seraturam novam falsam gariboldellum vel clavem (ad formam
        cere* in the text) and chapter 14- *Quantum solvere debeat fabricantes campanas.*

26   See for instance: Giulia Camerani Marri, ed., *Statuti delle arti dei Corazzai, dei Chiavaioli,
        Ferraioli e Calderai, e dei Fabbri di Firenze, 1321-1344, con appendice dei marchi di fabbrica
        dei fabbri, dal 1369* (Firenze: Olschki, 1957), 54.

27   In brackets the number of the correspondent paragraph in the vernacular version of 1592.
        5-(5) *De differentiis vertntibus inter ferrarios et carbonarios occasione saccorum a carboni-
        bus vel posture dictorumsaccorum* 6-(6) *De pena gubernantium (gubernantibus) carbones
        non mensuratos et de mensura paratici* 7-(7) *De manutentione mensure dicti paratici et
        qualiter debeant mensurari carbones* 8-(7) *De vendentibus carbones ad vallum et forma valli*

FIGURE 22   Assumption of Mary, patroness of the Cathedral of the city of Cremona, *front
cover of the statute of the* Paraticum et Ars Ferrariorum Civitatis et Districtus
Cremonae, *1529, parchment, cm 22.1 × 16.3.* The cover was added in 1529, when
Janello appears for the first time in documents concerning work on the clocks of
the Torrazzo. COURTESY OF THE BIBLIOTECA STATALE DI CREMONA. PHOTO BY
MINO BOIOCCHI.

FIGURE 23   The Saints Eligius and Anthony the Abbot, also called "of the Fire", the patrons of blacksmiths' guild, *back cover of the statute of the* Paraticum et Ars Ferrariorum Civitatis et Districtus Cremonae, *1529, parchment, cm 22.1 × 16.3,* COURTESY OF THE BIBLIOTECA STATALE DI CREMONA. PHOTO BY MINO BOIOCCHI.

commercialization of the finished items. Something similar can be seen in guilds encompassing physicians and painters. Both of these groups used the same powders for medicaments and colours and fell under the protection of Saint Luke, who was believed to have been a physician and at the same time the painter of a portrait of the Virgin Mary. In this way, the cult of Saint Luke provided a religious confirmation of this peculiar bond, which served to unify such a diverse range of professions: the taxonomy of professions moulded by the guild-system provided workshops with a religious framework to ritualise their corporative activity, divine protection from a celestial patron, and a rationalised activity within the common ethical and material systems of the city. This was reflected by the religious parades (hence called *paraticum*), which mirrored the structural bodies of the community.

Unfortunately, when it comes to the educational path in the *Paraticum Ferrariorum* of Cremona, the details are blurred.[28] There are no paragraphs (*rubricae*) concerning apprenticeship: it seems that this kind of regulation was based more on custom and private contracts than written norms. Most likely, the magistrates of the guild, the consuls, were duty-bound to examine the candidates for the title of master-craftsman. This was the case in the guilds of Milan, and it seems reasonable to assume that the same applied to Cremona.[29]

Between Public Clock and Private Workshop

We know that Janello Torriani took care of the public clock of his city at the time of Francesco II Sforza. What does this imply? We learn from the *Rubrica de custodibus Turracii et Turris Communis* that already in 1477 Duke Galeazzo Maria Sforza had stated that the community of Cremona had to pay a keeper to toll on the bells the hours of "*mattutino, terza and nona of the vesper*". Two more salaried keepers had to stay on the tower day and night to toll on the bells the traditional hours of the community and the hours at which the city council gathered and the heralds issued communications. With the Venetian occupation of Cremona in 1499, the administration of the Torrazzo tower and

---

Statuta paratici ferrariorum 1474-1590 (cover painted in the year MDXXVIIII), Biblioteca Statale di Cremona, Deposito Libreria Civica, B.B. 1. 7. / 17.

28   Antonio Padoa Schioppa has noted that guilds had many aims but not all of them were explicit in their statutes. Antonio Padoa Schioppa, *Giurisdizione e statuti delle arti nella dottrina del diritto comune*, [1964], Studia et documenta histriae et iuris 30 (Milano: Saggi di storia del diritto commerciale, 1992), 26 and following.

29   Santoro, *Collegi professionali e corporazioni d'arti e mestieri della vecchia Milano*.

of its clock had passed to military authority. After 1521, when Francesco II Sforza took back Cremona, he maintained the Torrazzo tower under military control. After things were eventually settled with the Emperor in 1529, the administration of the tower with its clock probably went back to the community. It was at this moment that, as previously mentioned, Torriani appeared for the first time in the documents as restorer of the public clock (Figures 24-25).[30]

The relation between knowledge and institutions is a crucial one. In the fifteenth century, in Torriani's region, administrative practices helped the process of professionalization of crafts such as clockmaking, creating specific offices for specialists who could run public clocks. The same was true of other mathematical professions, such as the public engineer, which, since the Middle Ages, had been part of the communal technical apparatus. In Torriani's Northern Italy, clockmaking and engineering were professions without institutionalised *curricula*, but they did have an institutionalised office. Indeed, two elements were key for the evolution and circulation of clockmaking-related knowledge during the Renaissance On the one hand, we have the office of public-clock keeper (often expressed with the titles of governor, keeper or moderator) in all major and secondary centres of Northern Italy. A public wage with public functions gave to this craft a relevant social position, putting the artisan close to political institutions, the economic and cultural elites. On the other hand, we observe the grounding of specialised dynasties of clockmakers in many of these cities.[31] In fact these offices were often held for life and, where the skills of family members were sufficient, they were even hereditary, as in the cases of Renaissance Italian clockmaking dynasties such as the Dondi, the Falconi, the Tezano, the Ranieri, the Sforzani, the Manfredi, the Mazzoleni, the della Volpaia and the Barocci. The lack of a codified *curriculum* for clockmakers and the possibility of winning a public hereditary position, together with the typical situation created by the necessity of transferring one's workshop to his heirs, encouraged the creation of professional dynasties. Scions of those families then increased competition by moving to other centres seeking appointments as public-clock keepers. Blood links allowed circulation of knowledge in terms of know-how and *ad hoc* specialisation.

---

30   Still in the middle of the sixteenth century, the Cremonese jurist Torresino, wrote that "there is a keeper adapted to the task, i.e., to take care of the clock, toll the bells at the time when divine offices are held, and to do the necessary". Giuseppe Galeati, *Il Torrazzo di Cremona* (Cremona: Emilio Bergonzi, 1928), 80-86.

31   Exemplar is the case of the Ranieri (Raineri, Rainieri) a famous clockmakers-family from Parma. See: Paolo Parmiggiani, "Ranieri," *Dizionario Biografico Degli Italiani* (Roma: Istituto della Enciclopedia italiana, 2016).

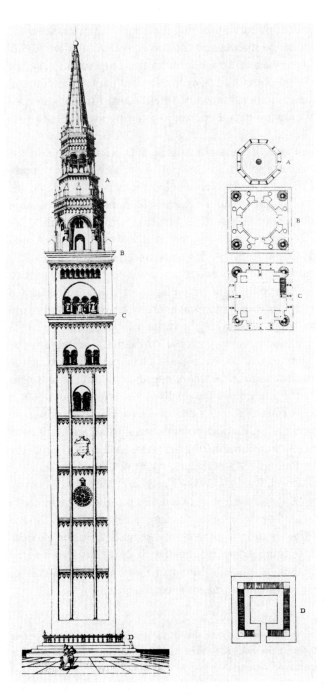

FIGURE 24
Antonio Campi
(*drawing*) *and David
De Laude* (*engraving*),
Torrazzo, Cremona
fedelissima, *1585*.
Depiction of how the
quadrant of the
Torrazzo's clock
probably looked at
the time of Torriani
before the interven-
tions of Giovanni
Francesco Divizioli in
1583.

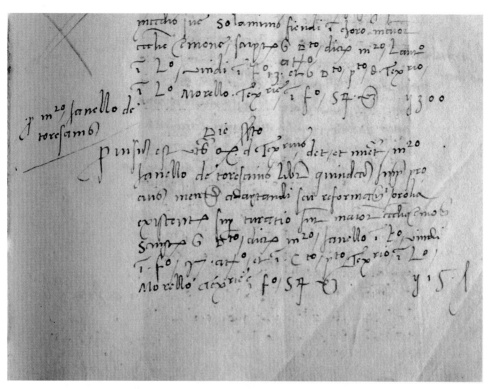

FIGURE 25   *First known document in which Janello appears mentioned.* It concerns a note
dated July 6th, 1529 of the Fabbrica della Cattedrale of Cremona to Janello
Torriani (here still called with his original name Janellus de Torresanis),
promising payment of fifteen imperial pounds for the adaptation and restoration of the Torrazzo's clocks: adaptandi seu reformandi horolia existentia super
Turratio. *ASDCr, Fabbrica del Duomo, Libri Provisionum, I, ca. 159r.*

The Renaissance market for technical knowledge did not depend only on
practical considerations made as a function of efficiency. Political support was
as important as competitive knowledge. It seems that artisans seeking such
appointments had to use patronage networks and their brokerage systems in
order to achieve their goals. The history of the office of keeper of the public
clock of Cremona can help us to see this. Public clocks had multiple functions:
they meant prestige and with their carillons they were also instruments of
entertainment, but first of all they performed a practical function: they marked
the important terrestrial and sometimes astrological hours of the day in relation to civil and religious activities. A document from the State Archive of
Milan well illustrates this function and also dates the existence of a public
clock in the city of Cremona to the early 1460s instead of the 1470s as it was

presumed till now. On the 27th of September 1462 the representatives of the community of Cremona wrote to the Duchess Bianca Maria Visconti, wife to Francesco Sforza:

> It has been a long time that this community of Yours has been badly served by the clock and by a master who is not able to make it work properly: to toll the hours of Your aforesaid city, that is a thing of great necessity and much importance; and with great inconvenience ... because when it is supposed to toll a hour, it tolls instead another one, so that the whole population is bothered and it has often manifested its discontent asking to us to spend the necessary to fix the problem.[32]

---

32   ASMi, Fondo Autografi: Artisti diversi (1447-1842), pezzo 93: Orologiai e Orologi: Letter of
      the 27th of September 1462 from the community of Cremona to the Duchess of Milan:
      "*Illustrissima dna ... Essendo za longo tempo questa v[ost]ra co[mun]itade male servita/ de
      horologio, & magistro apto a tal mestere p[er] sonare le hore de dicta v[ost]ra citade, / che
      nel vero è una cosa di grandisi[m]a necessitade, & importa[n]tia molta ; et cu[m] gra[n]de
      / incomodo, et posse dire più tosto essine senza, p[er] che qua[n]do debbe sonare una hora /
      ne sona una altra ; si che a tutto lo populo e molesto, i[n] modo che pure più fiate / ne ha dato
      lame[n]to & ricordo di degna p[re]visione neli presidenti. Et pare[n]doli / de necessitate
      questo. Pregano v[ost]ra Illustri[ssim]a S.ria volgia provedere di questo officio / ad mag[ist]
      ro Anto[n]io Tezano, apto et sufficie[n]te più che altro mag[ist]ro di questa citade / quale
      offerisse molti partidi che forse altro no[n] poteria ne saperia fare. Per ho / come optimo
      mag[ist]ro se obliga grati[s] qua[n]to p[er] manufactura a questa co[mun]itade / fare le
      campane necessarie suso lo turazio et cosi quelle che i[n] futuro ca[s]u aliq[u]o / fosseno
      rotte che sera co[m]modissi[m]o et utilissi[m]o ad la fabrica de D[omina] S[an]cta Maria
      Ite[m] de fare uno horologio ch[e] rebatta le hore, et de aptare questo o[p]u[r]e farne / uno
      p[er]fecto ad sue spexe. Et altre cose digne di laude et memo[r]ia ad v[ost]ra Ill[ustrissi]ma
      / S.ria et a questa co[mun]itade, a la quale ex corde se recoma[n]da, p[re]gandola ne con-
      ceda / questa n[ost]ra iusta richesta como i[n] quella firmis speramo. Da Cremone die 27
      septe[m[bris mcccclxii / Eiusdem celsitudinis v[est]re Servitores deputati presidentes
      negotiis Co[mmun]is Civitatis v[est]re Cremone cum reco[m]mendatione.*" ; Letter of the
      27th October 1462, from the community of Cremona to the dukes of Milan: "*Illustrisi[m]e
      Princeps &cS . Horologium quod turatio huius Vre Civitatis sup[er] / Positum est universam
      hanc civitate[m] & opifices eiusdem alias rogere co[n]suevit. Nu[n]c / Vero nescimus quo
      fato sive per negligentiam sive per rectoris ip[s]ius imperitia n[u]llo / ordine s[er]vato nu[n]
      c ad has extremitates modo ad alias fluctuat. Ex cui[us] fluxu / Multi as nos querelantes
      confugere, ut sup[er]inde oportune p[ro]videremus, quod / Usq[uam] modo distulimus.
      Nu[n]c vos oblata nobis co[m]moditate Mag.ri Anto.ii d[e] Teza[n]is / Concivis n[ost]ri i[n]
      eo magiste[r]io p[er]itissimi, qui oi [...] ad ... ibita dilige[n]tia se ad illud vaca[r]e / Permisit,
      & om[n]em errorem ab eo remove[re]. Ultra alias obligatio[n]es co[m]modissi[m]as, & /
      Huic civitati v[est]re honorabiles int[??] quas duas co[n]memora[n]do se obligat suo / Per-
      spicaci ingenio artificium unu[m] fabricare suis expensis, quod de p[er] se, sonatus / Horis*

But the keeper of the public-clock was not just in charge of programming the clock. He was also supposed to toll funerals and sermons, and in Cremona, some 40 years after the previous document, we have notice of Zanino da Clusone (keeper of the Torrazzo's public clock from 1499 to 1505) asking money for this service, provoking the discontent of scandalised citizens.[33] Another unpublished document from 1462 reveals a very interesting point: until that year, the keeper of the public clock of Cremona was actually also its owner. At the end of 1462 Antonio Tezano, supported by the local elite, was granted the office of keeper. Benedetto da Corte, Governor of the city (and therefore representative of the central government of the duchy), supported instead the old master Petro del Pena, and he complained to the dukes about their decision to confirm the election of the new master Tezano. In order to justify his concerns before the dukes, and to convince them to change their minds, Benedetto da Corte claimed that:

> With this letter I recommend him [Petro del Pena] to you, communicating that the clock, which strikes the hours, belongs to him and therefore he can take it away as he wishes, and taking it away, master Antonio [Tezano] would not be able to strike the hours and this city would rather turn into a village.[34]

---

as eas iterum resonabit. Ite[m] si configat aliquo casu campanas fabrice / D. S.cte Marie, sup[er] Turatio p[re]dic[t]o existentes infrangi, eas velle refice[re], op[or]tu[m] est / P[ro] factura, sine aliqua mercedis solutio[n]e.q[uo]d dicte fabrice co[m]modissimu[m] & / Et utilisimum erit. Ne igit[ur] i[n] tanto erro[r]e p[re]servarem[que] ex quo confusio oritur, / D[o]natio[n]i v. supplices exposamus, ut p[re]d[i]ct[u]m mag[ist]rum Anto[n]ium Tezanu[m] toti huic / Civitati v[est]re in eo magisterio comprobatum, ad ip[s]ius horologii gub[er]nacula / & curam, cu[m] salario & phemine[n]tus consuetis, eadem eligere et co[n] stituere / dignetur. Quod nobis o[mn]ibus, ac univ[er]se huic civitati v.re pergratissimum / erit. Da[tum] Cremone diexxvii Septe.bris mcccclxii / Eiusdem Cel.d. fid.lissimi s.vitoris deputati / P[re]sidentes negotiis co[mmun]is civitatis v[ost]re Cremone / Quem recomand" (my translation).

33   Bedini, "Falconi," fasc. 1, 46.

34   ASMi, Fondo Autografi : Artisti diversi (1447-1842), pezzo 93: Orologiai e Orologi: Letter of the 22nd of January 1462 (anno ab incarnatione) from Cremona to Milan: "Magniffice maior hon[orande]. Questi di passati scripsi ala ex.tia del S. de la suffitientia / de Petro del Pena, per lo offitio de lo oriologio quale fu dato a mastro Antonio / Tezano per che li diputati scripseno i[n] suo favore q[uondam] meastro [Crist]offoro da le / Balestre, et tunc no[n] fu no[m]i[n]at il dicto Pedro quale è homo da bene / Lo quale sel fusse stato no[m]i[n]ato haveria obtenuto intenderete da luy como / è passata la cossa ... per q[ue]sta mia ve lo aricomando advisandone / Che lo orologio quale sona le hore e suo lo quale po tore a sua posta / et tolendollo mestro Antonio no[n] poteria sonare et sic q.sta cita saria / Una villa. Me

Petro del Pena lost his place, he probably took his clock with him, and Tezano built a new clock for the city. From this controversy we can also infer that by the second half of the fifteenth century, public clocks were a compulsory facility for a city that desired to be recognised as such in its full dignity.[35] Furthermore, the vicissitudes of technical offices highlight political conflicts: the governor nominated by the central power often promoted different people for the same office in contrast with local elites. Subsequently, the prince, playing the role of the supreme arbiter and benevolent father, decided to which party to give satisfaction.

Technical offices constitute a field for political competition. At the time of Janello's childhood, another controversy had sprung from the appointment of the public clock's moderator: from 1499 to 1505 the above-mentioned Zanino of Clusone was keeper of the public clock. After the city fell into the hands of the Republic of Venice in 1499, Zanino was granted this office by the Venetian military and civil governors Domenico Trevisano and Nicolo Forscarini and confirmed in his post by the supreme office of the Republic: the Council of the X. However, in 1503, the Cremonese nobility protested to the Doge in Venice, claiming that, with Zanino's election, the ancient administrative customs of their community had been broken. Thus in order to get Zanino excluded from the post they proposed a different candidate. Probably, in order to make their claim acceptable, they supported the election of a master who was well connected in Venice and therefore not suspected of anti-Venetian spirit: the "fedelissimo" (most loyal) Falconi from Bergamo. The representatives of the City Council of Cremona, as we have seen, had accused Zanino of misbehaviour. It was likely only after 1505 that Zanino lost his place and the candidate of the local elite was elected.

In order to be appointed as keeper of the public clock, Janello Torriani was probably required to be seen as both loyal to the ruling power and sufficient in his craft. We shall try to understand in the next part of the book what kind of political support Janello might have found at the beginning of his career. For the moment we shall only consider his sufficiency as master clockmaker. Going back to Janello's practical education, as we notice above, the statute of the *Paraticum Ferrariorum* of Cremona does not explain the nature of the masterpiece which enabled an apprentice to become a master. However, from the

---

ricuman. a. V. aa. Cremone xxi Ja.ii 1462 / v.r. S. Benedictus de Curte / C[remone] Potestas et capitaneus".

35   At the time of the reign of Gaezzo Maria Sforza (1466-1476), the Duke will order that all cities had a public clock: Gregory Lubkin, *A Renaissance Court: Milan under Galeazzo Maria Sforza* (Berkeley: University of California Press, 1994), 75.

rubricae defining certain standards of quality we infer a constant monitoring of the production: for instance in the 1474 statute of the Cremonese guild of blacksmiths, the consuls were called to evaluate the good craft of making knives (a different art from bladesmithing, whose practitioner at this time still constituted a different guild, which would join blacksmiths in the following century). According to the statute, knives were to be made *"bene et laudabiliter"* (well and praiseworthy). In the later 1592 statute in vernacular, one can find clear descriptions of something similar to a master-examination. Paragraph 36, entitled *"those foreigners who want to enter into the guild have to provide guarantees,"* refers to the acceptance-procedure of foreign masters into the *paraticum*: the evaluation of their skills was supposed to be made by the consuls of the guild. Indeed, we learn from paragraph 37 that a bladesmith who wanted to be inscribed in the guild had to demonstrate to the consuls that he was able to make a sword and a dagger. Clockmakers do not appear in the statutes, probably because of their scarcity.[36]

---

36   In brackets the number of the correspondent paragraph in the vernacular version of 1592: *9-(8) Quod ferra incidentia vel pro incidendo debeant fieri ben azallata et bullita; 10-(9) De facientibus cultellos Statuta paratici ferrariorum* 1474-1590 (cover painted in the year MDXXVIIII), Biblioteca Statale di Cremona, Deposito Libreria Civica, B.B. 1. 7. / 17. *"Che i forestieri che vorranno entrare nel paratico diano sicurità."* Perhaps the higher skills of a bladesmith in respect to a plain blacksmith required a more serious and institutionalised examination-system: indeed, after the fusion of the bladesmiths guild into the *Paraticum Ferrariorum*, one sword-maker consul was to be added to the two traditional blacksmith consuls. But this may also testify to the desire of existing members of the guild to maintain some degree of identity within the new structure, a sort of a guarantee of representation for a wide and well-established group of craftsmen. *"1- dell'elezione delli nuovi consoli"*: Almansi Sabbioneta, "L'arte ferrarecia a Cremona tra i secoli XV e XVIII," 29 and 41. The guild of blacksmiths of Cremona had annual magistrates elected in two periods of the year: in August (fest of St. Laurence) two consuls were elected with a random extraction: *"Dell'approvazione delli spadari e che non si vendano arme nuove fuori delle botteghe"*. The consuls had to swear an oath of fidelity to the guild, to exercise control over the correct behaviour of the members of the guild, to judge the irregularities and to punish the guilty, to impose the respect of the statute, and to defend the members from external frauds. On the first of January three inspectors (*sindici*) and treasurer (*massarius*) were chosen with an open election. The inspectors had to investigate the consuls' deeds and the *massarius* had to keep the accounts of the guild and collect the money. It is curious that in the statute of 1592 the election system was changed to a secret ballot. The process of evaluation of the workmanship of the candidates applying for the title of master clockmakers likely involved the few clockmakers of the community: the consuls, to be elected by lot, might not have had the necessary competences to judge the specialised work of a clockmaker. This is just a supposition; after all, the clockworks inside a complex lock were not different from the ones used for timekeepers. Furthermore, metallurgical technical skills may

Although Torriani's apprenticeship remains mysterious in the absence of documentary evidence, we can grasp by analogy some information from two contracts that Janello Torriani himself signed between 1536 and 1550 with two of his own apprentices. It is likely that the rules Janello and his apprentices had to respect were not too different from those Janello and his unknown master(s) must have followed some years earlier. The notary note reporting a contract between Janello and a certain youngster Gianfrancesco Botti shows that Janello took on a young boy as apprentice with a two-year-contract. In the notary act Torriani committed himself to "teach the same Johannes Franciscus the art of clockmaking and similar things that the aforesaid master Janellus exercises and sells in his house and in his workshop". Torriani was also required to provide the boy with clogs, clothes, bed and board. In turn, the apprentice had to work for Janello, producing and selling for his master. Gianfrancesco's family had to pay the master four florins a year.[37] Fourteen years later, during his residence in Milan, *"magister Ianellus de Torrianis"* took into his house and workshop another apprentice with a notary act dated 25th of May 1550: the apprentice was one Sigismondo de Bacillieri from Ferrara.[38] This took place at the same time when Janello was elected abbot of the Milanese *Paraticum Ferrariorum*. The apprentice, by contract, could not leave the workshop without his master's permission. The same unilateral possibility of contract recession is explicit in the statute of the blacksmiths' guild of Cremona in the *rubrica* that deals with dependent work. A master could not assume a worker already employed at another workshop whose master did not agree.[39] This was a rule surely favourable to the category of the masters: the tendency was to maintain a strong hierarchical distribution of the contracting power. Torriani learned his profession within this very framework. His father had probably signed a contract not much different from the ones that later Janello signed as a master with his apprentices' fathers. It remains an open question whether or not Janello attended other people's workshops for more than a biennium, and

---

have been the most important features to be evaluated, such as the smelting process, smithing, production of good-quality steel, of wrought iron, of bronze and other worked-metals. It is not to be excluded that an informal custom of testing masterpieces involved some of the few members of the guild dealing with clocks. Indeed, it is possible that blacksmiths specialised in clockmaking or even literati familiar with mathematics advised ignorant consuls, but there is no evidence to add weight to this hypothesis.

37   Document reported in: Barbisotti, "Janello Torresani."

38   García-Diego, *Juanelo Turriano, Charles V's Clockmaker*, 148, n. 19.

39   *Satuta paratici et artis ferrariorum Civitatis et districtus Cremonae, Rubrica* IV, "*De labora-tore recedente a magistro in discordia vel ante tempore*", manuscript, Biblioteca Statale di Cremona, Deposito Libreria Civica, B.B. 1. 7. / 17.

if he attended a master's workshop in Cremona or in another city. Considering Janello Torriani's commitment with the public clock of Cremona, we may consider the hypothesis that he could have been trained at the workshop of some master who had previously performed such a function.

At Cremona, since the midst of the fifteenth century, the clockmakers who were employed as keepers of the public clock were: a certain master Petro de Pena, keeper of the public clock until 1462, and Antonio Tezano (1463-1473). This Tezano has been always confused with Antonio da Trezzano, keeper of the public clock in 1515. This probably happened because another Tezano, a certain Tomé, perhaps son of Antonio, was working in 1511 for the Cathedral. Antonio Tezano was said by the representative of the City Council to be the most skilled clockmaker of Cremona, showing that in the city there was more than a clockmaking-workshop.[40] Then in 1473 we know that a certain Giacomo Pezathis de Lagrave improved the public clock. We have already mentioned for the years 1499-1505 Zanino of Clusone (Bergamo) as keeper of the public clock,[41] followed by Falconi from Bergamo, probably keeper of the civic clock from 1505 to the end of the Venetian dominion in 1509. After that, we find in 1511 Tomè Tezano, and from 1515, once the French king Francis I took Cremona, a certain Antonio da Trezzano. The next artisan mentioned is Janello Torriani, who, between 1529 and 1533, at the time of the last Sforza duke Francis II, took care of the civic clock and of other things concerning his art within the complex of the Cathedral such as the already mentioned gilded-copper doors with iron locks in the Baptistery.[42]

---

40   *Libro delle Convenzioni della Cattedrale di Cremona*, 23, manuscript, Biblioteca Statale di
       Cremona, Deposito libreria civica, AA.3.72: in 1561, one Nicola Tezano, probably a descen-
       dent of this family, built the organ for the church of San Francesco in Cremona. For this
       period we have some obscure documents about a Bernardo of Cremona called Caravag-
       gio, or Caravaglio. He is one of the less clear figures of this art: indeed we do not know
       whether Caravaggio is the same as Caravaglio or Carovaglio, clockmaker to Luis XI of
       France (1480) and to the city of Paris. Morpurgo, *Dizionario degli orologiai italiani*, 1950;
       and Del Vecchio and Morpurgo, *Addenda al Dizionario degli orologiai italiani edizione 1974
       di Enrico Morpurgo*.

41   According to Bedini he was the same Zanino who built in 1457 a clock for the Sforza and,
       as already mentioned, the same who tried in 1478 (without success) to restore Dondi's
       Astrarium. If this is true, he must have been very old at this time. Bedini, "Falconi," fasc. 1,
       39-51.

42   After Janello we have notice of a master Benedetto Capironi who in 1548 reformed the
       public clock, changing the Italian hours for the German ones (from 24 to 12), and much
       later between 1582-1583, Gio. Francesco de Visioli (Dovitioli or de Vitioli) and his father
       Gio. Battista rebuilt the clock after the Gregorian Reform.

Falconi's activity in Cremona, if it occurred, overlaps with Fondulo's epistolary correspondence with Paolo di Trezzo, and makes of this master a very
interesting figure for our story. Unfortunately we have documents about
Falconi only for the span of a few years, from 1500 to 1507.[43] He was, like Janello,
a product of the superior craftsman's workshop, where literati and artisans
could unify their knowledge in a common task, under the encouragement
of political institutions. Falconi had been public clockmaker in Brescia and
perhaps in Cremona as well, and he had produced several astronomical instruments. He appears for the first time in 1501 in Brescia where the City Council
paid for a new public clock to be set within the tower of *Piazza della Loggia*.
Falconi was at that time running his business in Venice. He was said in a note
by the General Council of Brescia (22nd August, 1501) to be:

> Superbly skilled in the art itself, as they testify to have seen his works, he
> desires to make his residence in this city [of Brescia] and to practice his
> art for the honour of our city, and then in managing the said clock and
> radia [probably "dials"] and also in bringing [them] up to standard, and
> to practice diligently and faithfully everything in our city dealing with
> arithmetic and geometry ...

Falconi was thus hired to design the astronomical clock, and perhaps to take
care of practical problems involving applied mathematics, perhaps even engineering. Falconi had a son who worked as a clockmaker too: Giovanni Antonio
Falconi, who was also in charge of the public clock of Brescia until 1514. At this
time master Falco, said "*quondam*" in documents, was already dead.

When technical offices such as public clock keeper or engineer were
appointed to a master, his workshop became a kind of an institutional place,
where other technicians, decurions (i. e., the members of the City Council)
with public problems to solve, would go. Franco Franceschi argued for fifteenth
century Florence that the workshop was a stage for the interaction of different
social classes that did not usually have direct relations. Highly specialised
workshops, those of the "superior craftsmen", were open to all members of any

---

43   We have no record other than the name "Falconius" for this figure, perhaps his family was
      involved with goldsmithing and came from a valley near Bergamo. Bedini suggests a num-
      ber of hazarded hypotheses on the age and formation of master Falco; but unfortunately
      we do not have any evidence that would offer us direct information on these points.
      Bedini also claims that "likely ... he studied basic Latin, mathematics and astronomy
      probably at a local seminary or monastery, and he may have served a period as an assis-
      tant or apprentice to a maker of public clocks in the region ... " It is impossible at the
      present time to accept such a baseless hypothesis. Bedini, "Falconi." 39-51.

FIGURE 26   *Falconi,* Nocturlabe, *1505.* With tower engraved on the handle between the
initials "I" and "T" (Janello Torriani?). FOR A TIME IN THE EPISCOPAL COLLEC-
TIONS OF VERONA. PHOTO BEDINI, 2004.

class who had specific interests involved with that craft on both a theoretical
and practical level. I am not only talking about the numerous workers, the
master, his apprentice and the wealthy customers that were crowding a successful workshop (as we shall see through Stradanus' visual work), but I have in
mind other workers whose cooperation was necessary in a complex art such as
that of Falconi and Torriani, from the coal-seller providing the fuel for the
forge, to the carpenter who made the bellows, from the goldsmith and the
painter who provided for decorations, to the glassmaker who provided magni-
fying glasses, not to mention the official seeking solutions to practical problems
scholars in love with mathematics, and children fascinated by red-hot iron and
the wonders such a workshop could produce.[44] Both Vespasiano da Bisticci

---

44   It was quite moving, talking with clockmaker Alberto Gorla (born in 1939), to find a simi-
      lar pattern: Alberto Gorla is a blacksmith by training, with only an elementary school
      licence. Despite this, because of his ability in calculation and mechanics, he is one of the
      most esteemed restorers of ancient clocks, though respected historians of horology such
      as Beppe Brusa have pointed out Gorla's lack of phylological sensibility in his restoration.
      For instance the controversial restoration by Gorla of the clock of the Mori Tower in
      Venice has been at the centre of a harsh dispute. Gorla is also a capable reproducer of
      ancient lost clocks, such as Lorenzo della Volpaia's planetary clock at the Museo Galileo

and Vasari reported that workshops were places of encounter for "*molti homini singulari*", that is to say, for people with great intellectual interests and curiosity.[45] The urban workshop of the Renaissance superior-craftsman involved with applied mathematics can be considered as a place of knowledge and entertainment, a *Wunderkammern* for human *poïesis*.

Like apprentice Sigismondo de Bacillieri who came from Ferrara, Janello himself may have been educated at some workshop outside of Cremona. As his route to fame passed through the capital of the state, Milan, so might have his education. Indeed, many other clockmakers had been attracted to this centre of political and economic gravity, among them other Cremonese clockmakers, such as the Mascaroni. Milan, with its rich Sforza court, welcomed numerous clockmakers also from far places: at least four clockmakers from Germany and the Netherlands.[46] In addition to Milan, many other centres of Northern Italy offered a remarkable network of private and institutionalised workshops where, from the beginning of the sixteenth century, clockmakers could learn a most refined craft.

---

       in Florence. During a visit to his amazing workshop in Cividale, a small village on the borderline between the provinces of Mantua and Cremona, I asked him when he discovered his passion for metalwork. He said that when he was five years-old, he went with his grandmother, who had found a piece of iron on a countryside path, to the blacksmith's. She needed this blacksmith to refashion the iron into a tool to open the corn-cobs. The little boy was completely overwhelmed by the red fire in the forge. From that moment he knew he wanted to be a blacksmith. After elementary school, his father asked him what he wished to do, and the boy gave a clear answer. From the age of 12 to the age of 18 he worked as an apprentice at a blacksmith's workshop.

45   In fifteenth century Florence, workshops were places where people could get information and socialise, contributing to a common discussion with their specific knowledge. With reference to sixteenth century France, Natalie Zamon-Davis has shown that workshops also facilitated the circulation of books. We also know that in Venice books were produced to be read in workshops. Franco Franceschi, "La bottega come spazio di sociabilità," in *Arti Fiorentine: la grande storia dell'artigianato*, ed. Gloria Fossi and Franco Franceschi, vol. 2, Il Quattrocento (Firenze: Giunti, 1999), 65-83.

46   During the second half of the fifteenth century the following Italian clockmakers were active in Milan: Gio. Battista Albani, Mondello Lario, a certain Lorenzo, Master Zanino (perhaps the same Zanino of Clusone we have previously mentioned), Baldassarre Dordone, Antonio Balbo, Antonio de Lombardis, Isaac Milanese, Angelo Santo, Marco Reijna, Marco Mainieri. A Gasparo d'Alemagna was moderator of the clock of the court of Milan until 1470; A Giovanni Tedesco was also employed there between 1471-1473. We have also documents talking about a magister Jacobus Theotonicus for the year 1482, and the famous Zelandinus: Morpurgo, *Dizionario degli orologiai italiani*, 1950; Del Vecchio and Morpurgo, *Addenda al Dizionario degli orologiai italiani edizione 1974 di Enrico Morpurgo*.

Parma, some 53 kilometers from Cremona, gave birth to one of the most important dynasties of clockmakers: the Ranieri da Ramiano.[47] This family dominated the North Italian market for more than a century, from the 1440s to the 1570s. Parma at that time was under the control of the dukes of Milan, and therefore, the Visconti-Sforza employed the Ranieri. It seems that in 1441 Giampaolo Ranieri, called Zampaolo degli Horologi, the most talented of his family, moved to Reggio, a nearby city under Este rule.[48] From 1481 to 1483 he

---

47   ASMi, Fondo Autografi: Artisti diversi (1447-1842), pezzo 93: Orologiai e Orologi: fifteenth century plea to the duke without date. Bartolomeo and Antonio da Ramiano are well documented by historians of horology. In the State Archive of Milan I found this letter mentioning a further clockmaker, a member of the following family: David da Ramiano who was in the fifteenth century at the service of the Duke of Milan. Beside the Ranieri there were other clockmakers in Parma: Rolando Rogolli of Parma, who remade with B. Clavario the public clock of Reggio (1416), and Cristoforo Ponte da Parma. It is believed that Marchionne Toschi from Brescello educated the first members of the Ranieri da Ramiano during the first half of the fifteenth century. See: Paolo Parmiggiani, *Ranieri*, in Dizionario Biografico degli Italiani (2016).

48   Reggio, together with Modena and Ferrara, constituted the territory of the Este. This family was very active in clock patronage: beside the above-mentioned Rainieri and Sforzani-Parolari, the Este invited clockmakers from different areas of Northern Italy. From 1540s, but especially in the second half of the century, the dukes of Ferrara hired a growing number of German clockmakers. In a minor number, there were also French and Flemish ones. Bolzone deDonati from Bergamo, goldsmith and *intagliatore*, was made in 1435 keeper of the public clock of Ferrara. He also worked for the Este. In the year 1472 he was still living elderly and impoverished in Ferrara. Andrea and Marco da Crema were master clockmakers and crossbow-makers active in Ferrara from 1458-1506. In 1475 they asked to elevate scaffolding in front of their house so that people could behold the astronomical clocks they put there. In 1503, the king of England asked the Este ambassador for a clock made in Ferrara. A master Manfredo (2nd half of the fifteenth century) was a clockmaker in Ferrara, and Giovanni da Landinara was also a clockmaker of the Estensi (1450). Then we have a Gio. Stefano Martinelli from Reggio, a Pellegrino Canevaro 1503-1510, keeper of the clock of Rigobello of Ferrara (1503-1510); a Bernardino dalli Orologi, specialist in alarm clocks; and a Zampaloca of Modena, who was active in 1526. From the other side of the Alps we have a Nicolò Fenis, French (clockmaker in Modena in the 1540s); Corrado d'Innspruch (1542); Giovanni and Jacopo Marcoat; Marquatti or Marquart from Augsburg (1554-1582); a Michele from Germany; a Giovanni Fiammingo, who was asked in 1593 to go from Anversa to Ferrara with a carillon clock; a Daniele Micheletti from Germany († 1599); a Nicolò Clusar, German alabardier of the duke (1552-1580); and the trickster Bartolomeo Schneeberger or Schneberger (1589). For this character see the documents in appendix in: Giuseppe Campori, *Artisti degli Estensi: orologieri, architetti ed ingegneri : con documenti inediti ed indici* (Modena: Vincenzi, 1882); Morpurgo, *Dizionario degli orologiai italiani*, 1950; Del Vecchio and Morpurgo, *Addenda al Dizionario degli orologiai italiani edizione 1974 di Enrico Morpurgo*. Another trickster and charismatic figure, the Jew

rebuilt the public clock and then he became its keeper. The office was here
made hereditary. Another Ranieri, also named Paolo, was clockmaker in Reggio
around 1463, and he was brother-in-law to another clockmaker: Bartolo de
Comadri. Zampaolo degli Horologi, with his son Gian Carlo Ranieri, built the
public clock of St. Mark in Venice (1493-1499). Gian Carlo Ranieri had two
brothers working as clockmakers too, Gian Lodovico and Lionello, and a son
named Girolamo. Reggio also gave birth to another important dynasty of clockmakers: the Sforzani, alias Parolari. Just like the Rainieri, the Sforzani-Parolari
had a large web of patrons: Girolamo Sforzani spent all his life in Mantua working for the Gonzaga where the enthusiastic prince called him "dearest friend"
(*amico charissimo*). The most famous of the family was Cherubino Sforzani, as
celebrated as Torriani: he worked in 1518 for the Duke of Este, from 1524 to 1531
for Pope Clemens VII, in 1540 for the dukes of Mantua, for Emperor Charles V,
and also for the King of Portugal.[49]

The neighbouring small state of the Gonzaga of Mantua was also very active
in sponsoring clockmakers.[50] The Gonzaga's entanglement with the ruling
houses of Ferrara and Urbino created channels of technological circulation.[51]

Abramo Colorni from Mantua, worked for the Este during the second half of the sixteenth
century.

49   Already in the 1380s, Giacomo Sforzani had made together with Zilino Clavario the public
clock of Reggio. Giovanni Battista Sforzani was active between 1497-1525. He was *Magni-
fice Universitatis Artistorum Rector* (1520). Benedetto and his brothers, Francesco Sforzani
and Cherubino Sforzani, were famous clockmakers: Morpurgo, *Dizionario degli orologiai
italiani*, 1950; Campori, *Artisti degli Estensi*, 8-9.

50   Pietro Adamo Micheli produced in 1547 what is now the oldest existing print in the field
of horology, and in the same year Paolo deOrsi fixed with Giulio Rainieri the public clock
of Mantua. Francesco Filipono, expert in clocks, died at the age of 105 in 1575. Gio. Battista
Guidotto was in the years 1525-1544 keeper of the public clock in Mantua, Giovanni Tra-
versino was keeper in 1544, Gio. Battista Doni was a 16th century court clockmaker, and
Cesare Giacobiniwas keeper of a tower clock in Guastalla 1574. Morpurgo, *Dizionario degli
orologiai italiani*, 1950; and Del Vecchio and Morpurgo, *Addenda al Dizionario degli orologiai italiani edizione 1974 di Enrico Morpurgo*.

51   Ibid., Both Elisabetta (1471-1526) and Eleonora Gonzaga (1493-1550) married dukes of
Urbino. Titian's portrait of Eleonora Gonzaga della Rovere, painted between 1536-1538,
immortalised the duchess with a fine small table timepiece. In fact, this small duchy was
famous for its mathematical school and for a very important dynasty of clockmakers: the
Barocci. Ambrogio Barocci moved there originally from Milan in the 15th century. The
most famous clockmaker of the family was probably Gio Maria Barocci (†1593). Also his
brother Simone Barocci had clients from all over Europe, but it was in Rome and Florence
that they had their greatest fortune. Like Ferrara, Urbino was a fief of the Holy See. Rome
used to attract princes and artisans from all over Europe. Indeed, since Rome was the
centre of Latin Christianity, it hosted innumerable cardinals' courts in constant competition. For Rome we have documents referring to German clockmakers, such as friar Paolo

It should be noted that the Gonzaga controlled lands very close to Cremona; this, as we shall see, meant a great business opportunity for the artisans of Cremona. The fact that some historians consider neighbouring Mantua the birthplace of portable watches,[52] and Torriani as a specialist in micro-mechanics, makes the circulation of technical knowledge between these two centres an intriguing hypothesis. As mentioned before, also the controversial Bernardo Caravaglio da Cremona has been indicated as the inventor of the spring-driven watch.

Another important centre of attraction for clockmakers was Venice.[53] First of all, Venice was a rich market place and the foremost printing centre of Europe. Around 1501 Falconi (probably the keeper of the public clock of Cremona from 1505) had a workshop for astronomical instruments in Venice. Furthermore, Venice lies only 20 kilometres from the university centre of its dominions: Padua. This city could boast different clockmaking workshops and a brilliant tradition in the craft. Giovanni de Dondi, the most famous clockmaker of the Middle Ages, was active there. A university such as Padua, at that time the most famous in Europe for medical studies, offered a peerless source

---

Alemanno and Enrico d'Alemagna who restored the clock of the Apostolic Palace in 1470, and a certain Erasmo, master clockmaker in the second half of the 15th century. French clockmakers Giuseppe and Sebastiano were both employed by the cardinals Ippolito and Luigi d'Este at the beginning of the 16th century. We also know that Pietro l'Astronomo of Sweden went to Rome around 1513, and that French clockmaker Vincenzo was a member of the blacksmiths' guild in 1519. We have already seen how Cherubino Sforzani was employed at Clemens VII's court. We shall discuss the great number of clockmakers working in Rome at the time of Pius IV in the next part of the thesis, taking in account the possibility of employment for Janello in Rome at the beginning of the 1560s.

52   Ibid., 120-21 For instance it seems that Pellegrino Canevaro, active between 1503-1510 as keeper of the clock of Rigobello of Ferrara made a clock for Isabella d'Este marquess of Mantua (1509). Pietro Guido of Mantua, who had run a workshop at the Rotta di Revere since 1494, specialised in small portable clocks. Between 1506 and 1507 he made watches for Isablella d'Este Gonzaga, for the humanist Pietro Bembo and for the duchess of Urbino. The Manfredi of Mantua constituted another important dynasty of clockmakers. Bartolomeo was son of Giovanni dell'Orologio, who was brother to Nicolò. Nicolò was father to Galeazzo, who made in 1468 the public clock of Marcaria, a village between Mantua and Cremona. Brother to Galeazzo was then Gian Giacomo. Bartolomeo Manfredi († 1478), made the first documented portable watch in 1462.

53   Ibid., Morpurgo and del Vecchio mentioned some local and foreigner clockmakers active in Venice: Adamo Tellarolo, Corrado Rego (a German goldsmith who made a clock for the Gonzaga in 1508), Viviano Piccoli dell'Orologio 1491-1521, Austrian clockmaker Magnericus Reckinger 1560, Stefano Scaguller end of the sixteenth century, and Zuane Farina Montagnana 1563.

for astronomical knowledge. Beside the Dondi family, we have another local dynasty of clockmakers who enjoyed certain fame in Venice: the Mazzoleni.[54]

Besides Venice and Padua, the Republic boasted other rich urban centres, among them Brescia and Bergamo, with their iron mines and related manufacturing, suitable places for metal-crafts. Among the clockmakers coming from that region were Zanino da Clusone and Falconi, the two keepers of the public clock of Cremona at the time of Venetian rule, when Janello was a child (1499-1509).[55]

---

54   Ibid.. Beside the Dondi and the Mazzoleni for the first half of the sixteenth century at Padua we know of a certain Jacopo Veneziano who fixed Piazza dei Signori's clock in Padua 1530, the French Paolo Dujardin who invented some wheel- mechanisms for arquebuses, and a Bernardino Sabeo (1552). Deujarden arrived with his brother to Padua in the 1550s from Blois, an important centre for clockmaking. Martellozzo Forin reports some clockmakers resident in Padua and mainly from the Western Part of Norther Italy, such as Giovanmaria da Civenna. For the second half of the century, he mentions 11 master clockmakers active in Padua: Martellozzo Forin, *La bottega dei fratelli Mazzoleni, orologiai in Padova, 1569*, 97 and following. A member of this family was to cooperate with Galileo in the construction of scientific instruments. Before that time, one Giuseppe Mazzoleni had become keeper of the clock of St. Mark in Venice 1551, after winning a harsh competition. In fact, the previous year, as we have previously mentioned, the Republic gathered a large commission involving many clockmakers from different places to discuss the problem of the broken clock. This commission was formed by Annibale Raimondo, Giuseppe Mazzoleni, the Venetian blacksmith Ambrogio delle Ancore, Leonardo Olivier (perhaps the French clockmaker of cardinal Tournon), and by Bernardino degli Orologi from Padua. Except for Annibale Raimondo, all the others took part in the contest. We have already seen that the Rainieri of Reggio had nearly monopolised the office of keeper of this clock for half a century: Gian Paolo Rainieri with his son Gian Carlo built the clock of St. Mark in Venice (1493-1499), then Gian Carlo Rainieri († 1529) and his son Girolamo tried to establish a dynastic tradition. They controlled this office until 1549, exept for the years 1517 (when a German clockmaker was appointed), and between 1518 and 1528, when Bernardin Cardo was made keeper, and for the years 1531 to 1539, when Raffaele Pencino from Padua substituted for Girolamo who went to study the craft at his uncles' workshop in Reggio. Beside Venice and Padua, the Republic had many other important cities in which clockmaking was established.

55   South of the Apennines Florence was another important centre for clockmaking, where the dynasty of della Volpaia gained fame mainly thanks to Lorenzo della Volpaia's planetary clock. Beside the members of this family Florence was home to other clockmakers such as Carlo Marmocchi (until 1500 keeper of the clock of Palazzo della Signoria), carpenter Rizio from Lugano (1500-1503), Miniato Pitti (fl. 1558) and Giovanni Ventrossi.It is however difficult, because of the great distance and the lack of documentation about possible networks, to imagine an apprenticeship or journeymanship for Janello at Florence or at Rome.

PART 2

The Emperor's Clockmaker
(1540-1558)

∴

The Artisan Courtier

The Grand Tour in Reverse

During the fifteenth and sixteenth centuries, Europe's royal courts attracted a great number of highly specialised artisans. This means that the elites of Christendom (and to some extent of Islamic Istanbul as well) competed among themselves sharing the same artistic and scientific languages. That is why we find clockmakers from Northern Europe employed in the South and the other way around.[1] Janello Torriani, after working for the local community in Cremona, joined this circuit of European courts. When he was around 40, he moved to Milan, the capital of the duchy, the first step in a trajectory that led him from the periphery of the empire to the centre of power. Indeed, from the 1540s on, Janello lived mainly in key sites of power within the Habsburg dominions: besides spending several years in Milan, he served briefly at the imperial court in Worms, Ulm, Innsbruck and Augsburg. After that, he was attached to the imperial court in Brussels and in Yuste (the monastery where Charles v spent his last months after the abdication) and later to the Castilian court of King Philip II in Madrid, Valladolid and Toledo.

---

1   For instance the master bombardier and bronze-caster Guglielmo Monaco de Lo, "*egrege instructus arte horologiorum*," moved in 1451 from Paris to Naples where he was employed at the local court. He made the bronze gates of Castel Nuovo. For the same castle, King Alfonso I of Naples and Aragon commissioned him to build a clock for 1117 ducats. A couple of decades later, the Catalan clockmaker Antonio Buchet was employed in the household of the count of Calabria. Zelandinus from the Low Countries was also employed in Italy, like many other German, Flemish and French clockmakers in the following century. On the other hand we have several Italian clockmakers who, since the Trecento, moved across the Alps: for instance, around the middle of the fourteenth century we find a certain Antonio Bovelli, papal plumber at Avignon, who worked as clockmaker to the Aragonese king at Perpignan. Two hundreds years later, at the time of Janello we have between 1560 and 1571 a Lorenzo Fortuna in London, employed as helper of the French clockmaker Peter Dellamare; a Bartolomeo Campi from Pesaro who, after working for the kings of France and Spain, died in 1572 at the siege of Haarlem; and a Martino Belcampo, who was in 1576 court-clockmaker at Vienna. Cyril Frederik Cherrington Beeson, *Perpignan 1356: The Making of a Tower Clock and Bell for the King's Castle* (London: Antiquarian Horological Society, 1982); on the background of Renaissance clockmakers see also: Cipolla, *Le macchine del tempo*, 24-25; Morpurgo, *Dizionario degli orologiai italiani*, 1950; and Del Vecchio and Morpurgo, *Addenda al Dizionario degli orologiai italiani edizione 1974 di Enrico Morpurgo*.

© KONINKLIJKE BRILL NV, LEIDEN, 2017 | DOI 10.1163/9789004320918_006

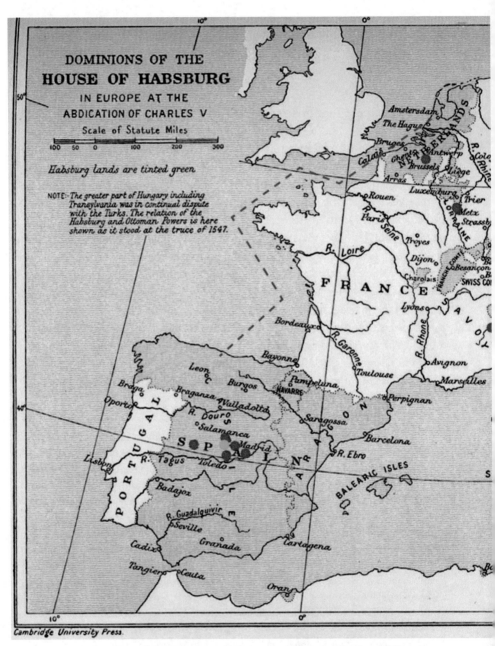

FIGURE 27   *Map indicating Janello Torriani's activities in Europe and the range of the Habsburg Empire in the mid-sixteenth century.* Highlighted in red are those places Janello visited in the course of his career; highlighted in blue are the places where Janello obtained privileges of invention for his hydraulic machines and mathematical instruments, and yellow indicates those places in which privileges were refused.

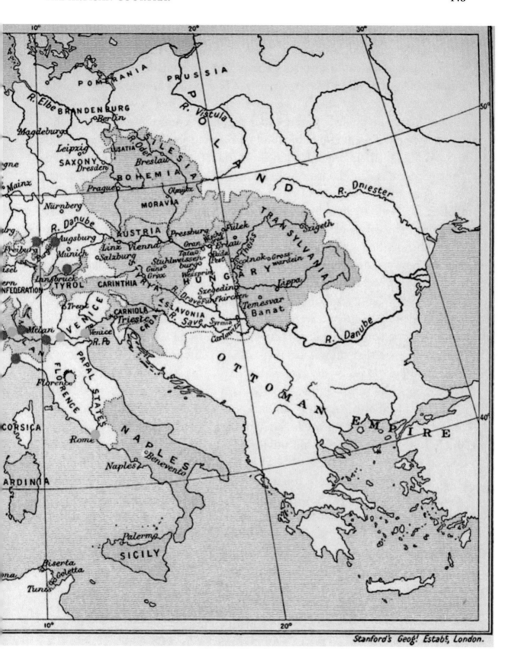

The passage from urban society to court society needs to be considered as not less relevant than the geographical passage from Italy to Brabant and Spain. During this multi-layered journey, Torriani's identity changed. This mutation, like any cultural process, needs to be seen as a function of its agenda and reception. In 1535, when Francesco II, the last Sforza duke of Milan died without heirs, the duchy entered under the direct control of Charles V, who later bestowed this state upon his son Philip, the future king of Spain. Moving his workshop to Milan offered to Janello Torriani the chance to live close to the "control room" of the duchy. All the most important characters of the state, first among them the senators, had to meet in Milan, where the economical, political and military decisions were discussed and taken in accordance with the Crown. The governor was living in Milan where he held a seigniorial court. In 1542 the city had a population of around 79,000 inhabitants, nearly double that of Cremona; despite the fact that there were fewer dwellers than before the Italian wars, Milan was a big city on a European scale.[2] What better situation for a skilled and ambitious artisan selling luxury products? Monasteries, nobles and affluent citizens were interested in clocks and astrological objects. The only signed creation still extant made by Torriani is indeed a small brass armillary sphere that shows, together with the date 1549, the fact that it was made in Milan: "*Ianellus 1549 Mediolani f[ecit]*". This scientific instrument was made for one unknown Hermes Delphinus, and it later entered in the Settala collection and afterwards in the Ambrosiana Library, where it is still held today.

There are not many known documents about Torriani in this long decade in Milan. However, the ones so far published give us some important pieces of information about the clockmaker. First of all, from this period in Milan, the family name "Torresani" was turned into "Torriani". Among the administrative documents, the first one known dates back to the year 1544, during the governorate of the Marquis del Vasto Alfonso III d'Avalos, and it is related to a payment to Torriani for "staying by His Majesty's army in Piedmont".[3] This payment is an important testimony of Janello's early employment in the executive structure of the duchy, and it is the first document that calls the clockmaker an "engineer". However, this document also raises some questions: first of all we do not know what Janello was actually doing as an engineer for the imperial army. Except from the construction of a ballista (i.e., a Roman

---

2 Domenico Sella and Carlo Capra, *Il Ducato di Milano dal 1535 al 1796*, Storia d'Italia 11 (Torino: UTET, 1984), 109.

3 "*Al primo d'Aprile 1544, in Asti. Un mandato al magistrato che facci pagar a Maestro Leonello Torriani ingegnero, qual ha da star' presso l'exercito de S. M.ta in Piemonte ...*": this is the first of a series of very important documents published for the first time by Silvio Leydi: Leydi, "Un cremonese del Cinquecento," 133.

catapult described by Vitruvius), an antiquarian more than a military enterprise, Janello has only been linked to civil activities. Moreover, the document
records Janello as *Leonello Torriani*. But, as previously mentioned, this may be
the sign of the same process of ennoblement of the clockmaker who changed
his plebeian patronymic and name for gentler ones. At the same time we cannot be sure about the wrong spelling by ducal officials who were not acquainted
with such a rare name. What makes us identify this Leonello with our Janello is
the frequency that from that moment on the Cremonese appears in Milanese
documents. In 1545, Torriani probably travelled first to Genoa, where he
brought the clock the Marquis del Vasto had ordered him to make for Andrea
Doria, and later to the imperial court held in Worms. Indeed, an order of payment is recorded in the name of a certain *"M.ro Jamello* (here is another
variation in the spelling of the name) *Torriano Ingegnero"*; the wife of the governor ordered the disbursement to Janello of 275 lire (amounting to the not
small sum of 50 *scudi*) in order to reach the Emperor in Worms, most likely
together with the Marquis del Vasto, who was indeed directed North of the
Alps. Unfortunately the document does not give further information about
this first voyage in Northern Europe. In the year 1547, the new governor of
Milan Ferrante Gonzaga, successor to Alfonso III d'Avalos, who had died in
1546, ordered the disbursal of money to Janello Torriani to construct a planetary clock for the Emperor. Ferrante's order followed Charles v's own command
given personally to Janello in Ulm in February 1547.[4] It is possible that the trip
to Worms was already connected with this task: in a letter written in the year
1628 by the mathematician Muzio Oddi from Urbino to a member of the
famous Milanese family Settala (in whose collection Torriani's armillary sphere
was preserved), it is said that it had been Alfonso III d'Avalos who had introduced Janello to the Emperor. Then, according to Oddi, Janello guaranteed to
Charles v something that nobody else in Germany had dared to promise: to
rebuild the Astrarium. An incredulous Charles v, thinking the task to be impossible, asked the Marquis, who had introduced the clockmaker to him, if the
Cremonese was a fraud. The Marquis, afraid of losing the favour of the Emperor,
threatened Janello's life if the Cremonese was lying. The clockmaker answered
that if he had said so, it meant that he was going to accomplish the task. Muzio
Oddi concluded that nobody could have built such a clock but Torriani.[5] We
cannot be sure if this encounter was taking place in Worms, or on some earlier

---

4   About Janello's clock made for Andrea Doria, see: Alvar Gonzáles-Palacio, Mobilile in Liguria,
    Banca Carige – Fondazione Cassa di Risparmio di Genova ... 37-38. I thank Angelo Landi for
    this reference. Gulielmus Zenocarus Snouckaert, *De republica, vita ... imperatoris, Caesaris,
    Augusti, Quinti, Caroli* ... (Gandavi: excudebat Gislenus Manilius tipographus, 1559), 149.
5   Enrico Gamba and Vico Montebelli, eds., *Le scienze a Urbino nel tardo Rinascimento* (Urbino:
    QuattroVenti : Distribuzione P.D.E., 1988), 190-93.

FIGURE 28
Francesco Porro, The
armillary sphere of Janello
Torriani, *Biblioteca Estense
Universitaria, Modena, Ms
gamma.h.l.21, f. 46.* Porro's
drawing, executed
between 1640 and 1660,
shows Janello's sphere
with a different base than
its current one. COURTESY
OF THE BIBLIOTECA
ESTENSE UNIVERSITARIA
OF MODENA.

occasion. Nevertheless, despite the fictional tone of the story, there are no good
reasons not to trust Oddi's passage.

Since the moment Janello started to work on the Microcosm (this is the
name the humanist Marco Girolamo Vida suggested for this imperial planetary
automaton), the clockmaker was strictly tied to the court of the governor.
However, his election in 1550 as abbot of the guild of the blacksmiths of Milan
reveals that Torriani still belonged to civic society. It would be different when
he became employed at the Habsburg court. From that time on Torriani did
not need to be enrolled in any guild.[6] When Torriani concluded the main part

---

6   Perhaps, the ambiguous status of the governorship in Milan discouraged the vicar of the duke
    from holding an official court, as it would have been by a hereditary lord. This did not prevent
    the Gonzaga acting as *Signore* with something very close to a court. Anyway, the juridical
    framework for his clients was different.

of his complex clock, he was invited to bring his creation to Charles V. A voyage to the imperial court was already planned for the year 1549, but it was just 20 months after the last payment, in August-September 1550, that Torriani brought his planetary automaton to the imperial court, at that time held in Augsburg. Perhaps this delay happened because Ferrante Gonzaga had needed Torriani on the occasion of Prince Philip's visit to Milan on the 10th of November 1549, and imaginably it was on this occasion that Philip entered in possession of a clock made by Torriani in 1547 in Milan.[7]

Despite the trip to Augsburg in summer 1550, the job was not entirely finished. Maybe, under direct suggestion of the Emperor, or of Ferrate Gonzaga, Torriani had started cooperating with the precious-stone worker Jacopo Nizzola da Trezzo, in order to install, beside other decorations, a rock crystal celestial globe on the top of the clock, in which the famous cosmographer Gerhard Mercator was later to place a paper-globe representing planet Earth. It is likely for this reason that Torriani, between 1550 and 1556, visited the Emperor in Augsburg, Innsbruck and Brussels four times, always carrying his work-in-progress clock with him.

Janello Torriani, Jacopo Nizzola da Trezzo, Giuseppe Arcimboldo, Giovanni Ambrogio Maggiore and Leone Leoni were just a few among many highly skilled Milanese artisans who were attracted by courts all over Christendom. Like Nuremberg or Augsburg, Milan was one of the most prolific manufacturers in Europe of luxury goods: Milanese armours, swords, damascened metals, silks, engraved and carved rock-crystals, hard and precious stones, turned ivory, ceroplastics, illuminations, bronzes and other precious metals together with Torriani's clocks were considered in European courts the best one could purchase and display. Written works such as Bugati's (1570), Lomazzo's (1587), and Morigia's (1592 and 1595) bestowed upon us a long list of Milanese highly skilled artisans who gained a broad fame with their crafts.[8] Silvio Leydi, analysing those texts, has advanced the hypothesis that the quality required to the artisans employed in European courts was not just craftsmanship (nevertheless yet a *conditio sine qua non*), but the capacity through *disegno* to constantly

---

7   Francisco Javier Sánchez Cantón, ed., *Inventarios reales: bienes muebles que pertenecieron a Felipe II* (Madrid: Real Academia de la Historia, 1959), item 4630.

8   Giovanni Paolo Lomazzo, *Trattato dell'arte della pittvra, scoltvra, et architettvra di Gio. Paolo Lomazzo milanese pittore* (Milano: Per Paolo Gottardo Pontio ..., a instantia di Pietro Tini, 1584); Giovanni Paolo Lomazzo, *Idea del tempio della pittura* (Milano: per Paolo Gottardo Pontio, 1590); Paolo Morigia, *La nobilta di Milano, diuisa in sei libri* (Milano: Nella stampa del quon. Pacifico Pontio, 1595).

a

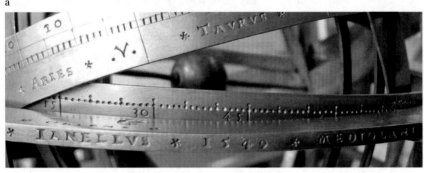

b

FIGURE 29a-b   *Janello Torriani's trademark at the 1549 Milanese Armillary Sphere (above).*
Latin signature of Janello with date and place of production (Milan): "Janellus
1549 Mediolani". COURTESY OF THE VENERANDA BIBLIOTECA AMBROSIANA,
MILAN.

stupefy with *inventio*: a new creation.[9] And this is in fact what unifies the production of those migrating Milanese artisans. Janello Torriani too should be
understood not only as a skilled engineer appreciated for his service, but also
in his function of *virtuoso* and "inventor".[10]

Thanks to the exceptionality of the Microcosm, in 1552 Charles granted
Janello with a privilege (Fig. 35): a long-life pension, later turned into a here-

9   During the Sixteenth century the concept of *inventio* could also be applied to literature,
      poetry and music. For a general overview of the use of this word in the Renaissance, with
      special regard to the work of Virgil Polidoro, see: Catherine Atkinson, "Inventing Inven-
      tors in Renaissance Europe: Polydore Vergil's De Inventoribus Rerum" (Mohr Siebeck,
      2007).
10   Silvio Leydi, "'Al fì, chi vol de tut cora a Milan': arti suntuarie milanesi del Cinquecento.,"
      in *Arcimboldo: artista milanese tra Leonardo e Caravaggio/ Palazzo Reale*, ed. Sylvia Fer-
      ino-Padgen (Milano: Skira, 2011), 51-55.

ditary one. Approaching Charles' abdication, Torriani's periods of permanence at the imperial court became longer, a clear indicator of the Emperor's fascination with his art. It was at this stage in Brussels that Charles took the decision to bring Janello with him to Spain in his retirement at the monastery of San Jerome of Yuste. In 1555, Torriani was granted 1,200 crowns by the Emperor to case the clocks and to return for the last time to the Duchy of Milan in order to take care of his family businesses, before embarking on a galley in Genoa heading to Spain. However, the Emperor's abdication and his gout delayed his departure, so that Janello had time to rejoin Charles in the Low Countries from where, in mid-September 1556, he embarked with Charles on the *Espiritu Sanctu* heading to Spain and never came back.[11] At this moment of the story Antonia, Janello's wife, reappears for a moment in the documents as a flighty yet lively shadow, showing how difficult it must have been for artisans to cope with choices that conflicted with their family's livelihood: the *condottiere* Giovan Battista Castaldo, who had to face his own wife's anger because he failed to win the golden fleece, the most prestigious military decoration given by Charles v only to a few of the military elite of Christendom, wrote in the 1550s that he was travelling with Janello and Leone Leoni, who were also fleeing their wives' fury. It seems possible Antonia was irked by her husband's apparently foolhardy decision to leave a respectable, if not comfortable life in Milan, where they had also invested in estates, for a Spanish adventure as he approached his 60s.[12]

In reality, some new documents recently found, help us to understand what was Janello's predisposition towards this migrating life.[13] He was not very keen on following the Emperor to Spain unless Charles v would guarantee him a truly advantageous economic treatment. As compensation for a move to Spain, Janello asked Charles v for a wage worthy of a prince. Through the priceless letter of the Mantuan ambassador in Brussels to a secretary of the Duke, we start to know better the picturesque character of Janello that we have previously glimpsed in Vida's description, and we learn more in the continuation of this book. On the 10th November 1555, the Mantuan ambassador wrote:

---

11   Federico Badoer, *Notices of the Emperor Charles the Fifth, in 1555 and 1556 Selected from the Despatches of Federigo Badoer, Ambassador from the Republic of Venice to the Court of Bruxelles*, ed. William Stirling Maxwell (London: Printed for the Philobiblon Soc., 1856), 20 and 57.

12   Marino Viganò, "Parente et alievo del già messer Janello," in *Leonardo Turriano, ingeniero del rey*, ed. Alicia Cámara Muñoz, Rafael Moreira, and Marino Viganò (Madrid: Fundación Juanelo Turriano, 2010), 214-15.

13   I thank Almudena Peréz de Tudela for sharing the next two archival finds with me.

Master Gianello of the Clocks, after lots of prayers of the bishop of Arras [i.e., the later nominated cardinal Antoine Perrenot de Granvelle, the powerful minister of Emperor Charles v and of future King Philip ii] and of the principle cavaliers of the court, he eventually settled to go and serve the Emperor in Spain, after the Emperor himself had to ask him many times, promising good conditions, but he used to answer that he requested a very clear contract, and a first payment of 2,000 *scudi* of gold, and 1,000 *scudi* of wage, and responding His Majesty that it was too much for Janello's condition, he replied that it was very little for an emperor!

Torriani's recalcitrance greatly exasperated the Emperor, so much so that he told him that he would push him to go but instead of bending to the will of his lord, Janello grew even more stubborn, without missing a note of sarcasm: he responded that:

It would certainly be a great honour for the Emperor to bring master Janello as a prisoner, however if he was to go by force, he would not work, so it was necessary to convince him with some resources, and the agreement was to give him 1,000 *scudi* of gold cash, and a 400 *scudi* salary, to be collected half in Spain and half in the state of Milan.[14]

---

14   The Ambassador of Mantua to secretary Francesco Catena, Brussels, 10 November 1555, ASMn, AG, b. 2988, ff. 166v-167. "*Mro Gianello dalli Horologli, doppo molti prieghi di Mons.r d'Arras, et dei principali cavaglieri della corte, s'è contentato di andar' a servir' lo Imp[erato]re in Spagna, cosa che no' si volse disponer' di fare per instanza che su M.tà le ne facesse più volte di bocca propia, perche se ella le faceva delle proferete assai, egli rispondeva che voleva fa' le Cap[ito]li chiari, et che voleva per la p.ma ii/m scudi in contanti, et mille di entrata, et rispondendo su M.ta che era troppo per un par suo, egli replicava che era ben poco per uno Imp[erato]re. Al fine ella disse che lo sforzaria ad andar, et a q[ues]to rispose, che le sarebbe in vero un grand' honore amenar' Mro Gianello prigione, pur che se andasse per forza no' lavorarebbe, et come dico conviene che si mettese sotto delli mezi per disporlo, et se acco'modò in mille scudi alla mano, qual'hebbe in continente, et 400 d'entrata, la metà in Spagna, et l'altra metà nello stato di Milano, Dicendoli Mons.r d'Arras perche cosa s'havea fatto tanto pregare, et che havendolo da fare, era pur manco male à farlo à prieghi dello Imp.re che no' di persone private, Rispose, che haveva voluto mostrar' a su M.tà ch'anch'ella ha bisogno di mezi, soggiongendole doppoi, che con le mille scudi haute poteva mo' andarsene à Milano allegram.te a far le feste, se come ne haveva dimandata, et ottenuta la licenza, replicò, che voleva anche l'ispeditione dell'entrata se haveva da tornare, et dicendo esso Mons.re, che egli in tanto la farebbe fare, et che stesse sopra di se, rispose ch'egli era Prete, et che non li voleva creder' Dicendoli un'altro ch pazzia era stata la sua non havendo fig.li a mettersi in una servitu tanto lontana da casa sua. Rispose, che una cosa lo confortava di doverne tosto uscire, perche se no' moresse il p[at]rone [Carlos V] morirebbe [tosto] l'Asino, et cosi havendo*

When Janello was reproached for his impudence by Granvelle, he responded almost mockingly that he did so in order to impart an economy lesson to the Emperor:

> Reproaching him, Monsignor d'Arras asked the clockmaker why he had had to be implored so much, and even worst having the Emperor himself to do that, and not just private people. Janello answered that he wanted to teach His Majesty that even the Emperor needs resources, adding later that with this 1,000 *scudi* he could have now gone happily to Milan to celebrate.

However, it was not enough for Janello, and once more he told Granvelle that:

> He wanted also the expedition entry [the payments of his wage] if he had to return, and saying it the Monsignor [bishop Granvelle], that he would personally take care of it in the meantime, responded [Janello] that Granvelle was a priest, and therefore, he did not want to trust him![15]

The unscrupulous and disdainful words of the clockmaker still leave us stunned, especially in times of Reformation, when every sign of anticlericalism could be called an act of heresy. This letter renders plausible the tale told by Muzio Oddi of the first meeting between Charles v and Janello. Returning to the rather enjoyable recollections of the Mantuan ambassador, we also read of somebody telling Janello that:

> It was a madness to leave for Spain without male heirs to leave in his house. He responded, that a thing comforted him in such business: that he would anyway come soon out of it, because if it were not the patron [Charles v] to die, it would die rapidly the donkey [i.e., Janello himself].

The liberties Janello took with the Emperor did not stop here. Before leaving for Milan from Flanders:

---

incassati tutti gli horogli da uno in fuori, et co'mandato allo Imp.re, che non le movesse, perche no' le guastasse, et promesso su M.tà di ubedirlo, se ne è partito dui di ha per Italia di dove pigliarà il camino per Spagna come meglio li parrà, la cosa è passata appunto come ve la descrivo, ne v'aggiongo niente del mio ...".

15   Ibid.

Having collected all of the clocks … he commanded to the Emperor to promise that he would not move them, because they could be damaged, and his Majesty promised to obey him[.] He [Janello] left two days ago for Italy from where he would later leave for Spain in the way he prefers. The thing went as I am describing to you, I do not add anything of my own.[16]

Even after the death of Charles V, Janello wished to return to Milan, as shown by Alonso de Santa Cruz's letter to King Philip II, Valladolid, 30 March 1559; but the doubling of all his income, and perhaps the transformation of his life-long pension into an hereditary one, convinced him to remain in Spain: the problem was that without Janello, no one would have been able to make use of his complex planetary automata: *"Porque los reloxes sin el no valen cosa y con el tienen algún valor".*[17]

During the years 1557-1558, Torriani was part of the small entourage of 50-70 servants who followed the retired Emperor in the small palace built attached to the monastery of San Jerome de la Vera de Yuste in Extremadura. The steward Luis Méndez de Quijada (*mayordomo*), the Flemish physician Mathys, the Basque secretary Martin de Gaztelu, and the chamberlain Van Male were the most important figures in this monastic court.[18] Together with Janello travelled

---

16   Ibid.

17   AGS, Estado, leg. 137, f. 98: *Juanelo Reloxero q[ue] fue del enperador esta aquí con los reloxes q[ue] su mg.t [Carlos V] le mando hazer y muy desabrido y descontento por le pareçer q no se haze con él lo q fuera razón y porq su estada es tan provecha sup[li]co a vra mg.t sea servido de le mandar escribir o hazer alguna memoria del en la carta del sec[retari]o Juan Vazquez de Molina diziendole q diga a Juanelo el servi[ci]o q[ue] v[uest]ra mg.t reçibe con su estada en esta corte hasta q[ue] venga en estos sus reynos y mandara vra mg.t q[ue] le sea mostrado este capitulo de la carta porque se yo q[ue] terná más sosiego en ver q[ue] vra mg.t se acuerda del. Porque los reloxes sin el no valen cosa y con el tienen algún valor.* In 1562 Philip II confirms Torriani in his service with his salary doubled, DGT, inv. 24, leg. 563 and leg. 564.

18   "*Quedaron en su servicio Luys Quixada por Mayordomo, Martin de Gaztelu seruia de Secretario; Morón, vn Cavallero de Borgoña, era Camarero, y Limosnero, porque el tenia el dinero que en obras pias gastaba el César; Henrique Matisio, Medico, Charles era Uxier, Matía, y Guillermo eran de la Camara, Iuan Gaytan seruia de Veedor, Iuanelo, el que hizo el ingenio de agua que sube al Alcaçar de Toledo, seruia de entretenerle con reloxes y otros ingenios, y al fin todos los otros oficiales necessarios, que eran en suma como setenta criados, y de la estofa que he dicho, casa de un honesta hidalgo, en comparación de aquella magestad primiera*": José de Sigüenza, *Tercera parte de la Historia de la Orden de San Geronimo Doctor de la Iglesia* (Madrid: En la Imprenta Real, 1605), bk. I, XXXVII.

his assistant Giorgio di Diana.[19] The latter was brother to a physician named Orfeo di Diana who had married Barbara Medea, herself the daughter of Janello.[20] Orfeo was then a member of a higher social level than Janello. This may be a further clue regarding the importance of astronomical and planetary clockmaking for Renaissance medicine, and the apprenticeship of Orfeo's younger brother at Janello's workshop is supplementary evidence for the same conclusion. Another clockmaker, the Flemish Joannes Vallin (Juan Valin), was with them, most probably as a skilled helper of Janello, his wage being inferior to that of the Italian engineer.[21] The group of clockmakers was not lodged at the monastery, but at the close village of Cuacos, were Gatzelu, Quijada and the young Don Juan, the beer-brewers and the two women in charge of the laundry were also quartering.

The role of Janello in Yuste is controversial. Many legends surround it.[22] Some later narrative records speak of mechanical birds that could fly, little soldiers fighting and dancing dolls, as we shall discuss in the chapter on automata.[23] What we know from a contemporary eyewitness such as Friar Hernando del Corral – one of the monks of Yuste – is that in the mornings, the first one to access Charles' apartments was Janello who put back in order his planetary clock. Afterwards, Charles spent a moment of prayer with his confessor Friar Juan Regla, before the physician Mathys and the barbers took care of his health. At ten o'clock Charles had to be ready for the mass.[24] This activity was run on a

---

19   AGS, Estado, Castilla, Leg. N. 121, published in the 19th century by Louis-Prosper Gachard, *Retraite et mort de Charles-Quint au monastère de Yuste.* (Bruxelles: C. Muquardt, 1854), L-LI; Don Modesto Lafuente, *Historia general de España,* vol. 13 (Madrid: Establecimiento Tipográfico de Mellado, 1869), chap. 33.

20   AHPT, escribano Gaspar de Soria, doc 15 nov. 1571, sin foliar

21   Cadenas y Vicent, *Hacienda de Carlos v al fallecer en Yuste*, 92-94.

22   In a visit to the monastery in late 2010, the guides told me that all the engineering works at Yuste, from the ponds to the sundials, fountain and mill, were Torriani's creations. Indeed, hydraulic works in Charles' palace in Yuste have been attributed to Torriani, with no documentary proof other than the fame that originated from his later achievements in Toledo. The documents related to pipe-works in the gardens seem to exclude Janello from hydraulic practice in that context. Antonio Perla and Fernando Checa Cremades, "Una visita al monasterio de San Jerónimo de Yuste.," in *El monasterio de Yuste* (Madrid: Fundación Caja, 2007), 64-65, 81 and 126-127.

23   Francesco Arisi, *Cremona literata*, vol. III (Cremonae: Apud Petrum Ricchini, 1741), 338.

24   "*Cada día, por la mañana, luego que se abría su aposento, entrava luego Janelo a ver y concertar el relox [el Planeterio] que tenía de assiento encima de un bufete. Entrava el padre fray Juan Regla, su confessor, a reçar con él. Y acavado de reçar, entravan los barberos y cirujanos, y hacían lo que era menester conforme a las yndisposiciones que tenía su majestad, juntos con el médico Mathiso. Entretanto los officiales davan bueltas por sus officios para que a las*"

FIGURE 30   *Antonio Lafrery,* Plan of Milan, *1573.*

daily basis. Friar Sigüenza (1544-1606), prior of the order of San Jerome (the very same order that was running the monastery of Yuste), historiographer and curator of the royal library at El Escorial, noted that Janello was used to entertaining the Emperor with clocks and other devices.[25] Perhaps, as later sources claimed, those other devices were automata, though the detailed inventory made at Charles V's death does not report any other machine that was not a planetary or plain clock.[26]

The choice of the monastery of Yuste has puzzled many historians: it is far from urban centres and set in an area infected with malaria. On the other hand, the climate of the southern side of the Sierra is mild and allows cultivation of oranges, lemons, almonds and cherries. The elevation of the place assures more gentle temperatures during the hot summers than the nearby plains of Extremadura and Castile la Mancha. Its isolation also complimented Charles' desire for a quiet and pious retirement. Some historians consider the choice of Yuste as a consequence of Charles' friendship with the great-commander of Alcántara don Luis de Ávila y Zuñiga, who had his palace in the closest urban centre: Plasencia. It has also been suggested that the selection of the Extremadurenian place had been encouraged by the Order of San Jerome itself, traditionally supported by the Emperor. Friar Prudencio de Sandoval reported that the prior of Yuste monastery once reported to Charles' daughter the old intention of the Emperor and his wife (evidently before the death of the Empress in 1539) to retire in two Extremadurenian Jeromitan convents: Yuste and Guadalupe. Charles V's cosmographer Alonso de Santa Cruz wrote in his chronicle that after Isabel died, the Emperor spent a certain period in isolation in the monastery of the order of Saint Jerome at Sisla. An ambassador anticipated the declaration of such intent even before that time, in 1535, after

---

diez estuviesse todo a punto y comiessen todos los que avían de assistir a la mesa de su majestad presidiéndoles el gentilhombre que aquel día era de guardia. Entretanto, se vestía su majestad y en acavándose de vestir, acavaban de comer los officiales y salían con su majestad a oyr missa, y los que le avían vestido, se yvan a comer. Entretanto que su majestad oya missa, ponían la mesa y aparejava cada official lo que era de su officio, para que en acavando de oyr missa comiesse. El gentilhombre que presidió en la mesa de los officiales, assistía con Su Magestad, quando oya missa, y en todo el día no se apartava de su vista": in Vicente de Cadenas y Vicent, *Carlos de Habsburgo en Yuste, 3-II-1557-21-IX-1558* (Madrid: Hidalguia, 2000), 62-63.

25   *"Iuanelo, el que hizo el ingenio de agua que sube al Alcaçar de Toledo, seruia de entretenerle con reloxes y otros ingenios":* Sigüenza, *Historia de la Orden de San Geronimo,* bk. I, chapt. XXXVII. Lafuente claimed instead that the emperor spent all his time and energies in devotional issues: Don Lafuente, *Historia general de España,* vol. 13, chaps. 33, 141.

26   Cadenas y Vicent, *Hacienda de Carlos V al fallecer en Yuste.*

the campaign of Tunis.[27] Among this array of hypotheses about Charles' choice of Yuste, there is one that attributes to Janello Torriani the picking of the place, as we shall soon see.

A Broth of Clocks for the Emperor

For economic, fiscal and cultural reasons, Renaissance princes promoted crafts and their practitioners,[28] who entered with new authority in the processes of science making, technological innovation and technological transfer.[29] The work of humanists, who rediscovered and popularised ancient technical knowledge, influenced this trend, so that the role of mathematical instruments and machines became more and more relevant, especially for those who exercised power.[30] Innovative technologies could increase the prince's wealth.

---

27   Manuel Fernández Álvarez, *La España del emperador Carlos V: (1500-1558; 1517-1556)*, 3a ed., Historia de España (Espasa Calpe, S.A.) 20 (Madrid: Espasa-Calpe, 1982), 915-26; Alonso de Santa Cruz et al., eds., *Crónica del emperador Carlos V* (Madrid: Real Academia de la Historia, 1920), 24-25; Luis Montañés, "Los Relojes del Emperador," in *Relojes olvidados: sumario de relojeria historica española ...* (Madrid: Artes Gráf. Faure, 1961), 20; Jesús Sáenz de Miera, "The Emperor's retreat from public life," in *Carolus*, ed. Fernando Checa Cremades (Toledo: Museo de Santa Cruz, 2001), 158-59.

28   Recently, Luca Molà has shown the important role of rulers in Medieval and Renaissance Italy in the conscious process of introducing and supporting the establishment of a certain technical knowledge through the invitation of foreign specialised practitioners. The aim of this process was to increase production, taxation and prestige. Luca Molà, "States and Crafts: Relocating Technical Skills in Renaissance Italy," in *The Material Renaissance*, ed. Michelle O'Malley and Evelyn Welch (Manchester: Manchester University Press, 2007), 133-47.

29   Mario Biagioli brings to our attention the role of the court in experimentation, making of science, and improvement of technology: "In particular, the court contributed to the cognitive legitimation of the new science by providing venues for the social legitimation of its practitioners, and this, in turn, boosted the epistemological status of their discipline". Biagioli's book on Galileo courtier provides an analysis of power-creativity relationships; Mario Biagioli, *Galileo, Courtier: The Practice of Science in the Culture of Absolutism* (Chicago: University of Chicago Press, 1993), 2.

30   Moran, "Princes, Machines and the Valuation of Precision in the 16th Century"; María Isabel Vicente Maroto and Mariano Esteban Piñeiro, *Aspectos de la ciencia aplicada en la España del Siglo de Oro* ([España]: Junta de Castilla y León, Consejería de Cultura y Bienestar Social, 1991); Klaas van Berkel, "'Cornelius Meijer Inventor et Fecit': On the Representation of Science in Late Seventeenth-Century Rome," in *Merchants & Marvels: Commerce, Science, and Art in Early Modern Europe*, ed. Pamela H Smith and Paula Findlen, 2002, 277-94.

More resources meant greater military power: as Cicero wrote: "*Primum nervus belli pecunia,*" which means that "Money is the first nerve of war". In the sixteenth century, the main field for state-investment was war, as shown by the huge costs of raising and maintaining armies, of paying mercenaries, of casting firearms and especially of implementing and evolving defensive-architecture. Any innovation that reduced expenditure or increased military potential, usually achieved thanks to the application of mathematics, was eagerly welcomed. The related field of navigation, as a function of military and commercial expansion, also attracted royal investments. Also within their dominions, rulers, and especially Charles V, were promoting the construction of infrastructures for communication and agriculture, and land-measurement in order to create a better land-register, a powerful tool to raise taxes.[31]

For its "global" dimension, the Spanish empire was in the front-line in the development of discourses involving applied mathematics. For instance, as we shall later see, Charles V was popularised by his librarian Zenocarus as the most knowledgeable in mathematics among all rulers, and Giuseppe Ceredi observed that it was a characteristic of the court of his son King Philip II to have mathematics as a central issue of discussion, as demonstrated by the establishment at court of a mathematical academy, and the famous adage "*A la espada y el compás más y más y más y más*" epitomised the sixteenth century Castilian program of dominion through military power and mathematical knowledge.[32]

Technology and science developed significantly within this imperial frame of reference.[33] Moreover, princes did not restrict their competition to the battlefield, to administration and to commerce, but they also contended for supremacy in all fields that could increase their prestige: the finest products of each art were necessary ornaments for any ruler who claimed superiority over his subjects and over his competitors, and the display of luxury goods visually represented the order of things emphasizing hierarchy within society. If clothes are a social skin, and the elite distinguishes itself by dressing in expensive fab-

---

31   Matteo Di Tullio, "L'Estimo di Carlo V (1543-1599) e il Perticato del 1558: per un riesame delle riforme fiscali nello Stato di Milano del secondo cinquecento," *Società e Storia* 131 (2011): 1-35.

32   "*By compasses and by the sword more and more and more and more*": motto appearing beneath the portrait of captain Vargas Machuca on the frontispice of his description of the Indies of 1599: John Huxtable Elliott, *The Old World and the New 1492-1650* (NY: Cambridge Univ., 1998), 53.

33   Carlo M. Cipolla, *Vele e cannoni* (Bologna: Mulino, 1983); Geoffrey Parker, *The Military Revolution: Military Innovation and the Rise of the West, 1500-1800*, 2nd ed. (Cambridge: Cambridge University Press, 2003).

ric, within the same social class knowledge could create further levels of respectability. In addition to spending large quantities of money on luxury products such as architectonical enterprises, jewellery, clothes, and visual arts, it became necessary for the ruling classes also to invest in libraries, collections of curiosities, scientific instruments, clockworks and other mechanical devices. Preciously crafted scientific instruments were an unavoidable part of sixteenth century rhetoric of power. The well-preserved collections of Renaissance scientific instruments of the courts of the Medici, of the Wettin and of the Hessen, illustrate this trend.[34]

In the time of Euclidean, Archimedean and Pythagorean revival and of the wide interest that the Aristotle of mechanics had provoked, precision-measurement and geometrical relations were at the core of the philosophical systems and methods of investigation. The advantageous results of their practical applications made them into a necessity. From the use of mathematical linear perspective to map-making, from ballistics to engineering, from architecture to clockmaking, from mathematical compasses to instruments of observation, everything practical was helped by measurement, and inside the heart of the Renaissance scholar, both the Christian and the humanist looked at mathematics as the key to unveil God's design in Nature. After all, in the Christian Bible the *Book of Wisdom* read: "Thou hast ordered all things in number, measure and weight".[35] This is why, as previously mentioned, that during the *Quattrocento* it was the Roman Curia which patronised a new Latin translation of Archimedes, and it was thanks to Cardinal Bessarion that Regiomontanus wrote a new translation of Ptolemy's Almagest, and that the Republic of Venice opened the first public library with all the most important mathematical texts from the Ancient Greek tradition, including Aristotle's *Mechanics*. A paradigmatic example of this entanglement of religious and philosophical reasons can be seen in an event occurred in Venice upon the 11th of August, 1508. A crowd of hundreds, both citizens and foreigners, packed

---

34   Michael Korey, *The Geometry of Power- the Power of Geometry: Mathematical Instruments and Princely Mechanical Devices from around 1600 in the Mathematisch-Physikalischer Salon* (Dresden: Staatliche Kunstsammlungen Dresden, 2007); Karsten Gaulke, ed., *Der Ptolemäus von Kassel: Langraf Wilhelm IV. von Hessen-Kassel und die Astronomie*, vol. 38, Kataloge der Museumslandschaft Hessen Kassel (Kassel: Museumslandschaft Hessen Kassel, 2007); Filippo Camerota and Mara Miniati, eds., *I Medici e le scienze: strumenti e macchine nelle collezioni granducali* (Firenze: Giunti: Firenze musei, 2008); Giorgio Strano, ed., *European Collections of Scientific Instruments, 1550-1750* (Boston: Brill, 2009).

35   *The Holy Bible: Containing the Old and New Testaments with the Apocryphal/Deuterocanonical Books*, New Revised Standard Version. (New York: Oxford University Press, 1989): The Book of Wisdom, 11:21.

into the church of San Bartolomeo where the mathematician Friar Luca Pacioli – probably the pupil of Piero della Francesca, and collaborator and friend of Leonardo Da Vinci, who illustrated his book *De Divina Proportione*, or *On the Divine Proportion* – gave an introduction to the fifth book of Euclid's *Elements*, focusing on mathematical proportions. Amongst those present, in addition to the printer Aldo Manuzio, the aristocrat Bernardo Bembo and the sculptor Pietro Lombardo (the designer and executor of Dante's tomb in Ravenna), was the Sienese practical and theoretical metallurgist Vannoccio Biringuccio, the cosmographer Francesco Rosselli, and Friar Giovanni Giocondo of Verona, author in 1511 of the first illustrated edition of Vitruvius' *De Architectura*, arguably the most important text of the Renaissance.[36]

One of the fields in which practical mathematics particularly thrived in the *Cinquecento* was hydraulic engineering. Janello will make the most impressive contribution to this field; however, his was not an isolated case. From the Duchy of Parma and Piacenza comes an interesting example of princely patronage of applied mathematics in antiquarian hydraulic engineering: in the introduction to his book on the Archimedean screw, published in 1567, the physician Giuseppe Ceredi wrote:

> The favour and incomparable liberality of the most illustrious and most excellent Duke and our Lord Ottavio Farnese, to whose Excellence, thanks to his kindness, I had familiarly shown my project, not just because he is very much acquainted with mathematics, and especially to those that belong to warfare, not just was he very convinced and praised it a lot, but he even ordered to his treasurers – who were at that time complaining because of the many expenses made by our illustrious Lady the Princess that left the state treasury nearly empty – to disburse to me a good amount of money. Thanks to this, without a big burden of my own money, and without giving up my main profession [medicine], I was able make an almost infinite number of small and big models [of Archimedes' water-lifting screw].[37]

---

36   Francesco Paolo Di Teodoro, "Pacioli, Luca," Dizionario Biografico Degli Italiani (Roma: Istituto della Enciclopedia italiana, 2014).

37   *"Il favore, et la liberalità incomparabile dell'Illustrissimo, & eccellentissimo Duca, et signor nostro Ottavio Farnese, alla cui Eccellenza havendo io molto familiarmente, per sua gran cortesia, manifestato il mio disegno; non solamente, per essere ella molto avvezza alle ragioni mathematice, & especialmente all'appartenenti all'uso della guerra, ne restò capacissima, & lo lodò assai, ma ordinò anco ai suoi Thesorieri, li quali allhora si dolevano, che le borse erano quasi vuote dalle molte spese fatte per l'Illustrissima Signora la Principessa nostra, che mi fusse sborsata una buona somma di soldi; con l'aiuto dei quali senza molto*

In order to create innovation, committed and skilled people needed to be able to devote time to challenging experiments. The prince, with his personal taste and individual interests, was able to distract public resources for mechanical specialities, creating a privileged condition for experimentation and innovation in a precise field. In the case of Ottavio Farnese and Cerdi, an inventive physician, thanks to princely patronage, could experiment in the field of applied mathematics without leaving his medical profession: probably Ceredi could pay some technician, with a long and therefore expensive practical education, to carry out the models under his direction,[38] showing how the princely support for innovation encouraged the cooperation among literati and craftsmen. We also have an example from the court of Wilhelm IV of Hessen, where this prince united a group of craftsmen and mathematicians to construct his planetary clocks.

Courts provided a stage on which to gain credit and prestige and to enter the market for pensions and inventor-privileges, a stage Janello Torriani entered in the 1540s. It was Charles V's passion for clockworks that made Torriani's career unpredictably fortunate. Though some historians, reacting to the excessive embroidery of narrative and poetic literature on this issue, rejected it as a "Pink Legend",[39] the Emperor's obsession with clocks was not a myth:[40] When he abdicated and found refuge in the monastery of Yuste, among the small court he took with him were two clockmakers and an apprentice: Janello Torriani, Jan de Valin and Giorgio de Diana. The Emperor considered clocks a serious matter: in a letter with instructions for his son Philip, dated 4th of May 1543, Charles found no metaphor more suitable to express reliability than a clock. Advising his son to place his utmost faith in a most loyal courtier, he wrote: "keep Don Joán Çuñaga as your clock and alarm, and be always ready to listen

---

discommo do (sic) *delle mie sostanze, & senza ritirarmi punto della mia principale profes-sioni, ho potuto fabbricare quasi infiniti modelli piccoli, & grandi"*.: Ceredi, *Tre discorsi*, 7.

38 This case recalls the model described for Boyle: Steven Shapin, "Invisible Technicians: Masters, Servants, and the Making of Experimental Knowledge," in *A Social History of Truth: Civility and Science in Seventeenth-Century England* (Chicago: University of Chicago Press, 1994), 355-407.

39 Montañes defines the myth of the emperor clockmaker as *"a pink legend"*: Montañés, "Los Relojes del Emperador".

40 It was not even part of Burgundian courtly tradition of providing patronage for automata as a practice of power- representation, as someone saw it Daniel Damler, "The Modern Wonder and Its Enemies: Courtly Innovations in the Spanish Renaissance," in *Philosophies of Tecnology: Francis Bacon and His Contemporaries*, ed. Claus Zittel, vol. II, Intersections 11 (Leiden: Brill, 2008), 432-33.

to him and to trust him too".[41] By the end of his reign Charles' passion for clocks had fully emerged: from the letters of the ambassadors of England, France and Venice, it appears that Charles V, especially after the failure of the siege of Metz (which ended in January 1553), could only find pleasure when dealing with his collection of clocks. Indeed, at that time, according to the English ambassador, the Emperor had entered a state of mental and physical prostration:

> Before Yuste, on his return from Metz, he spent long hours plunged in deep thought and weeping like a child. Nobody dared to proffer any comfort to him. Neither had anybody sufficient authority to dispel the sad notions which were so prejudicial to his health. He granted audiences to ambassadors, which lasted as long as it might take to say a Creed. His only occupation and his exclusive concern day and night was to care for his clocks and keep them all working in unison. He has many of them, and they constitute his greatest obsession, together with another kind of clock, which he may have invented and which he has ordered to be placed in the frame of a window. As he cannot sleep at night, he often calls together his servants and others and orders them to light torches and to help him to take certain clocks to pieces and put them together again.[42]

Another contemporary source, this time the official Emperor's biography written by his librarian and counsellor Zenocarus (Guill Van Snouckaert), gives us precious insights into Charles' relation to clocks. Zenocarus had a special interest in the science of stars; he knew personally Janello Torriani and Petrus Apianus (Peter Bienewitz) from Leipzig (1495-1553), the author of the famous *Astronomicum Caesareum*. Zenocarus was charmed by the measurement of the World in a deeply religious way. In the frontispiece of the biography of

---

41   My translation: *"Instrucciones de Carlos V a Felipe II ... tengays a don Joán Çuñaga por vuestro relox y despertador, y que seyas muy pronto a oyrle y también en creerle"*: Manuel Fernández Álvarez, *Corpus documental de Carlos V (1539-1548)*, vol. II (Salamanca: [Ed. Universidad], 1975), 102.

42   Garcia-Diego reports indirectly that Charles V, according to a record of a not better specified English Ambassador, between the unsuccessful siege of Metz (31st of October 1552-5th of January 1553) and his departure for Yuste in September 1556, or the abdication in October 1555, was in a grave state of prostration: García-Diego took this quote from an anonymous newspaper article (ABC) attributed by L. Montañes, an expert in Spanish historical horology, to Rafael Sánchez Mazas. In the 1550s the English ambassadors in Brussels were Sir Philip Hoby (1505-1558), Thirlby and Sir Richard Morysine. Perhaps these manuscripts belong with the Harvey Collection. García-Diego, *Juanelo Turriano, Charles V's Clockmaker*, 81-82.

Charles v, he adopted as a motto the above-mentioned biblical line from the *Book of Wisdom*. Once Zenocarus had written Charles v's biography, he sent it to the great-commander of Alcántara don Luis de Ávila y Zuñiga, who lived in Plasencia, a few miles away from Yuste, Charles' place of retirement. Zenocarus writes in his book that both the great-commander of Alcántara and the Emperor reviewed and considered acceptable the biography.[43] This makes of this document a very important source for the three pieces of information regarding Janello Torriani, and for how Charles v wished to have his image officially perceived by posterity. Zenocarus' account is consistent with the reports of Charles amusing himself in keeping all his clocks working in unison and later, once in the cloister of Yuste, refusing anybody's visit, except for Janello Torriani's:

> Charles admitted nearly nobody to his presence ... Count Ruy Gomez was admitted to treat public affairs ... For a similar reason, Don Diego de Azabedo was admitted ... Janello Torriani, the foremost among clockmakers (who had made with the greatest mastery the instrument of the motions of the eight sphere[s], and brought it almost every day to the emperor) because of Caesar's habit, with great familiarity was used to visit him, the only one among Charles' twelve officials.[44]

This report is consistent with what Friar Hernando de Corral and Friar Jose de Sigüenza wrote about the Emperor's retirement at Yuste, as we have already seen. Going back to Zenocarus' account, the Flemish biographer described Charles' love for numbers and mathematics, and he concludes that passage with the following sentence:

> But it was especially in sandglasses and clocks that Caesar manifested his love for numbers: nearly nothing made him more happy then when he

---

43   Zenocarus Snouckaert, *De republica*, 288.

44   Ibid., 289 : "*Carolus in solitudine, nullos ad colloquium admisit. Ad colloquium admittebat fere nullos. Perillustris vir Rodericus Gomezius a Sylva Meliti Comes, summus Regis Philippi Iusti Cubicularius, & Consiliarius ter ad Caesarem officii publici causa veniebat, cum in Hispaniis nuper esset : fuit & vir clarissimus, & humanissimus D. Diegonus ab Azeuedo, Oeconomus Regis simili de causa a Caesare ad colloquium admissus. Ianellus Turrianus horologistarum vertex (qui motum octavae spherae instrumento faberrime constructo Caesari quotidie fere ostendebat) familiarissime, consuetudine Caesaris utebatur, ut qui unus esset ex xii. Caesaris ministris.*"; "*Sed & ad horaria, & horologia hanc numerorum curam Caesar transtulit: nunquam fere magis hilarescens, quam cum omnium concordiam, consensionemque perspiciebat, nec mirum, si numeros tanti semper fecerit*".

was seeing all of them synchronised and in harmony, not a remarkable thing, if it was not for the fact they were so many.[45]

Indeed, the inventories of the courts of Charles v and of his son show a great collection of clocks, probably beyond any comparison with any other ruler of the time: at the death of Philip ii, the royal collection of scientific instruments hosted around 56 mechanical clocks, most of them precious. The total value of these clocks was 9,186 ducats of gold. Among them the most expensive ones, even after half a century, were still the two planetary clocks by Janello, which alone covered more than half of the value of the entire collection: 5,000 ducats.[46]

It is recorded that already at an early stage Charles manifested his curiosity about science and scientific instruments: the cosmographer Alonso de Santa Cruz, who wrote a chronicle about the Emperor, claimed that in the years 1538 and 1539, Charles spent time with him seeking to learn, together with navigation issues and cosmographical globes, astrology, *the Sphere*, and *the Theory of Planets*.[47] Planetary and astronomical clocks, like other mathematical instruments and Apianus' planetary paper-volvelles, were the *via regia* to Geometry once denied by Euclid to King Ptolemy i, the royal road to the mathematical reading of the World, easier than purely arithmetical calculation and direct observation. As we have seen, this was the purpose of the geometrical volvelles behind planetary clocks. In this regards, Zenocarus reports that:

> The Emperor enjoyed himself very much in the investigation of the sciences of the stars and of mathematics, and in the study of its theory: and in this matter, in 1541, in Regensburg, when Charles summoned the diet, he made use of the works of Petrus Apianus, famous mathematician. And in that moment, for the command of the same Emperor, not a little I satisfied His Majesty translating in French the words and the volvelles of his art: thus, in order to make it very clear to me, I studied this subject for a long time. Hence, already from that time, written in French by my own hand, there is this booklet on the subject, which I learned directly from the mouth of my preceptor Apianus. Also in the last time [after the abdication], the Emperor certainly did not give up studying [this subject]

---

45   Ibid., 142: *"Sed & ad horaria, & horologia hanc numerorum curam Caesar transtulit: nunquam fere magis hilarescens, quam cum omnium concordiam, consensionemque perspiciebat, nec mirum, si numeros tanti semper fecerit"*.

46   Sánchez Cantón, *Inventarios reales*, 309-332.

47   Santa Cruz et al., *Crónica del emperador Carlos v*, vol. iv, 24-25.

with calculations and orbits, in order to keep precise [note of] the motions of these heavenly planets, and ordered the trajectories of the stars, the eclipses of Phoebus and Diana [i.e., the Sun and the Mooon], where in the sign, and which one of the moving stars, and where will be in a certain span of time, which will arise and in which sign it will be placed in the period of a day, or which one will set down. For this matter, Caesar made use of very convenient, and very ingenious instruments made with great mastery, in which conjunctions and oppositions of the planets formed triangular and square configurations: the instruments clearly explained the differences of days and nights. He had a special one in which the motion of the eighth sphere was produced in seven times 7,000 years, according to Alfonso, King of the Romans and of the Spaniards and to Johan Regiomontanus ...[48]

As previously mentioned, a planetary clock like Torriani's was a mechanical equatorium driven by an engine. Its major medieval theoretician, Campanus of Novara (1220-1296) explained that the purpose of an equatorium was practical: indeed, it was made:

For those who, because of other occupations, or because of their lack of experience, or even due to their intellectual limitations, encounter difficulties in the solution of these problems [the calculation of the position

---

48 Zenocarus Snouckaert, *De republica*, 146-47: "*Valde quoq[ue] astror[um] & matematicaru[m] artium cognitione, eiusq[ue] doctrinae studio est delectatus: eaq[ue] in re anno quadrgesimo primo, cu[m] Ratisbone, in Germania conventus ageret, Petri Appiani insignis Mathematici opera est usus. Ac ego tu[m] ipsius Caesaris iussu, in gallice interpreta[n]dis nominibus, & circulis artis eius, no[n]nihil maiestati eius inservivi: sic ut plane perspectum sit mihi, eius ea in re longo tempore studium. Libellum enim iam ab illo tempore meis digitis gallice conscriptu[m] de ea re habet, quam ex ore Appiani preaceptoris mei didiceram. Atq[ue] hoc studium ne ultimo quidem tempore adhuc remisit, sic ut orbium illorum coelestium motus difinitos, syderumq[ue] cursus ordinatos, Phoebi Dianaeq[ue] defectiones, quo in signo quaeq[ue] errantium stellarum, quoq[ue] tempore futura sit, qui exortus quoq[ue] die alicuius signi, aut qui occasus futurus sit, numeris, & rotis persequeretur. Ad eamq[ue] rem explicandam Caesar commodissimis, & faberrime, ingeniosissimeq[ue] factis instrumentis utebatur, in quibus coniunctiones, oppositiones, trigoni, quadratiq[ue] aspectus planetarum: Dierum, noctiumq[ue] intervalla dilucide explicabantur. Habebat & unum praecipuum in quo septies septem chiliadibus annorum motus octavae sphaerae secundum Alphonsum Romanorum, & Hispanorum Regem, & Ioannem a monte Regio conficitur. Qui anni in unum ducti faciunt quadraginta novem milia annoru[m], qui numerus vestri ordinis Equitum est proprius. Quando unusquisq[ue] vestrum illustres Equites uni chiliadi praesset*".

Inside the figure:

Diebus 6 horis 1 2 ante
radicem Aftrologorũ, qui
anni initiũ fumunt in Ca
lendis Ianuarij, quæ funt
diebus 6 horis 1 2 poft
initiũ anni ecclefiæ. Sub
altitudine poli G 32.
Natiuitas Domini Noftri
I E S V C H R I S T I.

FIGURE 31

The controversial Horoscope of Jesus Christ, *by Girolamo Cardano, published in the appendix to his commentary on the Tetrabiblos of Ptolemy:* In Claudi Ptolemaei De astrorum Iudiciis, Aut (Ut Vulgo Appellant) Quadripartitae Constructionis Libri IIII Commentaria: ab autore postremum castigata..., *Basel, 1554.*

of the planets], in order to allow them to overcome these difficulties in the quest for numbers and so that they can always find the exact position of the planets and see it through the means of a practical instrument ...[49]

However, the didactical function to represent the motions of the heavens was not the most important task of these machines that were ancillary to the science of astrological influence. This was what really brought planetary clocks to Charles' attention. Astrology, as we have seen before, was an important part of the scientific paradigm of the time. Charles V used an astrological rhetoric to describe his political thought:

> [the Emperor] used to say that princes, like the sphere of Saturn, the highest of all seven planets, and the slowest in motion, they should not be hasty in their decisions and actions. And in the same way as the Sun, that is the same towards poor and rich, being the equal and common to everybody, in the same way those who rule have to equally show

---

49   My translation: '*Quindi, per tutti coloro che per alter occupazioni, o per mancanza di esperienza, o per limiti intellettivi hanno difficoltà nella soluzione di quei problemi, affinché essi, aggirate le suddette difficoltà nella ricerca dei numeri, trovino sempre l'esatta posizione dei pianeti, e possano vederla per mezzo di uno strumento pratico ...*': Alessandro Gunella, 'Campanus de Novara: un precursore del Dondi?', *La Voce di Hora*, 4 (1998), 58. The historian of science Bruce T. Moran noted this very idea in the *Astronomcum Caesareum* by Apianus, however he did not mention Campanus' tradition: Moran, "Princes, Machines and the Valuation of Precision in the 16th Century".

benevolence and justice to each one. And as the Sun eclipse is most of the times a sign of great turmoil, so each half-mistake that the king or lord makes, provides human kind with great disturbance. He also used to say that as the Sun melts the wax down and hardens the mud, so kings' liberality makes good people better and the evil ones ungrateful and worst. He also added that as the Moon especially moves inferior things, not because she is more powerful, but because she is the closest to Earth among planets, so the vicinity of the king is of a paramount importance in quieting war or in calming down turmoil that may be generated in peace.[50]

Moreover, on Charles v's relation to astrology, mathematics and clocks, the same Zenocarus also asks rhetorically:

Is it not true that this Caesar among all kings, emperors and monarchs was the greatest astrologer and mathematician? And that when he was in Ulm, in Swabia, and he prepared the funerary celebrations for his sister-in-law Queen Anne of Hungary, in that very time he summoned Janello Torriani of Cremona, who arrived in the very day of Charles v's forty-seventh birthday? And immediately after listening to him, he ordered the task to build the instrument of the motion of the eighth sphere [i.e. the Microcosm]? And that instantly afterward, there was a [military] progress in Saxony almost as a prophecy of the future victory?[51]

---

50   Zenocarus Snouckaert, *De republica*, 272; and Ludovico Dolce after him: Lodovico Dolce, *Vita dell'invittissimo e gloriosissimo Imperatore Carlo Quinto* (Vinegia: Appresso Gabriel Giolito de Ferrarii, 1561), 92-93: "*diceva, che si come la spera di Saturno, che è il più alto di tutti sette i Pianeti, è tardissima a moversi: cosi dovrebbono i Prencipi non esser frettolosi nelle deliberazioni & opere loro. E nella guisa, che'l sole è il medesimo cosial povero, come al ricco; ne è diverso, ma eguale e comune a tutti: cosi parimente quei, che reggono, debbono mostrar benevolenza e giustizia egualmente a ciascuno. E, come lo Eclissi del Sole è le piu volte segno di gran movimenti; cosi ogni mezano errore, che commette alcun Re o Signore, apporta gran disturbo a gli huomini. Diceva anco, che, si come il Sole liquefa la cera, & indura il fango: cosi la liberalità de i Re fa divenire i buoni migliori, e i malvagi più ingrati e peggiori. Nè taceva, che, come la Luna muove specilmente le cose inferiori, non per essere ella più potente, ma per esser più vicina de gli altri Pianeti alla terra: cosi è di grandissima importanza ad acquetare i movimenti della Guerra, o i sollevamenti, che si fanno al tempo della pace, la vicinanza del Re*"

51   Zenocarus Snouckaert, *De republica*, 149, my translation. "*Sed hic Caesar non omnium Regum, & Caesarum, & Monarcharum maximus fuit Astrologus, & Mathematicus? an non cum vlmae Sueuorum, Annae Reginae Pnnonum defunctae pompam funebrem duceret, exequiasq; faceret? (fuerat enim Ferdinandi fratris vxor) eodem illo tempore, anno*

Astrologia Judiciaria was one of the main functions of scientific instruments, and Charles v, though cautious towards the excesses of this discipline,[52] was not the only ruler to use mathematical devices for astrological practice. For instance, in 1574, Landgrave Wilhelm IV of Hessen, the "Ptolemy from Kassel", who made of his courtly astronomical observations and related production of scientific instruments a source of universal prestige, wrote a remarkable letter to his ally, Elector Augustus from Saxony. In this missive, he informed Augustus about some important and difficult astrological calculations he had to perform about the future maximum conjunction between Jupiter and Saturn, which was to bring about important historical changes. In the end of the letter he asked the elector to keep this secret: "since Your Grace knows how despicable is this art with those who do not understand it".[53]

But historical changes and military operations were not probably the only preoccupation that amplified Charles' attention towards astrology and clocks: the Emperor had indeed a very poor health,[54] being badly tortured by gout, asthma and haemorrhoids. The people and the objects that Charles took with him in his retirement may be considered as a reliable mirror of his interests and more pressing concerns: cooks, bakers, confectioners and beer-brewers could appease his avid desire for drink and sustenance. Political secretaries took care of his correspondence, which kept the retired Emperor in touch with political life. The confessor organised Charles' exercises of piety; the monks (selected apparently from the best of the Order of San Jerome) performed high quality music for the religious offices, and Charles' physician surveyed his weak health and entertained him with learned conversations.[55] The steward Quijada took care of the organisation of Charles' practical life and scheduled rare visitors' audiences. Finally, the clockmakers provided for one of the strongest interests of the retired Emperor. Moreover, it seems that Janello and the chamberlain van Male were used to play also chess with their lord.[56] At the

---

quadragesimo septimo vitae suae, Ianellum Turrianum Cremonensem as se accersivit? ac is quidem die natali Caesaris ad eum venit? ac statim illi (auditus cum esset) instrumenti de octavae spherae motu conficiendi curam, mandatumq; iniunxit? Mox in Saxoniam profectus est futurae quasi victoriae praescius? Quid autem habet astrorum cognitio cum belli tractatione sociale, quid coniunctum habebat? Magnus igitur Caesaris animus, qui humanas actiones semper contemnens: divinas, coelestesq; solas est admiratus".

52   Ibid., 147-149 and 200-201.

53   Moran, "Princes, Machines and the Valuation of Precision in the 16th Century," 218.

54   According to Monica Azzolini, medicine and politics were the two major fields of interest for Renaissance Astrology: Azzolini, *Duke and the Stars*, 11.

55   Badoer, *Notices of the Emperor Charles the Fifth*, 20-57.

56   Cadenas y Vicent, *Carlos de Habsburgo en Yuste*, 107.

moment of his death, Charles had an outstanding collection of scientific instruments: he had three watches to carry on the breast, furthermore a clock called *El Especho* (the Mirror) and one named *El Portal* (the Portal), plus Torriani's Microcosm and Crystalline, this last still under construction.[57] Charles also possessed more than twenty quadrants, and many other scientific instruments such as astrolabes, astronomical rings, sandglasses, the paper-astronomical instruments inserted by Apianus in his *Astronomicum Caesareum*, a manuscript on it by Santa Cruz, a book by Ptolemy and more. Besides the mathematical instruments were a certain number of golden rings and brooches with rare stones and bones set in them: these were all considered to be useful natural magic medications against Charles' chronic diseases and beyond, as in the case of the stone believed to cure the plague.[58]

According to Friar Jose de Sigüenza, Torriani told the monks of San Jerome of Yuste that he had chosen that very location as the most salubrious place for the Emperor's health. The choice was based on astrological calculations.[59] Sigüenza was well-connected at court and he knew many eyes-witnesses, including Torriani himself. Therefore, as concerns the use of astrological calculation in relation to health, this testimony suggests that attributing such a delicate role to Janello, even if not real, might have been perceived as likely. This is indeed an astonishing point: a craftsman was entitled to open scientific speculation! As we shall see, in at least other three occasions, the Habsburgs bestowed upon Janello Torriani a scientific authority, calling him a "mathematician" when granting him a life-long pension in 1552, when discussing cosmographic issues with Mercator in Brussels in 1554, and when participating in the Reform of the Calendar in the late 1570s.

During this period, physicians, according to astrological calculations, could organise medications and diet. In this regard, it is significant to note that in the inventory of the Emperor's belongings at the time of his death, among the instruments listed for the office of his four barbers Nicolás Benigne, Guillao Mebi Querlot, Diritacil, and Gabriel Banden Bosquen, were two astrolabes, an astronomical ring, a sundial, and two books with tables for the ephemerides

---

57   The *Christallinum* would be finished at the beginning of the next decade, and King Philip II would pay for it in 1566: 26th of May 1566, Bosque de Segovia: Cervera Vera, *Documentos biográficos*, doc. 15.

58   Cadenas y Vicent, *Hacienda de Carlos V al fallecer en Yuste*, 18 and following.

59   "*Y afirmaba Juanelo su ingeniero que tenía buen voto en esto, por saber mucha astrología, que con ser Vera de Plasencia de lo mejor de España para la habitación de los hombres, hacía el sitio de Yuste conocidas ventajas a todas las tierras vecinas*": Sigüenza, *Historia de la Orden de San Geronimo*, bk. I, XXXVII, 200.

for the years 1533 and 1548.[60] It seems to be likely that the barbers used such instruments to calculate when to bleed the Emperor, or for other similar practices. If Torriani and the Emperor set the clock each morning,[61] Charles' physicians and barbers would have had a useful instrument displaying the planetary alignments for the entire day: direct observation of the sky was difficult or even impossible in daylight hours and with cloudy weather. Giorgio Fondulo's friend Paolo da Trezzo, the professor of medicine and astrology at Pavia whose manuscript on the construction and use of the astrolabe is kept in the Biblioteca Ambrosiana of Milan, wrote in this very text that when the clouds prevent the direct observation of the stars, thanks to a well set clock, one could find one's position on the astrolabe,[62] and a planetary clock is an instrument that unifies the functions of the clock and of the astrolabe. Moreover, as already observed by Campanus and Zelandinus, consulting astronomical tables was less convenient than simply looking at the clock's displays. The planetary automaton provided comprehensive information about the movements of all celestial bodies at once. One could consider the proliferation of planetary clocks in this period as part of a consistent Renaissance cultural trend that attempted, through the mathematically constructed visual representation, to provide a more instinctive, easy and universal medium to artificially retrieve precise knowledge from Nature.

Ambassadors, King Philip and Charles' attendants in Yuste, all rejoiced at the renewed physical condition of the Emperor, improved by the salubrious air of Yuste. But no enthusiasm could purify the area from malaria; Charles died from this disease in September 1558.[63] If Janello was really the wise man who, thanks to astrological calculation, had identified Yuste as the most salubrious location for the Emperor's health, the clockmaker probably later preferred to keep his role secret. Charles v's lack of moderation in his meals increased the course of his diseases. Exactly this problem generated one day a comical incident that was recorded in an official report. This episode gives us a further insight into the Emperor's utter passion for clocks. Charles v loved his gargantuan meals, which were heavy and rich in condiments and spices. The Venetian

---

60   Cadenas y Vicent, *Hacienda de Carlos v al fallecer en Yuste*, 34-36.

61   It is for this reason that sundials are found close to tower-clocks and in the monastery of Yuste: Sixteenth century clocks needed be adjusted with the sun every morning because of the lack of precision in their winding systems.

62   *Canon 16: Ad Inveniendu[m] gradus Ascendentis tempore nebuloso*: Paolo Tritii, ms. I.20. Sup, Veneranda Biblioteca Ambrosiana, Milan.

63   William Stirling Maxwell, *The Cloister Life of the Emperor Charles the v*, second edition (Boston; New York: Crosby, Nichols & company; C.S. Francis & Co., 1853), 101; Julian de Zulueta, "The cause of death of Emperor Charles v," *Parassitologia* 49, no. 1/2 (2007): 107-9.

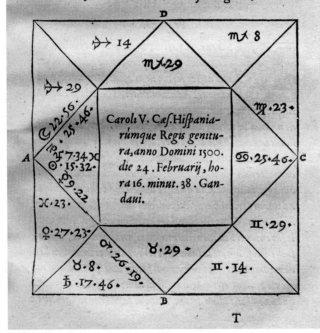

IN PLANISPH. LIB. III. 145

ortu ad occasum centum atque octoginta partes semper interiacent, non tamen nonaginta semper partes, hoc est, numeri totius media summa inter occasum & culmen, ac rursus inter culmen & ascendentem cardinem æquis portionibus interlabuntur. Quod item in hemisphærio inferiori ab occasu ad imum, ab imo ad horoscopum usu uenit. Hactenus Pontanus. Cuius uerba ut facilius intelligätur, diagrāma mathematicum hoc in loco à me subiectum est, diuinam tuā, Maxime Cæsar, cötinens geniturä: in quo, A litera ascëdens, cardo est, B imum cælum, C occasus angulus, D culmë.

FIGURE 32
Juan de Rojas,
Horoscope of
Charles V, *in*
Commentariorum in
Astrolabium, quod
Planisphaerium
vocant, libri sex,
Paris, 1551.

Federico Badoer, Ambassador to the King of the Romans, wrote to the Senate of Venice about Charles V. In this interesting document, Badoer observed that:

To what concerns the table, the Emperor exceeds everyday ... But this is not yet enough for him. With a bitter tone, one day he said to his major-domo Monfalconnet, that the latter did not show any judgment in the orders he gave to the cooks, because all the courses served to him were tasteless. "I do not know – responded the butler – what could I do more in order to please Your Majesty, unless I try a new dish for You: a broth made out of clocks!". These words made the Emperor laugh a lot, and for such a long time that nobody had ever seen before; those in the room did not laugh less: because there is nothing in the world, as everybody knows, that His Majesty loves more than [to] stand in front of clocks.[64]

In 1555, before the monarch abdicated, courtiers understood that Charles would soon leave for Castile, when he asked Janello back to pack the clocks that were his only amusement, before sending the clockmaker back to Lombardy. As noted above, Charles had to postpone his departure. Thus Janello had time to re-join his master in the Low Countries. In 1556, when the emperor had all his belongings finally cased for the last voyage to Spain, he waited until the last moment to have his bed and clocks packed up, too.[65]

But Charles v did not just watch clocks. As we have learned from the English ambassador, he both assembled them and took them to pieces. He synchronised them. The French ambassador noticed that even if Charles had three fingers of a hand amputated because of the grave form of gout, he was still amusing himself using the remaining two fingers to put clocks together.[66] He might have even contributed to their invention and design: Torriani's five voyages to the imperial court (1545, 1547, 1550, 1552, 1554) before definitely joining Charles v's household after a last trip to Italy in 1556, testify to a constant and interactive relation with the Emperor: Janello always travelled with his clock, and once back in Milan, he constantly performed some works on the clocks under imperial command.[67] Moreover, the above-mentioned clock "to be put in a frame of a window" to which the English ambassador referred might have been the Crystalline, the new marvel (smaller and less complex than the

---

64   "*Pour ce qui est de la table, l'Empereur a toujours fait des excès ... N'etant pas encore content de tout cela, il dit un jour à son majordome Montfalconnet, d'un ton de mauvaise humeur, qu'il ne montrait plus de jugement dans les ordres qu'il donnait aux cuisiniers, car tous les mets qu'on lui servait étaient insipides. "Je ne sais pas – lui répondit le majordome- ce que je pourrais faire de plus, pour complaire à Votre Majestè, à moins que je n'essaie pour elle d'un nouveau mets, composé de potage d'horologe." Ces paroles firent beaucoup rire l'Emperour, et plus longtemps qu'on ne le vit jamais; ceux de la chambre ne rient pas moins: car il n'y a chose en ce monde, comme on le sait, qui plaise autant à Sa Majesté que de s'arrêter devant des horloges*". Louis-Prosper Gachard, *Relations des ambassadeurs Vénitiens sur Charles-Quint et Philippe II* (Bruxelles: M. Hayez, imprimeur, 1855), 23.

65   William Stirling Maxwell, *The Cloister Life of the Emperor Charles V*, 4th ed. (London: J.C. Nimmo, 1891), 463.

66   French ambassador in Brussels; García-Diego, *Juanelo Turriano, Charles v's Clockmaker*, 81-82.

67   Leydi, "Un cremonese del Cinquecento".

Microcosm) Janello was creating since the year 1554, after Charles V employed him in his household. Although this clock was finished only some years after the Emperor had passed away, Janello had constantly worked on it under imperial supervision for several years. Ambrosio de Morales tells us that in the Microcosm:

> Despite the brass partitions leaving open the movements of the planets and much more, a lot more is hidden: all the internal movements of the gears. For this reason he made another square clock, a little smaller than the other, and with fewer movements, and there he put walls of crystal so that all the movements of the gears could be revealed. On this clock was engraved an ingenious philosophical phrase: VT-ME-FVGIENTEM-AGNOSCAM [to acknowledge myself as I flee away]. He says he made the movements so visible to better understand ... how fast you approach death.

Complex mechanisms were something to be admired, especially in relation to their virtuosic complexity: our post-industrial sensibility makes it difficult to perceive clockworks, cranks, and cogwheels aesthetically.

The imperial patron had most likely a strong part in the definition of the qualities of this clock, which cost the enormous sum of 3,000 ducats of gold. Charles V must have been familiar with Claudianus' poem on Archimedes' crystal-machine, and perhaps Cardano's 1554 edition of his *De Subtilitate* triggered this new mechanical enterprise. In this edition, the Milanese physician comparing Janello Torriani's planetary clock with Archimedes' crystal one, perhaps with a touch of malice, added that "the ancient machine [of Archimedes] was far more noble and beautiful than ours, but ours is far more durable to time" because it was made of metal.[68] This consideration must have tormented (and perhaps stimulated) the unappeasable spirits of both Janello and Charles V to emulate the ancient mathematician. Janello himself knew about this mythical Ancient Greek crystal machine: he had told Ambrosio de Morales that already with the first planetary clock, he wanted to emulate and surpass the Archimedean cosmic machine.[69]

---

68   Cardano quotes Honorato Juan of Valencia, tutor of Charles's son. Bishop Honorato Juan is said to have been writing about this issue: Girolamo Cardano, *Hieronymi Cardani De svbtilitate libri XXII* (Basileæ: per Lvdovicvm Lvcivm, 1554), 452-55.

69   "*Un relox con todo s los movimientos del cielo, asi que fuesse más que lo de Archimedes*": Morales, *Antiguedades de las ciudades de España*, fols. 91-94.

Climbing The Social Ladder

It has been previously observed that Renaissance political powers were interested in supporting the creation and growth of new crafts within their dominions for the prestige of manufacturing luxury products, implementation of the market, new tolls entering the state treasury, and sometimes even a sensibility towards social improvement.[70] But, if Renaissance rulers were interested in the flourishing of crafts on an official level, how did they relate to them in their private sphere? The status of aristocrats, both in republics like Venice as in principalities, excluded them from any profession but government (rule, army, and administration of justice) or non-mechanic but lucrative trades.[71]

However, since the Renaissance, there were many cases of members of the elite who delighted in activities traditionally exclusive to artisans, such as Charles v's passion for assembling clocks. The Medieval communes of Northern Italy and especially the guilds of the Republic of Florence (since the end of the thirteenth century) had created the base for a growing grey-zone between these hierarchical divisions. This process led to a progression in the dignity of wealthy but non-noble citizens. A complex mixture of traditional examples and political customs provided the ideological source for an alternative (or rather complementary) idea of social dignity: republican humanism, the mobility within the hierarchies of the Church, religious examples starting from the biblical Bezalel to God himself incarnated as the son of a carpenter, together with the innovative hereditary policy of Frederick II Hohenstaufen with his illegitimate sons and with the previously discussed humanists' fascination with crafts and practical mathematics are the most evident components of this cultural process. Under the masks of Aristotle, Archimedes, Apelles, Ctesibius, Phidias, and Daedalus, even high craftsmanship became during the fifteenth and sixteenth centuries a suitable object of fascination for the elite: personal interest was devoted to fields apparently very different from the traditional ones codified for the ruling class, such as hunting, tournaments, games,

---

70   For instance, in Venice, people like Girolamo Miani, founder of the Somaschi fathers, convinced the Republic to promote the attraction of new crafts to be taught to poor orphans in the city, so that this might become a future livelihood for them: Luca Molà, "Privilegi per l'introduzione di nuove arti e brevetti," in *Il Rinascimento italiano e l'Europa*, ed. Philippe Braunstein and Luca Molà, vol. 3, *Produzione e tecniche* (Treviso: Angelo Colla Editore, 2007), 533-72; Molà, "States and Crafts".

71   In a republic like Venice, inhabitants could aspire to be included among the elite of the burgers, the *cittadini originari*, only if they could demonstrate their family's distance from the profession of mechanical arts for at least three generations: Frederic Chapin Lane, *Venice, a Maritime Republic* (Baltimore: Johns Hopkins University Press, 1973), 151.

balls, and courtly intellectual activity in the fields of the liberal arts. My aim is to underline some of the most relevant factors that changed the traditional hierarchy of knowledge and practices, from the direct participation of the ruler in specialised manual arts to the ideological use of new and old ideas.

Virtus Vera Nobilitas Est[72]

Since the flourishing of a historiography devoted to the Renaissance, this new aristocratic trend has been integrated in the narrative of the breakdown of the old medieval system for the "rise of modernity" in Italy. In the nineteenth century, Jacob Burckhardt emphasised the unusual relation between the lords of Ferrara and the mechanical arts. Alfonso I d'Este (1476-1534), Duke of Ferrara, Modena and Reggio, was a patron of letters, but did not know much about them. On the contrary, he was an expert in mechanical arts. He was said to have travelled in Northern Europe driven by his curiosity for crafts, and he was also said to have amused himself working at the pottery wheel, creating ceramic furnishings, chess pieces, flutes and many other things. But especially, he was a most skilled expert in casting bronze cannons, obtaining an outstanding level of workmanship.[73] Before him, Sigismondo Pandolfo Malatesta (1417-1468) was said to have been the inventor of an iron-ringed wooden bombard.[74] Outside of Italy, Sultan Mehmed II (1432-1481), who adopted the expressive codes of Italian Renaissance, Greek and Timurid-Turkmen styles in order to represent himself as a universal ruler, found artisanal work suitable to his status. Mehmed II was said to have "whiled away his idle hours fashioning archer's thumbrings, beltbuckles – and scabbards. Patronage found its pendant in the Sultan's own handiwork".[75] In Saxony, at the court of Dresden, elector August of the Albertine line of the House of Wettin (1526-1586), spent numerous hours at the lathe, turning pieces of ivory, a craft that under his

---

72   Horace, Seneca, and Juvenal were supporting this idea. From the Thirteenth century this
      message was gaining in Italy new fortune, for instance in Brunetto Latini and later in his
      pupil Dante: Quentin Skinner, *Visions of Politics*, vol. 2, Renaissance virtues (Cambridge:
      Cambridge University Press, 2002), 132-33.

73   Jacob Burckhardt, *The Civilization of the Renaissance in Italy*, ed. Peter Burke and Peter
      Murray, trans. Samuel G.C. Middlemore (London: Penguin Books, 1990), 49-50.

74   Roberto Valturio attributes to his lord such an invention: Luisa Dolza, *Storia della tecnolo-
      gia* (Bologna: Il mulino, 2008), 92.

75   Gülru Necipoğlu, "From Byzantine Constantinople to Ottoman Kostantiniyye: Creation of
      a Cosmopolitan Capital and Visual Culture under Sultan Mehmed II," in *From Byzantion
      to İstanbul: 8000 Years of a Capital* (Istanbul: Sabanci University Sakip Sabanci Museum,
      2010), 262-77; Julian Raby, "A Sultan of Paradox: Mehmed the Conqueror as a Patron of the
      Arts," *The Oxford Art Journal*, 1982, 5.

patronage reached its peak. August justified his practical occupation by reminding the sceptic that, after all, he was just imitating God who was the first true moulder of the World.[76] In Florence, Francesco I de Medici (1541-1587) was even said to have trained in mechanical arts: contemporaries recorded his obsession with the experiments he was leading in the Casino of San Marco, a place where he spent entire days in pyrotechnic, chemical-alchemical and distillatory experiments.[77] The French philosopher and traveller Michel de Montaigne defined him as *"Dux mechanicus"*.[78] Precious stones and pottery attracted him, and from his workshop came the best western Renaissance imitation of Chinese porcelain before Meißen.[79]

In the centre of the ceiling of the *studiolo* of this Francesco I de Medici at Palazzo Vecchio, an emblematic image was painted. This fresco represents Nature giving rough stones to a chained Prometheus, who then transforms them into precious ones.[80] All the decoration of the room is based on the theory of the four elements and their interactions through human intervention. The room has no windows and had to be illuminated only by fire, recalling the cave and the gift of Prometheus. In this context, man was not just transforming Nature, he transformed himself as well, through the means of research and conquest of knowledge. The echo of Pico's *Oratio de dignitate hominis*, of Bovillus' *liber de Sapiente* (and of other writers of the Renaissance) was not just suitable for a fervent alchemist such as Francesco de Medici, but for the entire scientific Christendom, which, despite different methods and diverse theological choices, was absorbed in the process of scientific research and records of data, in the hope of improvement, and practical mathematics were the field where this hope became certainty.

These testimonies are more interesting for the discursive arguments they use in praise of princes than for the facts themselves. From a cultural perspective

76   Sabine Haag, "A Signed and Dated Ivory Goblet by Marcus Heiden," *The J. Paul Getty Museum Journal*, 1997, 45-59.

77   Luigi Biadi, *Notizie sulle antiche fabbriche di Firenze non terminate e sulle variazioni alle quali i piu ragguardevoli edifizj sono andati soggetti: operetta* (Firenze: Stamperia Bonducciana, 1824), 233.

78   Daniela Lamberini, *Il principe difeso: vita e opere di Bernardo Puccini* (Firenze: La Giuntina, 1990), 12.

79   Gino Benzoni, "Francesco I de Medici, granduca di Toscana," *Dizionario biografico degli italiani* (Roma: Istituto della Enciclopedia italiana, 1997).

80   For a bibliographical discussion about science and the symbol of Prometheus from the Middle Ages to the Modern period: *La nuova Scienza e il simbolo di Prometeo*, in Paolo Rossi, "La nuova Scienza e il simbolo di Prometeo," in *I filosofi e le macchine, 1400-1700*, Seconda edizione (Milano: Feltrinelli, 2007), 177-88.

it is relevant that these societies did not find any shame in the involvement of rulers with mechanical skills. Of course not all mechanical arts were considered equal.[81] The strictly hierarchical society of Renaissance Europe needed a clear cut between different social levels. The new and more worthy mechanical arts that provided for the social elevation of their practitioners comprised a small sector of the crafts. In the first part of this book we have seen how, from an intellectual standpoint, the legitimation of these crafts into the framework of aristocratic respectability took place through their identification with Classical categories. The Aristotelian category of *scientia media*, placed between natural philosophy and mathematics, and figures such as Vitruvius, Archimedes, Ctesibius, and Hero of Alexandria, were as much the noble fathers of the modern ambitious artisan as the code that the learned noble used to acknowledge and support these craftsmen's aspirations. Torriani entered in the Olympus of craftsmen not as a blacksmith, but as an "architect of clocks" and as a "mathematician". Moreover, the crafts referred to here, besides their noble ideal ancestry, involved precious materials, impressive design, alchemical transformations and invention.[82]

Renaissance Italy was a privileged workshop for cultural models for the rest of Christendom.[83] The continuous rediscovery and circulation of ancient texts

---

81   On the different meanings of "mechanical arts" during the Renaissance, see: Helbing, "La scienza della meccanica nel Cinquecento," 573-75.

82   We have already discussed how distinctive was, from the Quattrocento, a conscious idea of scientific progress. Among other things the mechanical clock provided evidence for this. For instance Bernardus Saccus wrote that it was unknown at the time of Pliny: "... *similiter horologii ... Plinii temporibus in usu non fuit*" in Bernardo Sacco, *De Italicarum rerum varietate et elegantia libri X* (Paiae: Hieronymus Bartholus, 1566), 151. In Torriani's professional life, this idea of novelty would return constantly. The imperial diploma issued to him in 1552 states that the Lombard clockmaker had been the first to craft such a planetary clock. Morales and others affirmed the same about the Toledo device. Torriani's participation in the market of ideas, that is the market related to patents for inventions, produces more evidence about the close relation between his professional career and the idea of invention, therefore innovation and progress – with the caution due to the philosophical problematic connected to this category.

83   Some influence in this process of redefinition of the suitability of mechanical arts for the ruling class, reached the court of Spain, since the time of Charles v the strongest political power in Italy. In the Castilian court, Philip II was said to show more interest in humble people than in nobles. Despite the reliability of this judgement, it is out of doubt that the Spanish king was an enthusiastic patron of architecture, and himself a practitioner of it. García-Diego collected the judgments regarding this characteristic of Philip from Justi and Anibal Alberti: García-Diego, *Juanelo Turriano, Charles v's Clockmaker*, 117; Philip, when still a prince, purchased a large quantity of books dealing with architecture and

offered new tools to unhinge traditional structures of knowledge. Humanism, antiquarianism, sophisticated urban and courtesan life and the religious milieu of the Renaissance were the cultural *humus* of this change. For instance, it seems that the term "mechanics" first underwent a process of decoupling in Italy and acquired dignity thanks to the impact exercised by the already mentioned Greek manuscript belonging to Cardinal Bessarion: Aristotle's *Mechanical Problems* or plainly *Mechanics*, which was well-received by scholars and believed to be a piece of work by the philosopher from Stagira himself. Aldo Manuzio included this Greek text in his edition of Aristotle (1497) and a printed translation in Latin was then made in 1517. In 1525, Leonico Tomeo edited a Latin version with explicatory drawings. The imperial ambassador to Venice, Diego Hurtado de Mendoza, asked Alessandro Piccolomini to prepare a periphrasis of this work in Latin, showing how agencies in the making of knowledge were not only Italians. Piccolomini also published another book that is relevant for our topic: in 1542, *De la instituzione di tutta la vita de l'homo nato nobile e in città libera*, where he includes mechanics in the gentleman's curriculum.[84]Alessandro Piccolomini, like other later sixteenth century writers such as Tommaso Garzoni or Bernardino Baldi, looked at Classical authors and found elements that could endow the mechanical arts with a noble status.[85] Discussions over machines and the natural laws behind their functioning flourished in the name of Aristotle and Archimedes, and their often-incongruent positions. This debate on statics, movement and gravity would become central for the creation of a "new science" in the end of the century.

If Christian religion and humanism opened a theoretical path to nobility for the Vitruvian artisan, traditional Italian social liquidity made this way practicable: indeed, the political situation of central and northern Italy provided strong examples of unorthodox social climbing. Even outside of Italy burgers like Nicolas Rolin could be appointed with key political positions, but what we see in Italy appears to be much more radical. The social mobility in these Italian states, even in the high sphere of politics, created the conditions for a reconsideration of the right to nobility. As Aeneas Sylvius Piccolomini, future

---

astronomy: David C. Goodman, "Philip II's Patronage of Science and Engineering," *British Journal for the History of Science* 16 (1983): 52, where he quotes an article by P. Guillermo Antolín, "La Librería de Felipe II (Datos Para Su Reconstitución)," *La Ciudad de Dios* 116 (1919): 42. Philip employed Torriani in this field of interest: architecture and engineering.

84 Helbing, "La scienza della meccanica nel Cinquecento," 573-92.

85 Alessandro Piccolomini, *De la institutione di tutta la vita de l'homo nato nobile e in citta libera: libri X* (Venetijs: Apud Hieronymum Scotum, 1545), 71; Tommaso Garzoni, *La piazza universale di tutte le professioni del mondo* (Venetia: Ad instantia di Roberto Meglietti, 1605), 757-64.

Pope Pius II, used to say: "In our change-loving Italy, where nothing stands firm, and where no ancient dynasty exists, a servant can easily become a king".[86]

And this was not a metaphor far from reality: the Sforza and the Corradi from Gonzaga were originally peasants, and the Medici merchants. Even the ancient nobility had to reach a compromise with this trend. The Savoy and the Este, households of antique aristocracy, intermingled with these powerful parvenus. The noble Duke of Ferrara married the daughter of an innkeeper, Lucrezia Borgia. This pragmatic approach tended to hide the lack of blood-based nobility where this did not exist, and to cynically substitute it with a discourse based on virtues. In Italy, illegitimate sons were also invested with titles legally the prerogative of lawful siblings. The line of Alfonso of Aragon, King of Naples, was not a legitimate one. The wife of Francesco Sforza, new Duke of Milan *de facto*, but not *de iure*, Bianca Maria, was illegitimate daughter to the last Duke of Milan of the Visconti stock and probably the real source of legitimacy in the taking over of the new family from the old Milanese line. The Este family gave great responsibilities to its bastards, and Federico del Montefeltro, duke of Urbino, and Alessandro de Medici, first Duke of Florence, were also illegitimate. Burckhardt preferred to emphasise the relevance of this phenomenon during the fifteenth century as a sign of novelty rather than searching for its medieval tradition. In fact, Otto, bishop of Frisinga and uncle to Emperor Frederic Barbarossa, had already noted how easily Milanese plebeians could be made knights, and the nephew of Barbarossa, Emperor Frederic II, in the first half of the thirteenth century devolved important titles and crowns to his bastards, chiefs of the Ghibelline party in Italy, like Manfred, King of Sicily, Enzo (or Heinz), King of Sardinia, and Frederic of Antioch. This likely set a precedent that was later widely accepted. Also on the Guelf side, political practice created a suitable environment for social mobility independent from blood and legitimate lines. For example in Florence, people with official roles in the government, such as Chancellor Leonardo Bruni (1370-1444), wrote in a public funeral oration: "For our city requires virtue and honesty in its citizens. It considers anyone with these qualities to be noble enough to govern the state".[87]

This discourse was followed by a *memento* of the *Ordinamenti di Giustizia* (Laws of Justice), a revolutionary law promulgated in Florence in the last decade of the thirteenth century that excluded old feudal nobility from the

---

86   Burckhardt, *The Civilization of the Renaissance in Italy*, 12.

87   Leonardo Bruni, "Funeral Oration for Nanni Strozzi, 1427," in *Major Problems in the History of the Italian Renaissance*, ed. Benjamin G. Kohl and Alison Andrews Smith (Lexington, Mass.: D.C. Heath and Co., 1995), 281.

government of the Republic. Since that time, only members of guilds were allowed to participate in the administration of the Commune.[88] The ideological routes of this radical shift had been there in early humanist culture for a long time. For instance, in the previous years, Guelf writers such as Dante's master, Brunetto Latini, had drawn upon Classical literature (Horace) in order to create a discourse promoting a different idea of nobility, no longer deriving from blood-connected rights, but from talent and virtue. This was taken to its logical consequence by the representatives of what Hans Baron has called "Civic Humanism": one of the most influential thinkers of this movement and a member of the Papal Curia, Poggio Bracciolini (1380-1459), agreeing with Cosimo the Elder's brother Lorenzo and with the other famous humanist Niccolò Niccoli (1365-1437), wrote that "there is no other nobility than that of personal merit".[89] These factors led to the development of a specific literary genre: beside Brunetto's work, part of Dante's *Convivio* dealt with the same subject, as did Buonaccorso da Montemagno's *Oratio de Vera Nobilitate* (1428), Poggio Bracciolini's *De Nobilitate* (ca. 1440), and *De Vera Nobilitate*, written in 1475 by the already mentioned fellow countryman of Torriani, Bartolomeo Sacchi, nicknamed *il Platina*. At the beginning of the sixteenth century these ideas were circulating outside of Italy, and Desiderius Erasmus, the most famous northern humanist, touched on the topic in his *The Praise of Folly* (1511) where he stated: "who, though they differ nothing from the meanest cobbler, yet 'tis scarcely credible how they flatter themselves with the empty title of nobility".[90]

In this context, ingenious mechanical arts were no longer considered a shameful activity for a gentleman.[91] On the top level, we observe that great princes with a strong fascination for a specific craft could spend huge quantities of money on specific projects that resulted in technological innovation, a trend well illustrated by Torriani's great mechanical accomplishments that

---

88   *The Ordinances of Justice of Florence*, 1295, in Benjamin G. Kohl and Alison Andrews Smith, eds., "The Ordinances of Justice of Florence (1295)," in *Major Problems in the History of the Italian Renaissance* (Lexington, Mass.: D.C. Heath and Co., 1995), 139-41.

89   Burckhardt, *The Civilization of the Renaissance in Italy*, 143.

90   Desiderius Erasmus, *The Praise of Folly*, trans. John Wilson (Rockville, MD: Arc Manor LLC, 2008), 47.

91   Antonio Clericuzio, *La macchina del mondo: teorie e pratiche scientifiche dal Rinascimento a Newton* (Roma: Carocci, 2005), 30-32. In the past it has been pointed out that the social milieu that made respectable the association between technicians and gentlemen was characteristic of European civilisation: Joseph Needham, *Science and Civilisation in China*, vol. 3, Mathematics and the sciences of the heavens and the earth (Cambridge: Cambridge university press, 1959), 154-55.

reflect his patrons' personal interests. Within the lower ranks of aristocracy, usually devoted to war, law, medicine, administration of agricultural or urban estates, and luxury production and trade, the practice of certain mechanical arts opened new possibilities. For instance, all the Anguissola sisters, and especially Sofonisba, the most talented among them (a fellow countrywoman of Torriani, and appreciated paintress at the court of Philip's wife during the 1560s), and Francesco Sitoni, a gentleman from Milan and a rather undistinguished hydraulic engineer at the same court, were nobles who, in order to gain a better share of public resources, performed mechanical and technical works for their prince. In the case of Sofonisba Anguissola and her sisters, it was their father who promoted her obsessively until the oldest daughter obtained such a position at the Spanish court. It was in turn Sitoni's personal ambition that took him to Spain, where there was need for hydraulic engineers and good money for the taking. Sitoni's case shows a member of nobility involved not with military engineering, which remained connected with the aristocratic practice of war, but with civil engineering. Sitoni's North Italian background probably had a role in the way he shaped his professional strategies: even a gentleman such as he could shamelessly practice an *agrimensor* or an *ingegnero* in exchange for a salary.

However, Aristotelian scholastics defended the importance of aristocratic lineage as a guarantee of wealth and virtue. According to this vision, public service needed sufficient resources to guarantee a full commitment to the office: something that except for what was happening in Venice and especially in Florence, only the status of nobility, with its *beneficia*, could constantly provide.[92] The growing prestige of the Vitruvian artisan does not have to be

---

92   The roots of these discussions on the fundaments of nobility are probably to be found in the tensions between great imperial feudal lords and their vassals in Italy around the tenth and eleventh centuries. The study of Roman Law at Bologna University bore signs of a political discussion in which ancient models were used and adapted to the contemporary situation. The adoption of elective titles such as *"consules"* is a clear sign of a political use of the image of the ancient Roman *Res Publica*. However, the old feudal nobility ruled the communes for the entire twelfth century, and it was only in the following century that the *Populus* (another classical loan-term) emerged as a political actor. Though still a controversial issue among historians, it seems that the *Populus* was a social group formed by the non-noble but economically relevant population of the cities. Professor Azo of Bologna (fl.1150-1230) maintained that according to Roman Law, the assembly of the people enjoyed a greater authority than the emperor, although the emperor's dignity and authority was superior to that of any of his single subjects. This interpretation of the *Lex Regia* (Digest) assumed in fact that it was the people who bestowed the power upon the emperor. Skinner, *Visions of Politics*, 2, Renaissance virtues:15 and 132-133; Piero Fio-

equated with their assumption in the higher layers of society: except for a few craftsmen, especially painters, who were elevated to the rank of knights, a noble (both of old or of new lineage) was always in a better position, at least in relation to honour, i.e., to social value.[93] For instance, the noble woman Sofonisba Anguissola, once she had to get married, received from the king an annual pension of a 1,000 ducats.[94] Francesco Sitoni, received a wage of 960 ducats when he was working, and 600 when not active, in this latter case a salary equal to that of the much more successful Janello Torriani when he was in his full functions.[95] Janello was even ordered by the King to evaluate the job of his fellow countryman Sitoni, judging it incorrect. This did not change the economic difference between the two Lombards: their social status was far more important in the redistribution of wealth than their competences. The experience of Torriani who, despite the satisfactory execution of his job, fell into debt because of the delays in payment on the part of the King, was likely the result of his comparatively low social status. A king had to protect the honour of his gentlemen before paying debts he owed to people of a lower status. The honour of a noble was to be measured with the grace granted him by the prince, which provided for his *decorum* through the confirmation of his benefits or by the provision of the necessary to maintain his prestige. This was a period of transformation of the tradition originated from the royal distribution of the resources to the vassals, which were supposed to perform in exchange military,

---

relli, "Azzone," *Dizionario biografico degli Italiani* (Roma: Istituto della Enciclopedia italiana, 1962).

93 Several Renaissance authors, such as Pier Paolo Vergerio and the *arbiter elegantiae* Baldassarre Castiglione reconsidered in a positive way the art of painting because of ancient authorship that considered it part of a noble education. Vergerio, "Pier Paolo Vergerio Defines Liberal Learning," 102-9; Castiglione, *Il Libro Del Cortigiano*, 44.

94 Even Philip's wife was involved with mechanical arts. In 1559 Isabel of Valois was married via proxy in Paris. The procured marriage between the king and Isabel was carried out by the third Duke of Alba, who had been governor of Milan in the years 1555 and 1556. During his stay in Milan, don Alfonso Álvarez de Toledo had noticed the female- painter Sofonisba Anguissola, a noblewoman from Cremona, intensively promoted by her father. The duke of Alba thought she might be good company and a teacher for the young queen. Indeed, Isabel would take great amusement in learning this mechanical art from Sofonisba: Orietta Pinessi, *Sofonisba Anguissola: un "pittore" alla corte di Filippo II* (Milano: Selene, 1998), 42.

95 Giovanni Francesco Sitoni, *Giovanni Francesco Sitoni: ingeniero renacentista al servicio de la Corona de Espanã : con su códice inédito, Trattato delle virtù et proprietà dell'eacque, en su idioma original y traducido al castellano*, ed. José A García-Diego and Alex Keller ([Madrid]: Fundación Juanelo Turriano : Editorial Castalia, 1990), 34.

juridical and administrative service. The practice had become a custom long before, and the noble, by right of blood, expected economical support from the Crown even without serving in administrative or military offices.

Vivitur Ingenio, Caetera Mortis Erunt!

The Knight Without a Horse[96]

Harpalus: *But it's not in our power to be born nobles*
Nestor: *If you're not so by birth, strive by your good deeds to establish a noble line*
Harpalus: *A long, drawn-out business!*
Nestor: *The emperor will sell you a title for a trifling sum.*
Harpalus: *Nobility got by purchase is commonly an object of ridicule.*
Nestor: *Since nothing is more absurd than faked nobility, why do you covet the title of knight so badly?*

Erasmus' moralistic arguments represent a highly hierarchical society where social climbing was sometimes possible. If aristocracy found it worthwhile to engage in the workshop with Vitruvian craftsmen, those artisans took their chance to climb the social ladder exploiting the credit they enjoyed. They used the fascination of the prince with their arts as an instrument to ennoble themselves together with their crafts whose products belonged to the courtly language of prestige. Nonetheless, artisans did not pursue this aim as a professional collective, but as individuals. Even these highly qualified artisans shared that desire for glory that Burckhardt described as a typical feature of Renaissance civilisation.

If we accept that Janello's employment at the imperial court was a sign of social success, which instruments did he adopt, if any, to refashion his role in this very different context? As we have seen, Charles v amused himself assembling and setting clocks. We do not know whether he ever touched the file, lathe, hammer or bellows or whether he just participated in the planning and

---

96   Translation in English from Desiderius Erasmus, *Collected Works of Erasmus*, ed. Dominic Baker-Smith, 1997 "ΙΠΠΕΥΣ, ΑΝΙΠΠΟΣ, sive Ementita Nobilitas ... Harpalus: At nobis in manu non est, ut nascamur nobiles. Nestor : Si non es, enitete benefactis, ut a te initium capiat nobilitas. Harpalus: Perlongum est Nestor: Exigua summa tibi vendet Caesar Harpalus: Vulgus ridetur emticia nobilitas. Nestor: Cum nihil sit magis ridiculum quam ementita nobilitas, quid est ut tantopere affectes nomen equitis?

FIGURE 33   *Obverse and reverse of the medal of Janello Torriani, bronze casting, diam. 80 mm,*
            ca. 1550-1552, attributed to Jacopo Nizzola da Trezzo. COURTESY OF THE MUSEO
            CIVICO ALA PONZONE, CREMONA.

composition of the clockworks. Nevertheless, what we do know is sufficient to
entitle us to call him an adept of clockmaking. We have seen how this did not
diminish his noble status. What impact did it have on Janello's rank? A series of
documents chart an interesting shift in his professional and social identity.

Over the 1550s, we face a change in Janello's image, which clearly tends to an
ennoblement of his character. In fact, as we mentioned earlier, this process was
likely begun when the clockmaker, moving to Milan, changed his old patronymic "*de Torresani/Torrexanis*" for "*Torriano/Torriani*". The humanist Giovanni
Musonio of Cremona, a scholar contemporary of Janello, wrote in 1551 that the
clockmaker took his name from the major tower of Cremona where the public
clock was set.[97] Janello, as previously seen, was in charge of it. The medieval
Torrazzo tower was famous because it was the tallest medieval brickwork
structure in Christendom (112 metres), and it was the pride of Janello's countrymen in their never-ending competition with the neighbouring cities.[98] It is
possible that Janello, once he moved to Milan, changed his name in order to

---

97   "*Janiculus decus Italicum, celsaque Cremonae, Cuius ab insigni traxit cognomina Turri*":
      "Janiculus pride of Italy, and of the high Cremona, Whose famous Tower he took his
      familiy name from", Johannes Musoni, *Apollo Italicus, nuper in lucem restitutus. His etiam*
      Emblemata accedunt, VIII. Ad Jacobum Albensem Juris consultiss. Ode I (Ticini: Ex typis
      Francisci Moscheni, 1551).

98   For instance there is evidence of a Guelf adage circulating in Italy since the fourteenth
      century that stated: "*Unus Petrus in Roma, Una turris in Cremona, Unus portus in Ancona*"
      (*Ther is only one Peter and is in Rome, there is just one tower, and is in Cremona, there is just*

make clear which *patria* was his. Both his controversial seal and the trademark on his armillary sphere, after all, shows a tower (Figs 10, 29a). Another plausible hypothesis is that, once he had moved to the capital city to work on the Astrarium, Janello – perhaps encouraged by his brokers – assumed a surname that evoked the ancient and noble Milanese family of the Torriani. This must have projected a more effective image of the clockmaker, now involved with the ducal administration, in the eyes of the people of Milan.[99] The fact that the Torriani family had long been in decline probably gave Janello the chance to adopt this prestigious name without stepping on someone else's toes. The similitude between the two surnames allowed a smooth transition from the previous name to the latter. At least from the year 1549, Torriani used a shield with a tower as trademark.[100]

Torriani was not alone in the process of fashioning his new image. His program must be shared with people ready to recognise the new identity, endorsing a common agenda and often adding their personal touch to it. As for the other celebrating artefacts (medals, paintings, poems, sculptures, etc.), it was probably not Janello who had them done, but most likely it was his patrons and friends. As we have seen, the first to put Torriani on a pedestal were Cardano and more emphatically the humanist Marco Girolamo Vida, who elected a craftsman as one of the defenders of a city's superior dignity over another.[101] Among all the most excellent artesans, Janello was chosen to repre-

---

one harbour and is in Ancona): Maria Teresa Saracino, *Il Torrazzo e il suo restauro* (Cremona: Banco popolare di Cremona, 1979), 20.

99   In the very same context, even the powerful Medici di Marignano used a similar strategy, playing on the homonymy of their family with the powerful Florentine house. In fact, the Milanese Medici claimed a common ancestry with the Tuscan stock. The latter instead superciliously refused any association, until the Milanese family became very important. From that moment, the Florentine Medici stopped denying their blood relation with the Lombard kin, who shamelessly adopted the Florentine coat of arms with the *palle medicee*. The Medici di Marignano were part of the family network that supported Janello throughout his long career. According to Pietro Verri it was after the victorious campaign of Siena that the imperial *condottiero* Gian Giacomo de Medici received from Cosimo I the *palle* crest: Verri, *Storia di Milano*, II:350; before Verri, there was another tradition circulating, which identified in Gian Giacomo's brother, Pope Pious IV, the one who first adopted the Florentine coat of arms. According to this version, the Florentine Medici did not protest, considering this appropriation prestigious for their kin: Pompeo Litta, *Celebri famiglie italiane* (Milano: P.E. Giusti, 1819), see Medici di Marignano.

100   I thank prof. Mario Biagioli for having suggested me to look for such a trademark.

101   Perhaps it was an unknown fellow countryman of Vida who provided him with all the necessary information on Torriani that the *Cremonese* bishop of Alba, appreciated humanist, afterwards transformed into a refined Ciceronian pamphlet, "*Orationes III Cremonensium adversus Papienses in controversiam principatus*". On the 21st of March 1549,

sent Cremona's dignity because he was already very famous among the audience of Vida's pamphlet, being the clockmaker already working for Governor Gonzaga and for the Emperor, as Vida emphasises.[102] Going back to Vida's oration we read that:

> As it would be a long – and for you a very tedious – matter to enumerate one by one those persons most distinguished in each of the arts, I believe it was worth the trouble – though attempting to summarise it – to bring to your consideration only one single personage, who is well known to you, so that he might be a kind of prototype who can guarantee the others by himself alone.[103]

But being employed by the Governor and the Emperor it was probably not enough to transform Janello into an instrument able to transfer credit to the entire city. As we have previously seen, Vida highlighted that Torriani was both the inventor and the executor of the marvellous clock. Here is the necessary condition for Torriani's social rise: he incarnated the Vitruvian idea of "archi-

---

the city council of Cremona sent to the bishop of Alba the material collected to claim Cremona's precedence over Pavia in the internal hierarchy of the duchy of Milan. Then, Vida personally curated the printing of his work and sent it to Ferrante Gonzaga and to the Senate. The senators and learned people from Pavia reacted with anger to it, and vigorously protested toFerrante. Vida wrote a letter to the governor of Milan mentioning the fact that fellow countrymen from Cremona told him how numerous *Pavesi* went to Ferrante to complain about the bishop of Alba as he was the author of the pamphlet. This might be interpreted as the claim of Vida to be merely the redactor of the *Orationes*, which the admired humanist only assembled and translated into good Latin. This letter, together with the fact that the pamphlet was not signed, has led some scholars to believe that Vida was not its author. Tiraboschi produced evidence demonstrating that the bishop of Alba received by 1549 the request from Cremona to write the pamphlet using materials written by others. Anyway, Vida knew about the clockmaker and his clock, being the one that, on request of some friends of Torriani, suggested the name Microcosm for it. From Antonio Campi's map of Cremona one can also see that Vida's houses were very close to the parishes of S. Silvestro and S. Agata, where Janello's father lived for many years. If letters Vida received from Cremona in 1549 were to be found it could help us to solve this puzzle. The governor Ferrante ended the tense situation between Cremona and Pavia with a decree (7th of August 1550) imposing silence on the two parties: Vida, *Cremonensium Orationes III*; Girolamo Tiraboschi, *Storia della letteratura italiana*, vol. 4 (Milano: Per Nicolò Bettoni e comp., 1833), 257; for the part of the Orationes quoted, I used the English translation in: García-Diego, *Juanelo Turriano, Charles v's Clockmaker*, 73.

102   See in this book the section: *Fashioning the Aura of the Genius*.

103   Vida, *Cremonensium Orationes III*, 53-57.

tect" mastering a *scientia media* demonstrating both theoretical and practical skills.

For the Vitruvian artisan Torriani, the Habsburg patronage opened up the path of fame, giving him the necessary resources to demonstrate his unique ability in both design and physical construction. The reception was exceptional: the artisan was mentioned in more than a dozen different books published during his lifetime.[104] These books were mainly printed in the cultural framework connected with the duchy of Milan and the Habsburg court. There is evidence that the clockmaker was perfectly aware of this process of ennoblement: as Estebán de Garibay (1533-1599), historiographer to Philip II and Torriani's friend, noted, the Cremonese, while showing to him the planetary clocks, had also exhibited the imperial diploma that was praising him as mathematician and prince among the architects of clocks. Moreover, the clockmaker said to Garibay that:

One day he told the Emperor that he owed more to him than to his natural parents, because they had only engendered him for a brief life, whereas the Emperor would make him immortal.[105]

---

104 Vida, *Cremonensium Orationes III*; Musoni, *Apollo Italicus, nuper in lucem restitutus. His etiam Emblemata accedunt, VIII. Ad Jacobum Albensem Juris consultiss. Ode I*; Cardano, *De subtitulate (1554)*, 1554, not all the various editions report the same quantity of information about Torriani; Cardano, *De libris propriis* [*2004*], 71, 148; Zenocarus Snouckaert, *De republica*; Gasparo Annibal Cruceius, "Epigramma in Ianelli Turriani Cremonensis horologium," in *Carmina poetarum nobilium Io. Pauli Vbaldini studio conquisita*, by Gio. Pietro Ubaldini (Mediolani: apud Antonium Antonianum, 1563), 12; Bernardo Sacco, *De Italicarum rerum varietate et elegantia libri X* (Ticini: apud Hieronymum Bartolum, 1587), 150; Stefano Breventano, *Istoria della antichita nobilta, et delle cose notabili della citta di Pauia, raccolta da m. Stefano Breuentano cittadino pavese* (In Pauia: appresso Hieronimo Bartholi, 1570); John Dee, *The Mathematicall Praeface to the Elements of Geometrie of Euclid of Megara (1570)* (New York: Science History Publications, 1975); Esteban de Garibay y Zamalloa, *Los XL libros d'el compendio historial de las chronicas y vniuersal Historia de todos los reynos de España* (Anueres: por Christophoro Plantino, 1571), bk. 36, chap. 20; Lamo, *Sogno non meno piacevole, che morale*; Morales, *Antiguedades de las ciudades de España*, fols. 91-94; Alessandro Lamo, *Discorso di Alessandro Lamo intorno alla scoltvra, et pittvra doue ragiona della vita & opere in molti luoghi & à diuersi prencipi & personaggi fatte dall'eccell. & nobile M. Bernardino Campi, pittore cremonese* (Cremona: Appresso Christoforo Draconi, 1584); Campi, *Cremona fedelissima citta*; Lomazzo, *Trattato dell'arte della pittvra*; Gasparo Bugati, *Historia uniuersale* (In Vinetia: Appresso Gabriel Giolito de Ferrarii, 1570), 1025-26.

105 "*Yo le envié algunos años antes esta obra para que la viese, señalando este su titulo particular, de que el no tenia noticia. Sobre lo cual me dijo despues el mismo que un dia habia dicho*

Janello told Garibay about this fact, because the Basque historiographer, reading Zenocarus' book about the life of the deceased Emperor, had found out that Janello and his planetary clock were mentioned there. Consequently he sent Janello the book which was yet unknown to the clockmaker. Another Spanish friend of Janello's, the famous architect Juan de Herrera, had in his library a "notebook with several epigrams in praise of Janello's clock".[106] Praise through the means of epigrams and poetry was a well-established habit in humanist culture, but still quite exceptional for an artisan who was not a painter, sculptor or architect, professions already objects of public admiration. Although this booklet is now lost, we still have some of the poems from Cremona and Milan.[107] The aim of these poems was both to provide an explanatory text,[108] a compliment to the artisan, and a means of promoting his person, and celebrating his creations and his patrons.

On another occasion, in the letter of petition dated 1567 held in the Archive of Venice, Janello, or someone writing on his behalf, confirmed the clockmaker's desire for fame: indeed, writing to the Doge of Venice in order to obtain a privilege for invention – i.e., the precursor of modern patents – the author of the letter stated:

> Most Serene Prince, His Catholic Majesty's clockmaker and mathematician Janello Torriani, who has always had the aim to be helpful to others

---

al Emperador que le debia mas que a sus padres naturales, porque ellos solo le habian engendrado para una vida breve, y el seria causa de inmortalizarle esta". and *"En el titulo latino que le dio en Alemania de maestro de relojes, refrendado por Gonzalo Perez, su segretario, le llama principe facilmente entre los maestros de hacer relojes, que el me lo mostró originalmente en Toledo con otros papeles, con este mismo reloj y con el que después hizo para el católico Rey Don Felipe II, su hijo, muy mejor que el pasado, con otras obras dignas a su ultimo ingenio":* Esteban de Garibay y Zamalloa, "Memorias," in *Memorial histórico español: colección de documentos, opúsculos y antigüedades que publica la Real Academia de la Historia*, ed. Pascual de Gayangos, vol. VII (Madrid: Imprenta de José Rodríguez, 1854), 420-21.

106   The parts of the inventory of Herrera's library related to Torriani, are published in: García-Diego, *Juanelo Turriano, Charles V's Clockmaker*, 46.

107   Cruceius, "Epigramma in Ianelli Turriani Cremonensis horologium," 12; Lamo, *Discorso di Alessandro Lamo*; see also the poems by Giovanni Musonio and Ubaldini, reported in: Leydi, "Un cremonese del Cinquecento," 139.

108   As in the case of his *Flora* (made in Milan and sent to Rudolph II) the painter Giuseppe Arcimboldo sent some poetic compositions together with the painting that illustrated his work: Giacomo Berra, "L'Arcimboldo 'c'huom forma d'ogni cosa': capricci pittorici, elogi letterari e scherzi poetici nella Milano di fine Cinquecento," in *Arcimboldo / Palazzo Reale*, ed. Sylvia Ferino-Padgen (Milano, 2011), 289.

EN MADRID,

Por Luis Sanchez, en fin de Abril.

Año M. D. XCVI.

FIGURE 34
Portrait of Esteban de
Garibay, *in* Esteban de
Garibay, Ilustraciones
genealógicas de Catholicos
los reyes de España, *Madrid,
1596.*

and to himself, and to leave behind some memory of himself, as he has
done so far, has invented new instruments and ways to elevate water of
flowing rivers, lakes, springs and dead rivers, whose effect of this invention, will bring, one can say, infinite utility to the world ...[109]

---

109   "*Serenissimo Principe, Giannello Turriano Cremonese Orologiaio, e Matematico di Sua Mae-
      stà Cattolica, havendo sempre havuto per fine di giovare ad altri, et a se stesso, e lasciare
      qualche memoria di se, come ha fatto per sino ad hora, ha ritrovato nuovamente instru-
      menti, e modi di estrazer acqua di fiumi correnti, di laghi, di fonti, o di fiumi morti, il quale
      effetto sera al mondo di giovamento, si può dir, infinito ...*": ASVe, Senato, Terra, filza 50, doc.
      2; I thank professor Luca Molà for his help in finding this document. The petition does not
      contain any signatures, and it is probably a transcription by some officials of the Republic.
      It was the Mantuan ambassador Girolamo Negri who acted as a promoter of Janello's
      device in Italy.

This desire to be known by the next generations fits perfectly within the humanist mentality of the time. The broad diffusion of the pseudo-Virgilian motto *"Vivitur ingenio, caetera mortis erunt"* during this century provides a consistent frame of reference for Torriani's idea of immortality: "Only the genius survives; all else is claimed by death".[110] Here we can see that cogs and toothed wheels could ensure glory just as literature, sculpture, painting and music can.

"Vivitur ingenio" was not a purely rhetorical formula: it could also be literally translated as *"one lives in ingenuity"*, a proper motto for those who made their living from their inventiveness that could be remunerative even after the inventor's death; as we shall seen in the chapter on Torriani's household, the clockmaker's inventions brought economic advantages to his heirs for generations. And for Janello feeding his family was a major problem. In an already visited anecdote entitled *Quesito elegantísimo di M. Gianello a Carlo v Imperatore* (*Very elegant question addressed by Master Gianello to Emperor Charles v*) the Mantuan Camillo Capilupi tells that an almost 60-years-old Torriani was not too happy about the life-long pension that the Emperor granted him after the delivery of the Microcosm: indeed, in the imperial privilege awards the annual pension of 100 golden *scuti "durante ipsius Ianelli vita"*, which means for the duration of the life of the same Janello.[111] Capilupi says that since Charles v and the clockmaker spent a lot of time together, one day Janello asked the Emperor to whom his clock was supposed to be handed over after Charles' death. Charles answered "certainly to my son Philip Prince of Spain" and Janello replied:

> Why will my clock not come back in my hands when Your Majesty will die, but my pension will go back to you if I die? The clock should come back to me in case of your departure as my pension will come back to you when I pass away.[112]

---

110 Dürer used the *"vivitur ingenio"* motto in a portrait, Vesalius in one of his anatomical tables, and Jean Errard in his theatre of machines published in the year 1584. Humanist intellectuals, physicians and technicians all found this phrase apt to describe their desire. Dolza, *Storia della tecnologia*, 108-10.

111 See the text of the privilege at p.

112 Biblioteca Nazionale Cenreale di Roma, manuscript Vitt. Em. 1009, cc. 152v-153v, and manuscript Vitt. Em. 1062 a c. 33v: *"perciochè non gli assignò altro per recompensa / che una pensione di dugento scudi nello stato di Milano et questa in vita sola di esso Gianello, il quale benche sentis= / se molto male di vidersi così leggermente gratificoato, tuttavia se / lo tolerò pacientemente sperando alla fine colla virtù di ac= / quistarsi la gratia dell'Imperatore ch'egli vedeva ... Et dilectarsi / di cose simili, et però fermatosi alla servitù dell'Imperatore et / ogni di fabricando nuove cose et donandogliele divenne in / pochi mesi gratísimo*

Capilupi concludes that the Emperor was convinced by the witty and subtle way Torriani had manifested his discontent, and he granted him the possibility of handing over his pension to his heirs, something, observes Capilupi, that was not customary, because of the risk that too many resources of the state would thus be drawn away. In this manuscript, there was praise not only for the audacity that brings rewards, but also for the credit that backs audacity. We cannot tell if this story is true; however, the pension did become hereditary. Perhaps it was Philip II who changed the pension's nature when he had to convince Torriani to stay in his service, doubling his wage.

Torriani's career, now within the orbit of the Habsburgs, led to the construction of a new ennobled identity. Humanist culture had in Archimedes the shining prototype of the mathematician-inventor and many contemporary authors saluted also Torriani as the *"new Archimedes"* as they did with several other talented engineers.[113] Such learned models were reflecting an aura of

---

> *all'Imperatore che poi non tanto per la virtù / dell'huomo, ma per la ... sua natura volentieri*
> *et spesso quan / do voleva ricrearsi stava con lui le hore intiere / ... a parlar seco, andando*
> *egli stesso a trovarlo alla sua / stanza a vederlo lavorare onde presa una volta tra l'altre*
> *Gia=/ nello oportuna occasione che vide l'Imperatore di buona tempra gli / chiedesse un*
> *poco di chi sarà l'orologio grande che vi donai / quando voi morirete? Et egli rispondendo che*
> *sarebbe del / Principe di Spagna suo figlio, Gianello gli soggiunse et per / che volete poi che*
> *la pension che mi deste delli dugento scudi / in recompensa dell'horologio s'estinguano*
> *quando morendo io et / non si possi lasciarla a miei figloli et heredi, sarebbe giusto dunque*
> *ch'anche / l'orologio ritornasse a me mancando voi, si come mancando io la pensione / ricade*
> *a voi, nel qual modo videndosi convinto dulcemente da / Gianello gli disse ch'era ben ragione*
> *che le cose fossero pari / et così gli fece un decreto per lo quale la pensione potes=/ se passar*
> *a suoi heredi, il che forse Gianello non havreb/be insperato così facilmente se havesse tenuta*
> *la via ordinaria / non essendo veramente solito il darsi quelle pensioni ad alcuno / per ben-*
> *emerito che sia della Corona se non in vita, acciochè l'/ entrate di quello stato in pochi anni*
> *non venissero ad assai / alienarsi affatto".*

113   Sebastián de Covarrubias y Orozco, *Tesoro de la lengua castellana o española* (Madrid: Luis Sanchez, 1611) , see the voice *"Ingenio"* where the exemple given is Torriani: *"Janelo, segundo Archimedes"*; perhaps Covarrubuas followed the tradition of Agustín de Rojas Villandrando, *El viaje entretenido*, ed. Jacques Revel, [1604] (Espasa-Calpe, 1977), 296, here the author says that Janello *"(...) mereció igual gloria con aquel Arquímedes de Siracusa"*. Marshall Clagett, *Archimedes in the Middle Ages* (Philadelphia: The American Philosophical Society, 1978); Eberhard Knobloch, "Les ingénieurs de la Renaissance et leurs manuscrits et traités illustrés," in *Engineering and engineers: Proceedings of the xxth International Congress of History of Science, (Liège, 20-26 July 1997)*, ed. Michael Claran Duffy, vol. 60, De Diversis Artibus, XVII (Tumhout, 2002), 23-65. Taccola boasted to have been nicknamed *"Archimedes of Siena"*, Ambrose Bachote called Agostino Ramelli *"the Archimedes of his age"*. The famous clockmaker Giovanni Maria Barocci from Urbino was seen as a modern Archimedes. Lawrence Fane, "The Invented World of Mariano Taccola:

classical charm not just upon the capable artisan, but also on his powerful patron. Other ancient Greek models were used to emphasise Janello's unique talent as a constructor: some authors, like Luis de Góngora, called him the "Daedalus from Cremona".[114] Indeed, both the medal and the sculpture representing Janello Torriani (Figs 33, 53) show the clockmaker wearing on his sixteenth century clothes what seems to be an ancient Greek chlamys, the best "uniform" to identify him as the new Archimedes or the modern Daedalus. However, the most significant transformation of Janello's dignity was quite other and it had in the clockmaker himself the first actor: when Ambrosio de Morales referred to the magnificent invention of the Microcosm, he added that:

> The Emperor asked Janello what he wanted to be written on the clock. Janello answered: "Janellus Turrianus *horologiorum architector[um]*". Because Torriani at this point hesitated, the Emperor added: "*Facile princeps*". Together this means "easily the prince [i.e., the first] among the architects of clocks".[115]

Indeed, as we said, in 1552 the Emperor granted Janello a life-long pension of a 100 ducats a year. The privilege reads:

> We Charles V, by the Grace of Divine Mercy, August Emperor of the Romans ... recognise and by the tenor of the present letters, make manifest to those whom it may concern that, concerning the praiseworthy artistic and practical work which for us, for Our Empire and for the lieges of the Empire itself has been executed by Our dear Janellus Turrianus, a mathematician of Cremona and, with no doubt the foremost among architects of clocks [*inter Horologiorum Architectos facile Princeps*], in

---

Revisiting a Once-Famous Artist-Engineer of 15th-Century Italy," *Leonardo : Journal of the International Society for the Arts, Sciences and Technology*, 36, no. 2 (2003): 135-43; Morpurgo, *Dizionario degli orologiai italiani*, 1950, 23; Knobloch, "Les ingénieurs de la Renaissance et leurs manuscrits et traités illustrés," 47.

114   Luis de Góngora, *Las firmezas de Isabela*, 1610, in Luis de Góngora y Argote, *Teatro completo: Las firmezas de Isabela ; El doctor Carlino ; Comedia venatoria*, ed. Laura Dolfi (Madrid: Cátedra, 2016).

115   "*Preguntole el Emperador, que pensava escrvir en el relox? El respondio que esto. Iannelus Turrianus Cremonensis horologiorum architector. Parando el aqui, añadio su Magestad, Facile princeps. Y assi esta puesto todo junto, y dize. IANNELLVS TVRRIANVS. CREMONENSIS. HOROLOGIORVM. ARCHITECTOR. FACILE. PRINCEPS*". Morales, *Antiguedades de las ciudades de España*, fols. 91-94.

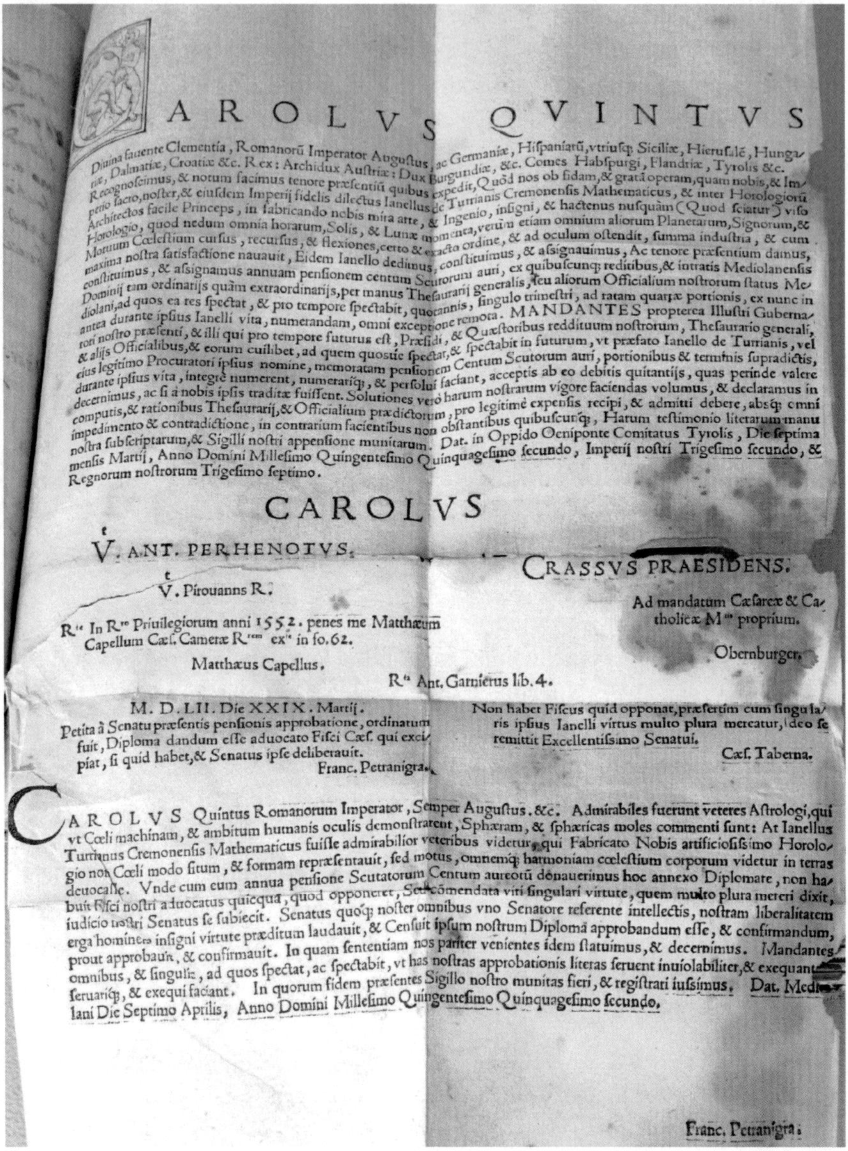

CAROLVS QVINTVS

Diuina fauente Clementia, Romanorũ Imperator Augustus, ac Germaniæ, Hispaniarũ, vtriusq; Siciliæ, Hierusalẽ, Hungariæ, Dalmatiæ, Croatiæ &c. Rex: Archidux Austriæ: Dux Burgundiæ, &c. Comes Habspurgi, Flandriæ, Tyrolis &c. Recognoscimus, & notum facimus tenore præsentiũ quibus expedit, Quòd nos ob fidam, & gratã operam, quam nobis, & Imperio sacro, noster, & eiusdem Imperij fidelis dilectus Ianellus de Turrianis Cremonensis Mathematicus, & inter Horologiorũ Architectos facile Princeps, in fabricando nobis mira arte, & Ingenio, insigni, & hactenus nusquàm (Quod sciatur) viso Horologio, quod nedum omnia horarum, Solis, & Lunæ momenta, verũm etiam omnium aliorum Planetarum, Signorum, & Motuum Cœlestium cursus, recursus, & flexiones, certo & exacto ordine, & ad oculum ostendit, summa industria, & cum maxima nostra satisfactione nauauit, Eidem Ianello dedimus, constituimus, & assignauimus, Ac tenore præsentium damus, constituimus, & assignamus annuam pensionem centum Scutorum auri, ex quibuscunq; reditibus, & intratis Mediolanensis Dominij tam ordinarijs quàm extraordinarijs, per manus Thesaurarij generalis, seu aliorum Officialium nostrorum status Mediolani, ad quos ea res spectat, & pro tempore spectabit, quotannis, singulo trimestri, ad ratam quartæ portionis, ex nunc in antea durante ipsius Ianelli vita, numerandam, omni exceptione remota. MANDANTES propterea Illustri Gubernatori nostro præsenti, & illi qui pro tempore futurus est, Præsidi, & Quæstoribus reddituum nostrorum, Thesaurario generali, & alijs Officialibus, & eorum cuilibet, ad quem quosuis spectat, & spectabit in futurum, vt præfato Ianello de Turrianis, vel eius legitimo Procuratori ipsius nomine, memoratam pensionem Centum Scutorum auri, portionibus & terminis supradictis, durante ipsius vita, integrè numerent, numerariq;, & persolui faciant, acceptis ab eo debitis quitantijs, quas perinde valere decernimus, ac si à nobis ipsis traditæ fuissent. Solutiones verò harum nostrarum vigore faciendas volumus, & declaramus in computis, & rationibus Thesaurarij, & Officialium prædictorum, pro legitimè expensis recipi, & admitti debere, absq; omni impedimento & contradictione, in contrarium facientibus non obstantibus quibuscunq;, Harum testimonio literarum manu nostra subscriptarum, & Sigilli nostri appensione munitarum. Dat. in Oppido Œniponte Comitatus Tyrolis, Die septima mensis Martij, Anno Domini Millesimo Quingentesimo Quinquagesimo secundo, Imperij nostri Trigesimo secundo, & Regnorum nostrorum Trigesimo septimo.

CAROLVS

V. ANT. PERHENOTVS.

V. Pirouanns R.

CRASSVS PRAESIDENS.

Ad mandatum Cæsareæ & Catholicæ M[tis] proprium.

Obernburger.

R[ta] In R[ro] Priuilegiorum anni 1552. penes me Matthæum Capellum Cæs. Cameræ R[rem] ex[te] in fo. 62.

Matthæus Capellus.

R[ta] Ant. Garnierus lib. 4.

M. D. LII. Die XXIX. Martij.

Petita à Senatu præsentis pensionis approbatione, ordinatum fuit, Diploma dandum esse aduocato Fisci Cæs. qui excipiat, si quid habet, & Senatus ipse deliberauit.

Franc. Petranigra.

Non habet Fiscus quid opponat, præsertim cum singularis ipsius Ianelli virtus multo plura mereatur, ideo se remittit Excellentissimo Senatui.

Cæs. Taberna.

CAROLVS Quintus Romanorum Imperator, Semper Augustus. &c. Admirabiles fuerunt veteres Astrologi, qui vt Cœli machinam, & ambitum humanis oculis demonstrarent, Sphæram, & sphæricas moles commenti sunt: At Ianellus Turrianus Cremonensis Mathematicus fuisse admirabilior veteribus videtur, qui Fabricato Nobis artificiosissimo Horologio non Cœli modo situm, & formam repræsentauit, sed motus, omnemq; harmoniam cœlestium corporum videtur in terras deuocasse. Vnde cum eum annua pensione Scutatorum Centum aureorũ donauerimus hoc annexo Diplomate, non habuit Fisci nostri aduocatus quicquã, quod opponeret, Sed cõmendata viri singulari virtute, quem multo plura mereri dixit, iudicio nostri Senatus se subiecit. Senatus quoq; noster omnibus vno Senatore referente intellectis, nostram liberalitatem erga hominem insigni virtute præditum laudauit, & Censuit ipsum nostrum Diploma approbandum esse, & confirmandum, prout approbauit, & confirmauit. In quam sententiam nos pariter venientes idem statuimus, & decernimus. Mandantes omnibus, & singulis, ad quos spectat, ac spectabit, vt has nostras approbationis literas seruent inuiolabiliter, & exequantur seruariq;, & exequi faciant. In quorum fidem præsentes Sigillo nostro munitas fieri, & registrari iussimus. Dat. Mediolani Die Septimo Aprilis, Anno Domini Millesimo Quingentesimo Quinquagesimo secundo.

Franc. Petranigra.

FIGURE 35 Imperial privilege. With this privilege Charles v grants Janello "mathematician … among the architects of clocks surely the prince" a life pension of one hundred gold ducats for the construction of a planetary clock "the likes of which had never been seen before", ASCr, Comune di Cremona, Miscellanea Jurium, vol. 10, f. 149. COURTESY OF THE ARCHIVIO DI STATO DI CREMONA.

constructing for Us, with admirable technique and talent, an exceptional clock and – so far as is known – never seen anywhere else.[116]

Garibay referring to the imperial document named it *"titulo"*, meaning both diploma and title.[117] Of course this was not a feudal title, but in practice it had a similar effect: this princedom of honour increased Janello's prestige and gave him a better social position, with the hundred ducats (which would double after the death of Charles) to add to his wage at court, a life-long pension that he later managed to turn into a hereditary one, a privilege quintessential to nobility. Both Torriani's granddaughter, who married a Milanese gentleman, and later his great-granddaughter, received this pension at least until the second decade of the seventeenth century.[118] It seems that the bombastic definition of "prince" was not extemporaneous in the field of clockmaking: we know that Janello's Microcosm was supposed to substitute for the Astrarium, and already in the fourteenth century Petrarch had addressed his friend Giovanni de Dondi, builder of that famous machine, as "with no doubt the foremost among

---

116   I do not agree with the translation that Garcia-Diego gave of the line *"inter Horologiorum Architectos facile Princeps"* as *"a mathematician of Cremona and, very probably, the foremost among the inventors of clocks"*. I indeed prefer to maintain the emphasis given by the imperial patent, which stresses the fact that Torriani is *"with no doubt the foremost among architects of clocks"*. García-Diego, *Juanelo Turriano, Charles v's Clockmaker*, 73.

117   *"En el titulo latino que le dio en Alemania de maestro de relojes, refrendado por Gonzalo Perez, su segretario, le llama principe facilemente entre los maestros de hacer relojes, que el me lo mostró originalmente en Toledo con otros papeles, con este mismo reloj y con el que después hizo para el católico Rey Don Felipe II, su hijo, muy mejor que el pasado, con otras obras dignas a su ultimo ingenio"* Garibay y Zamalloa, "Memorias," 420-21.

118   An interesting feature of this title is the fact that its pension is inheritable, or at least transmittable. In the year 1581 Janello transferred his 200 ducats pension to his granddaughter Emilia Filipa Diana Turriano, married to the Milanese Ludovico Besozzi or Besozzo. In 1612 the pension, reduced to its original amount of 100 *scudi*, was than inherited by her daughter Angela Magdalena Besozzo. The pension was doubled by Philip II in the 1560s and than reduced to its original quantity by Philip IV. Cervera Vera, *Documentos biográficos*, doc. 107; Adela González Vega and Ana Ma Díez Gil, eds., *Títulos y privilegios de Milán: siglos XVI-XVII* (Valladolid: Archivo General de Simancas, 1991). See the names in the index: *"Turriano, Juanello. Licencia a su favor para renunciar en Emilia Filippa de Diana, su nieta, los 200 escudos de pensión que tiene en el estado de Milán. Thomar, 15 de Mayo de 1581"*; *"Besoza Angela Magdalena. Merced de 100 escudos de pensión, en el estado de Milán, durante su vida, de los 200 que allí vacaron por su madre, doña Emilia Felipa de Diana, en atención de los méritos de su abuelo, Juanelo Turriano. El Pardo 1612"*.

astronomers".[119] Furthermore, Cesare Cesariano, in his comments on Vitruvius, used a similar terminology in order to define Giancarlo Ranieri of Reggio:

> Our Langobard [i.e., Lombard], who obtained the princedom making the very excellent clock for the Venetian Lords, placed in the great square of Saint Mark.[120]

Rulers such as Charles v promoted upward social mobility as a function of proven virtue in paintings, sculpture and astronomy: when Petrus Apianus offered to Charles v in 1540 the *Astronomicum Caesareum*, he was rewarded the following year by the Emperor with the title of Count. Was this title a real title and Torriani's simply a rhetorical gesture, despite the pension of a 100 golden *scudi*? Was not the official role of royal clockmaker a translation in courtly language of Janello's princedom within his profession? Apianus was the son of a shoemaker. Not so different from Torriani in terms of his background, Apianus, unlike the clockmaker, was however a literate, i.e., he knew the scientific language of the time: Latin. Since 1527, he had taught at the University of Ingolstad. Literacy and academic positions still had a higher dignity than that of the craftsman, even the most skilled one. In between one can find Vitruvian artisans involved with goldsmithing or with the visual arts, as in the case of Titian, Leone Leoni, Baccio Bandinelli, and later Giuseppe Arcimboldo, who were all elevated by the Habsburgs to the rank of knights.[121] A hundred years before, King Mathias Corvinus of Hungary had knighted Aristotile Fioravanti, a goldsmith and an outstanding engineer, and granted him the privilege to strike coins with his own effigy. After that, in contemporary documents from Bologna,

---

119   "*Mag. Joh. de Dondis phisicum, astronomicorum facile principem*": Morpurgo, *Dizionario degli orologiai italiani*, 1950, 61.

120   "Johanne Carlo Regiense nostro Longobardo obtene il Principato in fabricare lo Horologio excellentissimo facto a li Signori Venetiani collocato a la Magna Platea del Divo Marcho", Cesariano and Vitruvius Pollio, *De architectura libri dece*, CX.

121   Giorgio Vasari, *Le vite de più eccellenti pittori, scultori e architettori* (Novara: Ist. Geografico de Agostini, 1967); Michael Hirst, "Bandinelli, Baccio (Bartolomeo)," *Dizionario biografico degli Italiani* (Roma: Istituto della Enciclopedia italiana, 1963); Berra, "Al fì, chi vol de tut cora a Milan," 284; Bruce L. Edelstein, "'Acqua Viva E Corrente': Private Display and Public Distribution of Fresh Water at the Neapolitan Villa of Poggioreale as a Hydraulic Model for Sixteenth-Century Medici Gardens," in *Artistic Exchange and Cultural Translation in the Italian Renaissance City*, ed. Stephen J. Campbell and Stephen J. Milner (Cambridge: Cambridge University Press, 2004), 204.

Aristotile was called "a magnificent knight".[122] It seems that the first clock-maker to be invested with the title of knight was Giovanni Maria Barocci from Urbino in 1572, by Pope Pius v.[123]

As concerns the earlier ennoblement of goldsmiths, painters and sculptors, one has to make a few considerations: for instance a goldsmith, because of the material he used, had a privileged position in society. Even painting, within both courtly and humanist culture, enjoyed a high degree of respectability: in ancient Greek education painting was considered a suitable activity for a liberal education. Moreover, classical literature had provided powerful models to demonstrate the nobility of this gentle art: Pliny's famous account of Alexander the Great picking up Apelles' brush was projected into the sixteenth century with regard to Dürer's portrait of Emperor Maximilian, and Titian's of Emperor Charles v.[124] Moreover, Leon Battista Alberti and later Giorgio Vasari, thanks to their influential writings, stabilised this belief within their own society. This classical tradition was integrated into a practice that already in the first half of the fifteenth century saw the painter Jan Van Eyck performing as a diplomat for the Duke of Burgundy Philip the Good.[125] He was not the first visual artist to gain a position as a courtier of a Valois prince: Philip's grandfather, Philip the Bold (1342-1404), had already made in the last part of the fourteenth century sculptor Claus Sluter from Haarlem valet de Chambre.[126] His brother Jean Duke of Berry (1340-1416) had granted the same honour to Pol, one of the three Limbourg brothers, authors of the famous illuminated manuscript known as *Les très riches Heures du Duc de Berry*.[127] The brother of the dukes of Berry and Burgundy was the French king Charles v, who made *valet de Chambre* his court painter Jean Bondol of Bruges and the illuminator Jean d'Orléans.[128] We can trace even further back the fortune of painters at court: for instance, even if he

---

122   "*Magnificus eques*": Adriano Ghisetti Giavarina, "Fioravanti Aristotele (Fieravanti)," *Dizionario biografico degli Italiani* (Roma: Istituto della Enciclopedia italiana, 1997).

123   Morpurgo, *Dizionario degli orologiai italiani*, 1950, 23.

124   Gaius Plinius Secundus, *Storia delle arti antiche*, ed. Silvio Ferri and Maurizio Harrari (Milano: BUR, 2007).

125   Johan Huizinga and Eugenio Garin, *L'autunno del Medioevo* (Milano: Rizzoli, 1995), 353.

126   Susie Nash, "Claus Sluter's 'Well of Moses' for the Chartreuse de Champmol Reconsidered: Part I.," ed. Benedict Nicolson, *The Burlington Magazine*, 2005, 801.

127   Stefano Zuffi, *European art of the fifteenth century* (Los Angeles: J. Paul Getty Museum, 2005), 302.

128   William D. Wixom, "A Missal for a King: A First Exhibition ; an Introduction to the Gotha Missal and a Catalogue to the Exhibition Held at the Cleveland Museum of Art, August 8 through September 15, 1963," *The Bulletin of the Cleveland Museum of Art / Cleveland Museum of Art*. 50, no. 7 (1963): 162; Stephen Perkinson, "Engin and Artifice: Describing

was never granted with any title, Giotto between the end of the thirteenth and the beginning of the fourteenth century was highly esteemed and invited to paint for kings, popes, lords and, powerful religious orders and rich bankers.[129]

Even sculpture, which was engaged in a never ending and capricious humanist competition with painting in the field of *mimesis*, became a noble profession, despite the fact that the sweat, noise and dust it produced were seen as drawbacks, thanks to the experiences of Claus Sluter, to the revival of the classical tradition coming from Pliny the Elder, and to a milieu such as the Florentine one where admired sculptors such as Donatello, Ghiberti and Brunelleschi, and later Michelangelo, could proudly work in marble. And it was a sculptor, Leone Leoni, who contemptuously called Torriani "great king of blacksmithing, not of sculpture nor of statue-making",[130] in this way pointing out that the "title" granted by the Emperor, beside its encomiastic use, was a source of both jealousy and mockery. Garibay, who attended Torriani's funeral, would still call him "famous prince in all things he put his most bright ingenuity and his hands to"[131] and in 1584 Alessandro Lamo of Cremona would recall that Charles V made him by privilege "prince of blacksmiths".[132]

Something must have clicked in Torriani's mind from the time he received his imperial diploma. The self-confidence of the clockmaker, who called himself "architect of clocks", was now bringing Torriani to reinvent his identity, or as recent scholarship might have it, to refashion himself.[133] The modesty of the urban craftsman was abandoned for a new magnificent courtly image: thanks to the new language he adopted, we no longer see a skilled clockmaker among clockmakers, but the prince of them. Torriani's choice of the word "architect" is here of a crucial importance: it is a clear attempt to move from the category of

---

Creative Agency at the Court of France, Ca. 1400.," *Gesta / International Center of Medieval Art*, 2002, 61-67.

129   Dante Alighieri, "Purgatorio," in *La divina commedia*, ed. Natalino Sapegno (Milano: Ricciardi, 1957), XI, 91-96; Miklós Boskovits, "Giotto di Bondone," *Dizionario biografico degli italiani* (Roma: Istituto della Enciclopedia italiana, 2001).

130   "*gran Re di fabro e non di scultore o statuario*": Leydi, "Un cremonese del Cinquecento," 137.

131   "*Príncipe muy conocido en todas las cosas en que puso su clarísimo ingenio y manos*": Llaguno y Amirola, *Noticias de los arquitectos y arquitectura de España*, II:250.

132   "*Prencipe de i Fabri*": Lamo, *Discorso di Alessandro Lamo*.

133   The work by Greenblatt on Renaissance self-fashioning is considered to be the progenitor of this analytical trend. With these lenses the historian looks at people as they were all actors on a stage: any action or product is considered as a function of cultural codes. I have been inspired by Biagioli's work on Galileo as a courtier, and his self-fashioning as a court-philosopher from the mathematician he was; Greenblatt, *Renaissance Self-Fashioning*; Biagioli, *Galileo, Courtier*.

the *faber*, to be intended as craftsman, who makes things but does not design them, to the Vitruvian one of *architectus*, who was also inventor.[134] Janello's library has been unfortunately scattered, but in Spain two books still show Janello's interest in Vitruvius: an illustrated Latin edition of Vitruvius from 1552, edited by Philander, still displays the note in pen "This belonged to sir Janello" and a later Italian edition by Daniele Barbaro from 1567 display the *ex libris* "S[ir] *Ioanelo*".[135] This should be considered as evidence of Janello's proud conscious intellectual aspirations as a Vitruvian artisan: besides his construction of a ballista and of a Ctesibian pump, both described by Emperor Augustus's architect, the ownership of these editions of the *De Architectura*, and the choice of being called "architect of clocks" are all pointing in this direction. In the 1570s, Louis Hurtado, praising the great water device of Toledo, called the clockmaker "the most subtile Janelo Turriano from Cremona, prince of architecture",[136] thus showing that the definition of Janello as *architectus* in the Vitruvian meaning of designer and constructor was widely received at court. Archimedes was the best archetype of this Vitruvian idea of architecture, so different from what it became during the following centuries.

This helps explain why a craftsman, during his lifetime, beside all laudatory literature lavished on him, was also portrayed on his own planetary clock, on

---

134   Garzoni, *Piazza universale di tutte le professioni del mondo*, 757-64.

135   *"Fue del S[eñor] Janelo"* Marcus Vitruvius Pollio, De architectura libri decem ad Caesarem Augustum, omnibus omnium editionibus longe emendatiores collatis veteribus exemplis. Accesserunt, Gulielmi Philandri Castilionii, civis romani annotationes castigatiores, & plus tertia parte locupletiores. Adjecta est epitome in omnes Georgii Agicolae de mensuris & ponderibus libros eodem autore cum graeco pariter & latino indice locupletissimo, Lugduni, apud Joan[nes] Tornaesium, 1552. (Real Academia des Bellas Artes de San Fernando, Madrid. Archivo-Biblioteca). *"S.r Ioanelo"*: Marco Vitruvius Pollio, *I Dieci libri dell'Architettura di M. Vitruvio, tradotti et commentati da Monsig. Daniel Barbaro eletto Patriarca d'Aquileia, da lui riueduti & ampliati; & hora in piu commoda forma ridotti* (Venetia: appresso Francesco de Franceschi Senese et Giovanni Chrieger Alemano compagni, 1567); Fernando Marías, "Entre modernos y el antiguo romano Vitruvio: lectores y escritores de arquitectura en la España del siglo XVI," in *Teoría y literatura artística en España*, ed. Nuria Rodríguez Ortega and Miguel Taín Guzmán (Madrid: Real Academia de Bellas Artes de San Fernando, 2015), 199-233.

136   *"Sutilísimo Janelo Turriano de Cremona, príncipe de la arquitectura"*: Luis Hurtado de Toledo, "Memorial de Algunas Cosas Notables Que Tiene La Imperial Ciudad de Toledo," in *Relaciones Histórico-Geográfico-Estadísticas de Los Pueblos de de España Hechas Por Iniciativa de Felipe II*, ed. Carmelo Viñas Mey and Ramón Paz (Madrid: Reino de Toledo, 1951), nos. 103, 1958, 92.

medals, in a marble bust and in a couple of paintings.[137] If Torriani was not the one asking for the portraits to be made, he was certainly not opposed to being represented as the new Archimedes and as a vessel of virtues, as he was depicted on the fine bronze medal probably coined by his friend Jacopo Nizzola da Trezzo:[138] "Virtue never died" one can read.[139] It was the Emperor Charles v who promoted and confirmed Janello's metamorphosis, ordering his

---

137   Cristiano Zanetti, ed., *Janello Torriani, genio del Rinascimento* (Cremona: Comune di Cremona, 2016). Older works on the subject, often presenting mistakes, are: Leydi, "Un cremonese del Cinquecento," 153-56; Jaume Coll and Ivana Iotta, eds., *Realismo y espiritualidad: Campi, Anguissola, Caravaggio y otros artistas cremoneses y españoles en los siglos XVI-XVIII* (Alaquàs: Ayuntamiento de Alaquàs, 2007); and Angel del Campo y Francés, *Semblanza iconográfica de Juanelo Turriano* (Madrid: Fundación Juanelo Turriano, 1997). The best analysis of the problem of Torriani's portraits is to be found in: Leydi, "Un cremonese del Cinquecento," 153-56. An extremely refined medal, attributed to Jacopo da Trezzo, friend to Torriani and servant as well to the king of Spain, was probably cast for this occasion. Zaist referred to another one, but there is no evidence to verify its existence. There is a marble bust portraying Torriani in the Museo Santa Cruz of Toledo. The sculpture, which was probably meant to be placed somewhere on the Toledo device, has been attributed to Pompeo Leoni, son of Leone. Two paintings of Torriani are stil lextant, one in the library of the Escorial (Patrimonio Nacional inv. 10034530) and the other, badly damaged and heavily restored, in the museum of Cremona. They are both painted, or largely restored in the 17th century, and both come from a common model. There was another portrait of Janello made by Bernardino Campi in the 1550s, and brought to Spain by Torriani himself in 1556: Lamo, *Discorso di Alessandro Lamo*, 52. Another portrait of Janello belonged to Juan de Herrera. In the inventory written at the time of his death, there were only five portraits in his house: his own , that of his wife and those of Raymondo Llul, Michelangelo and Torriani. A token of Herrera's affection and consideration for the clockmaker, the other portrait is probably the one that senator Filodoni sent to the town of Cremona together with a model of the Toledo device after Janello's death. See for the images of the portraits: García-Diego, *Juanelo Turriano, Charles v's Clockmaker*; one can find coloured pictures of the portraits, medals and sculpture in the following books, where there are as well some pieces of news about this topic, though with some mistakes: Coll and Iotta, *Realismo y espiritualidad*; Campo y Francés, *Semblanza iconográfica de Juanelo Turriano*.

138   From a document published by Babelon, it appears that in 1582, Torriani and Jacopo da Trezzo were cooperating on some business, within the framework of their court service, concerning the cast of medals. Jacopo da Trezzo asked for the payment of a certain sum for two wooden models of coins and medals ordered by the king and now in Torriani's hands:"*de dar reali 215 per tanti spesi in Toledo in doii modelli de legno, l'uno per far moneda, l'altro per far li gitoni et medaglie quali sono in mano de M. Janello et fatte fare de ordine de Sua M-ta*": Jean Babelon, *Jacopo da Trezzo et la construction de l'Escurial: essai sur les arts à la cour de Philippe II, 1519- 1589* (Paris: E. de Boccard, 1922), 64.

139   "*VIRTVS NUN[QUAM] DEFICIT*"

marvellous planetary automaton to be inscribed with the above-mentioned sentence, and with a portrait of Janello, most probably the prototype for the medals. The Latin motto said: "Thou shall understand who am I, only if thou can accomplish a work equal to mine"[140] and on the later creation he made for Charles v, the *Crystalline*, as we have already seen, he had engraved the motto "to acknowledge myself as I flee away".[141] Though German clocks often had mottos engraved on their cases, usually related to the themes of Death and Temperance with which clocks were symbolically associated, we can consider the mottos on Janello's clocks and medal as a point of encounter between this tradition and the Cremonese's aspiration to social advancement.[142] Among nobles, the adoption of mottos went back to medieval times, and by the sixteenth century, mottos were compulsory for high-ranking figures. Charles v had the Herculean motto *PLUS ULTRA*, meaning "further beyond". His loyal vassal Ferrante Gonzaga choose a motto referring to the same heroic myth: *FINIUNT RENOVANTQ[UE] LABORES* (*a labour ends, a labour comes*). This record, together with others related to the entourage of Charles v and the governors of Milan, is taken from a manuscript of the Biblioteca Ambrosiana of Milan, which also houses the lectures of the *Accademia degli Inquieti*,[143] the institution that, since the late sixteenth century, had taken Torriani's Toledo water-machine as its own device, accompanied by the Three Musketeers-like motto: *LABOR OMNIBUS UNUS* (a single effort for all), referring to the fact that the hundreds of mechanical components of the giant Spanish hydraulic structure were moving in perfect concert thanks to one single motor.

Still in 1774, at Cremona, distant nephews of Janello would jealously and proudly keep an exemplar of the Cremonese's celebrative medals, as many

---

140   "*En otra parte donde esta su retrato de Ianelo, dize:* QVI.SIM.SCIES.SI.PAR.OPVS.FACERE. CONABERIS. *No podra tener en Castellano toda la lindeza que en el latin: mas todavia se puede traslatar assi. Entenderas quien soy, si acometieres hazer otra obra ygual desta".*: Morales, *Antiguedades de las ciudades de España,* fols. 91-94.

141   VT.ME.FUGIENTEM.AGNOSCAM

142   Leydi, "Un cremonese del Cinquecento," 146.

143   "... *col motto* FINIUNT RENOVANTQ[UE] LABORES *impresa di Don Ferrante Gonzaga, scoprendo l'honestà dell'eroiche fatiche già per lui superate, il rappresentava pronto di ... a nuovi per condursi con la scorta di tramontana tale nel porto della felicità dove voglio credere, ch'egli finalmente riposi. le navi del Granvela sbattute da contrarii venti con iscrizione* DURATE, *mostrando l'honestà nella resistenza ch'ei faceva a colpi di fortuna lo confortavano tacitamente a perseverar ... Il miglio della marchesa di Pescara con l'arma* SERVARI, ET SERVARE MEUM EST *addita l'honestà nelle attioni nelle quali s'havea ...*": Veneranda Biblioteca Ambrosiana, Ms. Y 93 sup. Ercole Cimilotti, "Lezioni Tenute Presso l'Accademia Degli Inquieti, in Casa di Muzio Sforza, Marchese Di Caravaggio" n.d., 38 v, Y 93 sup.

museums and private collections still do today.[144] Medals were considered to be powerful means of advertisement and celebration.[145] The tradition of medals was well established at the Gonzaga court: the lords of Mantua had been the greatest patrons of Pisanello, who reinvented this fashion, starting with a medal-portrait of the Greek-Roman Emperor John VIII Paleològos when he visited Italy in 1439. The medal, inspired by Roman coins displaying the heads of gods, emperors and members of their house, now represented a modern sitter to whom, with the honour of taking the place belonging to a Caesar, was transferred the nobility of the Ancients.[146] It seems that Janello was the first

144   Giambattista Zaist, *Notizie istoriche depittori, scultori, ed architetti cremonesi* (Cremona: Nella Stamperia di Pietro Ricchini, 1774), 154-55; Giuseppe Grasselli, *Abecedario biografico dei pittori, scultori ed architetti cremonesi* (Milano: Co' Torchj D'Omobono Manini, 1827), 249.

145   The use of medals in humanist culture had several functions: they were buried in the foundations of important buildings in magical and symbolic manner, they were worn on hats to represent a statement or a motto, or they served to remember someone or to promote him. This type of medium passed from the celebration of rulers and aristocrats to other respectable categories, such as scholars, painters, sculptors and even engineers and clockmakers: Janello was the first blacksmith to be celebrated in this fashion. An interesting work on medals made in these years, giving some information on the concepts related to this object, such as memory and celebration, in the sixteenth century, can be found in the following book: Sebastiano Erizzo, *Discorso di M. Sebastiano Erizzo sopra le medaglie de gli antichi Con la dichiaratione delle monete consulari & delle medaglie de gli imperadori romani. Nella qual si contiene vna piena & varia cognitione del' istoria di quei tempi.* (In Vinegia: Appresso Gio. Varisco & Paganino Paganini, 1559).

146   A painting in the style of Jean Bruegel the Elder and workshop, depicting "*Linder Gallery Interior*" (ca. 1622-1629) is a good example of the celebrative and didactic use of medals. The Milanese gallery was belonging to Peter Linder, a rich merchant of German origins. His gallery appears crowded with paintings, sculptures and astronomic books and instruments, and on the table at the centre of the painting are displayed a number of medals which portray his friend Mutio Oddi (1569-1639), together with Girolamo Cardano (1501-1576), Andrea Alciati (1492-1550), Albrecht Dürer (1471-1528), Michelangelo (1475-1564), and Donato Bramante (1444- 1514). This collection, put on show, has the function of demonstrating the patron's will to celebrate this group of characters, whose ingenuity and fields of competence reflect his own interest, as a clear program of intellectual self-promotion. Indeed, it has been demonstrated that medals were also pierced and worn as a representation of one's intellectual aspiration, or as tokens of affection. Alexander Marr, in his book on Mutio Oddi, gives us some more insights on Linder's story providing us with an interesting example of the use of medals in the Early Modern Period. Linder was Oddi's friend and enthusiastic pupil. Positioning Oddi's medal at the centre of the picture was a claim: Linder shared Oddi's scientific ideas and values. Besides that, positioning this medal among those that celebrated a carefully selected array of famous characters,

blacksmith and clockmaker to be celebrated with medals, something that had a fast influence on other similar craftsmen: for instance, the Sienese sculptor Pastorino in 1553, only one year after the imperial diploma and the most probably contemporary medal celebrating Janello, made a medal celebrating the clockmaker Francesco Parolaro Sforzani from Reggio, city of the Este duchy.[147] The famous clockmaker Cherubino Parolaro Sforzani, was instead celebrated with an epitaph on his tombstone composed by his son Teofilo and sculpted by Prospero Spani. It reads: "To an almost unique clockmaker, who in his own craft several times was enhanced with rewards and honours by the Supreme Pontiff and by Emperor Charles v".[148]

Another record of Janello is to be the first clockmaker to have been celebrated during his lifetime with paintings and with a marble bust (Figs 36, 53). Janello's portraits should not be considered as naturalistic representations of the sitter, but as cultural documents of a rhetorical kind.[149] Janello and his

---

mainly practitioners of *disegno* and mathematics, elevated a still living Mutio Oddi into a *pantheon* of illustrious men in the disciplines he admired the most. Close to the table one can see an allegorical representation of Arts and Virtue resting on *Disegno*. Marr convincingly suggests that the portrait of old and long-bearded *Disegno* reproduces the features of Oddi's fellow-countrymen Federico Barocci from Urbino, mathematical instruments maker. This interpretation together with the presence of Bramante's among the medals, seems to Marr a celebration of the mathematical Urbinate tradition in design. We have seen how scientific traditions were in Renaissance Italy also shaped around the cult of one's tiny *patria*. Mutio Oddi refers to the medal in a letter, giving us some more information on the collective process of these celebrating objects: "... *the* impresa *which has been made on the reverse of the medal made in Milan to honour me, which is a heaven ... with a motto taken from Dante ... ; this very* impresa, *was commended to me by Monsignor Paolo Aresi with one of his letters, and it seemed good*". Paolo Galluzzi et al., *Galileo: immagini dell'universo dall'antichità al telescopio*, Giunti arte mostre musei (Firenze: Giunti, 2009), 388; Luke Syson, "Holes and Loops: The Display and Collection of Medals in Renaissance Italy," *Journal of Design History Journal of Design History* 15, no. 4 (2002): 229-44;Marr, *Between Raphael and Galileo*, 180-209.

147   Morpurgo, *Dizionario degli orologiai italiani*, 1950, 177.

148   My translation from the original Latin:"*Horariis conficiendis prope singulari ut cuius ea in arte exc. a sum. Pontt. et a Carolo v Imp. praemiis et dignit. aucta non semel fuerit*": Campori, *Artisti degli Estensi*, 8-9. A special thanks to Simone Manfredini for his help in translating this epigraph.

149   Burke has dealt with the problem of portraits in the chapter "*The presentation of self in the Renaissance portrait*", affirming that these paintings are to be considered historical documents not because they reflect an image but because they can be considered works of a rhetoric representing a system of values in a social context: a portrait is a means of communication: "This is not to reaffirm the view, rejected earlier, that the painter is a mirror or camera. On the contrary he (or occasionally she, as in the case of Sofonisba

FIGURE 36   Portrait of Janello Torriani, *Spanish painter,* oil on canvas, *last quarter of the*
                     sixteenth century. The painting was highly damaged and badly restored at the
                     beginning of the twentieth century. See for instance the blunders in transform-
                     ing the year MDLXXXVII (1587) in MDCXXXVII (1637) of Horologioru[m] in
                     Philologioru[m]. COURTESY OF THE MUSEO ALA PONZONE, PINACOTECA
                     CIVICA, CREMONA.

Anguisciola), is a rhetorician. The point is that rules of rhetoric changed as the wider
culture changed, so that they too must be studied as a historical source. Historians not
only can but must use portraits and other paintings as part of their evidence, because

patrons and friends with these portraits tried to bestow upon the figure of Janello a dignity that was previously accorded only to holy characters, nobles, and urban and courtly élites. Pietro Aretino, in a letter written to his fellow countryman Leone Leoni, denounced this new trend: "It is the disgrace of our age that it tolerates the painted portraits even of tailors and butchers".[150] The fine portrait of a commonly alleged tailor (1565 ca.) painted by Giovan Battista Morone from Bergamo and now held in the National Gallery of London provides a very appropriate sample of this much detested fashion.

The wage Torriani received was a considerable amount of money, suitable for a gentleman. Even the fact that in the majority of administrative documents issued within the Spanish Court Torriani was addressed with his simple first name *Juanelo* might be considered as a sign of his status of "special character" within this social system, recognizable without mentioning his function. It is difficult to understand when Torriani consciously started to pursue a more elevated social status. Around the year 1530, when Janello Torriani was freshly married and his daughter was born, we find his deceased father-in-law addressed as *dominus*, meaning he was most likely a gentleman. Moreover, Janello's daughter's name, Barbara Medea, reflects a humanist trend of Italian urban centres. Classical names were a constant reminder of the refined knowledge of one's family. Some of the most powerful houses continued to use the Germanic and Christian-related names of their ancestors, but emerging families and decayed ones made a great use of this erudite practice. In the very same milieu of Torriani, the noble but decadent family of the Anguissola had among its members one Asdrubale, two Annibales, and one Sofonisba (the famous female painter 1535-1625), who had two sisters named Minerva and Europa. Another famous woman from Cremona, the humanist Partenia Gallarati (1526-1572), bore a remarkable name too.[151] Unlike her contemporaries Partenia and Sofonisba, Barbara Medea was not a member of the aristocracy, and despite her humanist name, she did not even know how to write. It was only after the death of her father that she learned to write her name in order to sign documents connected with the property she had inherited with a trem-

---

      images often communicate what is not put into words" : Peter Burke, *The Historical Anthropology of Early Modern Italy: Essays on Perception and Communication* (Cambridge [Cambridgeshire]; New York: Cambridge University Press, 1987), 150-67.

150   Luba Freedman, *Titian's Portraits through Aretino's Lens* (University Park, Pa.: Pennsylvania State University Press, 1995), 41.

151   Rachele Farina, *Dizionario biografico delle donne lombarde: 568-1968* (Milano: Baldini & Castoldi, 1995).

bling *ductus* that moves the reader.[152] We cannot say whether Torriani chose this name for his daughter in order to present himself as someone who spoke the same humanist language of the refined elite that promised to open the purse, or if he (or his wife Antonia) was a consumer of the literary or musical production on the Argonauts. What one can say is that Torriani was part of a cultural milieu dominated by humanist literary models, and he, his admirers, supporters and friends, used this language to reaffirm and strengthen Torriani's credit as a learned craftsman.

---

152 Cervera Vera, *Documentos biográficos*: see the documents about Torriani's inheritance after his death. The myth of Medea is connected with the popular legend of Jason and the Argonauts and it also provided the name for one of Euripides' tragedies. In 1503 Aldo Manuzio published Eurpides' *Medea*. The name had in fact been in use already in the previous century: for instance the daughter of the famous condottiere Bartolomeo Colleoni was named Medea († 1470). In fact since the time of the Paduan early humanist Lovato deLovati (ca. 1240-1309), Seneca's tragedy also titled *Medea* was circulating. Moreover other ancient authors such as Ovid had elaborated on the myth (*Metamorphosis*). Benjamin G. Kohl, "Lovati, Lovato," *Dizionario biografico degli Italiani* (Roma: Istituto della Enciclopedia italiana, 2007).

CHAPTER 5

Networks and Technology in Habsburg Europe

From Commoner to Courtier

This chapter aims to highlight the relation between the concepts of patronage and brokerage and the successful career of a Renaissance inventor and the making of scientific authority in context. Without stepping into the spotlight of the Habsburgs, Janello Torriani would be just a name of a few documents, nothing more than an obscure clockmaker from the province, and the Cremonese was well aware of that.[1] Habsburgs' patronage gave him the chance to challenge his own skills with tasks that no one was able to complete. On the stage of the most important court of Christendom, Janello, like a *deus ex machina*, successfully accomplished them, winning public praise. This admiration was expressed in several media that, together with the rich administrative documentation from a sophisticated court like the Spanish one, produced an imposing historical source. The court gives a special visibility to those actors who go on its stage. Janello Torriani succeeded in this venture thanks to his talent, with a range of impressive design abilities, successful mechanical executions, and constant commitment to his tasks, together with a charismatic character and distinctive physical presence. These factors made of Janello a courtly wonder and a literary figure. As just seen in the previous chapter, once he entered the court-system, Janello Torriani started to adopt its language, and was celebrated with written works, medals, paintings and sculpture. After Charles v passed away, his son Philip, the King of Spain, employed him at his court with a doubled wage. This position as clockmaker royal later led to him receiving the commission for an enterprise that would win Janello immortal fame and, alas, economic disaster.

At the time of Janello's employment at the service of the governors of Milan, the Cremonese appears and almost magically disappears from Girolamo Cardano's works. The great number of reprints of Cardano's books has created some confusion on the reasons for this phenomenon. I think we can read this

---

1 We have seen what Janello had told Garibay about the Emperor making him immortal. Moreover, in his late years, in the letters directed to King Philip II, Janello stressed his proud courtly identity of servant (*criado*) that put him in a direct personal relation with the monarch. To be a *criado de su* Majestad, i.e. a servant of His Majesty, was for Janello the supreme accomplishment. Cervera Vera, *Documentos biográficos*, doc. 130.

© KONINKLIJKE BRILL NV, LEIDEN, 2017 | DOI 10.1163/9789004320918_007

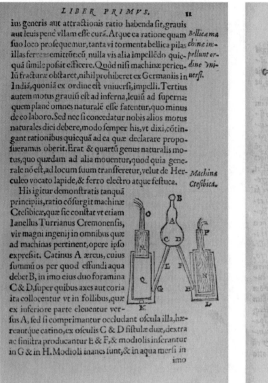

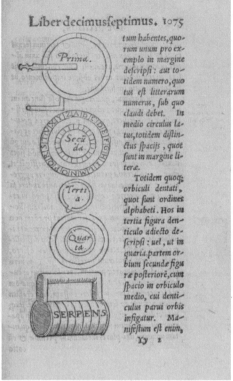

FIGURE 37   *Girolamo Cardano,* Machina Ctesibica made by Janello Torriani, *from* De
subtilitate, *book xxi, Lugduni, 1550 (left), and* Seven-letter combination
padlock by Janello Torriani, *from* De subtilitate, *book xx, Basel, 1554 (right).*

shift as a function of the dedications of the different editions of two of his
works to a group of patrons that Cardano addressed in different times.[2] For
instance, the first edition of *De Libris Propriis* (1544) was dedicated to the
Cremonese Francesco Sfondrati (1493-1550), certainly a crucial character in
Cardano's fortune, and perhaps as important in Janello Torriani's success. In
this book, printed in Nuremberg in 1544, Janello is said to have restored
Zelandinus' planetary instrument. In a following manuscript edition from
1550, the very same year Sfondrati died, Cardano added a new detail about
Janello's important role: "We had written about Zelandinus' machine, but
Janello of Cremona set me free from this labour. Indeed, having constructed
that divine – so to say – instrument for our Caesar, he described its structure

---

2   Elisa Andretta, "Dedicare libri di medicina: Medici e potenti nella Roma del xvi secolo," in
Rome et la science moderne: entre Renaissance et Lumières, ed. Antonella Romano, Collection
de l'École française de Rome, 403 ([Rome]: École française de Rome, 2008), 207-55.

with very careful writings".[3] Cardano then dedicated the next printed edition (1557) to Niccoló Secco, captain of Justice of Milan. Despite still mentioning to have started to write about Zelandinus' instrument, in this edition Cardano completely omitted the name of Janello.[4] What, at first glimpse, seems to be just a minor change, assumes a totally different meaning if observed in a broader picture: in the three editions of Cardano's bestseller *De Subtilitate* (1550-1554-1560) this process of obliteration is even more evident: the first two editions were dedicated to Ferrante Gonzaga, Governor of Milan, who died in November 1557 after suffering defeat at the Battle of Saint Quentin. In the first edition Cardano attributes to "Janello Torriani from Cremona, a man of great ingenuity in anything that concerns machines," a model of a Ctesibian pump.[5] In the book XVII, dealing with crafts, Janello is considered the inventor of a mechanism for clocks, and of the gimbal, employed by the Cremonese to sustain a litter (or sedan-chair) for gouty Charles V and some oil-lamps.

In the second edition of 1554, still dedicated to Governor Ferrante Gonzaga, one can observe the references to Janello doubling in number: "Janello Torriani of Cremona, previously mentioned, a man of sharp ingenuity"[6] is mentioned four times in connection to inventions and technological innovations. Cardano added now the description of the wonderful planetary clock made for the Emperor and a combination lock. In the biennium 1557-1558, Ferrante Gonzaga and Emperor Charles V passed away. The 1560's new edition of *De Subtilitate* was now dedicated to the new governor of Milan: Consalvo Fernandez de Cordoba, Duke of Sessa.[7] Any reference to Janello has disappeared except for the secondary passage regarding the invention of a combination lock, where the name of Janello appears without any laudatory adjective at all![8] However,

3   "*Scripseramus de Zelandinis, sed Ianellus Cremonensis nos ab hoc labore liberavit. Cum enim instrumentum illud divinum, ut ita dicam, Cesari nostro composuisset, illius structura, diligentissime litteris mandavit: itaque eum ...*": Cardano, *De libris propriis* [*2004*], 9, 10, and 148.

4   Girolamo Cardano, *Hieronymi Cardani Mediolanesis Medici Liber de libris propriis* (Lugduni: Apud Gulielmum Rovillium, 1557), 77.

5   "*Ianellus Turrianus Cremonensis vir magni ingenij in omnibus quae ad machinas pertinent*": Girolamo Cardano, *De subtilitate*, ed. Elio Nenci, vol. I, Libri I-VII (Milano: F. Angeli, 2004), 67, footnote (a): Editiones 1550 and 1554.

6   Cardano, *De subtitulate* (*1554*), 1554, 452-53: "*Ianellus Turrianus Cremonensis, cuius etiam supra meminimus, vir acris ingenii multa talia aut excogitavit, aut ab aliis excogitata in melius traduxit*".

7   Twice in charge as governor of Milan: 1558-1560 and 1563-1564.

8   Geronimo Cardano, *De subtilitate libri XXI: ab authore plusquam mille locis illustrati nonnullis etiam cum additionibus ; addita insuper Apologia adversus calumniatorem, qua vis horum librorum aperitur.* (Basileae: Ex Officina Petrina, 1560), 1074: "*Unius tamen exemplum subijciam, quam Ianellus construxit*".

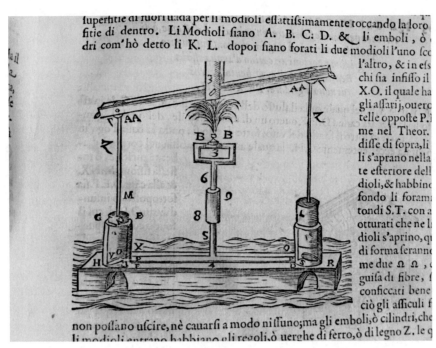

the descriptions of the inventions once attributed to him are preserved in the
text. The *damnatio memoriae* was so sudden and rough that the sentences previously praising Torriani now look mutilated, presenting the Latin verb still
conjugated in the third singular person, but showing no subject. In another
sentence, where the name of Janello appeared in the first edition, is now substituted *"artifex noster"*: a craftsman of ours.[9] Why this eradication? Was there
a motive of animosity between Cardano and Torriani? Was the powerful
patronage of Francesco Sfondrati and Ferrante Gonzaga the reason why
Cardano had to (or decided to) quote Janello in the first editions of his books?
Most likely it was.

     Despite Cardano's attempt of *damnatio memoriae*, upon Janello's entrance
at court, his personality became relevant in chronicles and later even in anecdotal literature On the one hand, these accounts highlighted the magnanimity
of the prince, in the figure of Charles v; on the other, those stories provided
readers with an exemplum of the ingenious inventor and his wit. These anecdotes are especially relevant because they describe the relation between power

---

9   Ibid., 1031.

and scientific knowledge in a crucial moment for the evolution of technology: the wit of Janello reaches often the grade of insolence, but it is kindly accepted from the monarch. The reason for this tolerance must be found in the "truth" spoken by the clockmaker: the respect for his talent in applied mathematics is the key to understanding this unusual relation. It seems that the crowned patrons recognised through Torriani's creations his knowledge (*scientia*) and his wisdom (*sapientia*), which expressed themseves through Torriani's wit. Torriani's savoir-fare was the demonstration of a divine genius, which six-teenth century society acknowledged. Spanish elites especially believed in that, and the Crown invested in the teaching of mathematics, stressing its necessity.[10] He who can control nature or imitate it, as Torrini did in his hydrau-lic device and in his planetary clocks, can do that because he can understand, through mathematics, Creation. Therefore, mathematical knowledge gives wisdom. As we are soon going to see, Ambrosio de Morales' account of Janello, praising Saint Augustine's idea that "the one who perfectly knows everything related to numbers, will have access to marvellous things that will be like mir-acles" seems to move in this direction.

We know from several influential Spanish court employees such as Juan de Herrera, Ambrosio de Morales, Esteban de Garibay and others, that Janello was very much respected. Esteban de Garibay, when he attended Janello's funeral, sketched a short but vivid portrait of the man:

> He was tall and massively built, not talkative, hardworking and of great generosity. He had something fierce in his expression and he was some-how rough in speech: he never spoke Spanish properly, and because of his old age, the lack of teeth created big problems for the use of Italian too.[11]

---

10 See the fist chapter in this book on the idea of the necessity of mathematics in the Span-ish *Siglo de Oro* and Vicente Maroto and Esteban Piñeiro, *Aspectos de la ciencia aplicada en la España del Siglo de Oro.*

11 "*Solo fue de mi voto Joanelo Turriano, natural de Lombardia, el que habia hecho la admira-bile fabrica de la subida del agua del Tajo al alcázar de Toledo. Este insigne varon, antes de ver acabada esta navigacion murió en la misma ciudad en 13 de junio de 1585 á los ochenta y cinco años de sue dad, poco mas ó menos, y fue enterrado en la iglesia del monasterio del Cármen de ella en la capilla de nuestra Señora del Soterraño, siendo yo presente, no con el debido acompañamento que merecia quien fue príncipe muy conocido en todas las cosas en que puso su clarísimo ingenioy manos, Fue alto y abultado de cuerpo, de poca conversacion y mucho estudio, y de gran libertad en sus cosas: el gesto algo feroz, y la habla algo abultada, y jamas hebló bien en la española; y la falta de los dientes por la vejez le era aun para la suya italiana de grave impedimento. Túviole en mucho el dicho católico Rey D. Felipe, y le regaló y*

Bishop Vida, talking about his enormous hands, already noticed the huge physical structure, his monstrous look (covered with soot), as if he was one of the
Cyclops-helpers of Vulcan.[12] Besides his amazing talent, this huge build, his
crusty and silent attitude, only interrupted by caustic and witty sentences,
enhanced the charisma of the clockmaker, and likely made the few things that
he said in front of the Emperor the subject of some interest. As we have seen in
the Mantuan ambassador's letter from Brussels, and also in the anecdotes
reported by Mutio Oddi and Capilupi, this was happening since the time an
apparently over-self-confident Janello was introduced to the Emperor. Thanks
to this combination of features, roughly on the model provided by Diogenes,
the clockmaker become a moral character assuming the mask of the witty and
laconic old wise man with no fear for expressing his opinions in front of powerful people thanks to the credit he received from the admiration provoked by
his creations.

This role of a new Diogenes is well presented in a book by Cardinal Federico
Borromeo, the founder of the Ambrosiana Library. The Cardinal, cousin to
Carlo Borromeo, who had the plan to build a colossal planetary clock in the
Ambrosiana Library in Milan in the 1620s,[13] in a book on the morality of the
princes and their courts, reports two short stories about Janello. The first is
about honesty and directness. Cardinal Federico Borromeo argues that many
people say that the thing the prince most lacks is to be told the truth, but that
those at court do not dare speak honestly in a proper moment unlike the
famous artisan Janello, who:

> Among the fine things he was used to say there was the following: "Three
> very beautiful and very good things, and very much valuable and beloved
> by each man are denied to the princes: to be told the truth, to behold
> dawn, and to feel hungry".[14]

---

honoró siempre, como quien sabia bien lo que él merecia, imitando lo que en esto habia hecio
con él el preclarísimo Emperador D. Cárlos su padre": Llaguno y Amirola, *Noticias de los
arquitectos y arquitectura de España*, II:250.

12   Leydi, "Un cremonese del Cinquecento," 137.

13   Marr, *Between Raphael and Galileo*, 160-166.

14   "... *di nuna cosa hanno maggior carestìa i Principi, che di persona, la qual dica loro il vero.
Ma sarebbe pur tollerabile usanza questa, e men degna di esser ripresa dalle savie persone,
se essi, in pronunziando tali detti, si valessero di qualche bella ed opportuna occasione, come
pur'avvenne già ad un famoso artefice lombardo, chiamato Maestro Giannelli, e molto ne'
suoi dì celebrato per l'esquisito ingegno nel fabbricar diversi artificiosi ordigni e lavori. Egli,
tra le altre belle cose, che soleva dire, haveva in costume di dir questa, che di tre cose bellissime e bonissime, e degne di essere apprezzate molto, ed amate da ogni huomo, erano privi i*

The last part of this caustic saying curiously recalls what Tommaso Garzoni, in his encyclopaedic work on professions, defined to be "the best clock", that is the belly of the peasant: "Without any mistake, this clock tells admirably to the workers the right times for breakfast, lunch and dinner!"[15]

The second story highlights the obstinacy and cheek in Torriani's character: Cardinal Federico Borromeo tells that one day Janello stubbornly refused to do something required by the Emperor; therefore Charles gently reproached him asking: "What does he deserve, who refuses to obey the Emperor?" implying that one should always obey his orders, "to whom the master promptly and without losing his spirit answered back: "to be paid and to be told goodbye!"[16] Janello's insolent answer well represents the condition of the court-artisan, who had often to face monstrous delays in getting his wage paid, a sad and frustrating condition that left Janello in troubles for the last twenty years of his life. This was the price a court-employee had probably to pay in order to join the protected and prestigious household of the ruler.

Beside the printed work on the "grace of the princes", Cardinal Federico Borromeo memorialised Janello in a manuscript, still kept in the Ambrosiana Library. It was probably part of the preparation material for the printed book: in this manuscript one can find a third unpublished story. In this anecdote the stage is Toledo and the story probably refers to the year 1569, when King Philip visited for the first time the finished water-lifting device built by Torriani. On this occasion, a knight of the royal entourage criticised Torriani's machine because, in his eyes, it brought little water to the *Alcázar*. Janello, feeling attacked by this comment, could not bite his tongue, and stepping forward said: "I am not like Moses, who could make abundant fountains spring out from the stone". Everybody present well understood and appreciated Janello's

---

Principi; cioè di chi dicesse loro la verità, del riguardare l'aurora, e della fame": Federico Borromeo, *Il libro intitolato La gratia de' principi di Federico Borromeo ...* (In Milano: [s.n.], 1632), 167-68.

15   Garzoni, *Piazza universale di tutte le professioni del mondo*, 627.

16   *"Tale, e tanto dee poi essere l'amore della virtù, quando ella in eccellente grado si trova in alcuno, che non tutte le sue colpe si hanno da ricevere per gravi offese; ma gli si deono di buona voglia, e liberalmente perdonare. Né si reputa per falsità e menzogna quello, che si racconta di maestro Gianello di sopra mentovato, e cotanto caro all'Imperadore Carlo Quinto. Ricusò un giorno il buon Gianello, per certa ostinazione, di far una cosa, che l'Imperadore voleva, che egli facesse nella sua arte. Laonde l'Imperadore con piacevol modo gli disse. E che meriterebbe uno, il qual non volesse obbedire all'Imperadore? Cui il maestro prontamente, e senza perdersi d'animo, rispose. Pagarlo mandarlo condio".*: Borromeo, *Gratia de' principi*, 178.

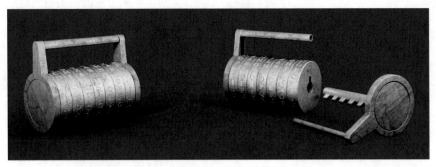

FIGURE 39   *Andrea Bossola, Reconstruction of the seven-letter combination padlock (SER-*
              PENS) of Janello Torriani.

wit, the writer says, because the criticising knight had a quarter of Jewish blood.[17]

In addition to the previously mentioned letter of the Mantuan ambassador from Brussels, the image of a stubborn and irreverent man also emerges from some other epistolary material. In a letter dated 16th of May 1556, sent by Ferrante Gonzaga to the ambassador of Mantua Annibale Litolfi, the Count of Guastalla wrote:

> For your own life recall once again to M.ro Janello to answer to the letter that I sent to him, and in case he does not want to do so, at least make him say something to you about what he has decided to do about that clock of mine.[18]

---

17   *"Perché habiamo ragionato del Gianello narreremo alcune cose di lui, dimostranti prontezza*
      d'animo et ingegno. Egli fece quella tanto nominata Machina, che portava l'acqua insino alla
      somità della Rocca di Toledo et essendo Filippo Secondo andato a vedere questa nuova
      meraviglia, et laudandola, un Sig.e che era officiale suo, e favorito in quell tempo disse che
      era poca acqua p[er] il bisogno. Gianello sentendosi punto, incontinente si trasse innanzi e
      disse che egli non era come Moise, che poteva far scaturire le fontane abbundevoli, e cavarle
      dale pietre. Fu inteso subito il detto e fu stimato acutissimo, perché questo Cav.re che coleva
      più acqua teneva un quarto di giudeo, ed il suo sangue non era netto, come si dice volgar-
      mente": in De Fabricam olim typis orbium caelestium, Veneranda Biblioteca Ambrosiana,
      manuscript. G.9.Inf.4, fol. 56 and following.

18   *"Per vita vostra tornate a ricordare a M.ro Gianello la risposta de la letter ache io gl iscrissi,*
      et quando non voglia rispondere, procurate almeno, che dica a voi quell che egli si risolve di
      fare intorno a quell mio horologio": Giuseppe Campori, *Lettere artistiche inedite* (Modena:
      Erede Soliani, 1866), 49.

Sculptor Leone Leoni, like Janello working between Milan and the imperial court, gave several harsh judgements on his colleague: Leoni's descriptions of Torriani go from ungrateful to obstinate; he was described by the sculptor as expressing himself by grunting. Leoni was famous for his choleric nature and his judgements are perhaps too harsh; however, even if distorted by a bad temper, these images are somehow fitting with the representation other sources give of Janello Torriani's temper. On other occasions Leone Leoni seemed to be on good terms with Torriani. In a letter to Ferrante Gonzaga, for instance, he wrote that Janello, perhaps with one of his sharp remarks, had made him laugh, and in another missive written on the 2nd of April in the year 1583 from Milan to the old common friend Jacopo Nizzola da Trezzo, he even quoted a witty saying of the clockmaker.[19]

A couple of years before, Janello had just finished his treatise and mathematical instruments for the reform of the Calendar. On this occasion he clearly demonstrated he was probably not an easy person to deal with if one did not have his trust: the Apostolic Nuncio could not convince Janello to deliver to him the material he had prepared for the reform of the Calendar, because the 80 years-old clockmaker would deliver these items to none other than the King himself. We shall come back to this fact. Also on another occasion the Apostolic Nuncio Sega underlined a certain anarchic tendency of the clockmaker, when he wrote to Rome that:

> Here is Janello with his treatise, which, in my opinion, will deserve to be carefully studied, because with the help of an instrument he made, it is possible to illustrate the reduction [of the calendar] and its reasons very easily, that no one could desire a better method. And because he is a man who does not do things until he feels like, it is necessary to be patient. And because this is such an important business that will have an influence for the time to be, it seems to me that there are many people who hope he will not quit to work on the instrument. I will with dexterity follow from a short distance this erudite brain, which no king, queen nor tower can compete with.[20]

---

19   Leydi, "Un cremonese del Cinquecento," n. 4, 137; Babelon, *Jacopo da Trezzo et la construction de l'Escurial*, 319.

20   "*Aquí se encuentra Juanelo con su discurso que, a mi juicio, será digno de ser estudiado con detenimiento, porque, con un instrumento que él ha construido, se va demostrando fácilmente que no se puede desear mejor modo para esta reducción y sus razones. Y, como es un hombre que no hace las cosas sino cuando le viene en gana, es necesario que se tenga un poco de paciencia. Y, como se trata de un negocio de tanta importancia y que influirá en el futuro, me parece que hay muchos que esperan que no deje de trabajar en este instrumento. Yo seguiré de cerca, con destreza que conviene, a este erudito cerebro con el cual no puede ni*"

The name Torriani coming from the Italian word for tower, i.e., "*torre*", inspired Sega to make a playful analogy between Janello's cleverness and the challenging game of chess, in which the rook is called in Italian *torre*. As mentioned, Torriani was probably a skilled chess player. It is astonishing that also Ferrante Gonzaga, a quarter of a century before, referencing a certain freedom in action of Torriani in his profession, had written that:

> He is the man that needs a lot of reminders, or even spurs in order to have him to do things he does not want to do.[21]

This certain anarchism should be probably seen in light of the treacherous and highly competitive framework of the court, where Torriani spent most of his time absorbed in numerous projects on behalf of the Crown. Within the numerous tasks absorbing him at court, Janello probably gave priority to the ones he considered most interesting and more suitable to his fame as the new Archimedes.

The clockmaker felt honoured, even proud of his position at court. If we detect an unusual level of freedom in his relation to authority it is probably due to the trust and admiration, which issued from what Herrera had called *maestria* or craftsmanship. When put on a task, Janello took his duty in the service of the Crown very seriously, almost obsessively: the administrative material coming from the time when Janello was directing the construction of the water devices in Toledo shows how he could not tolerate inefficiency or delay. The royal officials were often stressed about situations that might provoke Torriani's discontent, and in Toledo there were many occasions for it, as we shall see. The royal bureaucratic machine was slow, and the King's officials of the *Alcázar* of Toledo, fearing Torriani's reaction, got restless: once they wrote to the royal secretary Gatzelu that Janello:

> Is carrying on so fast the construction of the device that there is an immediate need of the money scheduled for the year; otherwise Janello will be so sad not to get them that we [officials of the *Alcázar*] will have to appeal to all our diligence in order not to get him very upset – *que él no se disguste.*[22]

Rey, ni Reina, ni Torre (del Ajedrez), y espero que lo tendremos en breve": Ángel Fernández Collado, *Gregorio XIII y Felipe II en la Nunciatura de Felipe Sega (1577-1581): aspectos políticos, jurisdiccional y de reforma* (Toledo: Estudio Teológico de San Ildefonso, Seminario Conciliar, 1991), 245.

21   Giuseppe Campori, *Lettere artistiche inedite* (Modena: Erede Soliani, 1866).
22   30th of May 1576, Cervera Vera, *Documentos biográficos*, doc. 74.

Torriani's commitment to his tasks and their successful outcome were the basis of his self-confidence. This certitude backed his belief that he was entitled to write directly to the Emperor, to the King or even to the Pope, when he was not happy about some issues involving the payment for his work. The reason was always a request of money for some work left unpaid. Similar to the tone of the surrealistic debate on his wage he had with Charles v witnessed by the Mantuan ambassador in Brussels, in some of the letters addressed to Philip ii, Janello, who usually manifested a humble devotion, when he was not satisfied because he felt he was suffering an injustice, could be impudent or even impertinent: for instance, he once reminded the King that the latter had given his word, and was therefore duty-bound to keep it:

> Our Majesty needs to be informed of the great necessity in which I find myself for being employed for His Majesty's order and for his royal mandate in the project of the water-lifting device of Toledo ... Given that Your Majesty was all the cause and reason that pushed me into this business, I thought always that ... justice would have guaranteed that I would be given all that was promised to me, and with this firm hope, doctor Velasco reassuring me with letters or words in your royal name every day, and because I have been showing my good disposition in the service and obedience of Your Majesty, and because I trusted these words, I did not constrain myself from continuing to spend all of my wealth and even that of my friends ... And still, at the end of six years that the said water runs and that I have complied [to the contract] how also Our Majesty has seen and confirmed with his own royal mouth all of the times that he has visited the Device and that no one can claim the opposite, instead of paying me what I deserve according to justice, including the rent, one proposed to me instead to sign new agreements outside the contract ...[23]

Similarly, after he received delayed payment from Philip ii, he underlined that the King had paid "how it was supposed to be" – "*como era justo*" – and he even dared to define what were the moral duties for the monarch in the controversy around the first Toledo Device: "His Majesty and the City of Toledo are obliged to repair this scorn of justice and conscience".[24]

Torriani's consciousness of having accomplished outstanding works is striking: of the hydraulic machines of Toledo, he thought he had made "such an exceptional thing", in the Spanish original reads "*haver ... hecho obra tan*

---

23   Ibid. Toledo, doc 49, 15th of February 1575.
24   Ibid. doc. 44, 1575?

insiñe",[25] and the clockmaker, as just seen, reminded the King that Philip II himself visited the device several times and on these occasions he expressed satisfaction "*de su real bocca*", meaning "with his own royal mouth".[26] The prince's satisfaction was the highest confirmation of his technological success. However, there are clues that testify to Janello's strong self-confidence even before accomplishing his first great creation. As appears in the above-mentioned letter by Mutio Oddi, when the Emperor had asked the Marquis of Pescara if the clockmaker was a fool, promising to do something so ambitious that everyone else had declined to do, and the Marquis, feeling under pressure and fearing Janello's vainglory could damage him, threatened the Cremonese for his life, the latter did not lose his temper, and according to Oddi's account he securely stated that if he had said so, he was going to do it.

This planetary clock named by Vida "Microcosm", also called "Cesar's Sky," or, in the Spanish inventories simply the "Big Clock of the Emperor", was the source for Janello's fortune. His second planetary clock, the Crystalline, strengthened it. Ambrosio de Morales, discussing with Torriani these complex machines, could not help but feel awe and ask him how he had accomplished similar enterprises:

> And he answered in this way: "This is why: have you seen all I made in the clocks? Well, I have seen men who know very well and even more astronomy and geometry than me. But so far, I have not met anybody who knows as much arithmetic as I do". Thus, I told him that I was not surprised by what Saint Augustine used to say: that the one who perfectly knows everything related to numbers, will have access to marvellous things that will be like miracles. He was happy to hear that, and he considered that the Saint had to know a lot of arithmetic to reach such a level of knowledge.[27]

Janello Torriani's multi-layered knowledge typical of a *scientia media*, together with a natural talent interpreted as divinely inspired and with a picturesque appearance, made him a desirable beast for the seraglio of the court. For our

25   Ibid. doc. 44, 1575?

26   Ibid. Toledo, doc 49, 15th of February 1575.

27   "... *y respondendiome desta manera. Assi es. Porque veis todo lo que he hecho en los relojes? Pues hombres he visto que saben tanta y mas astronomia y geometria que no yo. Mas hasta agora no he visto quien sepa tanta arithmetica como yo. Entonces le dixe, que ya no me espantava lo que dezia santo Augustin; que quien supiesse perfectamente todo lo que se puede saber en los numeros, haria cosas maravillosas, y que fuessen como milagros. Holgose de oyrlo, y creyo que el Santo supo mucho de arithmetica, pues llego a tal conocimiento*": Morales, *Antiguedades de las ciudades de España*, fols. 91-94.

FIGURE 40   *Andrea Bossola, Reconstruction of the gimbal lamp, attributed to*
            Janello by Cardano. This reconstruction was inspired by a lamp
            drawn in one of the illustrated catalogues of the Settala collection
            in Milan, where was also held the armillary sphere by Janello
            Torriani, as of now the only surviving object signed by the
            Cremonese master.

own sake, thanks to his employment at court, there are so many traces left
behind to reconstruct the trajectory of the most successful machines-builder
of the Renaissance. But how could Janello jump from Cremona to the
Governor's court, and from Milan to Charles v's *entourage*?

The Artisan's Apotheosis

Access to Power

Could a craftsman from the "periphery" reach the favour of the Emperor by
chance, or by his own force? It is clear that Janello Torriani was, a prodigy,
as Cardano wrote, "a man of great ingenuity in anything which concerns
machines".[28] Nevertheless, the system he was living in was a strictly hierarchical one with its own severe etiquette, and according to which, court-employees

28   "*Ianellus Turrianus Cremonensis, vir magni ingenii in omnibus quae ad machinas pertinet*":
      Cardano, *De subtitulate* (2004), I, Libri I-VII:67, footnote (a).

were privileged individuals. They could reach a position in such a system only thanks to the acceptance of members of the very court, after prior presentation by somebody connected to the very sphere of power. Brokers and recommendations provided this protection, and this took place on the local level, where those nobles who were involved with the court had their ancestral roots, their resources and their clients.[29]

The grandiose patronage the Habsburg rulers provided Janello Torriani with has so far attracted the attention of scholarship, neglecting the necessary link that connected in the first place the clockmaker with the ivory tower of the court. Making a census of Torriani's effective or potential patrons during his long career enables us to sketch a picture resembling a consistent structure of power. What emerges is a network of intermingled Ghibelline families well rooted in the territory around Torriani's hometown: Cremona. Torriani's *patria* was placed between several centres of political gravity: among them especially emerged Milan and Gonzaga's Mantua exercised a force of attraction on Cremona, a city with no court, and its diocese. The courts present in those two cities often attracted the most accomplished virtuosi produced by the refined urban culture developed in the many centres of the plain around them, the largest of these being Cremona. The most renowned artists and musicians from this city and from its dioceses, moved first towards Milan or Mantua (with its satellites: Sabbioneta, Guastalla, etc.). In this process of attraction scholars like Jacobus Cremonensis and Platina, painters such as the Campis and Sofonisba Anguissola, the musician and composer Claudio Monteverdi, the painter Michelangelo Merisi da Caravaggio, together with Janello, are the most famous examples. Milan and Mantua drew also the excellence of production, like the musical instruments of the Amati workshop.

The first step on Torriani's path towards the imperial court was his migration from the political periphery to the capital of the duchy. In the case of a craftsman like him, trust had to be transmitted first through the recognition by urban institutions, such as crafts-guilds and the *massari* of the Cathedral, through remarkable skills, and finally through self-promotion. There are not known documents that refer to the patrons who introduced Torriani to the court of the governors of Milan. However, a contemporary parallel experience provides us with a useful matter of comparison, which illustrates the model for similar social ascents. In his autobiography, Girolamo Cardano (1501-ca.1575) recalls how he gained protection from his patrons:

---

29   See Cristiano Zanetti, *Juanelo Turriano, de Cremona a la corte: formación y red social de un ingenio del Renacimiento*, Colección Juanelo Turriano de historia de la ingenieria (Madrid: Fundación Juanelo Turriano, 2015).

FIGURE 41
*Agostino Carracci
(engraving) after a
design by Antonio
Campi,* Cardinal
Francesco Sfondrati,
Cremona fedelissima,
Cremona, 1585.

Thanks to the benevolence of apothecary Donato Lanza, I became friend
to Senator Francesco Sfondrati from Cremona, who was afterwards
elected cardinal: Sfondrati's friendship provided me with Giambattista
Speciario's, who was criminal prefect and a man of singular virtue and
knowledge. He was from Cremona too. Thanks to him Alfonso Davalo,
governor of the province and general captain of the imperial army, came
to know me.[30]

---

30   My translation: "*per Donatum etiam Lanzam Pharmacopolam, amicitiae Francisci, Sfon-
     drati Senatoris Cremonensis, qui postmodum Cardinalis evasit: per hunc autem Prefecto
     criminum, & ipsi Cremonensi Jo. Baptistae Speciario, viro erudito, & virtutis singularis: per
     quem innotui etiam Alfonso Davalo Provinciae Proconsuli, & militiae Caesarae Duci. Per
     Sfondratum quoque munus profitendi Papiae medicinam adeptus sum*". Girolamo Cardano,
     Hieronymi Cardani Mediolanensis, De propria vita liber ..., ed. Gabriel Naudé (Amste-
     laedami: Apud Joannem Ravesteinium, 1654), 50; see also: Nancy G. Siraisi, *The Clock and
     the Mirror: Girolamo Cardano and Renaissance Medicine* (Princeton, NJ: Princeton Univer-
     sity Press, 1997).

Cardano was able to reach the top of the hierarchy thanks to four factors: he was a graduated physician, he was successful in his diagnoses and cures, he was ambitious and he was in the right place at the right moment. Indeed, Cardano was effective in curing the pharmacist Donato Lanza, who introduced him to Senator Francesco Sfondrati, whose son fell ill in 1533 or 1534. Cardano was able to cure the boy and Senator Sfondrati introduced the physician to the governor Alfonso d'Avalos (end of 1536),[31] and promoted Cardano in the Senate, in the Milanese *collegium* (guild) of physicians, where Cardano had some troubles due to his condition of having been born an illegitimate – though later recognised – son, and in Pavia. In 1536 Cardano had among his patients the powerful Borromeo family, of whom Carlo (future archbishop, *cardinal nepote*[32] and eventually saint) later orchestrated Cardano's employment at the University of Bologna.[33]

Francesco Sfondrati and Giovanni Battista Speciano were powerful gentlemen from Cremona, members of the Senate of the Duchy.[34] As with Cardano,

31   Girolamo Cardano, *Hieronymi Cardani,... De Sapientia libri quinque. Ejusdem de consolatione libri tres... Ejusdem de libris propriis liber unus...* (Norimbergae: apud J. Petreium, 1544), 427.

32   I.e. "cardinal-nephew [of the Pope]", a custumary role that consisted in being a kind of prime minister of the Patrimonium Sancti Petri.

33   Cardano, *Hieronymi Cardani Mediolanensis, De propria vita liber ...*, 50-51, 105 and 133-134.

34   The Senate was the highest collegial institution of the State and Luis XII created it when he became Duke of Milan. Nevertheless, we refer to the ducal counsellors before this time as senators. The governors of the towns of the Duchy were chosen exclusively from its members. The most important offices of the Milanese Senate were: *Presidente del Senato, Gran Cancelliere, Presidente del Magistrato delle entrate ordinarie. Capitán de justicia* and senator Giovanni Battista Speciano (†1545), cardinal and senator Francesco Sfondrati, and Giovan Battista Schizzi who, from 1546, was elected senator after Speciano's death, were all paramount representatives of Cremona; Francesco Sfondrati had also been the governor of Pavia in the 1520s at the time of Duke Francesco II Sforza, when, perhaps, one decided to replace the rotten Astrarium with a new one. There are also other influential nobles from Torriani's city which aspired to a place in the Senate of Milan such as a certain Paolo Ala. *La Corte de Carlos V: Corte y Gobierno*, segunda parte, vol III, dir. José Martínez Millán, ed., *La corte de Carlos V* (Madrid: Sociedad Estatal para la Conmemoración de los Centenarios de Felipe II y Carlos V, 2000), 392; rich bankers such as the Affaitati might have had a role of promoters too. Waiting for new documents, the possibilities remain numerous. For the topographic reference of Sfondrati's house see the map by Antonio Campi (1583). For what concerns nobles from Cremona who served the Emperor at war: M. Rizzo, "Ottima gente da guerra: Cremonesi al servizio della strategia imperiale," in *Storia di Cremona : l'età degli Asburgo di Spagna (1535-1707)*, ed. Giorgio Politi (Bergamo: Bolis ed., 2006), 126-45. In 1530, Gherardo, Janello's father, was living in the very parish where (by the early 1580s) a member of the Sfondrati kin possessed some houses. Beside

they could have also promoted in the highest hierarchies of the Duchy of Milan their fellow-countryman Torriani. As previously mentioned, in the 1543-44 edition of his *De Sapientia*, dedicated to Francesco Sfondrati, and published together with *De Consolatione* and *De Libris propriis*, Cardano mentioned Janello as the restorer of a planetary instrument, a reference that had probably pleased Senator Sfondrati.[35]

Janello's house in Cremona around the year 1530 was in the parish of Sant'Agata. His first residence was probably at his father's place by the parish of San Silvestro, only a hundred meters away from the Church of Sant'Agata, and not far from compounds of bishop Vida, the local literate who exalted Torriani in the year 1550. Beside the Sfondrati and Vida, another powerful family had palaces in the same area: the Trecchi. Their palace is located on the square of the church of Sant'Agata. During the sixteenth century the Trecchi family enjoyed a near-monopoly on the office of abbot of the collegiate and its rich prebends, and their role as hosts for the institutional guests of the duchy in Cremona could have easily pushed them to display luxury goods such as clocks and automata in their residences.[36] From 1526, after the battle of Pavia, which determined the end of French supremacy in Lombardy and opened the region to the Imperial-Spanish forces, the following guests are recorded as having stayed in their palace. First of all, Duke Francesco II Sforza had stayed in Cremona, and in this palace, for long periods of time between 1526 and 1529, the year when Janello was appointed for the prestigious position of restorer of the public clock, which lasted at least until the year 1534.[37] Having already an important public position, it is possible that the last Sforza, interested in restoring clocks, had already been interested in him, as we have previously suggested.[38]

the Trecchi, there were many other imperial vassals who played important roles in Charles' army, for example the Dovara, the Picenardi, the Persico, etc. who could have provided a link between the clockmaker and the higher spheres of power: Guido Sommi Picenardi, *Luigi Dovara, gentiluomo cremonese, agente Mediceo alla corte di Filippo II* (Firenze: Tipografia Galileiana, 1911), 5.

35 Cardano, *De Sapientia libri quinque*, 427-29.

36 The Trecchi kin, originally from Milan and connected to the Trivulzio, enjoyed the utmost influence during the Sforza rule. Both Jacobus and his son Antonio were made ducal counsellors. From the time of Antonio Trecchi († 1540s) the estate of the family was exempted from taxation: in exchange, the Trecchi had the honour, and the burden, to host in their splendid brand-new palace all the official visitors who spent the night in Cremona.

37 For example, he was in the city for eight consecutive months between 1526 and 1527; see Antonio Campi in relation to these years.

38 See in part 1 of this book the section: *Between Public Clock and Private Workshop*.

Besides the Duke, several other important political characters took residence at the Trecchi palace: the constable of Bourbon, Don Antonio de Leyva, the great chancellor cardinal Gattinara, the president of the Senate of Milan Giacomo Sacco, and that very captain of Justice Giovanni Battista Speciano mentioned by Cardano.[39] Moreover, we have Charles v himself who sojourned in Palazzo Trecchi three times: in 1533 (5th of March), in 1541 (18th-20th of August) and in 1543 (15th-23rd of June).[40] On the last two occasions Charles was accompanied by the governor of the state: the Marquis del Vasto and Pescara, who was allegedly the one who introduced Janello to him. We do not know when Torriani moved to Milan; however, of the three visits of Charles v to Cremona, during the one of the year 1533, the clockmaker was certainly still living in his hometown where he held a workshop at least until the year 1537.[41]

---

39   The Duke of Milan Francesco II, living there for 8 months between 1526 and 1527, Violante Sforza, Renata di Francia, Alessandro Bentivoglio, etc.: Achille Giussani, *L'ospitalità ai Principi nel palazzo Trecchi*, Seconda edizione (Cremona: Ind. Grafica Editoriale Pizzorini, 1992), 21-26; *Karl v imperatore: L'imperatore Carlo v conferma a Giacomo Trecchi e ai sui fratelli I privilegi loro concessi da Francesco Sforza, duca di Milano*: Biblioteca di Stato di Cremona, Ms. Gov. 273.

40   Grandi, in the middle of the nineteenth century, wrote that during the visit of the Emperor to Cremona in 1543, which lasted for eight days, Torriani was creating the mechanical apparatus to entertain Charles and his court at night time. The source from which Grandi acquired this data is unfortunately not clear. It is likely that it was in fact guesswork. In fact there are descriptions of magnificent feasts taking place in the very same palace to entertain in 1563 Rudolph and Ernst, the sons of Emperor Maximilian II, with great numbers of men-at-arms jousting after being vomited by monsters spitting fire. This record plus the proverbial relation between Charles v and Torriani might have pushed the historiographer to such a supposition, but it is not impossible that archives will also offer documents on this uncertain event in the future. "*Giulio Campi ... vi rappresentò le imprese di Ercole, per fare così allusione a quelle dell'imperatore Carlo v, allorchè passando per Cremona ebbe ivi ad alloggiare negli anni 1541 e 1543 per otto giorni, nella quale occasione diede prova del suo matematico ingegno certo Lionello Torriano cremonese in disporre sì maestrevolmente meravigliosi e scariati meccanismi producenti veramente una magica illusione, onde accrescere il notturno divertimento*": Angelo Grandi, *Descrizione dello stato fisico, politico, statistico, storico, biografico della provincia e diocesi di Cremona*, vol. 1 (Cremona: Ed. Turris, 1856), 250. For the visit of the sons of emperor Maximilian II, see: Campi, *Cremona fedelissima città*, xliiii.

41   It was the counsel of the municipality to elect among its number the three *massari* of the Cathedral of Cremona, in charge of the keeping and decoration of the *Duomo*, Baptistery and major tower, including its clock. Their duties as such lasted for one year and they were chosen from the elite of the city: the decurions. The Trecchi family was not alien to this office: for instance in 1479, the father of Senator Antonio Trecchi, Senator Jacobus, was one of the two *massari* of the Cathedral who chose the governor and keeper for the clock of the Torrazzo tower: Carlo Bonetti, *Memorie: la fabbrica della Cattedrale: laica od*

The gentleman Francesco Grasso or Crasso from Pavia represents another possible link among Torriani, the Emperor, the Duke and the later Spanish Governors of Milan. Francesco Crasso had direct contacts with Charles V since 1529. After being accused and condemned for corruption by the imperial administration, he was made cardinal by the Milanese Pope Pius IV Medici di Marignano, the pontiff that had wished to have Torriani at his service in the early 1560s in Rome. Before his deposition, Francesco Crasso filled important institutional offices in the duchy of Milan as one of the permanent 60 decurions of the city, as senator, as great chancellor, etc. He was also appointed twice *Podestà* (civil governor) of Cremona in the years 1536-1537 and 1544-1545. The last document testifying to Janello's activity in Cremona is the interruption of the contract with his apprentice in 1537, when Crasso was *Podestà* of Cremona. The suspension of the contract has been interpreted as a possible proof of one of Torriani's voyages to Milan.[42] Moreover, during Crasso's second mandate as governor of Cremona, we find Torriani's first records at the service of the governor Alfonso d'Avalos, Marquis del Vasto and Pescara.[43]

---

ecclesiastica? (Cremona, 1936), 6. Recently Marubbi has claimed that the *massari* were elected evry two years, but this seems to be a mistake: Mario Marubbi, "Le 'Storie del testamento Nuovo': cronaca di un cantiere," in *La cattedrale di Cremona Alessandro Tomei. Saggi di Francesco Gandolfo ... Fotografie di Pietro Diotti.*, ed. Alessandro Tomei and Francesco Gandolfo (Cinisello Balsamo (Milano): Silvana, 2001), 86; Giussani, *Ospitalità ai Principi nel palazzo Trecchi*, 12.

42 As Rita Barbisotti has suggested in her article.

43 Moreover, Torriani was living in the same quarter of Milan where Crasso had been elected: Porta Romana. When Torriani received his privilege in 1552, it was Francesco Crasso to sign it as president of the Senate. Perhaps the fact that Giorgio Fondulo, as we have seen, was mentor, beside Torriani, to a certain Aurelio Grasso, a well-off citizen of Pavia was only a coincidence. Unfortunately I was not able to find any clear relation between these two figures with the same names and origins. Even if all these geographical and chronological coincidences between Crasso's and Torriani's trajectories were but an accident, they well represent possible chanels of communication between peripherical Cremona and the centre of power in Milan at the time of Janello's translation to the capital. Other institutional figures that may have performed as linking actors between Torriani and the governors of Milan were the military commanders of the city with their office of captains of the castle. In the year 1537, when Crasso was *podestà* and Torriani mysteriously suspended the contract with his apprentice Botti, Pier Antonio Gargano, nephew of the governor of Milan Cardinal Caracciolo, was commander of the castle of Cremona. Torriani's employment in the imperial army at the service of the governor Alfonso d'Avalos as engineer in the 1540s may suggest that Janello had also worked with the military administration of Cremona in the 1530s. But without clear documentation, this remains for the time being a plain supposition. Barbisotti, "Janello Torresani," 263; Franca Petrucci, "Francesco Grassi (Grasso, Crassi, Crasso)," *Dizionario biografico degli Italiani* (Roma: Istituto della Enciclopedia italiana, 2000); Leydi, "Un cremonese del Cinquecento," 138; Guido Sommi

FIGURE 42
Portrait of the Milanese
physician Girolamo
Cardano, *in the
Nuremberg edition of his*
De subtilitate, *1550.*

Another likely hypothesis of brokerage for the clockmaker brings us back to the very beginning of Janello's story, and precisely to Giorgio Fondulo, his mentor. The Fondulo were a noble family and one of its scions had also been lord of Cremona and Castelleone at the beginning of the fifteenth century, before he was beheaded at Visconti's command.[44] Despite his tragic ending at the hands of the Visconti, many members of the Fondulo family maintained a respectable status in Cremona. It is not unwise to think that Giorgio himself may have introduced Janello into the entourage of the governors of Milan. The Fondulo were gentlemen by status, and it is difficult to believe that Giorgio, if he really

---

Picenardi, "Tentativo fatto dai Francesi per impadronirsi del castello di Cremona nel 1537," *Miscellanea di Storia Italiana* XXIV (1885): 7. Leydi quotes two Milanese administrative notes for the year 1544: "*Maestro Leonello Torriano ingegnero, qual ha da star presso l'exercito de S.M. in Piemonte*" and for the year 1545: "*M.ro Jomello Torriano ingegniero*": Leydi, "Un cremonese del Cinquecento," 133.

44   This Cabrino Fondulo was famous because once he hosted both pope and emperor in Cremona. According to a curious anecdote, took them to the top of the Torrazzo tower, where he barely managed to control his desire to throw them both off the tower, and with this double killing to win eternal fame. Jacob Burckhardt made famous this fact in the chapter dedicated to the despots of the fifteenth century in his *The Civilization of the Renaissance in Italy.*

admired young Janello's bright mathematical mind sufficiently to become his mentor, would not have created all the conditions to promote his protégé among high circles. Antonio Campi, who gave us the only details about Torriani's mentorship, was well-informed, and had direct knowledge of Giorgio Fondulo's family: indeed, as we have already seen, the very son of Giorgio was trained as a painter at Antonio Campi's workshop. His name was Giovanni Paolo Fondulo, and in 1569 he had followed the governor of Milan, Francesco d'Avalos, Marquis of Vasto and of Pescara, son of Alfonso d'Avalos (first known patron of Torriani when he was governor of Milan in the years 1536-1545), when he was appointed Viceroy of Sicily. If members of the Fondulo family had access to the governors of Milan, Janello could have easily made use of their brokerage.

Stairway to the Imperial Court

The Habsburg's main strategy to consolidate their victories in Italy was to bind the local rulers to their blood. During the sixteenth century the lords of Milan, Parma, Ferrara, Mantua, Florence and Turin all shared their nuptial beds with Habsburg women. The building of a ramified family network was one of the most efficient features of Habsburg policy with consequences that would last for centuries. Furthermore, during Torriani's lifetime, Milan, Naples, Sicily, Sardinia and the *Stato dei Presidi* in Tuscany became the definitive possessions of the Spanish Crown. In the context of this web of power, I trace clues about Torriani's professional life, trying to define whether there was a consistent patronage network behind his remarkable career.

Once entered into the orbit of the Emperor, Torriani immediately consolidated his position, becoming Caesar's favourite entertainer. Already in 1552, the very year he had received the privilege from Charles v for his planetary clock, Torriani dared to write directly to the Emperor to have his pension paid by an excessively slow and unwilling Milanese bureaucracy. On the 31st of December, Charles, at the inglorious and unfortunate siege of Metz, immediately found time to satisfy Janello's petition, ordering Governor Ferrante Gonzaga to put pressure on the administration in order to have Torriani's pension granted without any further delay.

Janello Torriani seems to have found supporters within the court-system already in the first years of employment: in 1556 the feelings of Leone Leoni, who documents suggest had thus far enjoyed a friendly relationship with the clockmaker, suddenly changed into hate. In his letters to Ferrante Gonzaga he accused Torriani and "his clique" of damaging his image at court and of denigrating him. Insulting indirectly Torriani could be interpreted as a sign of jealousy in a very competitive and demanding environment in which the Emperor's obsession for clocks inflated Torriani's authority to the detriment of

Leoni's: the sculptor could not accept the fact that this "grand king of black-smiths, and not of sculptors" was entitled to check out the quality and state of progress of his sculptures as imperial inspector. Leoni's excessively slow deliveries and Torriani's bigger success at the Habsburg court were probably behind this bitterness.[45] The difference of value attributed to the clockmaker's and the sculptor's best pieces may explain Leoni's attitude: Torriani's most precious clock, the Crystalline, surpassed Leoni's masterpiece, the bronze sculpture portraying the Emperor enchaining the personification of Fury, by 1,000 ducats of gold![46] The Tuscan sculptor's allusion to a *clique* can be read as a sign of Torriani's settled position within the court where he likely created some alliances which Leone Leoni considered worrying. It is hard to understand who were in Leoni's eyes the members of Torriani's *clique* at this early stage: certainly the clockmaker was admired by several courtiers such as the precious stones worker Jacopo Nizzola da Trezzo, the multifaceted Juan the Herrera, and the historiographers Ambrosio de Morales and Esteban de Garibay.

Undeniably, together with Charles V in his retirement at the monastery at Yuste, the only members devoted to a luxury craft admitted to the company of the Emperor were Janello and his helpers. Philip II will always remember this service, and Torriani will serve him until the end of his days for more than a quarter of a century, a long period that allows us to analyse the structure of power Janello was belonging to. If we list all the patrons and promoters of Torriani, we will see emerging a consistent political network. At the top of the pyramid of Janello's patrons were of course the Habsburgs: Charles V was not the only member of his family to own clocks and automata made by Janello Torriani, and the clockmaker was not the only member of his family to be favoured by the Habsburgs: Philip II and his half sister, *la Madama* Margaret of Parma (1522-1586), former Duchess of Florence and now Duchess of Parma and Piacenza, also owned Janello's clocks. Duke Ottavio Farnese, her husband, granted in 1572 a privilege for invention for elevating water and digging machines to a certain Bernardino Torriani of Cremona.[47] This Bernardino was a relation to Torriani.[48] Bernardino Torriani was the father of Leonardo

---

45   Rosario Coppel Areizaga, "Carlos V y el Furor," in *Los Leoni (1509-1608): escultores del Renacimiento italiano al sefvicio de la corte de España.* (Madrid: Museo del Prado, 1994), 102-9.

46   Sánchez Cantón, *Inventarios reales.*

47   ASPr, Archivio Farnese, Patenti, 3 1564-1575, f. 279: *Bernardino Torriani di Cremona: patente per macchine per cavar terra ed elevar acque*

48   A few years later a certain Bernardo Torriani from Cremona, "*nephew and pupil of Janello*", was proposed (without success because he was judged insufficiently able) for the position of court architect of the Gonzaga of Mantua. After the death of Zelotti, the duke asked around for a new architect. Tintoretto and Palladio were said to have suggested an array

Torriani, nephew of Janello, who, by the end of the century, will be entitled *ingeniero mayor* of the Kingdom of Portugal, after its acquisition by Philip II of Spain.

The Imperial Admiral Andrea Doria and the Governor of Milan Ferrante Gonzaga had at least one clock each made by Janello.[49] After Ferrante's death in 1556, his son Cesare Gonzaga (1530-1575), tried to promote Torriani's hydraulic devices before the Grand Duke of Tuscany Francesco I de Medici (1541-1587), and his cousin Guglielmo Gonzaga (1538-1587), Duke of Mantua, granted Torriani a privilege for invention in 1568.[50] In the same year, the Republic of Venice granted Torriani with another privilege for the same machines.[51]

---

of names. "*Somebody else*", is said in the letter kept in the State Archive of Mantua, suggested to the duke this Bernardo Torriani. The names Bernardo and Bernardino were interchangeable; therefore there is a fair chance that the two architects mentioned by the Farnese and the Gonzaga were the same person. A certain Gio. Battista Lamo, also brought news about Bernardino back from Spain to Cremona. Lamo, who was housesteward to the Cremonese Count Broccardo Persico, wrote to have met Bernardo Torriani in the household of Janello in Spain: Viganò, "Parente et alievo del già messer Janello," 217. Another Lamo, Alessandro, wrote about Janello in two different books and referred that the clockmaker took a religious painting by his friend Bernardino Campi to Spain, most probably to promote him at the Castilian court. Bernardino Campi became court painter of Vespasiano Gonzaga prince of Sabbioneta: Lamo, *Discorso di Alessandro Lamo intorno alla scoltura, e pittura, doue ragiona della vita, ed opere in molti luoghi, ed a diuersi principi, e personaggi fatte dall'eccellentissimo, e nobile pittore cremonese*, 52; Leydi, "Un cremonese del Cinquecento".

49   In 1545, Alfonso D'Avalos ordered Janello to deliver a clock to Andrea Doria: Alvar Gonzáles-Palacios, *Il mobile in Liguria* (Genova: Sagep, 1996), 37-8. The clockmaker, according to Leone Leoni's letter, was planning to deliver a clock to Ferrante on the 10th of July 1556. In the same year, from Mantua Ferrante asked Janello – who was in Brussels – why a clock of his (probably not a work of Torriani, it seems from the letter) did not work properly. Janello answered to the Gonzaga to send it to him for repair. This gives an idea of a continuing relationship between the clockmaker and Ferrante, even after the end of the governorship of the state in 1554. S Leydi, "Un cremonese del Cinquecento," 137-142.

50   In the Archive of Mantua I found a privilege for invention granted to Torriani in 1568 and all the letters of petition for it, as discussed in the following footnotes: ASMn, Archivio Gonzaga, indici decreti busta 6, Liber 48 L.T, 10 Junii 1568, "*Turriani Joanellus. Concessio construendi duo edificia pro irrigands pratis siccitate laborantibus aliaq. Faciendi de eum prohibitione ne quis preter eum proposit .... aedifitia construere*", fol. 140 tergo. This is discussed below in relation to the role of ambassadors in the market for privileges for invention.

51   ASVe, Senato, Terra, filza 342, f. 187r.: 7 February 1567 (More Veneto): "A Giannetto Turriano privilegio per un edificio da estrazer aqua" and Archivio di Stato di Mantova: Indici Decreti Busta 6, Liber 48 L. T., f. 140 tergo "*10 Junii 1568 Turriani Joanellus. Concessio con-*

FIGURE 43   *Janello Torriani's network of patrons,* Panel from the exhibition: Janello Torriani
genio del Rinascimento, September 10, 2016-January 29, 2017. AUTHOR'S PHOTO.

Beside the Habsburgs' and Gonzagas' circuits, three successive popes
showed an interest in Torriani: Pius IV, Pius V and Gregory XIII. In 1563, Pope
Pius IV (1559-1565) tried without success to have Torriani at his service, as a letter from the great-commander of Alcántara to King Philip II shows.[52] The

struendi duo edificia pro irrigands pratis siccitate laborantibus aliaq. Faciendi de eum pro-
hibitione ne quis preter eum proposit ... aedifitia construere".

52   The great-commander of Alcántara was Don Luis de Ávila y Zuñiga, an old friend of
Charles V. Don Luis was used to visit Charles in Yuste, his palace being in the closest town
to the monastery: Plasencia. It is possible that the commander of Alcántara met Torriani
on these occasions. See: Francisco Antonio González and Juan Tejada y Ramiro, eds., *Col-*
ección de cánones de la Iglesia Española (Madrid: Imprenta de Don José María Alonso,
1849), 635; Fernández Álvarez, *La España del emperador Carlos V*, 915-26.

great-commander, agent to Philip in Rome, transmitted Pope Pius IV's desire to have Torriani on loan for two years in order to accomplish some works and for the Pope's amusement.[53] The great-commander thought that this gift would have pleased the Pope in a moment of tension between Philip and the Holy See.[54] His fame as amazing clockmaker and his common regional identity with the Pope would have made of him a suitable decoration for the Apostolic Chamber. But King Philip II did not rent out his precious prince of clockmakers. Pius IV's successor, Pius V (1566-1572), granted a patent for invention to Torriani with a *motu proprio* dated 19th of January 1567. The next pope, Gregory XIII (1572-1585) would indirectly (through Philip II) involve Torriani in the reform of the calendar, and Torriani would write directly to him asking for a pension and a privilege for invention in relation to the publication of a book

---

53   12th and 13th of May 1563, Rome: Cervera Vera, *Documentos biográficos*; this way to circulate a technician was well established in Europe: for instance, in the fourteenth century, Pope Innocent IV released his *plomberius* to King Peter of Aragon to build a clock: Landes, *Revolution in Time*, 194.

54   We do not know what Torriani was supposed to do in Rome during this biennium: Pius IV was investing great energies in his projects of *renovatio Urbis* and Janello was already known in Milan, the Pope's fatherland, for his Ctesibian pumps. However, Janello at that time was much more famous for his clocks. Rome, as already seen, was a centre of attraction for clockmakers. The Apostolic Chamber and the many cardinals' and diplomatic courts formed a lively market for luxury goods. For instance we know that at the beginning of the 1560s, in addition to different clockmakers employed in the city, a certain Francesco Bassano (or Bassiano), a Flemish clockmaker from Cambrai, signed a contract in 1560 in which he committed himself to provide within a month the Apostolic Chamber with a 30 *scudi* clock. Francesco Bassano also offered a 2- year warranty for this timepiece. Then a certain Pietro Filanetto from Lodi was the keeper of the Apostolic Palace's clock in 1567. In 1569 Francesco Bassano was still in Rome at Borgo Vecchio. Transalpine clockmakers lived in this area: around 1565 the French clockmakers Pietro Deslino and Uberto Muj called Paternostri had their houses and workshops there. A certain Adriano, Flemish clockmaker, was living at Borgo too around 1558. Some other clockmakers from Flanders and Northern Italy lived in other parts of Rome, such as Giovanni Fiammingo (documented in Rome around 1557), and Govanni Morone from Bergamo (documented in Rome about 1569): Morpurgo, *Dizionario degli orologiai italiani*, 1950; and Del Vecchio and Morpurgo, *Addenda al Dizionario degli orologiai italiani edizione 1974 di Enrico Morpurgo*; Long, "Hydraulic Engineering and the Study of Antiquity," 1110; Maria Losito, *Pirro Ligorio e il casino di Paolo IV in Vaticano: l'"essempio" delle "cose passate"*. (Roma: Palombi, 2000); Maria Letizia Gualandi, "Roma resurgens: fervore edilizio, trasformazioni urbanistiche e realizzazioni monumentali da Martino V Colonna a Paolo V Borghese," in *Roma del Rinascimento*, ed. Stefano Andretta and Antonio Pinelli, Roma del Rinascimento 3 (Roma: GLF editori Laterza, 2001), 152-53.

dealing with mathematical instruments connected with this reform, a privilege that the Pope granted him.[55]

The powerful patrons who supported Janello and his family were strictly intermingled: Ferrante Gonzaga, cadet brother of the first Duke of Mantua, had spent some of his youth in Madrid, where he became a most dear friend to Charles V.[56] Later, the Emperor made him Viceroy of Sicily and Governor of Milan. Ferrante managed to create strong ties with the Milanese world, and especially with two powerful families: the Medici di Marignano and the Borromeo. The Medici di Marignano produced a Pope: Pius IV. His name was Giovanni Angelo Medici (as noted above, not related to the Florentine Medici), brother to Gian Giacomo Medici, marquis of Marignano, nicknamed Medeghino (Little-Medico), a ruthless imperial general. It was possible for Pius IV to be elected in 1559 just after the official suspension of Ercole Gonzaga's candidature, because of Philip II's veto.[57] Pius IV's sister Margherita Medici

---

55   Felice Zanoni, "Un brevetto pontificio d'invenzioni del 500: Janello Torriano e un documento dell'Archivio segreto vaticano," *Bollettino Storico Cremonese* X, no. settembre-dicembre (1940): 145; Turriano, *Breve discurso*, 1990; See also in part 3 of this book the section: *Invention and the Practice of Secrecy.*

56   The Court of Madrid pullulated with members of the Gonzaga family: Vespasiano Gonzaga was in Madrid between 1545-1548. He was page of honour to Prince Philip. Vespasiano went back again to Spain in the 1560s, where he met a young Rudolf who would later make the Gonzaga Duke of Sabbioneta. Vespasiano was one of the geratest patrons of Bernardino Campi. Ferrante Gonzaga of Castiglione delle Stiviere was as well at the Spanish court. His son St. Aloysius/Luigi Gonzaga was in Madrid between 1581-1583. His clock made by master Jorge, Torriani's helper, is still at the monastery of Castiglione today. Even the papal court, especially at the time of the three popes connected to Carlo Borromeo, was crowded with members of the Gonzaga family, both cardinals and laymen. Scipione Gonzaga (1542- 1593) was *cameriere segreto* to Pius IV and was then made Latin patriarch of Jerusalem and cardinal. He was a friend of San Carlo Borromeo and S. Filippo Neri. Jesús Escobar, "Francisco de Sotomayor and Nascent Urbanism in Sixteenth-Century Madrid," *The Sixteenth Century Journal* 35, no. 2 (2004): 357-82; Raffaele Tamalio, "Tra Parigi e Madrid. Strategie famigliari gonzaghesche al principio del Cinquecento," in *La Corte di Mantova nell'età di Andrea Mantegna, 1450-1550: atti del convegno : Londra, 6-8 marzo 1992, Mantova, 28 marzo 1992 = The court of the Gonzaga in the age of Mantegna, 1450-1550,* ed. Cesare Mozzarelli, Robert Oresko, and Leandro Ventura (Roma: Bulzoni, 1997), 69.

57   Giovanni Drei, "La politica di Pio IV e del Cardinale Ercole Gonzaga," *Archivio della R. Società Romana di Storia Patria* XL (1917); Mia J. Rodriguez Salgado, "Terracotta and Iron: Mantuan Politics (ca. 1450-1550)," in *La Corte di Mantova nell'età di Andrea Mantegna, 1450-1550: atti del convegno : Londra, 6-8 marzo 1992, Mantova, 28 marzo 1992 = The court of the Gonzaga in the age of Mantegna, 1450-1550,* ed. Cesare Mozzarelli, Robert Oresko, and Leandro Ventura (Roma: Bulzoni, 1997), 15-57; Raffaele Tamalio, *Francesco Gonzaga di*

di Marignano was married with Count Gilberto II Borromeo, another noble Milanese. In 1560, just one year after the election of Pius IV, his niece Camilla Borromeo (daughter to the Pope's sister) married Cesare Gonzaga, son of the recently deceased Ferrante. Camilla was sister to Carlo Borromeo, the *cardinal nephew*, and champion of the Counter-Reformation. Another nephew of Pius IV, Federico Borromeo, married Virginia, daughter of the Duke of Urbino and of Isabella Gonzaga, sister to Ferrante and to Duke Federico of Mantua. In 1575 the last son of Ferrante Gonzaga, Ottavio, married another Medici di Marignano: Cecilia.[58] Giulio Cesare II Borromeo, Saint Carlo's uncle, married Margherita Trivulzio, mother of the future cardinal Federico, author of the above-mentioned book in which Torriani is mentioned several times. The Borromeos had also direct ties with the territory of Cremona. Between the second half of the fifteenth century and the end of the sixteenth century three women of this family married members of the Trecchi family.[59] The Trivulzios were also tied to secondary branches of the Viscontis, with the d'Avalos and besides the Gonzagas of Guastalla and Sabbioneta, with the Ganzaga branch of Vescovato, a few miles outside the city-walls of Cremona.

If we look more closely at Janello's relations, we can find more evidence about his commitment to this network of power. A granddaughter of Torriani, Emilia Felipa Diana, married a certain Ludovico, a member of the Besozzo (or Besozzi) household, a noble family of Milan. Emilia Felipa would become the heir of the imperial pension of Charles V. From 1581 she obtained the rights to the 200 ducats of her grandfather, which a certain Giovanni Battista Besozzi had to collect in her name. The Besozzo family was entwined with the Borromeos. Antonio Giorgio Besozzo was a diplomat in the early 1580s, employed by Carlo Borromeo and Gregory XIII in the attempt to coordinate the Duke of Savoy and the King of Spain against Geneva and the King of France. At that time, six members of his family were at war in the Habsburg's army. Later, Antonio Giorgio Besozzo, after Cardinal Carlo Borromeo's death, entered in service of

---

Guastalla, cardinale alla corte romana di Pio IV: nel carteggio privato con Mantova (1560-1565) (Guastalla: Biblioteca Maldotti, 2004), 15.

58   As a last note, it should be recalled that already in 1555, count Camillo I Gonzaga of Novellara, Venetian Nobleman, married Barbara Borromeo, daughter to count Camillo Borromeo of Arona. Furthermore, in 1561 the duke Guglielmo Gonzaga of Mantua married Eleonore of Austria, daughter to Emperor Ferdinand I. There are other strong connections between the Borromeo-Gonzaga-Medici di Marignano to the Serbelloni and to Altemps, but I think this brief summary is sufficient to highlight the politically relevant side of the family network.

59   ASCr, fondo Trecchi cart. 1: carte genealogiche.

FIGURE 44   *Bernardino Campi,* Holy Family with Saints. Janello himself brought to Spain a
Holy Family painted by Campi, in order to promote his countryman at the court.
COURTESY OF THE MUSEO CIVICO ALA PONZONE – DEPOSITO FONDAZIONE
CITTÀ DI CREMONA.

his cousin Cardinal Federico Borromeo (1564- 1631).[60] In 1594, Antonio Giorgio became a member of the *Accademia degli Inquieti* of Milan, whose emblem was Torriani's Toledo Device. Antonio Giorgio Besozzo wrote two (now unfortunately lost) books: the *Libro delle Invenzioni* (*book of inventions*, probably inspired by late fifteenth century Polidoro Virgil's bestseller) and the *Trattato degli uomini illustri di casa Besozzo* (*Treaty of the illustrious men of the House of Besozzo*). When he died his library entered into the hands of his patron, Cardinal Federico, the founder of the Ambrosiana Library.[61] Cardinal Federico published in 1632 the already mentioned book about the behaviour of the princes and their relationships with their courtiers, where Torriani appears as a witty character. It is possible that Torriani's memory was cultivated in the bosom of the Borromeo's family, thanks to the Besozzo's relations with the *Accademia degli Inquieti*.[62]

These witty tales with Torriani as a protagonist also have a parallel in the manuscripts by Camillo Capilupi (1531-1603) preserved in the Biblioteca Nazionale of Rome. Camillo Capilupi was part of one of the richest families of Mantua. The members of the Capilupi family were closely involved with the Gonzaga rule, and on one occasion, one Alessandro Capilupi, ducal secretary, had a direct contact with Janello when, on the 10th of June 1568, he issued the privilege for invention for the clockmaker in the name of the duke.[63] A certain Muzio Capilupi was secretary to Vespasiano Gonzaga, the Duke of Sabbioneta. Camillo Capilupi's father was ambassador at the imperial court, and his uncle Ippolito Capilupi was a member of Ferrante Gonzaga's entourage. Ippolito was imprisoned by the anti-imperial Pope Paul IV Carafa, and received for his fidel-

---

60   Remo Ceserani, "Besozzi, Antonio Giorgio," *Dizionario biografico degli italiani* (Roma: Istituto della Enciclopedia italiana, 1967).

61   Gherardo Borgogni, *La fonte del diporto* ... (Bergamo: Per Comin Ventura, 1598), 26-27; Paolo Morigi and Girolamo Borsieri, *La Nobilità di Milano, descritta dal R.P.F. Paolo Morigi... aggiunte si il supplimento del Girolamo Borsieri....* (Milano: appresso G.B. Bidelli, 1619), 297-99: conte Muzio Sforza of Caravaggio, Giorgio Besozzo, Ludovico Settala, conte Matteo Taverna, conte Don Andrea Manriche, conte Nogarola and many others. They chose the hydraulic device of Gannello Cremonese as coat of arms, together with the explicatory motto: "*LABOR OMNIBUS UNUS*". In a 19th century bibliographic repertoire I also found a reference to another prolific writer of this family: a certain Giovan Giacomo Besozzi, author of works amounting to a list of 502 pages, one claimed. In the nineteenth century, this index and many other manuscripts were still possessed by the Besozzi family. Among them, one can find two manuscripts entitled: *Istorie memorabilia di casa Besozza* and *Theatrum suae gentis genealogicum* : Francesco Predari, *Bibliografia enciclopedica milanese...* (Milano: Tipografia M. Carrara, 1857), 501-2.

62   Borromeo, *Gratia de' principi*.

63   ASMn, Archivio Gonzaga, Indici Decreti, Busta 6 Liber 48, f. 140 v

ity (thanks to the diplomatic activity of the nephew Camillo) a pension of 400 ducats on the archbishopric of Cuenca by a retiring Charles v. Ippolito had introduced young Camillo to the court of the governor of Milan. Camillo followed Ferrante on his unfortunate trip to Brussels after being accused of incorrect conduct as governor. After the end of the Carafa pope, Camillo found new fortune at the courts of Pius IV, of Pius V, of Gregory XIII, of Sixtus V and of Clement VIII, fulfilling important offices for them and always performing as an unofficial diplomatic support to the Manutan court. Sixtus v made him *protonotarius apostolicus*, the acme of his Roman career. He carried out delicate diplomatic missions at the imperial court and on behalf of the Guisa. Camillo was a successful pamphleteer devoted to the counter-reformist cause, and he aspired to write a historical work about the relevant events of his time, rich in anecdotes and diplomatic first-hand knowledge. The book was never finished, and the two accounts concerning Torriani are a part of its preparation material.[64]

Ties with the fatherland seem to have played an important role even many years after Janello's emigration to Spain: when in 1569 Janello Torriani had to borrow money in Toledo, he went to ask for a loan to the banker Ippolito Affaitati from Cremona, who was running his business in Spain. It has to be noted that this banker was taking care of Cardinal Carlo Borromeo's interests in Toledo. Ippolito Affaitati belonged to what was probably the richest family of Cremona. The Affaitati opened different branches of their commercial company and bank in Portugal, Flanders, Spain and elsewhere. Ippolito Affaitati was probably brother to Gian Carlo, the fortunate beginner of the branch of Antwerp, who was invested of fiefs in the Low Countries: Charles v made him Count of Inst, Baron of Ghystghem and Lanachensache. The same happened to his cousin Ludovico in Lombardy, who was made Count of Romanengo, probably after he had lent large quantities of gold to the imperial troops during the battle of Pavia (1525). His son was made Marquis of Grumello and was married to the daughter of Don Emanuel de Luna, a castellan of Cremona and noble Spaniard. Ludovico II and Ottavio Affaitati were for a period at the service of Philip II in Spain. Ottavio served as a captain in the Portuguese War. In 1586, the Duke of Mantua and his daughter-in-law kept at baptism by proxy the daughter of Count Ottavio Affaitati, sending to Cremona their secretary.

---

64   During the same period another member of the Capilupi, a certain Giulio, natural son of Ippolito, dealt with horology and in 1590 published a book on the topic: Campori, *Lettere artistiche inedite*, 28; Gustavo De Caro, "Capilupi, Camillo," *Dizionario biografico degli Italiani* (Roma: Istituto della Enciclopedia italiana, 1975); Giulio Capilupi, *Fabrica et uso di alcuni stromenti horari universali* (Roma: Giliotti, 1590).

Ottavio Affaitati was also envoy for the Grand Duke of Tuscany Francesco I at King Philip II's Court.

The Affaitatis link with the Gonzagas, with the Borromeos and with the Habsburgs, are instrumental to understand the trajectory of the only other member of Torriani's family who had fortune at the court of Philip II: Leonardo. In the 1580s, King Philip II made him *ingeniero mayor* of the Kingdom of Portugal. Leonardo was the son of Captain Bernardino Torriani and Juana (Giovanna?) Carra. It seems that Emperor Rudolf II had recommended Leonardo to Philip II, and that the engineer came to Lisbon with Rudolf's wife, Empress Maria.[65] Maria, on her way to the court of her brother Philip II, passed by Cremona and stopped in Soncino (county of Cremona) for a couple of days. On this occasion (as Antonio Campi reports) Ottavio Affaitati went to pay her a visit. It might be suggested that in this situation Leonardo Torriani was attached to Maria's train.

In the year 1600, Leonardo Torriani married well: he took as wife a certain Juana de Herrera, daughter of a former *regidor* of Madrid, Pedro de Herrera and Juliana Osorio. The Osorio (if the same family) were relatives of Janello Torriani: a granddaughter of Janello, a certain *doña* Catalina Turriano Ossorio, had been indeed married to a member of this family.[66] It is still not clear to what extent the family Herrera of Leonardo Torriani's wife is related to Juan de Herrera, friend of Janello, and *aposentador mayor de palacio*, which one may translate as "first sergeant of the Royal Palace", of Philip II. Some clues suggest that Vespasiano Gonzaga, Duke of Sabbioneta, fief in the diocese of Cremona, may have played a role in the brokerage of the Bernardino and Leonardo Torriani. The Duke of Sabbioneta had very good relations to both Emperor Rudolf II, who had made him duke, and to Philip II, who decorated him with

---

65   Cámara interprets the information that *"Leonardo was sent by the emperor Rudolph II to those kingdoms (Philip II's)"* as evidence that the young Leonardo Torriani was employed at the Imperial court. We cannot assume Leonardo stayed in Vienna or Prague just because of this letter. Some Italian vassals to Rudolph could have asked their lord to promote one of their servants at the Spanish court. For instance, a letter written in 1579 by Rudolph's wife (Maria, sister of Philip II), to the governor of Milan, expresses the empress's thanks to the governor for the positive response to her recommendation for a certain Pietro Paleari to be employed in the service of the king of Spain. Alicia Cámara Muñoz et al., "Leonardo Turriano al servicio de la Corona de Castilla," in *Leonardo Turriano: ingeniero del rey* (Madrid: Fundación Juanelo Turriano, 2010); Paolo Emilio Marcobruni, *Raccolta di lettere di diuersi principi, & altri signori: che contengono negotij et complimenti in molte graui & importantissime occorrenze* (In Venetia: Appresso Pietro Dusinelli, 1595), 190.

66   Cervera Vera, *Documentos biográficos*. doc 150.

the prestigious Golden Fleece. This petty prince was also patron to Leone Leoni and to Bernardino Campi. The latter is the friend of Janello who became court painter of Vespasiano in the 1580s. Moreover, Vespasiano, who spent a considerable period of time in Spain in the service of Philip II, had a secretary of Spanish ancestry, a certain Antonio de Herrera. It is not clear yet if this Antonio de Herrera was related to the wife of Leonardo, or to Juan the Herrea. Perhaps Vespasiano was also the anonymous person who had suggested to the Duke of Mantua Bernardo Torriani as new court architect. New investigation is needed.[67]

In conclusion, if one looks at all the records so far discovered on Janello Torriani and his relatives, it emerges that all their patrons belonged to a well recognisable group of power, based on the axis Milan-Madrid-Mantua, and for Mantua we need to understand especially the western territories close to Cremona ruled by the secondary branches of Guastalla and Sabbioneta. Torriani moved along the nerves of this political structure, and in tracing his movements we mark this very alliance. That is why Janello's fame and memory survived in the Spanish Habsburgs' dominions and within the Gonzagas'-Medicis' di Marignano-Borromeos' circuit. With the role played by the Mantuan ambassador Girolamo Negri, Janello's connection with the Gonzagas of Guastalla will appear even clearer.

This is not to ignore the connections and blood alliances between the Gonzagas-Medicis di Marignano-Borromeos and other networks. However, a strong suggestion of political determination appears from the timing, the geography, the quantity, and quality of the links between the Gonzagas (especially in the cadet branches of Guastalla and Sabbioneta) and the two powerful families of Milan. Of course family alliances were not carved in stone: as in any structure, negotiation, depending on contingency, was a constant practice, but we can be quite confident that since the time of Ferrante Gonzaga and his son Cesare, the reproduction of specific links created a fairly stable configuration, a political structure. Family-links, political interest, and feudal duties contributed to make the game very serious. As emerges from the correspondence of

---

67   Some further signs seem to suggest Vespasiano's relationship to Leonardo Torriani: on the
      one hand, in a letter to don Diego Sarmiento, an important character in the Portuguese
      military set, Vespasiano gave a negative judgment on the engineer Spannochi (perhaps, in
      order to improve the reputation of Leonardo). On Leonardo Torriano's hand, the engi-
      neer, during a military survey in Northern Africa, supported all of Vespasiano's thesis
      about what was best to do with some Spanish outposts in the region, against another
      engineer who had a different opinion from the Gonzaga. Alicia Cámara Muñoz, "Immag-
      ini della Orano e della Mazalquivir di Vespasiano Gonzaga in un manoscritto inedito di
      Leonardo Turriano," *Civiltà mantovana* 3 (2010): 130.

BERNARDINVS CAMPVS PICTOR
CREMONENSIS.

FIGURE 45   Portrait of Bernardino Campi, *from* Discorso di Alessandro Lamo
intorno alla scoltura, et pittura, doue ragiona della vita, & opere
in molti luoghi, & . diuersi prencipi, & personaggi fatte
dall'eccell. & nobile m. Bernardino Campo pittore cremonese ...,
Cremona, 1584.

Cesare Gonzaga, it was stressed that it was strictly forbidden to the members
of this family to have as friends any *persona non grata* to his uncle Cardinal
Ercole Gonzaga and to the Borromeo clan.[68] This network of power was locally
rooted in an area with Cremona and its territory as a crossroad, and it exercised

68   ASPr, fondo Gonzaga di Guastalla, Busta 4-27 Doc. 3, 56 *"di non pigliar per amici et confi-*
denti suoi altri che quelli che saranno tenuti da H.G. et dalli S.ri Borromei per amici et confi-
denti ... et procurar loro gratie et favori da sua E,za et dalli ditti S.ri Borromei"

for more than two decades a strong control on both the State of Milan and on the Papacy. This happened thanks to a relatively stable alliance with the Habsburgs. This family-network was then in competition with other centres of power such as Ferrara, Florence and Parma, whose ruling families had their territorial clients to employ, and to promote in key centres of power such as Madrid or Rome, using often Vitruvian craftsmen such as Janello as instruments to gain credit. The Cremonese clockmaker was not alone in this game; the Gonzagas'-Medicis' di Marignano-Borromeos' circuit cultivated its proper army of skilled artisans, and sometimes promoted them successfully at court, gaining prestige for their *patria* and for themselves. Even the same artisans were trying to support each other, as Janello, Jacopo di Trezzo and Leone Leoni often did. Janello also tried to promote Bernardino Campi in Spain, taking a painting by him in his final voyage to the court.[69]

The Mantuan Ambassadors and their Brokerage of Janello's Inventions

Between the 1560s and 1580s, Janello received several privileges for invention from different European powers. In the juridical context of medieval and early modern Christendom, a privilege for invention was a gracious concession to an inventor who desired to have the originality of his creation economically and intellectually acknowledged. In a feudal system, this gracious concession came from the supreme legislative authority of a certain state structure, e.g., a prince, his vicar or a collegiate body (such as a senate). As with all other privileges,[70] it technically responded to a supplication. From a juridical standpoint the privilege positioned the supplicant in a special status "outside the law": indeed,

---

69   Bernardino Campi, like Jacopo di Trezzo and Leone Leoni, had worked extensively for this network of power: he portrayed Cecilia Borromeo, as he did many other members of the Gonzaga and Medici di Marignano families, Zaist, *Notizie istoriche depittori, scultori, ed architetti cremonesi*, 208. Leone Leoni, sculptor to Charles V and to Philip II, worked for Ferrante Gonzaga, for his son Cesare and for Vespasiano Gonzaga of Sabbioneta. He also worked for Pius IV, and made the sculpture of his brother for his tomb in the Cathedral of Milan.

70   Common types of pleas were requests to obtain a special status (as in a citizenship), or to be paid for a performance or to carry weapons in a certain area where it was not allowed, or to be dispensed from the payment of certain taxes. For a brief story of inventor privileges, see Marius Buning, "Discovering Inventions: A Short History of Inventor's Privileges," in Cristiano Zanetti (ed.) *Janello Torriani, a Renaissance Genius* (Cremona: Comune di Cremona, 2016), 59-60.

privilege derives from the Latin *"priva lex"*, which literally means "special law".[71] After the end of the Italian Wars with the peace of Cateau-Cambrésis in 1559, the market for invention privileges expanded all over Christendom with a new intensity.

The Republic of Venice, since the previous century, had the first well-established bureaucratic system with its proper juridical framework and its technical praxis of evaluation. The Venetian law on privileges for inventions of 1474 read that:

> There are in this City ... men coming from different backgrounds and with very sharp intellects who are able to invent different ingenious devices. If there was a way to stop other people to steal their honour, imitating their inventions after having seen these devices, these sharp intellects would exercise their ingenuity discovering things that are of great utility and benefit for our Republic. For this reason, by the authority of this very Council, anyone who will create in this City any new ingenious device, never seen previously in our dominion, once it is ready and functioning, must be registered at the Office of the *Provveditori de Comun* [an important communal office created in the thirteenth century]. Thus it will be forbidden up to ten years for anyone to replicate it in our dominion without a previous agreement and licence by the author. Otherwise, the above-mentioned author and inventor will have the right to sue at the tribunals of this City anyone who has imitated without said permission, and he will have to pay one hundred ducats and he will suffer the immediate destruction of the device. Our government will be instead free to take and use for our needs any of these devices and instruments, however, with this condition, that no-one but their creators must operate them.[72]

Even in Venice, the legal procedure to plead for a privilege was reminiscent of the traditional method that saw the supplicant addressing personally the prince begging for a gracious concession, as Janello Torriani did writing to the head of the maritime Republic, the Doge. Then, the offices of the Ducal Palace forwarded the plea together with a model of the machine to the *Savi* [i.e., Wise Men]: a technical office in charge of evaluating the novelty and the functional-

---

71   Daniela Lamberini, "Inventori di macchine e privilegi cinque-seicenteschi dall'Archivio Fiorentino delle Riformagioni," *Journal de la Renaissance*, 2005, 177.

72   Republic of Venice, year 1474 First law for protection of intellectual property (privileges for invention): ASVe, Senato Terra, registro 7, carta 32r (my translation).

ity of the machine. Consequently, the *Savi* sent their positive response to the Senate, which voted on the issue, granting Janello his privilege. Thus, it seems that a well-structured bureaucratic system such as the Venetian one was open to any good idea providing (zero-cost) opportunities of gain for the Republic.[73] However, it is noteworthy that after the positive response of the technical committee, during the voting for the granting to Torriani of a privilege, six senators expressed their vote against the general acceptance of the other 150. Moreover, the senators made the privilege effective for a span of 25 years, whereas the *Savi* had suggested 40 years.[74] These two details open up quite interesting questions about the interaction of technical knowledge and political agendas in bureaucratic procedures in early modern Europe.

One may expect that the market for innovative technology, because of the utility-factor, has always been driven by a positive appreciation of functionality. Daniela Lamberini on the contrary, studying the case of Florence, has described how important were personal loyalties in the processes of granting privilege for invention during the sixteenth century. Janello Torriani seems to have faced this protectionist attitude when, in the very Florentine Grand Duchy, he attempted without success to obtain an inventor privilege. Torriani's experience confirms that beside categories such as *utility*, *innovation* or *economic vantage*, personal favour and political factors were also variables in the market. State officers had the task of judging the projects, but the last word was the prince's or the senators'.

In the specific case of Janello, we can observe how the Habsburgs-Gonzagas-Borromeos-Medicis di Marignano network functioned as an infrastructure connecting different centres of power and circulating people and knowledge. Within this structure, the vehicle Torriani used to seek for privileges for invention was, beside the credit provided by his position at the Habsburg court, the Gonzaga diplomatic network. Other inventors wandered around circulating explanatory models for their inventions: for instance Giuseppe Ceredi, court

---

73   Indeed, beside an implementation of production connected most of the time with the practice of sixteenth century invention, the Republic owned an archive of models and ideas that anyone in search of new and more lucrative technologies could access. The adoption of any idea protected by privilege was implying the payment of a fee both to the inventor and to the Republic. Any stealing also implied a sanction to be shared by the inventor and the state.

74   This is an interesting and still open question offered by Torriani's story, concerning the different spans of time the privileges for invention covered in different places. In Janello Torriani's case, his water-lifting machine was granted a privilege of the duration of: – 30 years in Philip II's dominions, – 25 years in Venice (though the Office of the Savi proposed 40 years); – 20 years in Mantua; – 15 years in Rome.

physician of the Farnese, mentions a man who in the year 1566 had brought around Italy a model of a pump that could perform more than 80 different movements, activated by a system of little bellows.[75] Torriani, once in Spain, by contract could not leave the court, and he adopted another strategy to circulate models of his invention. The Mantuan ambassadors to the Spanish court, Girolamo Negri and Emilio Roberti, had a fundamental role in this process. In the letters asking for the granting of a privilege for Janello Torriani, both in Mantua and in Florence, Girolamo Negri appears a most enthusiastic promoter of the clockmaker, exalting Torriani's skills and ingenuity.[76]

In the case of the letter sent in 1565 to his lord Duke Guglielmo Gonzaga, he claimed the machine Torriani was building in Toledo to be very useful for the Duchy of Mantua as well. However, he assured the Duke in this letter, he had not yet talked to Torriani about the possibility to sell his invention in Mantua. Negri wanted first to know what the Duke his master was thinking about such an action. In order to assure the quality of the machine, although it was not yet constructed, ambassador Negri confirmed that Torriani had already obtained an inventor privilege from the King of Spain protecting his invention for 30 years, a very interesting detail:

> This master Gianello [Janello] Cremonese, who is so good with clocks, has to take water from the river of Toledo above a very high mountain; having made a working model, [he] will now make a full-scale version without failing and very easily. Those of Toledo give him 8,000 ducats one time, and 1,900 ducats as a perpetual pension with the obligation that ...

---

75   Ceredi, *Tre discorsi*, See the 1st *discorso*.
76   ASFi, Mediceo del Principato pezzo 526, ca. 297: "*Ill.mo et ec.mo Prencipe mio O.mo/ E' tanta la volontà e divozione con che/ scrivo a v. eccellanza Ill.ma ch'io ho preso ardire/ di supplicarla si come fò humilissimamente/ ad esser servita di conceder la grazia/ che si domanda in nome di Giovanello/ horologiero, nel particular di quei edifici/ di acqua; stando che io son' interessato/ in quel negotio più assai, che non è esso/ et non havendo scritto questa ad altro fine/ qui si fine co'l basciarle mani a v eccellentza/ Humilissimamente. In Mantua a 7 di Febraio 1567/Humilissimo et devotissimo servitore/Hieronimo Negri*". ASFi, Mediceo del Principato pezzo 526, ca. 821: "*Illustrissimo et eccellentissimo signor mio osservantissimo/ io mi era molto prima rallegrato tra me stesso come servitore/affittionatissimo di v. eccellenza della figlila che le è nata. Hora me ne/ rallegro co' lei co' ogni affeto, et insieme le bacio le mani/della parte che le è piacciuto darmene co' la lettera sua e/ VIIII. per risposta della quale no' me le occorre dir' altro/seno' che di tutto quello che a' vostra Eccellenza piace in quel particular di maestro/Gianello resterò sodisfattissimo et co' obligo infinito a' la vo=/ lontà che v. Eccellenza mostra di far' a' me sempre gratie, et favori/Et baciandole da capo le mani le porgo ogni felicità. Di/ Mantova 28 marzo 1567/ Cesare Gonzaga*".

the device will constantly work and he has to maintain it and provide for the costs. The Majesty of the King has granted a privilege to him that states that others can not do similar work in his kingdom for 30 years except for Janello, or anyone he himself will permit. ... I have been told that, after finishing this one, he is planning to make many other devices and that he will lower greatly the price, and much more easily and with less expense will it be possible to lift water; when it was true that it was said to me we could bring up water from the river Po, or from the river Mincio, and from the river Oglio, with not much expense; based on that which they tell me with certainty, this device would serve very well in those rivers. I did not want to speak to master Gianello yet ... he affirms that in six months he wants everything to be set in place. If ... Your Excellency commands me to do more about this, I will profusely obey, and with this to your Excellency, I humbly kiss your hand. From the Court ... xxy April MDLXV humble vassal, and servant. Girolamo Negri.[77]

Girolamo Negri was the Mantuan ambassador; however, it emerges that his brokerage of Torriani was not only related to his official duties in the interests of his prince, but it went beyond them: once back in Mantua, having ended his diplomatic mission in Spain, Negri wrote directly to the ruler of Florence seeking for a privilege for Torriani's invention. In order to enforce his request he managed to have an influential "friend", in the person of the Count of Guastalla

---

77   ASMn, Archivio Gonzaga, Estero 3 Spagna e Portogallo, b. 593, April 22nd 1565: "... *Quel Mastro Gianello Cremonese così raro nel fare/ horologi ha da tirrar l'acqua del fiume di Toledo/ sopra d'un monte molto alto; havendo fatto un modello/ con che si tocca con mano, che lo farà senza fallo,/ et molto facilmente. Quei di Toledo gli danno per /questa opera m/8 ducati alla mano, et /1900 ducati /perpetua pension che esso si obblighi (come ha fatto) che questo/ edificio ha da durar sempre, il qual edificio esso/ mantiene ancho a sue spese: et la M.tà del Re fa/ un privilegio a lui, che alcuno non possi far la/ detta opera ne suoi Regni, et stati per 30 anni/ fuor che esso, o chi vorra esso. Si come è ragione/ vuole, così mi vien affermato anche, che fatto, che/ habbia questa fattione, penserà di farne molte/ altre, et che abbassarà molto il prezzo, et tanto più facilmente, et con meno spesa, quanto, che me-/no si havrà da tirar su l'acqua; onde s'è vero quello/ che mi vien detto si potrebbe tirrar su l'acqua/ del Po, del Mincio, et dell'Oglio, con non molta/ spesa; percioche mi dicon certo, che questo edifico servirebbe/ in quei fiumi molto bene. Io non ne ho voluto parlare/ al detto Mastro Gianello aspettando, ch'esso ne parlasse /a me, tanto più che non è cosa questa, che importi molta/ fretta, parendomi anche, che sii sa ... a vedere come/ riuscirà in Toledo, poi che afferma esso, che in sei mesi/ nel/ vol haver ispedito ogni cosa. Se ... v. Ecc.za/ mi commanderà alcuna cosa intorno a ciò, io la obbedirò/ incontinente, et con questo a v. Ecc.za bascio humilissima-/ mente le mani. Dalla Corte Castigliana (?) xxy d'Aprile MDLXV/ humiliss. Vassallo, et serv. / Hieronimo Negri*" (my translation).

Cesare Gonzaga, to write a letter to the Grand Duke of Tuscany, to speak for him and Torriani.[78] Cesare, son of Ferrante Gonzaga, Governor of Milan at the time of the construction of the Microcosm, promptly helped Girolamo Negri, but Francesco I de Medici refused anyway to grant a privilege for invention to the clockmaker.[79] According to his own words, he had to refuse this privilege because he wanted to protect his own servants who were devoted to similar inventions.[80] The most important of those servants was probably Bernardo Buontalenti, who competed with Torriani not only in the field of hydraulic engineering, but also in the art of miniaturised clockmaking.[81] This situation fits very well in the circumstances of Francesco I de Medici's protective policy

[78]   Morpurgo, in his dictionary of the Italian clockmakers, misunderstood this process of sponsorship: Garcia-Diego and Leydi were probably led astray by Morpurgo, mistaking the reading of the correspondence between Cesare Gonzaga and Francesco I de Medici as well: these scholars claimed that Cesare asked Francesco for a machine made by Torriani, which the Florentine prince refused to give. In reality, the letters tell a completely different story: García-Diego, *Juanelo Turriano, Charles v's Clockmaker*, 4; Silvio Leydi reported correctly the epistolary exchange between Cesare Gonzaga and Francesco de Medici but then he inexplicably introduced among the senders, together with Cesare Gonzaga and Girolamo Negri, an imaginary Cesare Negri, attributing to him the authorship of Girolamo's letter: Leydi, "Un cremonese del Cinquecento," 142.

[79]   ASFi, Mediceo del Principato, 526, ca. 300 (for a full transcription of the letter see: Leydi, "Un cremonese del Cinquecento," 142 n. 51): letter by Cesare Gonzaga to Francesco I de Medici: "*Illustrissimo et eccellentissimo signor mio osservantissimo/So che v. Eccellenza è stata supplicata per parte di Maestro/ Giannello Torriano horologgiaro di S. M.ta Cat.ca che, havendo esso nuovamente trovati due istromenti di cavar acqua, l'uno dai fiumi correnti, et l'altro da laghi i fiumi morti ... di concedergli un privilegio ...*".

[80]   9th of March 1567 (*1566 more Fiorentino*), answer by Francesco I de Medici to Cesare Gonzaga "*Ill.mo Sig.re, non restava defraudato M.ro Gianello Orologero a pensare che nessuno altro potesse disporre di me più che v. S. Ill.ma s'io havesse potuto consolarlo, perch'io amo tanto lei et desidero si di farle servitio che nissuna cosa m'è più grata che l'esser ricercato da quella. Ma ell'ha da sapere ch'io ho alcuni istrrumenti, che è forza sieno simili, nondimeno sieno come si vogliono io sono per metterli in opera in alcuni luoghi miei, che sono molto opportuni. Si che per non impedire li proprii, et chi me gl'ha proposti, come industriosi et d'ingegno mi è forza denegargli il privilegio che domanda, e dio sa quanto me ne dispiaccia si per non poter satisfare a V. S. Ill.ma come per non poter gratificare M.ro Gianello, al quale in ogni occasione desidero di far benefitio*": Campori, *Lettere artistiche inedite*, 49.

[81]   Buontalenti made a trip to Spain in 1562 when Torriani was employed at the court. As we shall see in the chapter on clockmaking he was said to have produced for Philip similar things that other sources attribute to Torriani: Amelio Fara, *Bernardo Buontalenti* (Milano: Electa, 1996); Andrea Iovino et al., *Acqua, continuum vitae: ...il divenire Mediterraneo nel racconto dell'arte e della scienza.* (Salerno: Artecnica production, 2000); Cristina Acidini Luchinat et al., *La Fonte delle fonti: iconologia degli artifizi d'acqua* (Firenze: Alinea, 1985).

towards his technicians. Franceso I sponsored his favourite engineer Bernardo Buontalenti in his quest for privileges for invention all over Europe. Cesare Gonzaga, the Grand Duke of Tuscany, wrote to several princes in order to ask for privileges for invention to be granted upon his protégé, and this sponsorship was extremely successful: pope, emperor, the kings of Spain, France, England and Poland, the viceroys, governors, dukes, archdukes, and cardinals of Italian and German states, the Republics of Genoa and Lucca, the Thirteen Swiss Cantons and the Great Master of Malta responded positively to this pressure, granting privileges for invention to Buontalenti.[82] On the other hand, Francesco I de Medici adopted a protectionist policy against foreigners inventors; for instance, in 1587, when the Swiss Angelico Risi pleaded for a privilege for a machine he had invented to clean the harbour of Livorno, the administration contemptuously answered that "His Highness thanks the supplicant, but when he will decide to empty and clean up his harbour, he has people who can do it properly".[83] With the death of Francesco I and the enthronement of his hated brother Ferdinand, things changed. Privileges were granted more liberally to foreigners. However, it should be noted that the first group of recipients consisted of Germans of his personal guard, the Lanzi, and of Austrians, Lorrainians, Burgundians and Flemish, whose presence at court had increased by the time the Grand Duke had married Christine of Lorrain.

The rich attraction that the flamboyant court of Florence exerted on creative artisans encouraged Janello to attempt to gain favour still another time: indeed, his friend Jacopo di Trezzo, who had been working as a precious-stones worker for the Medici for years, would attempt once more in 1572 what seems a timid promotion of Torriani by Francesco's old father, Grand Duke Cosimo I. In this case Jacopo di Trezzo, in order to promote Janello, sent the clockmaker's greetings to Cosimo and at the same time mentioned the successful construction of Janello's giant water-lifting device in Toledo.[84] Jacopo's skills in turning rock crystal and precious stones was not probably to be found in Florence at such a level of refinement, and this created a special market for his works. In Florence, Janello had instead fierce competitors, both in clockmaking, as della Volpaia's family, and in hydraulic engineering, first of all in the person of

---

82   I found in the State achive of Milan the privilege granted to Buontalenti by the governor in 1578 for a new type of mill that could elevate water: ASMi: Acque Parte Antica, Molini.

83   Lamberini, "Inventori di macchine e privilegi cinque-seicenteschi dall'Archivio Fiorentino delle Riformagioni," 184-86.

84   1572, 10th of January: Letter from Jacopo de Trezzo to Cosimo I de Medici "... *y con deseo de venirla a vedere e servirla en su terra prima che muoya, de bon cor a V. Ecc-a mi raccomando e così fa M. Gianelo el qual sta contento per eser reusito a onor del suo edificio de subir l'acqua a Toledo*" in Babelon, *Jacopo da Trezzo et la construction de l'Escurial*.

Buontalenti. Thus, Jacopo da Trezzo's endorsement of Janello seems to have been fruitless.

Ambassador Negri's commitment as a broker for Janello's inventions has some interesting but mysterious sides: in the summer of 1566, when the project for the construction of the first water-lifting device was already started, the ambassador and the clockmaker signed a contract:[85] In the agreement one can read that Janello Torriani committed himself to make two wooden models of his water device for ambassador Negri.[86] However, it is not specified why. From

---

85   Garcia-Diego, in one of his unfortunate quotes without reference, claimed that Negri and Janello created a business-society in Toledo in order to share the profits from the market of Torriani's inventions. Garcia-Diego probably took this data from an article by Garcia Rey, who did not give any archival indication either. Verardo García Rey, "Temas de arte: Juanelo Turriano ; matemático y relojero," *Arte español*, 1929, 525.

86   I was lucky enough to find this contract at the Archivo Historico Provincál de Toledo: Archivo Histórico Provincial de Toledo, Gaspar de Soria, prot. 1980, año 1566, sin index, f. DCCCXL, 31st of August 1566: "*Scriptura de Juanelo. † En la muy noble y muy leal çiudad de Toledo, treinta e un dias del mes de agosto, año del señor de mill e quinientos e sesenta y seis años, en presençia de mi el escrivano publico e testigos yusoescriptos, parescieron presentes de la una parte el señor Geronimo de Negri, enbajador de Mantua y vecino de la dicha çib-dad de Mantua, y de la otra parte el señor Juanelo Turriano, vecino de la çibdad de Cremona, que es en el reyno de Ytalia, estantes al presente en esta dicha çibdad de Toledo, e dixeron que por quanto ellos son convenidos e conçertados en esta manera: que el dicho Juanelo se obliga de hazer al dicho Geronimo de Negri dos modelos de madera con sus herramientas y ader-ezos con çiertas condiziones y en çierta forma que ante mi el presente escrivano presentaron firmadas de sus nombres, que su thenor de las quales es este que se sigue: Aquí las condiçio-nes: Con las quales dichas condiçiones y con cada una dellas, los dichos señores Geronimo de Negri, enbajador, e Juanelo Turriano se obligaron de haçer los dichos modelos, según e de la misma forma y manera que de suso va declarado, las quales según que en ellas se contiene y cada una parte por lo que le toca se obligan, se obligaron de lo cumplir e guardar, pagar e aver por firme obligaçión sus personas e bienes muebles e raiçes avidos y por aver, e por esta carta dieron poder cumplido a las justiçias de Su Majestad de qualesquier partes, a la juris-dicción de las quales e de cada una dellas se sometieron, e renunçiaron su propio fuero, juris-diçion y domiçilio e la ley sid convenerid de jurisdicione, porque por todo rigor de derecho e bia executiva les conpelan e apremien a lo ansi cumplir con costas como por sentencia pas-ada en cosa juzgada y dada a entregar. E renunziaron todas leyes, fueros e derechos e otras cosas que en su favor en este caso sean, y espeçialmente renunçiaron la ley e derechos en que diz que qualquier renunçiaçion fecha de leyes non vala. En testimonio de lo qual anvas par-tes otorgaron esta carta ante mi el escrivano publico e testigos yusoescriptos en el dicho dia mes y año susodicho, que fue fecha y otorgada en la dicha çiudad de Toledo treinta y un dias del mes de agosto de mill y quinientos sesenta y seis años. Testigos que fueron presentes Pablo Alvarez y Pedro Loçano, alguazil, y Cristoval de Astorga, vezinos de Toledo, y los dichos otor-gantes lo firmaron de sus nonbres en el registro desta carta. Y el dicho Cristoval de Astorga e*

the letters I found in the archive of Mantua, we know that a model was sent to Italy most probably as a specimen to enrol for a bureaucratic procedure in order to obtain a privilege for invention in that city. Perhaps the second model was directed to Venice or to Rome. We know of another model, perhaps one of these two, or a third one, that together with a portrait of the clockmaker would reach Cremona after Torriani's death as a memento to his ingenuity and success.[87] It is not clear what Negri was expecting to gain from the agreement of Toledo. If Torriani had to produce the two models, it means that Negri probably paid for them. Nevertheless there is no mention of economic details in the document. We only know that the models had to respect certain requirements and that they were to be made in wood and metal. The ambassador substituting Negri in Madrid in 1566, Emilio Roberti, also had a part in the story: in a letter sent to Mantua in 1567 the new ambassador talked about a model of the Toledo device which Janello gave him and which Emilio's brother was taking to Mantua.[88] The year after, Duke Gulielmo granted Torriani with the privilege.

Why did Girolamo Negri promote Torriani in Mantua and Florence, and perhaps elsewhere? Negri was a gentleman from the court of Mantua.[89] He appears for the first time in 1542, as a young and resourceful courtier, Ippolito

---

Juan Anbrosio Virago, milanés, juraron en forma de derecho conozer al dicho señor Geronimo de Negri, ser el propio que otorga la escriptura, sin cautela alguna. Juanelo Turriano, Hieronimo Negri".

87   It was the great chancellor and senator of the duchy of Milan Danese Filodoni who sent these two relics to be worshiped by Janello's fellow-countrymen: ASCr, Comune di Cremona, Litterarum 38, ca. 65, 24th November 1587.

88   ASMn, Archivio Gonzaga, Estero 3 Spagna e Portogallo, b.594, *"Illustrissimo et Eccellentissimo Sig. et padron mio Oss.mo/Mio fratello viene in Italia e oltre che porta il modello che si è havuto qui/Da questo valent'huomo di m.ro Gianello per cavar acque de fiumi/E lagune per adequar terreni e far ontane nelle città [sic], come/ Il Negri, a chi s'ha da consignar, doverà haver deto a v. Ecc.za/ Egli ancor seco ... et perchè in questi otto anni che sono stato qui, dove ho esposta/ la mia persona a molti pericoli, le poche mie sostanze paterne/ sono andate a male, da che segue che mi trovo con molti de-/biti alle spalle, supplico con ogni riverenza l'Ecc.za vostra sia/ servita comandar che s'habbi riguardo a miei servigi,/ et che si come gli altri servitori del Cardinal Hercole di fe: me:/ sono stati riconosciuti delle servitu loro, che a me si dia/anco qualche satisfatione, che ... dalla Corte Cattolica, alli 27 di Otobre 1567/humilissimo vasallo et servitore Emilio Roberti".*

89   It has been difficult to obtain a clear picture of Girolamo Negri: indeed, there were several homonymous literates in sixteenth century Italy. For instance, in the Ambrosiana's archives there are several letters and other documents related to a certain Girolamo Negri from the Republic of Venice (1496-1580). Other contemporaries with the same name are to be found in Genoa and Piedmont.

Capilupi,[90] sent a letter to Ferrante Gonzaga describing a carnival feast where Girolamo Negri, together with some other courtiers and a Jewish musician, were dressed up like satyrs by Giuglio Romano, performing a *moresca* dance with musical instruments and sticks.[91] This attitude shows a charismatic personality, quite suitable to what would be his crucial office of orator to be sent to key places such as Madrid and Milan. Girolamo Negri, like many other scions of the local gentry, probably followed a *cursus honorum* consisting of training in local offices before being later launched in diplomatic missions abroad.[92] He was appointed Mantuan resident ambassador in Spain for the years 1559-1566 and 1589-1590.[93] In between these Spanish periods, in the year 1567, he was appointed ambassador in Milan, and in the year 1578, Girolamo was serving as preceptor of Vincenzo, the Prince of Mantua. The Duke called him "*a principis filii nostri gubernator*": i.e., "governor of our son the prince".[94] It seems that his personality led to his recognition with singular honours, as when the Duke of Alba helped him to be invested as a knight of Alcántara.[95]

From a Milanese document dated 26th July 1568, we are informed that during the year 1567 Janello had applied to have confirmed his Spanish inventor privilege in Milan:

His Catholic Royal Majesty, Our Lord [Philip II], has conceded to Joanello Torriano [Janello], loyal servant of Your Excellency, a privilege that no-one except the supplicant can use the two devices he invented to take water from the rivers, fountains, ponds, marshes, or, other low waters to water lands or, for use of the needy populace until thirty years. They cannot use the said devices under penalty to lose the instruments, and a fine of 200 ducats, as one can more extensively read in the privilege given in

---

90   As we have previously seen, he was uncle to Camillo Capilupi, who wrote about Torriani and was related to Alessandro, who signed the clockmaker's Mantuan privilege.

91   Carlo D'Arco, *Delle arti e degli artefici di Mantova*, vol. 2 (Mantova: Agazzi, 1857), lett. 168: Lettera scritta al 25 di febbraio del 1542 da Ippolito Capilupi a Ferrante Gonzaga.

92   Alessandro Luzio, ed., *L' Archivio Gonzaga di Mantova: la corrispondenza familiare, amministrativa e diplomatica dei Gonzaga*, vol. II (Mantova: Mondadori, 1993), 80.

93   ASMn, Archivio Gonzaga, registro degli ambasciatori.

94   Ibid. footnote 5.

95   From a letter written in 1565 to the Duke of Mantua, Negri tells how the *duque de Alba*, who ten years earlier in 1555 had been Governor of Milan after Ferrante Gonzaga, recommended him to acquire the investiture of knight of the military order of Alcántara: Archivio di Stato di Mantova, Archivio Gonzaga, Esteri 3, 593: 3 sett. 1565? 1566?: "*ho pigliato lo habito dell'ordine della cavaglieria di Alcantara*" and 3rd of Novembre 1565, where he talks about the help received from the duque de Alba.

Madrid on the 15th of May 1567 of which a copy is attached [unfortu-
nately lost]. On this, the supplicant, wanting the approval from the
excellent Senate in conformance with the laws of this State, has found
himself to be past the said year. Consequently, he comes to your excellency, supplicating that you may order to the said Senate, notwithstanding
that the year has passed, to approve said privilege, disregarding every
order, law or statute that they did on the contrary and this he hopes.[96]

For the moment one has not found the original plea, but just this petition that
one year later asks the Governor of the State of Milan to put pressure on the
bureaucracy of the Milanese Senate to speed up the procedure. Nevertheless, it
is difficult to imagine that Girolamo Negri was not involved in this business
too, having been just nominated resident Mantuan ambassador in Milan in
that very year, 1567.

In order to be endowed with all those delicate duties, Girolamo Negri had to
be more than a talented entertainer, as the record from Capilupi showed; he
must have been a man well-respected for his capabilities and loyalty, and here
is a fundamental feature of Negri's social position, a trait that substantially
explains his brokering of Torriani. The Mantuan ambassador was more than a
simple friend to Cesare Gonzaga, as the count of Guastalla had addressed him
in his letter to the Grand Duke of Tuscany. According to a late record, Girolamo
Negri married Cesare's half-sister Silvia, natural daughter to Ferrante Gonzaga.[97]
This suggests that the ambassador's loyalty was based not just on personal
trust, but also on blood ties. From this complicated web we can say that the
link connecting Torriani to Negri and Cesare Gonzaga was anchored in precisely the period when Ferrante Gonzaga († 1557) was Governor of Milan
(1546-1554).

In the case of the papal privilege granted to Janello in 1567, we do not have
for the moment any document that leads to Negri's agency. However, we can
observe that the secretary who signed Pius v's *motu proprio* was a certain
"Melchior B.", most probably that Milanese apostolic proto-notary and diplomat Melchiorre Biglia (1510-1571), friend of Carlo Borromeo and of his uncle,
the deceased Pope Pius IV.[98] Melchiorre Biglia performed several diplomatic

---

96   Petition to the Milanese Senate for the granting of a privilege for invention to Janello Tor-
        riani, July 26, 1568, ASMi, Correspondence of Chancelleries of the State, 227.

97   ASMn, Manuscript Carlo D'Arco, "Famiglie mantovane e mille scrittori mantovani" n.d.,
        285-95, b. v.

98   Gerhard Rill, "Biglia, Melchiorre," *Dizionario biografico degli Italiani* (Roma: Istituto della
        Enciclopedia italiana, 1968).

missions for Pius IV and Pius V at the imperial court. We cannot say if he had any influence on the Pope's decision to grant this *motu proprio*; however, people like Melchiorre Biglia are an example of the permanence in Pius V's curia of figures strictly connected with the Borromeos'-Gonzagas'-Medicis' di Marignano circuit. It would be interesting to investigate possible relations between this apostolic proto-notary and diplomat, and the ambassador Girolamo Negri.[99]

To return to Venice and to the idea that recommendations and political issues were behind the shortening of the span of time for the privilege duration, and the six votes against Janello, we can see that the Gonzaga family was equally well-connected in this city: first of all, the vicinity of the two states imposed a defensive diplomatic policy of Mantua in the political life of the Republic, and members of the Gonzaga-kin were honoured with the title of Venetian patricians. Besides them, other prominent Mantuan figures possibly connected with Torriani had their influence in Venice: for instance, the scheming Ippolito Capilupi, who also knew Girolamo Negri, had the office of apostolic nuncio in Venice, where he was well connected and on good terms with the Doge Girolamo Priuli (1559-1567), who supported his election to the cardinalship. Unfortunately for him, his lack of vigilance in fighting heresy and his former attempt to elevate to the pontificate Cardinal Ercole Gonzaga, while Pius IV was ill but still alive, alienated him from the benevolence of Carlo Borromeo, whose veto squelched the purple aspirations of the Capilupi. During the last months of Giralamo Priuli's *dogado*, Torriani's petition for a privilege was starting its bureaucratic procedure at the office of the *Savi et Essecutori dell'offitio sopra le Acque*, the technical office dealing with water. Could Torriani have used these connections for his patent granted in Venice?[100] It is difficult to know for sure. What we do have evidence of are some practicable Mantuan channels that Janello might have used. May this support have provoked those six senators who voted against the resolution? Were they filo-Florentines who supported the interests of Francesco I's protégés? After all, in the long list of authorities produced by Lamberini that granted Buontalenti with a privilege, Venice is missing. Janello Torriani's experience in the market for the privileges shows that procedures for granting such patents were not based on plain merit

---

99   A further investigation should also look at a possible diplomatic role in this issue of the Apostolic Nucio in Madrid at this time: Gio. Battista Castagna. Franco Molinari and Daniele Montanari, "Rapporti con i vescovi italiani," in *San Carlo e il suo tempo: atti del Convegno Internazionale nel IV centenario della morte (Milano, 21-26 maggio 1984)*, ed. Danilo Zardin, Studi e fonti su San Carlo Borromeo 2 (Roma: Edizioni di storia e letteratura, 1986), 325.

100   Gustavo De Caro, "Capilupi, Ippolito," *Dizionario biografico degli Italiani* (Roma: Istituto della Enciclopedia italiana, 1975).

and technical evaluations, but that they also took into account personal relations. His case demonstrates that if two parties were competing for a privilege offering a similar performance, the authority tended to protect the interests of his personal relation. In this system, whom one knew counted as much as competence.

The problem of the duration of an inventor privilege remains an open question. Going back to the description that Girolamo Negri gave to the Duke of Mantua the 25th of April 1567 about the contract that Janello had previously signed in Toledo with the local City Council and with King Philip II for the construction of the first Toledo Device, the ambassador adds that the King granted Janello with the privilege "that nobody could build this machine in his reigns and state for the next 30 years".[101] This is probably why after 1598, 30 years after the contract, once Juanelo Turriano de Diana, nephew of Janello, had died, the administration of the Toledo Devices passed to royal officials others than Janello's heirs.[102] However, the petition written perhaps by the very Girolamo Negri to the Governor of Milan in 1567 dates the privilege to the 15th of May 1567 (a date different from the above-mentioned contract), meaning that it was not the very contract to work as a privilege, but perhaps a specific privilege now lost. This chronology is getting even more complex to decipher after the discovery of another Spanish privilege that dates back to the summer of 1563. This document demonstrates that at that early date, Philip II had already granted him a privilege for the invention of the new machine "to dig and raise [water], with said instrument in a straight or oblique way with tubes open and closed, from rivers to high places thanks to a system of fastenings".[103]

---

101 If one looks at the huge sum of golden ducats (57,000) that the contract signed in Toledo in 1567 by Janello, the City and the King's representative assured to the clockmaker and his heirs, we see that it fits with the description that Negri had given to the Duke of Mantua: as previously seen, Negri wrote that Janello had received 8,000 ducats one time, plus 1900 ducats per year for the next three decades (the result of the multiplication is indeed 57,000, as written in the contract). Although the contract from Toledo does not explicitly express protection from imitation, as we shall see when discussing secrecy and technological innovation, no representations of the Toledo Devices, the marvels of macromechanics of the Renaissance, survive.

102 Cervera Vera, *Documentos biográficos*, doc. 139 and 140.

103 Thanks to Almudena P.rez de Tudela for having showed me this document: *"de sacar y subir con el dicho instrumento derecha y oblicuamente caços abiertos y cerrados de ríos a lugares altos por vías de subjecion"*, AGS, Cámara de Castilla, libro de cámara 134, f. 534, Madrid, 13th of july 1563. In the margines: *"Juanelo Relogero: El Rey/ Por quanto por parte de vos Juanelo n[uest]ro relogero, nos ha sido hecha rel[aci]on q[ue] vos haveis imbentado un artifiçio nuevo q[eu] antes de agora no ha sido hecho de sacar y subir agua con el d[ic]ho instrum[ent]o derecha y oblic[u]amente con caños abiertos y cerrados de ríos a lugares altos*

We cannot say if this was a machine slightly different from the one later built in Toledo.

Janello's story, together with Ambassador Negri's letter to the Duke of Mantua, enables us to see that during the sixteenth century the act of protecting one's authorship could have at least three different juridical forms. First, the clockmaker was granted with what we might call "privileges for invention" *tout court* (the ones he obtained in Spain, Rome, Mantua and Venice); second, he signed the contract for the Toledo Device in Toledo (1565), which gave him 30 years of monopoly on the hydraulic machine; and finally, he received the imperial diploma for the Microcosm (1552). Among these three different ways to reward inventions, the most advantageous for the granting authority seems to be the privilege for invention. This may be the reason of the latter success of this juridical practice.[104]

---

por vías de subjecion/ suplicandonos y pediendonos por m[e]r[ce]d que haviendo considera-çion a q en la imbençion del d[ic]ho ingenio haveis travajado y gastado mucho fuesemos servido de daros liçençia para q vos, o quien v[uest]ro poder huvi[er]e y no otra pers[on] a alguna podais hazer y usar el d[ic]ho ingenio en estos n[uest]ros reynos de la corona de castilla / o como la n[uest]ra m[e]r[ce]d fuese/ y nos acatando lo susod[ic]ho / y por os hazer m[e]r[ce]d/ por la pres[en]te no haviendose usado del d[ic]ho ingenio por otra per-sona alguna en estos d[ic]hos n[uest]ros rey[n]os y señorios de la corona de castilla, damos liçençia a vos el d[ic]ho Juanelo para que por t[iem]po de treynta años contados desde el dia de la hecha desta n[uest]ra çedula en adelante vos o quien el d[ic]ho v[uest]ro poder para ello huvi[er]e y no otra persona ni personas algunas podais y puedan hazer y usar el d[ic]ho artificio como arriba va declarado de sacar y subir agua de rios a lugares altos en todos los n[uest] ros reynos y señorios de la corona de casti[ll]a y defendemos que persona alguna de qualquier estado o condicion que sea no pueda hazer ni usar del sobred[ic]ho ingenio en ellos ni en ningu[n]a p[ar]te dellos durante el t[iem]po susod[ic]ho sin speçial poder v[uest] ro/ sopena que pierdan los instrumentos y aparejos que para ello tuvieren y de çinquenta mill m[a]r[avedie]s, la qual d[ic]ha pena se reparta en esta ma[ne]ra la t[e]rçia parte para mi camara y la otra t[e]r[ci]a parte para la Justi[ci]a que lo señ[al]are y executare/ y la otra t[e]r[ci]a p[ar]te para vos o de quien de vos tuvie[r]e y causa/ y mand[am]os a los de n[uest] ro q.o [consejo] esta tomando la razon della ant[oni]o de arriola n[uestr]ro criado fecha en madrid a treze de jullio de 1563/ yo el rey/ refr[endad]a de Erasso/ Sep.a de menchaca y v[elas]co".

104   The differences, from an economic standpoint, are that in the first case the authority granting the privilege did not spend anything; on the contrary, in the case of a fine for he who had stolen the idea of the patented machine, the state was earning money. Moreover, after the period of duration of the privilege, the state and its subjects could use the technology for free. In the second case the authority promised to pay for the invention only when a specific machine was finished and well-functioning. In this case, all the risk was on the inventor's side, who had to cover all the costs during the process of creation; the King and the City of Toledo were however promising to pay for the machine if

Janello Entrepreneur

As art historians have firstly shown, famous artisans, while perceived by the
market and by historiography as individuals, usually worked more as a collective entity.[105] Painters and sculptors made great use of apprentices, journeymen,
masters and servants. Clockmakers were no less involved in such collective
endeavours.[106] An engraving from the second half of the sixteenth century
made by Stradanus (Jan Van der Straet, 1523-1605) helps us to visualise the
structure of a clockmaker's workshop (Fig. 46). The engraving, entitled *Horologia ferrea*, shows the interior of such a workplace in the midst of its activity.
The preparatory drawing,[107] with its sketchy technique, suggests that it may
have been made in the workshop itself. In the drawing as in the engraving the
same number of people crowd the indoor space of the workshop, but in the
engraving one new character stands in the foreground: he is a gentleman,
depicted as a rich customer. His short damask cloak and dagger point to the
considerable value of the objects produced in this setting.[108] Sitting on the

---

the construction was successful. In the third case the invention was financed during its
making and rewarded with a perpetual pension. For the invention of the Microcosm,
between 1552 and 1623, Janello Torriani and his heirs received from the duke of Milan,
who was since Philip II the King of Spain, more then 12,000 *scudi* of gold. In 1612, though
reduced to its initial amount of 100 *scudi*, the pension was bestowed upon Janello's great-
granddaughter, who enjoyed it until 1623, when she sold it to Girolamo Marzorati. For
what concerns the Toledo Devices, the King of Spain had to push Torriani into a life-long
legal litigation with the City Council of Toledo, in order to pay less than he was supposed
to pay. Instead of granting him with a life-long pension as his father had done with the
Microcosm, Philip II gave Torriani and his heirs a 30 years monopoly on the invention.

105   For what concerns the world of Science see: Alessandro Biral and Paolo Morachiello,
       Immagini dell'ingegnere tra Quattro e Settecento: filosofo, soldato, politecnico, ed. Antonio
       Manno (Milano, Italy: F. Angeli, 1985); Shapin, "Invisible Technicians: Masters, Servants,
       and the Making of Experimental Knowledge"; Pamela O. Long, "Trading Zones: Arenas of
       Exchange during the Late-Medieval/Early Modern Transition to the New Empirical Sci-
       ences," *History of Technology* 31 (2012): 5-25.
106   We shall see in detail in the chapter where we discuss Torriani's innovation in the field of
       Renaissance planetary horology, how certain projects of complex clocks required the
       increasing collaboration of different specialists.
107   Cooper Union Museum for Arts and Decoration: Michel N Benisovich, "The Drawings of
       Stradanus (Jan van Der Straeten) in the Cooper Union Museum for the Arts of Decora-
       tion, N.Y.," *The Art Bulletin / Ed. John Shapley [U.a.]*. 38, no. 4 (December 1956): 249-51.
108   Martellozzo Forin, in his interesting book on the Mazzoleni, lists their customers, high-
       lighting their elevated status. For instance, within this group one can find Venetian patri-
       cians such as Alvise Foscari, Angelo Contarini, Agostino Foscarini and Paduan patricians
       such as Gio. Battista Santasofia, Alessandro Maggi da Bassano, Alessandro and Francesco

FIGURE 46   *Jan Van der Straat, known as Stradanus (drawing), Luigi Alamanni (patron),*
             Philip Galle, Theodor Galle and Jan Collaert (engravers and printer), Horologia
             Ferrea *(Iron Clocks), From the series of engravings entitled* Nova Reperta, *Antwerp*
             1587-1589, cm 31 × 23.

only chair, a serious and authoritative character with glasses, clearly the eldest
present, checks the teeth of a freshly filed geared wheel: he is the master
clockmaker. Around him a team of men and boys are busy at the forge, anvil
and clocks-assemblage, and putting the finishing touches to timepieces. The
master's role was managerial.[109]

Borromeo (of a branch of the Milanese family settled in Padua after a French victory
against the imperial party). Among the customers of the Mazzoleni, besides monasteries,
one can also find members of the urban professional elite, such as notaries, physicians
and jurists, professors and students of the Studium and servants of important households. The clocks produced by the Mazzoleni workshops cost between 15 and 20 ducats:
Martellozzo Forin, *La bottega dei fratelli Mazzoleni, orologiai in Padova, 1569,* 72-93.

109   See Molà, "Privilegi per l'introduzione di nuove arti e brevetti"; and Patrick Boucheron,
      "L'artista imprenditore," in *Produzione e tecniche,* ed. Philippe Braunstein and Luca Molà,
      vol. volume terzo, Il Rinascimento italiano e l'Europa (Treviso: Fondazione Cassamarca :
      Angelo Colla Editore, 2007)

The first thing to point out is that entering the court and leaving the guild-system meant that Janello stopped being subject to fiscal pressure. Now, Torriani had to work full-time for the Emperor and later for his son the King, but he could still hunt for other revenues, as with the privileges for invention he was seeking. On the level of master-workshop relations, we can see that Torriani was constantly surrounded by a group of people who depended on him and who performed different duties under his supervision. For the period in Cremona and Milan, poor in sources, we have seen that at least in two occasions, apprentices were included in his workshop. More data is available in the documents related to his Spanish period.

At the time of his employment in Toledo (1565-1585), we know of three officials salaried by Torriani who were daily running the water-lifting devices.[110] Though it is not clear what kind of technical knowledge those employees had, it is evident that they were assisting Torriani in his complex service for the Crown. When Juan de Herrera made a technical inspection of the Toledo Devices after his friend Janello had died, he recorded that the ordinary personnel needed for the daily running of the first and second machines were a carpenter, a blacksmith and welder, two unskilled labourers and an official. For this post Herrera recommended a certain master Jorge, servant of the late Janello. He was the qualified Giorgio di Diana, brother to the physician Orfeo, Torriani's son-in-law.[111] Giorgio, as an apprentice, was already in Yuste with the clockmaker; at that time (1556-1558), Torriani had two attendants: a Flemish master and the young Italian relative. Herrera highlighted that "Jorge" was the most experienced that one could ever find because he was a servant of Janello and had assisted him for the construction of the two devices. Herrera also recommended Jorge for the maintenance of the two planetary clocks, because, even in this case, he was the one that knew them best.[112] Indeed, afterwards,

---

110   Cervera Vera, *Documentos biográficos*, doc. 44, 1575 (?).. 1575 (?): doc. 44.

111   Giorgio was brother to Janello's son-in-law, the physician Orfeo di Diana. These documents from the Archivo Histórico Provincial de Toledo and from the Archivo Historico de Protocolos de Madrid make finally clear that Giorgio (o Jorge/Xorge) de Diana was brother to Orfeo, who was a physician: AHPT, Gaspar de Soria, prot. 1980, año 1566, sin index, f. DCCCCXL, 31st of August 1566 Gaspar de Soria, prot. 1982 sin index doc 15th nov. 1571 sin foliar: notary act of Giorgio di Diana who wrote: "... *al dotor Orfeo de Diana mi hermano que ests absente de / la villa de Sale en Italia en el estado de Milan ...*"; AHPM, prot. 2766, folio 31 (Luis Suarez): 1602, 22nd of January: "*Testamento otorgado por doña Barbara Medea, viuda del doctor Orfeo de Diana, vecino que fue de Toledo, hija legitima de Juanelo Turriano.*"; prot. 841 (Pedro de Zuola) ff 725r 734r: 1588: "*D.a Barbara Medea Turriano viuda del doctor Orfeo de Diana, medico milanes*".

112   Cervera Vera, *Documentos biográficos*, doc. 129, 6th of March 1586.

Jorge de Diana moved to Madrid together with the two clocks, and he was appointed as their moderator. From letters and administrative documents we know that Janello Torriani had also two other *criados*, or servants: the Italian Giovanni Domenico Malaspina and the Spanish Pedro de Almaguer (most probably from Toledo).[113] The tasks of these workers probably beside the administration and maintenance of the waterworks included taking care of the clocks, of technical inspections (bell smelting, hydraulic engineering, astronomical surveying, etc.), and the writing of manuals for the King. Especially when he was involved in large-scale works, like the Toledo Devices, Janello must have had a greater number of employees. Jacopo da Trezzo, working at the construction site of the Escorial, provides us with a good term of comparison: in 1585 he had to pay some 50 workers in his household for the construction of the famous altarpiece.[114] Among Torriani's employees, besides Jorge, or Giorgio di Diana, there were several members of his family.

Family links were probably perceived to be the strongest base for an efficient business because they were based on trust and they focussed the aspiration of individuals on the kin's common good. The family was considered the most natural place for solidarity in an intergenerational pact according to which the eldest provided education, resources and credit for the younger in exchange for respect and support. For instance, Janello Torriani's nephews Giorgio di Diana, Giovanni Antonio Fasolo and Diego Juffré de Soria, assisted Janello in legal affairs, and Janello helped them in their careers. The potent institute of nepotism was a stabiliser in a very fluid system of power.[115]

---

113 See documents dated 9th of July 1580, Toledo: 19th of July 1580, Toledo; 29th of July 1580, Toledo, and perhaps 13th of November 1570, El Escorial: Cervera Vera, *Documentos biográficos*; in 1639, the keeper of the Toledo Device died: he was Juan del Castillo, married to a certain *"doña Felipa de Almager"*. The widow passed the office of her deceased husband with a notary act to Luis Maestre, goldsmith and clockmaker of the Cathedral of Toledo. Although I undertook no prosopographic investigation on the Almaguer family, it is likely that Felipa was somehow related to Janello's servant. The family-occupation of the posts connected with the keeping of the hydraulic devices and with the clocks would offer further support for this hypothesis: Julio Porres Martín-Cleto, "El Artificio de Juanelo en 1639," *Anales Toledanos*, 1982, 175-86, see inventoty at page 182.

114 Babelon, *Jacopo da Trezzo et la construction de l'Escurial*, 58-61.

115 Nepotism had its best-known application in the Roman curia: following elections when a new Pope needed a loyal governor, who could be trustworthier than a member of his own family? The biggest issue related to this practice was the attempt of many popes from the second half of the fifteenth century to transform offices into hereditary ones in those administrative, juridical and military districts within the *Patrimonium Sancti Petri*. Criticism of this practice has given nepotism an entirely negative connotation that does not

The one who had a remunerative business equally had the moral obligation to redistribute it among his community, in which blood relations represented the first link. Consequently it was normal practice to involve relatives in one's successful business: living standards were strictly connected to family wealth. Modern Western standards in welfare are very far from any public warranty of social care in the sixteenth century. At this time the former was synonymous with the notion of charity, a very different concept from the idea of right, especially because alms often required gratitude to wealthy private subjects. The step towards *clientelismo* was very short. The idea that one's heir might risk falling into indigence was a most serious fear. In a bitter letter written by Torriani on his deathbed and delivered posthumously to Philip II, he claimed that he moved to Spain because he had wished to give his daughter and her heirs a better life.[116]

Let's have a closer look now at Janello Torriani's family. Janello's only son died still young in Milan. The only other child was Barbara Medea, born around 1531. She married the physician Orfeo de Diana in Milan, and had in all five children, two males and three daughters: Juanelo Turriano de Diana, Friar Domingo Turriano of the Order of the Minims, Emilia Felipa de Diana Turriano, Laura Antonia Turriano, and Maria Turriano. In the Duchy of Milan, a brother or a cousin of Janello had a family too that would later be in contact with the Spanish branch: the names we know so far are the one of Janello's nephew Bernardino, of his son Leonardo, who later became Philip II's *ingeniero mayor del reyno de Portugal*, and the one of a less qualified relation answering to the name of Marco Antonio Torriani, citizen of Milan. In Cremona, a branch of the family will survive for centuries, keeping alive the memory of the clockmaker. Except for Domingo Turriano, all the other four siblings got married, and at least three of them had children. Also Giorgio di Diana, who served Janello for 30 years, had a son with a Spanish woman, as we shall see. Janello involved most of his relatives in his business, and even after Torriani's death, it was the son of his daughter Juanelo Turriano de Diana who took his post as official of the Toledo Devices, and it was his nephew-in-law Giorgio di Diana, his helper since the time of Yuste, who was entrusted with the care of the two planetary clocks. This was a customary solution.[117]

---

take in account the social reasons for its existence. Sandro Carocci, *Il nepotismo nel Medioevo: papi, cardinali e famiglie nobili* (Roma: Viella, 1999).

116 "... de donde las hize venir pensado hazer por ellos lo que ya no puedo ... pues en toda mi larga vida no he podido dexarles más que una pobre casa ... en la Villa de Madrid y el artificio de que he suplicado a Vuestra Majestad se sirva": Cervera Vera, *Documentos biográficos*, doc. 130.

117 A similar situation is offered by Florence, Venice and Reggio nell'Emilia. Lorenzo della Volpaia had built a remarkable planetary clock by the beginning of the sixteenth century.

We have already seen that in 1585, shortly after Janello's death, Juan de Herrera was sent to assess the value and functionality of the water-lifting machines and to see what among Janello's possessions were to be taken for the service of the Crown. Herrera valued the two devices, which could lift more than 43,000 litres of water a day, together with the help of some other experts, in the sum of 37,000 ducats. The only thing that had not been valuated, concludes Juan de Herrera bitterly, had been Janello's craftsmanship. And actually, wrote Herrera, the clockmaker's heirs were dying of hunger – "*se mueren de hambre*" – and this was not Janello's fault because he respected all the contracts he had signed for the construction of the machines. Five wooden boxes were made in order to transport to the Royal Palace of Madrid all Janello's belongings related to his profession: books, papers and mathematical instruments. At the beginning of the following century, eleven machines and instruments of Janello would still be listed together with the two planetary clocks in the royal inventory. The heirs were meant to be compensated economically (and they would be with the significant sum of 500 ducats) for the material that once belonged to Torriani's workshop, and also with offices connected with the keeping of their deceased patriarch. It seems that Janello's heirs, or their successors as keepers of the Toledo Devices, charged a fee to curious visitors that wanted to visit them, using a royal possession to make extra money.[118]

The two imperial planetary clocks, the Microcosm and the Crystalline, were taken to the Royal Castle of Madrid. In 1591, the royal cosmographer Pedro Ambrosio Onderis, who lectured mathematics at court, was put in charge of

---

After his death, his son Camillo was put in charge of the instrument. After his death it was then the turn of Lorenzo's grandson Girolamo. The same scheme was repeated in Venice and Reggio for the offices of public clock keeper. As we have previously seen, Gian Carlo Rainieri, who built with his father the clock of St. Mark in Venice, received the office of keeper. This office was hereditary. This kind of solution was quite common: his father Gian Paolo Rainieri, who had also rebuilt from 1481 to 1483 the public clock of Reggio, was later made its keeper, and this office was hereditary and was handed over to Gian Carlo's brothers. It is remarkable how conflicts between property and possession of clocks with a public function could bring quite unusual situations. For instance, at the beginning of the 1460s, the civic clock of Cremona was the property of its builder and keeper. According to the Governor of Cremona, it was the clockmaker's right to take with him the civic clock at the moment he lost his office. The clock was then public in its function, more than in its physicality. In this period, public offices were probably perceived as specialised feudal functions that, unless problems of loyalty or insufficiency should arise in sons and nephews, had to be bestowed upon one's heirs. For references see chapter 3.

118   AGS, *Casas y Sitios Reales*, Leg. 271, fol. 248: Ladislao Reti, *El artificio de Juanelo en Toledo: su historia y su tecnica* (Toledo: Diputación Provincial de Toledo, 1967), 7.

regulating Janello's clocks, which "must not leave the palace of Madrid".[119] In 1594, the son of Giorgio di Diana, Jácome de Diana, was appointed to the administration of his granduncle by marriage's clocks, and to the clock of the palace with the wage of four *reales* a day – one of which was to be given daily to his mother while she was still alive. However, it was one Céspedes, servant of the "*illustrisimo Car[los] archiduque*", who was granted the task of moderating the clocks.[120] This was the mathematician and royal cosmographer Andrés García de Céspedes. Jacome de Diana was ordered to only clean them, but not to synchronise and set them according to the celestial motions, a task reserved to the royal cosmographer. In the year 1606, something odd happened: Jácome de Diana stole the manuscript written by Janello used to govern the clocks.[121] At the moment we cannot say whether Jácome took the book because he wanted to study it or because he was on bad terms with Céspedes. What we know is that the clockmaker felt he was entitled to take the manuscript, but he was the only one to believe that; therefore, he was arrested and the book was confiscated, together with other objects used in his profession.

The hereditary right to occupy an office was an *Ancien Régime* feature, and perhaps a general anthropological tendency too. At that time it was mirrored and strengthened by the noble family system. The anecdote by Camillo Capilupi, where Janello Torriani wittingly complained about the fact that the property of his clock was hereditary for Charles V, but the pension received as a reward was not, shows the tendency of the non-noble to adopt this hereditary model provided by the aristocracy. Janello actually succeeded in transmitting his pension down his family tree for generations, at least until 1623, when, after the death of his granddaughter Emilia Felipa de Diana (1612), her daughter Angela Maddalena Besozzo was confirmed in this right until she sold the pension (1623).[122]

119   Cristóbal Pérez Pastor, *Bibliografía madrileña, ó, Descripción de las obras impresas en Madrid.* (Madrid: Tip. de los Huérfanos, 1891), 112-13.

120   13th of August 1594, San Lorenzo: Cervera Vera, *Documentos biográficos*, doc. 138.

121   García-Diego, *Juanelo Turriano, Charles V's Clockmaker*, 138-39.

122   González Vega and Díez Gil, *Títulos y privilegios de Milán*, 365, n. 4375: "*Turriano, Juanello. Licencia a su favor para renunciar en Emilia Filippa de Diana, su nieta, los 200 escudos de pensión que tiene en el estado de Milán. Thomar, 15 de Mayo de 1581*". N. 482: "*Besoza Angela Magdalena. Merced de 100 escudos de pensión, en el estado de Milán, durante su vida, de los 200 que allí vacaron por su madre, doña Emilia Felipa de Diana, en atención de los méritos de su abuelo, Juanelo Turriano. El Pardo 1612*". N. 487: "*Besozo, Angela Magdalena – Concesión a dicha Angela Magdalena, nieta de Juanelo Turriano para que pueda vender a Jerónimo Marzorati, deudo suyo, la pensión de 100 escudos que goza en Milán por los servicios de su abuelo. Madrid, 30 de julio de 1623*".

For the heirs, like Jácome de Diana, it was often hard to keep up with the splendour of their glorious ancestors. In the same years Jácome was arrested, a nephew of Jacopo Nizzola da Trezzo, a certain Jácome de Trezzo, was arrested too.[123] Unlike his name, the skill and fortune were not hereditary items. This was also the case for Juanelo Turriano de Diana's son, Gabriel Juanelo Turriano de Diana, who died in Flanders on the battlefield in 1616 as the result of a harquebus' bullet. When in 1598 the King had placed Juan Fernández de Castillo, a non-member of Janello's family, in charge of the Toledo Devices, a life-long yearly pension of 50,000 *maravedis* (around 133 gold ducats) was given to Gabriel Juanelo Turriano de Diana for the sake of the memory of his great-grandfather's service.[124] The attempt of declining families to pass down the name of a respectable ancestor was clearly an attempt to turn their honour and respectability into hereditary prestige, and the offices with it, but it was as much a token of affection and respect towards a deceased loved one and a tool for maintaining the identity of a family group transplanted a thousand miles away from its roots.

To respond to the problem of territorial eradication, Janello actuated a smart familiar strategy in the places where he had fortune: Milan and Castile. To strengthen his household and its economic basis, Torriani intermingled his family with local ones, the more powerful the better. His daughter, in Milan, was married to a physician, a member of a highly respected professional group. Then, Janello and his daughter organised the wedding strategy for her sons and daughters. If, on the one hand, Emilia Filipa Diana Turriano was married to a Milanese gentleman named Ludovico Besozzo, on the other hand, Laura Antonia Turriano de Diana was married to Diego Juffré de Soria of a family connected to the milieu of the notaries of Toledo. In a letter to the King, Janello Torriani claimed that since his early times in Toledo he had been on friendly terms with the de Soria family,[125] and indeed, many of the notary acts written in Toledo on Torriani's behalf were made by a member of this household.[126] Maria Turriano was instead given to a certain Domingo Bravo, who would later become very important in the family economy. Moreover, the daughter of

---

123   AHPM: Prot 1457, fols. 116, 117 and 118, 1594, nov. 15th: Jacome de Trezzo, (nephew) jailed for
       a debt of 700 ducats. Pompeo Leoni, (Juan) Pablo Cambiago (son of Luca Cambiago?) and
       Trezzo's wife committed themselves to pay the debt and have Jacome set free.

124   Cervera Vera, *Documentos biográficos*, doc. 140.

125   Ibid., see the doc. 130 dated to the 19th of April 1586 (but written by Janello the year
       before)

126   This is one of the clues that I followed in order to find new notary acts concerning Torri-
       ani, such as the contract with ambassador Girolamo Negri: AHPT, *Escrivano Gaspar de
       Soria*.

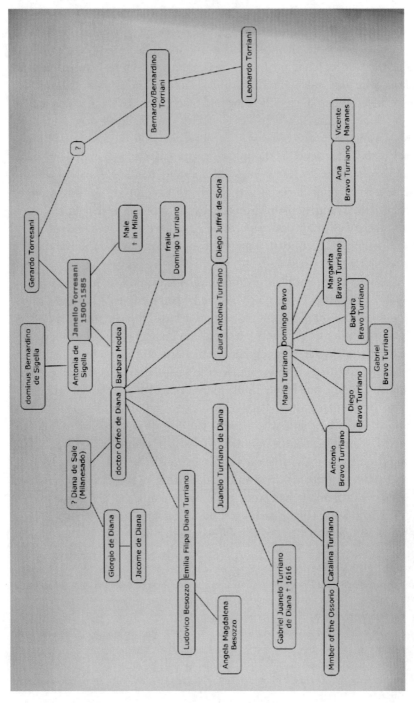

FIGURE 47 *Family tree of Janello Torriani.*

Juanelo Turriano de Diana married a member of the Ossorio (or Osorio) kin, another respectable family.

The hypothesis of a consistent wedding strategy planned by Torriani seems to be reinforced by documents related to Philip II's entourage: in the year 1563, Philip II wrote a note about some royal projects, among them the raising-water device in Toledo. In this note the King referred to a couple of characters, both appearing to be members of two of the three Spanish families later entangled with Janello's: Francisco de Soria and don Pedro Osorio.[127] Moreover, the mother-in-law of Leonardo Torriani, son of Bernardino, Janello's nephew, and future royal engineer of the Kingdom of Portugal, had this very surname, and perhaps was also a member of this family.

Being at the head of this complex household, Janello tried, when he could, to promote his nephews that were the sons of his only living daughter Barbara Medea, the husbands of her daughters, and obscure people like the nephew Giovanni Antonio Fasolo, perhaps the son of some female member of the family back in Italy. As regards the latter, it is clear that the year after Janello Torriani died, his old friend Jacopo da Trezzo wrote to the King's secretary Ibarra communicating that he had employed the clockmaker nephew at El Escorial. This may have been the consequence of a direct request by Janello, or a kind gesture of his friend Jacopo *in memoriam* of the old fellow.[128] In the case of Diego Joffré de Soria, married to his granddaughter Laura Antonia, Janello's brokerage is very explicit. In a letter posthumously directed to King Philip II by Barbara Medea the year after the clockmaker's death, Janello was asking for royal protection of his family because the debts he had accumulated – "thanks to the bad behaviour of the city of Toledo that did not treat me correctly", he wrote – were forcing him and his family to beg for alms in order to collect the money necessary to bury him. Janello adds a recommendation:

> Seven months ago I gave as wife one of my granddaughters to Diego Joffre, official of the scribe Gaztelu. Diego is a native of Toledo and is the

---

127 *"me parece ... que la de Francisco de Soria seria la major, aunque se habría de dar algo a los criados de don Pedro Osorio ..."*: Cervera Vera, *Documentos biográficos*, doc. 6. My suspicion that Torriani planned a canny marriage-strategy is strengthened owing to the fact that the Osorio family (if the same) was involved with the administration of the *Reales Sitios*, as one can see from the position of a certain Luis Ossorio, as royal *"gouvernador de Aranxuez"*: Ibid., doc 117, year 1583.

128 *"Yo tomado Antonio Fasol porque me ayude en estas obras de las armas que es muy el proposito; yo le ordenado a siete real al dia y ne merece mas de diez ..."*, letter written in Madrid on the 15th of January 1586; document published in Babelon, *Jacopo da Trezzo et la construction de l'Escurial*.

son of people that I know since I arrived here. I beg Your Majesty to give him any post of scribe. Diego has the skills for this as secretary Juan de Ibarra knows, having seen him working for Gaztelu for more than 6 years, before he died, and in the *contaduria* of Barcelona as major official and scribe in the building of 80 galleys (1570-1573). Diego worked other three years in the *Receptoría General de Penas de Cámara*, but he did not get a post.

Janello also begged the King to give Diego the chance to repair the misdeed that the clockmaker believed had been done to his own daughter whom he left in poverty after asking her to migrate from Lombardy to Spain. What Torriani desired the most, it is written in this last letter to Philip II, was to maintain the dignity of being a King's servant – *criado* – the best gift he could dream of.[129] During the same year, Janello had already tried to promote Diego with Juan de Ibarra: when Janello wrote to Ibarra about the functioning of the two devices, he added that Diego Joffré was kissing Ibarra's hands, and that he was going to pay him a visit.[130] Secretary Gatzelu was an old acquaintance of Janello since the times of Yuste, and it was thanks to this old friendship that Gatzelu had employed Diego. Now that Gatzelu was dead the family needed a new replacement within the system with Ibarra. After Janello's death, Barbara Medea, the clockmaker's universal heir and the new head of the family, lived together with her daughter and her son-in-law Diego Juffre. This was a sign of recognition of Diego's new position as male head of the family: indeed, Diego also helped the other grandchildren of Janello.[131] Diego, together with Ludovico Besozzo and Juanelo Turriano de Diana, are named executors for Torriani's testament.[132]

The administrative documents and the letters sent by Torriani and by his friends to the King and to the Pope reveal a constant need for money: the problem was not just greed or the burden of heavy responsibilities towards the family. A central issue was the extreme inability and unwillingness of the court system of the time, and especially of the Spanish monarchy, to administrate the economy and to redistribute the wealth of the state. Three bankruptcies of the King of Spain in less than 40 years (1557-1575-1596) offer proof of this disastrous situation.[133] This is not the place to analyse the reasons for this continuous

---

129   19th of April 1586: Cervera Vera, *Documentos biográficos*, doc. 130.

130   Ibid. 17th of April 1585, Toledo: doc. 121.

131   Ibid. See the documents related to Barbara Medea and Diego after the death of Janello Torriani in the year 1585.

132   Ibid. 11th of June 1585, Toledo, doc. 124.

133   Bartolomé Yun Casalilla, "Economical Cycles and Structural Changes," in *Handbook of European History, 1400-1600: Late Middle Ages, Renaissance, and Reformation*, ed. Thomas

economic crisis. Nevertheless, it is important to keep in mind the general economic framework in order to better understand Janello's poverty, together with the exasperating attitude of the aristocrat to procrastinate to settle their debts. Anyone who has browsed through sixteenth century court epistolary material knows that. Nevertheless, despite the lack of money, the King demanded the job to be done. Keeping his servants under pressure was not accidental. Federico Borromeo criticised the habit of certain princes who distributed their wealth to their servants little by little in order to keep them in a state of constant dependency.[134] Desperate court artisan-entrepreneurs like Janello Torriani or Jacopo da Trezzo had to pay in advance wishing perhaps to be one day paid back; indeed, their salaries were paid with at least a year of delay,[135] and sometimes, as with the first Toledo Device or the Crystalline, the payment was made on delivery, many years later.[136] Only a good artisan-entrepreneur could deal with such delays. For a period Torriani fell into the jaws of loan sharks, and for the rest of his life he had to struggle with his finances.

Begging for a debt to be paid and crying about one's misery was part of the rhetorical requirement for artisans who had the fortunate misfortune to work for the aristocracy; however, Janello's situation was truly disastrous. When he passed away, everybody at court was aware of the tragic situation in which he had unwillingly left his heirs: secretary Ibarra, the historiographer Garibay, and the *aposentador mayor* Juan de Herrera were all well aware of that, and they urged the King to help the engineer's family. Barbara Medea as well wrote to Philip II to remind him once again about her father's great efforts and misfortunes, and about the fact that she did not have anybody else to ask for

---

A. Brady, Heiko Augustinus Oberman, and James D. Tracy, vol. 1 (Leiden: Brill, 1994), 113-46; Fernand Braudel, *Civiltà e imperi del Mediterraneo nell' età di Filippo II*, vol. 1 (Torino: Einaudi, 2002), 546-48.

134   "Ma non è per questo d'approvarsi la bassezza dell'animo d'alcuni Principi, i quali donano poco, e molto a minuto, non già perchè riguardino alla conditione naturale di chi riceve il dono, ma sì perchè vogliono, che il soggetto dipenda sempre da loro, e stia sospeso con l'animo, e come appeso in aria.": Borromeo, *Gratia de' principi*, 153.

135   17th of June 1577, San Lorenzo el Real, Cervera Vera, *Documentos biográficos*, doc. 77.

136   Ibid. A special royal decree dated 1566, provided the payment for the crystal clock. A thousand *ducados* were granted for Torriani's occupational hazards plus his ordinary salary. To this sum of money were to be considered added 2,000 *ducados* that Janello had already received. Since the clock was valued by one evaluator for 3,000 and by another for 2,500 *ducados*, it is written that Janello had then to consider himself satisfied with the payment of 3,000 *ducados* (the price of the clock plus his salary consisting of four hundred *ducados*). The considerable sum of two 2,600 *ducados* should however be reckoned as the reward for almost a decade of hard work.

protection. The King was not insensitive to her prayer; on the exterior of the supplication one can read: "to be paid to Barbara Medea 6,000 ducados".[137] However, this was not enough: after Janello's death the King still owed him the enormous sum of 10,000 golden ducats. The royal help probably came also because Torriani played his last card on his deathbed: the King's debts were so large that according to the last contract the clockmaker's reparation was supposed to be to keep one of the two enormous water-devices for himself and his heirs; however, he decided to put his device, as a gift, in the hands of the King. He had no choice since he had a long-lasting and bitter story of litigations with the city council of Toledo, and King Philip II was the only possible party to sell the water to. This formal generous present – worth, according to the royal envoy Juan de Herrera, 20,000 ducats – required something in return. A note written on this posthumous letter with which Torriani delivers his own device into the hands of the King, reads as follows: "the King orders that this case has to be considered as soon as possible in the *junta*".[138]

The situation recalls battlefields where the imperial generals were often held in check by their empty purse. Jacopo da Trezzo once wrote that "one cannot do anything because of lack of money, and with those who work in my household, I do like I was a captain that cannot pay his own soldiers".[139] And Leone Leoni, complaining about the same problem, in a letter to Jacopo da Trezzo quoted something Torriani used to say:

> Our bad luck ... allows one to push, spur on, and the most of the time beat us in order to make us trot and gallop until we run, like one makes money run. This money (as your master Janello has yet said) is running so fast that it is almost impossible to reach or even to catch it.[140]

Since his early career in Cremona, Janello had a public wage. According to the communal rules the guardian of the Torrazzo tower was supposed to urge the master clockmaker in charge to visit often the clock, to keep at his own expense bells, ropes, timber and other materials in a good state. We infer from some

---

137   Ibid., doc. 125-126. Year 1585 (?): *terminus non ante quem.*
138   Ibid.
139   "... *ma no se puede aser nada por falta de dineros, y la jente que travajan en mi casa, ago comte [sic] un capitan que no puede dar la paga a los soldados* ..." in Babelon, *Jacopo da Trezzo et la construction de l'Escurial,* 278.
140   Ibid., 319: "*ma per la mala nostra ventura che permette così che siamo stimolati, solicitati et quasi sempre bastonati per farsi trottare o galoppare et fino a correre, si come essi fanno correre a denari i quali (come disse già Gianello v-o) corrono tanto che non si possono o di raro aggiungere a pena* ...", Leone Leoni 's letter to Jacopo di Trezzo: Milan, 2nd of April 1583.

bills of payment that Janello had this task, at least in the years 1529 and 1534. The wage for that job was of 12 lire and 10 *soldi* per month, about 2 ducats per month (or 24 per year).[141] Moreover, Janello had an income connected to the fee paid by his apprentices, and the sale of clocks, scientific instruments like the armillary sphere of the Ambrosiana, and locks like those made for the Baptistery of Cremona.[142]

Employment at court could provide a much better income for a clockmaker: for instance, already in the previous century, King Alfonso I of Naples, who had commissioned to the French Guglielmo Monaco de Lo a clock for 1,117 ducats, paid him an annual wage of 400 ducats (around 33 ducats per month).[143] But courtly wages, though more prestigious, were not always as generous as that of Guglielmo Monaco, the already mentioned Cherubino Parolaro Sforzani, who worked from 1524 to 1531 for Pope Clemens VII, receiving a monthly wage double that of the keeper of the Torrazzo's clock and a ducat less than the keeper of St. Mark's clock: four ducats per month.[144] However, his acquaintance with important courts gave him the chance to collect religious prebends as gracious gifts: in 1556 the Duke of Este granted him the income of the archpriest of Carpi, and later the Pope added the prebends of canon of Reggio Cathedral and the stipend of apostolic *protonotarius*.[145]

---

141   Bonetti, *Memorie*, 27. From the time of Charles V the imperial pound or *libra imperialis* (in Italian *lira imperiale*) suffered a progressive devaluation, so that the relation between golden coin and value in silver constantly changed: Carlo Capra and Claudio Donati, eds., *Milano nella storia dell'età moderna* (Milano, Italy: Franco Angeli, 1997), 73.

142   As terms of comparison one can quote some similar situations around Northern Italy: for instance, Gian Carlo Ranieri († 1529) who made with his father Gian Paolo the Venetian clock of St. Mark (1493-1499) for 2,701 ducats, was later made keeper of this machine at a salary of 60 ducats per year.

143   Morpurgo, *Dizionario degli orologiai italiani*, 1974; and Del Vecchio and Morpurgo, *Addenda al Dizionario degli orologiai italiani edizione 1974 di Enrico Morpurgo*.

144   Later clockmakers employed as keepers of the Roman Chamber's clocks slightly improved their remuneration: in 1543 a certain Domenico de Pedacchia had a monthly wage of seven golden ducats per month, and 24 years later, in 1567, the keeper of the Apostolic Palace's clock Pietro Filanetto from Lodi, received 6 *scudi* per month. But there were also those who gained less than that: around 1550, a certain Barnaba, master of the Palace clock in Rome, had a salary of 28 *scudi* per year, not much different from that which Janello had likely earned in Cremona. In the second half of the sixteenth century, at the court of Ferrara, the duke of Este paid Giovanni and Jacopo Marcoat (Marquatti or Marquart) from Augsburg a monthly wage of five *scudi* each. However, it was also likely that these figures would have been able to increase their income with extra sales, as it seems they did when they sold timepieces to Cardinal Luigi d'Este in 1560. Ibid., 10.

145   Campori, *Artisti degli Estensi*, 8-9.

This solution of payment was not an exception: during the long employ-ment of Janello at Philip II's court, the King had tried to pay his debts with the clockmaker in different ways. Some of the royal payments were in cash. Some others came in the form of offices. For instance, in 1579, the King granted Torriani the *alcaldía* (so to say, the office of director) of the jail of Ocaña, an honour that the clockmaker sold immediately to a certain Alonço García de Haro. After Torriani's death, in order to help the clockmaker's daughter Barbara Medea, the King granted her the revenue of a toll to pay on a card game in *Castilla la Vieja*, Old Castile.[146] Janello's homologue in Florence, Bernardo Buontalenti, bloodied as well by a long controversy related to a water-lifting mill, was granted by the Grand Duke a privilege on the revenues connected with the sale of snow and ice. By 1608, when he was ill and close to death, Buontalenti wrote to Gran Duke Ferdinando I de Medici, asking to be allowed to bestow this revenue upon his daughter Eufemia, who was poor: she had four daughters to take care of. In the following months Buontalenti died and Ferdinando decided to ignore the supplication of his hated brother's dearest servant. The privilege was passed instead to Ferdinando's secretary: Francesco Paulsanti.[147]

In Charles V's service Torriani received more than 300 ducats: 200 as a pen-sion and 173 as wage.[148] Since 1562, King Philip doubled the wage of Janello Torriani, which was now 400 ducats, plus the 200 ducats of pension. The spe-cial royal decree that defines Janello's new position at Philip's court states that the clockmaker had the obligation to permanently stay at the royal court.[149] A court artisan was paid only after finishing his job and after his creation was

---

146   See documents dated 22nd of December 1579, 29th of August 1580, and doc. 133-134: Cer-
       vera Vera, *Documentos biográficos*. Previously, in the first half of the fifteenth century, the
       last duke of Milan of the House of Visconti had awarded the clockmakers Bartolomeo and
       Antonio Rainieri da Ramiano with a privilege that exempted them from taxation. The
       same thing happened with the city of Reggio, which exempted another member of the
       dynasty of clockmakers, the famous Zampaolo degli Horologi Rainieri, constructor of
       Saint Mark's clock in Venice, from taxes. In 1506 his son asked for permission from the city
       of Reggio to open an inn exempted from taxation: Morpurgo, *Dizionario degli orologiai
       italiani*, 1950; and Del Vecchio and Morpurgo, *Addenda al Dizionario degli orologiai italiani
       edizione 1974 di Enrico Morpurgo*.

147   Lamberini, "Inventori di macchine e privilegi cinque-seicenteschi dall'Archivio Fioren-
       tino delle Riformagioni," 188.

148   Wages at the time of Yuste: Gaztelu 150.000 mrv. = 400 ducados; Physician Henrique
       Mathys 189.000 mrv. = 504 ducados; Janello: 65.000 mrv. (173,333 ducados) and 200 *escudos*
       of pension: Cadenas y Vicent, *Hacienda de Carlos V al fallecer en Yuste*, 92.

149   16th of July 1562, Madrid: Cervera Vera, *Documentos biográficos*, doc. 5.

evaluated. This means that usually the cost for any project made in the King's service required a starting investment from the craftsman himself. As we shall see in the construction of the Toledo Devices, it was this economic practice together with the usual royal delays in payment that ruined Janello.

A yearly income of 600 ducats was a very significant revenue, 10 times greater than that of a master mason.[150] To get a better idea of the level reached by Janello at court, we can attempt a comparison. In a document from the files of the Gonzagas of Guastalla, there is a list of expenses dated 1568, that a gentleman was supposed to incur in a year living at the court of Madrid. The costs were supposed to cover his needs: two servants, a horse, clothes and food. The total sum was 429 ducats, far less than Janello's salary.[151] However, important diplomats had very different standards. The apostolic nuncio received 400 golden ducats per month (4,800 per year); because of the splendour of the Spanish court, it was the richest remuneration among all apostolic nuncios. Such a wage allowed the nuncio to pay for all the offices of his *Nuntiatura* and for a personal household of 45 servants, without taking into account all the expensive apparatus (garments, stable, kitchen, presents, etc.) that the dignity of such a prelate required.[152] The total sum of Janello's yearly salary was set to become an impressive one had the city of Toledo honoured the contract of 1565: indeed, 2,500 ducats a year (the courtly wage of 600, plus the 1,900 promised in the contract) were a revenue suitable to a high-ranking gentleman. For instance, the title of *Prince of Eboli* brought an annual income of 3,000 ducats.[153] Unfortunately, Janello's economic expectations remained illusions. Faced with such an attractive but uncertain employment at court, the clockmaker thought to diversify his income by investing in real estate in the Duchy of Milan[154] and

---

150 Leydi, "Un cremonese del Cinquecento," 134. Sofonisba Anguissola, a noble woman, had the same pension of 200 ducats on the wine-toll of Cremona (year 1561); Giovanni Muto and Rossana Sacchi, eds., "Tra le carte degli Anguissola," in *La città di Sofonisba: vita urbana a Cremona tra XVI e XVII secolo ; mostra documentaria ; (Centro culturale "Città di Cremona" Santa Maria della Pietà, 17 settembre-11 dicembre 1994* (Milano: Leonardo, 1994), 129.

151 ASPr, Gonzaga di Guastalla, 1568: "*Notta di quello che poteria spendere un Gentilhuomo nella corte di Spagna con un par di servitori et una cavalcatura casa et vestire, e, mangier per la sua persona*".

152 Fernández Collado, *Gregorio XIII y Felipe II en la Nunciatura de Felipe Sega (1577-1581)*, 28.

153 Badoer, *Notices of the Emperor Charles the Fifth*, 34.

154 In 1582, Janello handed over to a relative, a certain Marco Antonio Turriano the right of collecting the money he gained from the rents of some houses he has in Milan and from his pension (613 *libras*, 12 *sueldos* and 9 *dineros*). Cervera Vera, *Documentos biográficos*, doc. 115, 19th May 1582, Toledo.

in Madrid, where the street in which he had a house still bears his name. This must have been a quite important residence if it was purchased after his death from Don Pedro de Medici, the controversial Florentine prince.[155]

Unfortunately, we do not know if Janello ever earned money from the privileges for invention he was granted in Spain, Venice, Rome and Mantua. What we know is that the moneylenders immediately devoured the payments he received in a lump sum for the devices. What was not devoured of his courtly salary was needed to support his household, which also comprised his workshop. Torriani's despair seems sometimes to turn into sarcasm. When in 1579, very old and ill, he wrote the treatise on the reform of the calendar, he added some mathematical instruments such as volvelles and tables. In the part of the treatise addressed to the King he said:

> If I had the gold and silver of Alexandrines and Babylonians, in gold and silver I would have carved them for Your Majesty ... However, it is more suitable that I made them out of paper, so in case they would not match their aim, it will be possible with no worries to tear them up and turn them into pieces.[156]

It was for ten years that Philip left Janello in this misery, and writing these words the clockmaker might have aimed to underline with witty but bitter irony the sadness of his situation.

This new image of Janello as a Renaissance artisan-entrepreneur who was running a complex family-company opens up quite problematic reflections. If we consider the entire body of a workshop such as Torriani's, it becomes apparent that it is still not possible to exclude the possibility that the famous *Los veinte y un libros de los ingenios y máquinas de Joanelo* were actually a production – perhaps just a partial one – of Torriani's workshop, as the title of the manuscript, rearranged by Juan Gómez de Mora in the first decades of the fol-

---

155   Ibid., doc. 135 and 136.

156   19th of June 1579, Janello's letter and treatise on the reduction of the calendar sent to the King: "*io che nè in me conosco quella facilità di dire, che seria necessaria per dichiarare difficoltà tanto importante, ho pigliato per espediente con una superficie delle Matematiche, che sono gli istrumenti delle operationi, di proponere a V.M.tà non solo il modo della riduttione, ma le Tavole, che la dimostrano ancora et queste sarano tre; Due in forma circolare ... ed una quadra. Le quali s'io havessi l'oro et l'argento de gli Alessandrini et Babilonici, d'oro et di argento le haverei scolpite a V.M.tà ... Ma forse sarà più accertato che io le habbia poste in carta, acciocchè no riuscendo al proposito di quello si desidera, si possano con manco pensiero et lacerare, et ponere in pezzi*" (my translation); Turriano, *Breve discurso*, 1990, 73-74.

lowing century, suggests.[157]. After Juan de Lastanosa, the probable Aragonese author of the text of the manuscript, died in 1576, Janello and members of his household could have continued composing the manuscript. Or perhaps some other engineer from Aragon replaced Janello after his death. More investigation is needed on this issue.

---

[157] Juanelo Turriano, *Los veínte y vn libros de los ingenios, y maquínas de Iuanelo: los quales le mando escribir y demostrar al Chatolico Rei D. Felipe Segundo Rey de las Hespãnas y Nuevo Mundo*, ed. José Antonio García-Diego and Alex Keller (Madrid: Fundación Juanelo Turriano, 1996). In the beginning, when Ladislao Reti published in 1967 for the first time this notable manuscript, no one doubted the attribution of the title of the same to the Torriani. However, the careful analysis of José García-Diego and Nicolás García Tapia showed that the language of the manuscript led back to the Aragon region, and that personal references that the author inserted into the narrative reported a sixteenth-century engineer who had had experiences in areas that Torriani never seems to have visited. Putting the two things together, it attributed the writing to the engineer Juan de Lastanosa. See: Nicolás García Tapia, *Pedro Juan de Lastanosa: el autor aragonés de Los veintiún libros de los ingenios* (Huesca: Instituto de Estudios Altoaragoneses, 1990); Nicolás García Tapia, "Y sin embargo es Lastanosa," *Técnica industrial*, no. 203 (1991): 54-61; Nicolás García Tapia and Jesús Carrillo Castillo, *Tecnología e imperio: ingenios y leyendas del Siglo de Oro: Turriano, Lastanosa, Herrera, Ayanz* (Madrid: Nivola, 2002).

CHAPTER 6

The Microcosm

The First Machine-Tool to Cut Gears

Janello Torriani's fortune is tied to the first planetary clock, the fourteenth century Astrarium made by Giovanni de Dondi for Gian Galeazzo Visconti, count of Virtú, and later first Duke of Milan. This clock was a possession of the Duchy of Milan, and it was kept in a tower of the castle of Pavia.[1] In 1566, Bernardo Sacco of Pavia, whose sources are unknown, reported that when in 1529-1530 Charles V was in Bologna for his coronation, he received Dondi's clock as a gift.[2]

---

[1] Giuseppe Brusa, following Cardano, considers Janello's planetary clock to be an improved copy of a planetary clock allegedly made by Zelandinus: *"Cardano's words establish beyond doubt that Torriano restored the planetarium of the Flemish maker and not that of Dondi as other less informed authors wrote afterwards. Indeed Cardano was so interested in the later clock that in 1539 he began to write about it ..."*. Brusa's claims are to be taken with caution, as we shall see in this section. Brusa, "Early Mechanical Horology in Italy," 512. It seems that just after the death of Duke Gian Galeazzo Sforza in 1494, his uncle Duke Ludovico il Moro had the clock moved to the castle of Rosate. The clock remained here for some time between November 1494 and March 1495, when the clock was on display in the castle of Pavia. In 1493, the fief of Rosate had been given to Ambrogio Varesi da Rosate, physician and controversial astrologer of Ludovico il Moro. Bedini and Maddison, *Mechanical Universe*, 25-37.

[2] *"Dominante deinde Transpadanis* [not Padua like Garcia-Diego reports!] *Joanne Galeacio Vicecomite, fabricatum fertur eiusmodi horologium, non solum horam, sed etiam syderum expressis notis, atq; temporibus ac Solis, Lunaeque; meatibus cuius operis autor ignoratur collocatumque illud horologium in arce, vel castello Papiae fuit, ubi defuncto principe tam mirabile opus despectum iacuit, circulis etiam a suo loco sublatis. Exacto postea saeculo anni millesimi & quingentesimi, circa annum vigesimum nonum, quo CAROLVS QVINTVS Bononiae Imperialem coronam suscepit, allatum eidem Imperatori illud horologium incompositum (ut erat) situ, ac rubigine foedatum, fuit quo conspecto, machinam admiratus, curari tanti operis instaurationem fabris undique evocatis iussit. Quibus circa opoficii restitutionem frustra laborantibus, unus accessit Ioannes Cremonensis, cognomento Ianellus, aspectu informis, sed ingenio clarus qui tantum opus speculatus, refici posse machinam dixit: sed nequiequam profuturam, ferris rubigine atritis, exesisque, nisi novum instrumentum ad illius vetusti similitudinem, ac symmetriam componatur. Aggressusque opus, sive priorem artificem immitando, atque aemulando, sive exaquando diuturno labore opificium absolvit: quod deferri in Hispaniam Imperator voluit, magistro Ianello simul deducto"*: Bernardo Sacco, *Bernardi Sacci Patritii Papiensis De Italicarvm Rervm Varietate Et Elegantia Libri X ; In Qvibvs Mvlta Scitv Digna Recensentur, De Populorum vetustate, dominio, & mutatione ; Item de Prouinciarum proprietate, & Ro. Ecclesiae amplificatione ...* (Pavia: [s.n.], 1565), 150-51. One can recognise the use of Vida from the line "aspectu informis,

---

At that time Francesco II Sforza was duke of Milan, and after a period of some serious tension with the Emperor, on this occasion he finally made peace with him. Unfortunately, the clock was broken and unrepairable, thus Torriani built a better version of it: the Microcosm. The antecedents of Torriani's project to create this planetary clock are still quite mysterious: first of all, even if Sacco's account is the most commonly accepted by historians, Silvio Leydi, with some reason, has observed that a broken and rusted clock was perhaps not an altogether apt ducal gift for the new Caesar.[3]

What we know about the genealogy of Torriani's Microcosm is that in the winter 1547, after having been summoned, as Gulielmus Zenocarus recorded,[4] Janello reached the Emperor in Ulm, arriving at court on the very day of Charles V's birthday, something that was received as a good *omen*. On this occasion, Charles gave him the go-ahead, and consequently Ferrante Gonzaga (Governor of Milan since April 1546) gave the order to the treasury of the duchy to provide for the payment of Janello's new machine.[5] In the following years, a globe of rock crystal, made by Jacopo Nizzola da Trezzo, which contained a paper-globe made by the famous Flemish cosmographer Gerhard Mercator, had to be placed on the top of the clock. Janello and Jacopo Nizzola da Trezzo eventually delivered the finished Microcosm in March 1554. However, as the eyewitness Gasparo Bugato wrote, even without the crystal sphere, the *Cesar's Sky*, as some people called it, was already working, and it was shown to the resident

---

sed ingenio clarus"; however, Vida never talked about the clock being given to Charles V as a present in 1529-30. Even the attribution to Cardano of this idea was a misconception, as we are going to see in this section.

3   Indeed, already in the previous century, the Astrarium went out of order, and it seems that nobody could fix it. Perhaps Gulielmus Zelandinus managed to repair it for a while, or even to build a new version of it. It seems that the dukes of Milan, in the years 1440, 1457 and 1488, had him moving to the court for this purpose. We have already met Zelandinus: he was a physician and constructor of astronomical instruments from the Low Countries who later emigrated to Carpentras in Province, and who worked for the crowns of France, Sicily (King René), and for the dukes of Milan. He is the author of the *Liber Desideratus*, a manual for the use of a planetary instrument; in the year 1494/5, the second edition of *Liber Desideratus* had been published in Cremona. Bedini, "Falconi," fasc. 1, 44.

4   We have previously met the Flemish Willem Snouckaert van Schauwenburg, Charles V's librarian and official biographer who had personal knowledge of Torriani.

5   Silvio Leydi has recently published a series of five tranches of payment (600 *scudi* in total) performed by the treasury of the duchy of Milan (April 1547-October 1548). The Microcosm cost 500 *scudi*, plus an extra hundred for the decorations in gold and silver. This amount of money was considerable: it was equivalent to 10-years' wage for a master mason: Leydi, "Un cremonese del Cinquecento," 134.

ambassadors in Milan.[6] Indeed, by 1550, the Microcosm, as bishop Vida suggested to name the machine, was already in full function. Janello had once told Philip II's historiographer Ambrosio de Morales that he had been studying on the project for the Microcosm for a good 20 years, before physically creating it. After this careful study, Janello needed only three years and a half to make with his own hands the clock with its 1500 wheels. Morales calculated that, excluding holidays, Torriani must have crafted three wheels a day: a terrific achievement, acknowledged Morales in awe. The clockmaker assured that he had not made use of anyone's help; on the contrary, he had crafted everything alone. The most impressive feat for Morales was that Torriani proudly assured he had never made the same wheel twice.[7] This was not just a matter of *sprez-*

6   *"Ne è punto da tralasciare di non far memoria dello stupendo horologio di Gianello Torriano, nato a Cremona, & cresciuto in Milano, chiamato Celo [sic] di Cesare: però che dedicato fu da lui a Carlo Imperatore. In questo horiolo, non solamente sono le rote, & le tavole dell'hore di tutte le nazioni, & calendario delle feste mobili, dell'epatta, delle lettere Dominicali, dell'aureo numero, del bisesto, con l'entrata desegni celesti; ma tutto il moto del Cielo di sfera in sfera, co'l moto contrario della trepidatione: ogni cosa condotta ordinata, & mobile per rote d'acciaio, & d'ottone, indorate, o inargentate secondo il bisogno: il che veramente è una meraviglia vedere in eserciaio continuo mille, & cinquecento rote, tutte rotando al movere d'una sola, come il Cielo dal primo mobile, mostrando tutti i pianeti, tutti gli accidenti, ordini, & moti d'esso Cielo, col tempo ben misurato d'ogni sua revolutione, & massimamente quella della stella di Saturno, che sta tanto tempo a compiere il suo. A questo pare che altro non manchi se non lo spirito, & l'armonia che ogn'uno afferma esservi: & come per me assai credo, se differenti temperature si dessero ai metalli, che s'havessero a comporre con buono studio, come per essempio si compone il metallo della quantità delle campane, & qualità anchora: affine che accordasse le grandi rote, con le mezzane, & con le picciole a guisa delle corde di Clavicembalo, di cetra, o leuto, esprimesero concento, esprimendosi tutte le proportioni, & l'altre, con ogni numero, ogni voce, & ogni consonante della musica Diatonica, Enarmonica, & Cromatica, o al modo Lidio, o Dorico, o Frigio, o Gionico, dandole spirito, & anima con tali temore, come solemo noi dare spirito (proportionalmente & per comparatione) al flauto, al piffero, all'organo, al corno, & alla tromba. Questo horologio non è più alto d'un braccio, & manco largo, s'apre nel mezzo per lo traverso, come l'aperse il proprio me presente, prima che lo portasse in Ispagna, mostrandolo ad alcuni Ambasciatori. Ma essi non videro, come io vidi poi, la palla di cristallo di montagna che v'andava in cima in forma del mondo, tutto cosmograficamente intagliato, diviso, compartito, & segnato deClimi, deParalleli, degradi, demari, dell'Isole, decontinenti, delle provintie, deRegni, co' monti, con le selve, co' fiumi, co' pesci, con gli animali, con gli huomini, con le donne, & con le battaglie navali, & di pietre preciose, forse deprimi che usasse col Diamante macinato intagliar l'istesso Diamante non che l'altre gioie. Era questo Horiolo opera da fare altrui doventare statua chi ben lo considerava per maravigliare a' movimenti, & alle contemplationi d'esso (restato poi in Spagna presso al re Filippo) assai si trastullava la felice memoria di Carlo Imperatore"*: Bugati, *Historia uniuersale*, 1570, 1025-26.

7   *"Tardo, como el me ha dicho en imaginarlo y fabricar con el entendimiento la Idea, veynte*

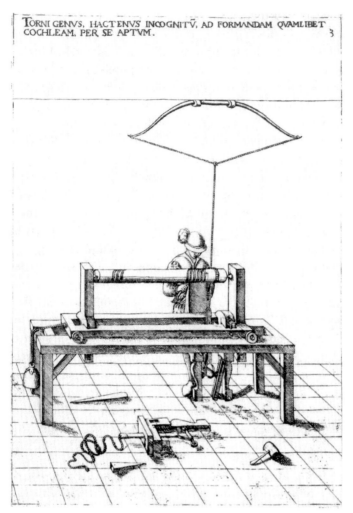

TORNI GENVS, HACTENVS INCOGNITV, AD FORMANDAM QVAMLIBET
COCHLEAM, PER SE APTVM. 3

FIGURE 48   Lathe, *from Jean Errard de Bar-Le-Duc,* Le premier livre des
instruments mathematiques mechaniques, *Nancy, 1584.*

años enteros: y de la gran vehemencia y embeveciniento del considerar, enfermo dos vezes en
aquel tiempo, y llego a punto de morir. Y aviendotardado tanto e nel imaginarlo, no tardo
despues mas que tres años y medio en fabricarlo con las manos. Es mucho esto, pues tiene el
relox todo mill y ochocientos ruedas, sin otras muchas cosas de hierro y de laton que entrevi-
enen. Assi fue necessario, que (quitando las fiestas) labrasse cada dia mas de tres ruedas, sin
lo demas, sindo las ruedas differentes en tamaño y en numero y forma de dientes, y en la
manera de estar enexadas y travadas. Mas con ser esta presteza tan maravillosa, espanta
mas un ingeniosissimo torno que invento, y lo vemos agora, para labrar ruedas de hierro con
la lima, al compas y a la igual dadde dientes que fuere menester. Y con todo esto, y con
entenderse que lo labro todo por sus manos, no causara asmiracion el dexir Ianelo, como

zatura: this remarkable achievement in speed and precision was made possible thanks to a special lathe Janello Torriani had invented, considered to be the first known rotary file cutter, able to mould wheels with equal teeth.[8] Zenocarus, in a passage of his first book on the life of the Emperor, published in 1559, just one year after Charles V passed away, states that Janello spent seven years constructing the clock. Zenocarus was obsessed with the number seven and its holy symbolism, in which he tried to include anything that had relevance. However, even if apparently different from Morales' chronology, the two versions could actually fit: the three years and a half (1547-1550) of the crafting of the mechanisms claimed by the clockmaker in Morales' report can be integrated into a longer span of time highlighting Charles V's agency and addenda: in this case we have as limits the year in which the Emperor ordered the construction of the clock (1547) and the year the clock, now adorned with the two spheres by Jacopo di Trezzo and Gerhard Mercator was finally delivered (1554).[9]

But how did Janello Torriani get involved in this project? Sacco was the first one to mention (in 1565) Charles V's coronation in Bologna (1530) as the moment when Charles received the old Astrarium. Not even Zenocarus, who was profoundly interested in Torriani's clock, mentions this gift taking place during the coronation. In Zenocarus' narrative, the only point of contact in the coronation and in the order to start the construction of the clock, taking place respectively in the years 1530 and 1547, is that they both happened during Charles V' birthday: on the 29th of February.[10] So far, what documentation tells us is that Torriani's employment in the service of the governors of Milan dated at least, to the year 1544, at the time of Ferrante's predecessor: Alfonso III

---

8   *dize, que ninguna rueda se hizo dos vezes, porque siempre de la primera vez salio tan al justo como era menester. Y sino precediera todo lo dicho, esto se tuviera por una estraña maravilla."*: Morales, *Antiguedades de las ciudades de España*, fols. 91-94.

8   Robert S. Woodbury, *History of the Gear-Cutting Machine: A Historical Study in Geometry and Machines* (Cambridge, Mass.: Technology Press, Massachusetts Institute of Technology, 1958), 45-46.

9   *"Horologium Caroli Maximi: Eius motus instrumentum aeneum habebat Caesar quod Ianellus Turrianus Cremonensis illi faberrime septennio confecit. Habet illud instrumentum mille quingentas rotas: est bipedale.Septem planetarum motus, & octavae sphaerae motum complectitur. Est lamina aenea inaurata, octagona, conclusum instrumentum. Interrogatus a me nuper Ianellus an septies septem chiliadibus annorum exactis : motus octavae sphaerae ad principium suum reverentur : respondit Deo id se iudicandum relinquere. Calculum enim secundorum, & tertiorum, esse tam immensum, ut id numerari a solo Deo, propter eius numeri immensitatem, & eorum qui primo hos motus observarunt vitae brevitatem, posit"*: Zenocarus Snouckaert, *De republica*, 203-7.

10   Ibid., 47.

d'Avalos d'Aquino d'Aragona, Marquis del Vasto and Pescara, imperial military governor of Milan from 1536 to 1546.[11] From Ambrosio de Morales, we learn that the marquis had "very much esteemed and favoured" Torriani, and Mutio Oddi's anecdote previously quoted seems to confirm this.[12]

Another witness, the famous physician, mathematician and bestseller Girolamo Cardano, confirms that Janello Torriani was already involved with planetary horology even before 1543: in his *De Libris Propriis*, in a section already written by the year 1543 and published in Nuremberg in 1544, Cardano said that by the end of the year 1536 he had made friends with the newly invested governor of Milan, the "most illustrious prince Alfonso d'Avalos".[13] In the following paragraphs he writes that in 1538, the "most illustrious prince [evidently Alfonso d'Avalos][14] upon procuring the noble instrument which

11   However, Janello may have encountered marquis Alfonso d'Avalos much earlier than 1536.
        Since the Battle of Pavia, the marquis had many previous occasions to stay in Lombardy.
        However, we have to keep in mind that while d'Avalos was said to have introduced Janello
        to the Emperor, he did not necessarily commission the reconstruction of the Astrarium in
        the first place. As we have already mentioned, the temporary interruption of six months
        that Janello took with his apprentice in the year 1537 (perhaps because of some extraordi-
        nary commission related to the planetary clock, as Barbisotti has suggested) is an interest-
        ing piece of information. The predecessor of del Vasto in the office of governor of the
        duchy of Milan was Cardinal Caracciolo, whose nephew was in that very year 1537 captain
        of the castle of Cremona. However, as we shall see, the commission for the restoration of
        Dondi's clock may have been assigned to Janello even earlier.

12   "*Ianelo Turriano que oyo esta platica, como muy estimado y favorido que era del Marques,
        començo luego a pensar (segu el a mi me ha contado) en como se podria subir el agua a
        aquella tan immensa altura, y fabricando con el entendimiento la suma de la Idea y modelo
        de su machina, lo dexo estar reposado, por andar entonces muy embevecido en la fabrica de
        su relox. El marques lo assento despues con el Emperador, y venido a España su Magestad, y
        retirado en el monesterio de Iuste, ninguna cosa humana llevo alli para su recreacion, sino a
        solo Ianelo y su relox, y alli lo tuvo hasta la muerte*": Morales, *Antiguedades de las ciudades
        de España*, fols. 91-94.

13   As we have previously seen, it was Senator Francesco Sfondrati of Cremona who intro-
        duced Cardano to the governor, and Cardano dedicates the collection of books published
        together in Nuremberg in 1544 to him: Cardano, *De Sapientia libri quinque*.

14   This passage from *De Libris Propriis* has created a lot of confusion: Ian Maclean thinks the
        "*illustrissimus princeps*" to be Charles v, but he was probably misled by the tradition: a few
        lines before, one can find the governor Alfonso d'Avalos called "*illustrissimus princeps*";
        therefore, the adjective must have referred to the Governor of Milan: Cardano, *De libris
        propriis* [*2004*], 132-133; Marino Viganò claims that Cardano had written that, before sell-
        ing it to Charles v, he had received the clock as a present from Francesco II Sforza (who
        was duke of Milan from roughly 1521 to 1535), and believes that both Cardano and Zeno-
        carus gave some clues that may suggest that the clock was given to Charles v at the time

represented the motion of all of the stars, which at another time had been sold by the author [is here Cardano referring to himself, or to the author of the instrument?][15] for twenty gold talents and was now dismantled with no hope of being restored, achieved its restoration thanks to the industriousness of Janello of Cremona. This [the apparition of this instrument] made me wish to describe the admirable symmetry, the mathematical relations, the movement and the shape of the machine, and so I therefore attempted this endeavour, seeing as I had already written about a minor instrument two years previously. I wrote something but I did not finish the work: sure enough, these things required a spirit free from public duties and the presence of the machine which, at that time, was not there being that I had not wanted it in my house".[16] Cardano does not attribute the authorship of this planetary instrument, restored by Torriani, to Dondi: "But the instrument was not engraved with letters, thus concealing the name of the first creator Guglielmus Zelandinus". Perhaps Cardano is here blandly accusing Torriani of not acknowledging Zelandinus' alleged authorship: "Indeed, it is a good practice so that even if you were to discover something by yourself which had already been discovered by someone else, even if you didn't benefit from any help from that, that you do not omit his name".[17]

Was Janello Torriani really indebted to Zelandinus' work? Because of their interests, and because of its Cremonese second edition (1494/5), it is highly probable that Torriani and his mentor Giorgio Fondulo knew Zelandinus' *Liber Desideratus*, as Cardano did. Zelandinus, in his book, stated that in the year 1488 he had left for the duke in Milan "a royal mirror of the celestial motions ... in which all the seven planets are moved with their own trajectories upon one single zodiac, with the motion of a single motor".[18] This instrument could be understood as both a real planetary display with a clock-motor, or as an equatorium where the volvelles of all seven celestial bodies were geared together, so that if one rotates one of them, all the others would follow. It seems that

---

of his coronation; but these passages are misunderstood, in as much as Cardano in *De Libris Propriis* and Zenocarus in his biography of Charles v are not giving this information: Viganò, "Parente et alievo del già messer Janello," 213.

15   Cardano claims in both his *De Libris Propriis* and in his *De Subtilitate* that he had been the one who brought the damaged instrument back to light..

16   This is an obscure passage: why did Cardano not want to have the machine in his house? Cardano, *De libris propriis* [2004], 133.

17   Ibid.

18   Zelandinus Aegidius Guillermus, *Liber desideratus super celestium motuum indagatione sine calculo*, Ed: Boninus de Bonini, [Lyon 1494] (Cremona: Carolus de Darleriis, 14 feb. 1494/95), f. a 7v

Cardano and his contemporaries acquainted with Zelandinus' book believed that Dondi's Astrarium and Zelandinus' Royal Mirror were the same thing. After all, it seems that it was just in the late 1550s that Giovanni de Dondi's authorship was restored.[19] Cardano's second edition of his *De Subtilitate* (1554) made clear that this first mysterious machine was the inspiration for Janello's Microcosm:

> Janello completely restored it [Zelandinus' machine]. Taking it as a model, he made, for the Emperor Charles V another machine that ... really depicts the whole universe.[20]

There is a further interesting particular emerging from this story: it seems that before 1543, when Janello Torriani made his first known intervention in planetary horology, the clockmaker had also written a treatise about it, as Cardano clearly stressed in his unpublished second edition (1550) of *De Libris Proriis*:

> We had written about Zelandinus' machine, but Janello of Cremona set me free from this labour. Indeed, having constructed that divine – so to say – instrument for our Caesar, he described its structure with very careful writings: therefore, you should refer to him.[21]

This treatise was probably a manual to build and govern the planetary clock, and it could be the one he brought together with the Microcosm to Spain, and that by 1575 he was ordered to finish for the functioning and management of the two clocks he had created for the Emperor, and that Jacome de Diana stole in 1606.[22] It seems reasonable to assume that Janello Torriani had also used this manuscript, most probably enriched with explanatory drawings and volvelles, in order to convince his patrons of the feasibility of his planetary clock, as

---

19   Bedini and Maddison, *Mechanical Universe*, 37.

20   Ibid., 37. Bedini and Maddison made here a mistake, writing at page 37 that this passage was from the 1550 edition of the *De Subtilitate*. Instead, this passage was added in the second edition from the year 1554.

21   "*Scripseramus de Zelandinis, sed Ianellus Cremonensis nos ab hoc labore liberavit. Cum enim instrumentum illud divinum, ut ita dicam, Cesari nostro composuisset, illius structura, diligentissime litteris mandavit: itaque eum ...*": Cardano, *De libris propriis* [2004], 148. Gio. Paolo Lomazzo, probably after Cardano, will remember this mathematical treatise: Lomazzo, *Trattato dell'arte della pittvra*. 4.

22   Cervera Vera, *Documentos biográficos*, doc. 50: Madrid, 20th of March 1575.

seems to have happened in Mutio Oddi's report on the Marquis Alfonso d'Avalos introducing Torriani to the Emperor.[23]

Going back now to Torriani's claim of having been involved with the project of the Microcosm for about two decades before physically constructing it, and considering that he started to craft it from the spring of 1547, we can push back the beginning of his involvement with the clock to the years around 1527. At this time, the context seems to be consistent with a project of restoration of the old glorious device. The Astrarium was not working since the previous century, when several clockmakers failed to repair it.[24] Moreover, during the Italian Wars the castle of Pavia, in whose library the Astrarium used to be kept, had been heavily damaged, especially in the course of the famous battle of 1525, when the King of France Francis I was captured, and during the siege of 1527, when the artillery of Lautrec managed to destroy two of the four towers of the castle – but not the one of the library. Once restored to power, it might have been in the interest of the last Sforza, Francesco II, to repair the Astrarium to its ancient splendour. In the State Archive of Milan, in a folder collecting letters by clockmakers, it becomes clear that in the year 1531, Duke Francesco II Sforza was looking for valiant clockmakers who could fix his clocks. Indeed, the Duke had ordered his servant Cavallerotto Fontanolla to seek capable craftsmen in Reggio, which, as previously seen, was the home of some of the most famous dynasties of clockmakers. The servant spoke with a member of the Ranieri family, an old man that had worked at the construction of St. Mark clock in Venice (1493-1499), perhaps the only public planetary clock of these times. The servant informs the Duke that the only two clockmakers considered worthy for the unknown ducal project of restoration were at that time not present in Reggio, because they were employed in Rome. The old clockmaker told the ducal servant that he had been their master. I wonder if this man was Gianpaolo Ranieri, father to Gian Carlo, Leonello and Ludovico. In these very years the Parolaro-Sforzani were also active in Reggio; it is interesting that the old clockmaker does not even mention them. Instead, the old master suggested a young clockmaker who was working in Parma, who had previously worked with him for 13 years.

If Francesco II Sforza was looking for clockmakers even outside of his duchy, it means that the craftsmanship required for the task had to be quite rare. Considering how difficult it had been for the previous century to find somebody able to fix the Astrarium, we may infer that the Duke was searching for

---

23   For Janello's numerous paper and cardboard mathematical volvelles, see: Sánchez Cantón, *Inventarios reales*, item 4736.

24   Bedini and Maddison, *Mechanical Universe*, 26.

clockmakers to give this very assignment. The Ranieri, Gian Paolo and his sons Leonello and Ludovico, were in charge of the public clock of Reggio.[25] The period in question coincided with the period when Janello was keeper of the public clock of Cremona. It may be possible that the Duke's first choice in his quest was to get in touch with those clockmakers that held the office of public clock keeper. If this was the case, it is possible that Janello was recognised as the most suitable to the task. In this case it is possible that Charles V took over, together with the duchy, the Astrarium-project from Francesco II Sforza, after the latter had died in 1535. Of course for the moment we cannot be sure if Janello started to study the clock as a private citizen around 1527 or later, or when he was officially put in charge of its restoration. More archival finds may shed new light on the story. In absence of certain data, we can accept the Battle of Pavia (1525) and the delivery of the Microcosm in 1550 as general chronological limits for Janello's study of the old planetary clock.

Anatomy of the Microcosm[26]

Since about the middle of the sixteenth century, the Kingdom of Germany, and especially the cities of Augsburg and Nuremberg, had been acknowledged as

---

25   ASMi, Fondo Autografi : Artisti diversi (1447-1842), pezzo 93: Orologiai e Orologi: "*Ill.mo et oss.timo Sig.re et ver patrono mio / Per una di vostra ecc.tia de tredici d[e]l presente che mi / Comen[dò] che io devia ricercar uno che sapia aconciar / li horologgi di vostra ex.tia et che subito fu gionto / qua fece la pratica ricordandome che altre volte /Vostra ex.tia havermelo dito : ho ritrovato che queli / Doi boni stano cu[n] el papa : et li altri no[n] ce che / Volia : qua ce uno vechio che me ha dito che / ce uno in Parma giovene suficientisimo et ha / lavorato cu[n] lui ani tredeci et questo vegio e de / quali che fecero lo horologgi de Venecia et / mastro de queli che fu a Roma : che penso di / haverlo et condurlo cu[n] meco che sera al più / presto sia posibile che mi par mili anni no[n] / Habia visto vostra ex.tia : no[n] ho ancora concluso / La cosa de mia fiola p[er] una cosa che poi diro / A vostra ex.tia : et humilmente me li recoma/nde : i Reggio die 26 novenbris 1531/ de vostra Ill.ma Sig.ria/ minimo servitor el cavallerotto Fontanolla*"

26   In this part of the book I address the same topic I previously analysed in my article: Cristiano Zanetti, "The Microcosm: Technological Innovation and Transfer of Mechanical Knowledge in Sixteenth-Century Habsburg Empire," *History of Technology* 32, no. III (2014): 35-65. This article was a revised part of a chapter from my doctoral dissertation. Despite arriving at the same conclusion, a new reading of the sources and of the secondary literature has brought me to reconsider some problems concerning the interpretation of Zelandinus and Cardano. Thanks to my postdoctoral fellowship at the Max Planck Institut für Wissenschaftgeschichte, I could focus once again on the topic, collecting new material that helped me to get a deeper understanding of the problem.

the main centres of clockmaking. Contemporary writers on technical issues, such as the German Johann Neudörffer (1497-1563),[27] the Frenchman Petrus Ramus (1515-1572)[28] and the Italian Tommaso Garzoni (1549-1589) were all in agreement on this matter. According to Garzoni, who published his book in 1585, the Germans:

> Are now the pride of this profession, because all the most beautiful and correct clocks are coming forth from their regions. Among these clocks, the one that Emperor Ferdinand sent to Soleiman, King of the Turks (as Bugatto writes) was a miracle.[29]

Following the example of Paris, which was most probably the first city to have a guild of clockmakers (established in 1544), Augsburg and Nuremberg set up their own guilds of clockmakers, the earliest in the Kingdom of Germany, in 1564 and 1565, respectively. To create a guild of clockmakers, one needed numerous specialized workshops. This can be interpreted as a sign of a shift in strategy regarding the production of clocks and watches. Traditionally, clocks were luxury goods that few could afford and so their production would usually be in response to a specific commission. These new guilds of clockmakers testify to a boost in production, probably enhanced by the specialisation of certain workshops in crafting pieces in series and the development of a specific knowhow in machinery. This process made it possible for a customer to enter a workshop and buy there a finished timepiece that he had not commissioned in advance.[30] Furthermore, due to the increasing manufacture of off-the-shelf timepieces, German products could also be easily exported, flooding other European markets. However, the superiority of sixteenth-century German production did not necessarily mean that its timepieces were also superior in terms of innovation. It seems that the cities of Italy remained cen-

---

27   Neudörffer, *Nachrichten von Künstlern und Werkleuten.*

28   Petrus Ramus, *Petri Rami Scholarum mathematicarum, libri unus et triginta.* (Basileae: Per Eusebium Episcopium, & Nicolai fratris haeredes, 1569), 31.

29   Gasparo Bugatto was a sixteenth-century Milanese domenican historiographer. Bugatto does not talk about this German clock for the Sultan but he describes instead Janello Torriani's planetary clock. Bugati, *Historia uniuersale*, 1570, 1025-26. Garzoni, *Piazza universale di tutte le professioni del mondo*, 625: "*hoggidì portano il vanto in questa professione, venendo tutti gli horologii più belli, e più giusti dale parti loro, ove sopra tutti fu miracoloso quello, che mandò Ferdinando Imperatore (come scrive il Bugato) a Solimano Re deTurchi*".

30   David Thompson, "Lo sviluppo dell'orologio meccanico: il contesto europeo," in *La misura del tempo: l'antico splendore dell'orologeria italiana dal XV al XVIII secolo*, ed. Giuseppe Brusa (Trento: Castello del Buonconsiglio. Monumenti e collezioni provinciali, 2005), 112.

tres for the highly specialized production of clocks and were the starting point for the dissemination of new technologies.[31] A manuscript written by a contemporary of Garzoni, the Mantuan nobleman Camillo Capilupi (1531-1603),[32] affirms that Germans, French and Italians of the time had learned from a Lombard master how to build the most complex clocks.[33] The Lombard master was Janello Torriani.

Between 1365 and 1588, around 29 planetary clocks had been created in Europe, mainly in the context of wealthy courts. From the moment Emperor Charles v put the Microcosm, his brand-new planetary clock, on display at court in the early 1550s, around 14 planetary clocks were crafted in just 28 years, whereas it seems that the same number had been made in the previous 200 years.[34] This impressive figure, together with the analysis of certain technical characteristics of these late 14 planetary automata, will illustrate the major role played by Torriani's planetary clock in the production of such instruments.

The Microcosm, also named Cesar's Sky, or the Emperor's Big Clock, is unfortunately lost. However, we have many descriptions of this object coming from inventories and contemporary accounts that can help to examine its structure. First of all, the Microcosm was more a "cosmometer" than a chronometer, providing a mechanical representation of the motions of the entire known Universe at once. The first description of this clock comes from Bishop Vida in the year 1550, who describes it as:

---

31   Some relevant examples can be found in the following articles: Ludovico Magistretti, "Geni della scienza e straordinari progressi nella misura del tempo: l'eredità di Galileo," in *La misura del tempo: l'antico splendore dell'orologeria italiana dal xv al xviii secolo*, ed. Giuseppe Brusa (Trento: Castello del Buonconsiglio. Monumenti e collezioni provinciali, 2005), 189-200; Silvio A. Bedini, "L'Orologio notturno: un'invenzione italiana del xvii secolo," in *La misura del tempo: l'antico splendore dell'orologeria italiana dal xv al xviii secolo*, ed. Giuseppe Brusa (Trento: Castello del Buonconsiglio. Monumenti e collezioni provinciali, 2005), 189-219.

32   As previously seen, Camillo Capilupi was connected with the Milanese circle of Ferrante Gonzaga, Lord of Guastalla, who, as imperial governor of Milan, authorized the funding for the construction of the Microcosm, Janello Torriani's clock, which was made for Emperor Charles v.

33   Camillo Capilupi, *ms. Vittorio Emanuele 1009*, cc. 152v-153v, and *ms. Vittorio Emanuele 1062*, ca. 33v, Biblioteca Nazionale di Roma.

34   Emmanuel Poulle, *Les instruments de la théorie des planètes selon Ptolémée: équatoires et horlogerie planétaire du xiiie au xvie siècle* (Genève; Paris: Droz ; H. Champion, 1980), see especially vol. 2, 777-84; Giuseppe Brusa, "L'orologio dei pianeti di Lorenzo della Volpaia," *Nuncius* 9 (1994): 645-69.

A clock of admirable, unusual and incredible artifice that measures not
only the daily course of the sun as it goes from rising to setting, and
divides the day into 24 hours of equal duration by means of mechanisms
and hidden energy, either from propelling weights or from a steel spring
strongly rolled up and wound round an axis – and always, because of the
natural stiffness of the steel, resistant and ready to unwind – just as hap-
pens in ordinary clocks, but figures out the order of the heavens and the
constitution and form of the world by means of its continuous motion; so
that, relying on it, we can always be perfectly aware of the state in which
the heavens are set and what their disposition is and not only in which
sign of the zodiac but even in what part of them are to be found the Sun
… and also every star, and know the rise and the setting of the aforesaid
signs with exactness and in their unalterable sequence each one at its
time; we can know also the summer and the winter solstices … the beginning of spring and the remaining parts of the year, apart from the
variations of days and nights, both of the vault of the heavens in motions,
and in the diversity of the regions of the earth itself, besides midnight
and noon in each period of the year, as well as the days of the Fasti, both
those determined by the course of the sun and by the course of the moon
– which we call movable feasts; also the approaches and withdrawals of
the sun and of the moon and their diverse eclipses; even at which moment
each of the stars rises and sets – in relation to us; I believe, senators [Vida
is addressing the senators of the state of Milan], that there is not one of
you who has not seen this admirable, extraordinary and, in a certain way,
portentous work already finished … in all its parts and numbers, in which
its egregious artificer, with his eminent talent and goaded eagerness for
investigation, has emulated the divine … until this moment, inimitable
activity of God Himself in the construction of the Universal World and
Nature entire; to such a point he succeeded in moulding all the revolutions of the firmament and the conjunction of the motions of the celestial
spheres, in which time itself consists, into his skilful and subtle construction of bronze and iron! You have seen how plastically he has represented
all the stars, both the fixed as well as the wandering ones … in such a way
that the journey of a single star occupies several years for all the signs …
When I was asked, some time ago, by some friend of his, by what name a
fabrication of such work could be designed … I could not agree to the
opinion that it should be called a representation or image or copy of the
world … but, given that this may appear a second Universe … it should be
called a small world or rather, perhaps, and more conveniently, in Greek,
by a single word, *mikrocosmon*.[35]

---

35   García-Diego, *Juanelo Turriano, Charles v's Clockmaker*, 54-58.

Besides Bishop Marco Girolamo Vida, the list of the eyewitnesses who described the Microcosm includes the officials who wrote the imperial privilege of 1552, Girolamo Cardano and Gerhard Mercator (1554), Guglielmus Zenocarus (1559), Paolo Casale, ambassador to the Duke of Urbino (1556), Domenicano Gasparo Bugato (1570), an anonymous secretary to the Venetian ambassador Antonio Tiepolo (1571), Ambrosio de Morales (1575), the Tuscan painter Federico Zuccaro (1586), the Flemish gentleman and member of the court Jehan L'Hermite (1599), and the authors of the royal inventories from the years 1602, 1603, circa 1760 and 1773, and many other less detailed writers.[36]

36   Vida, *Cremonensium Orationes III*; Musoni, *Apollo Italicus, nuper in lucem restitutus. His etiam Emblemata accedunt, VIII. Ad Jacobum Albensem Juris consultiss. Ode I*; Cruceius, "Epigramma in Ianelli Turriani Cremonensis horologium," 12; Sacco, *Italicarum rerum* (*1566*), 150-51; Breventano, *Istoria della antichita nobilta, et delle cose notabili della citta di Pauia, raccolta da m. Stefano Breuentano cittadino pavese*; Lamo, *Sogno non meno piacevole, che morale*; Lamo, *Discorso di Alessandro Lamo*. Besides these books, in the very year of Torriani's death, two other works referring to him were published: Campi, *Cremona fedelissima città*; Lomazzo, *Trattato dell'arte della pittvra*. The broad dissemination of Cardano's books, made the description of Torriani's clock circulate even in England: Dee, *Mathematicall Praeface*; Zenocarus Snouckaert, *De republica*, 142, 203-7, and 263. Ludovico Dolce followed Zenocarus, though with some misunderstandings: Dolce, *Vita dell'invittissimo e gloriosissimo Imperatore Carlo Quinto*, 80-88; Morales, *Antiguedades de las ciudades de España*, fols. 91-94; Jehan Lhermite, *El pasatiempos de Jeham Lhermite: memorias de un gentilhombre flamenco en la corte de Felipe II y Felipe III*, ed. Jesús Sáenz de Miera, trans. José Luis Checa Cremades (Aranjuez: Doce Calles, 2005), 509: "*This admirable device [the waterworks] was made by a certain master called Master Janello, a man of 60 years old, who had already made for His Majesty Charles V, the Emperor, a clock of 1,500 wheels, which figured all the movement of the spheres, in due proportion to the Heavens. It still exists in a very good state. His Lordship the Ambassador saw it and so did we. It is a foot and a half high; and of the same breadth in diameter and with eight faces; it can be taken to any place; because it is without ropes and counterpoises; it is wound with a single key. It shows the hours in Spain, Italy, France and of many nations; it gives the phases of the Moon, the growth and decrease of the waters; and to sum up contains so many things that there is not sufficient time to see it ... or memory to remember it or to describe it ...*" Libreria Marciana di Venezia: *Manoscritto di un servitore dell'ambasciator Antonio Tiepolo 1571* reported by: García-Diego, *Juanelo Turriano, Charles V's Clockmaker*, 59; Guido Panciroli, *The History of Many Memorable Things Lost, Which Were in Use among the Ancients: And an Account of Many Excellent Things Found, Now in Use among the Moderns, Both Natural and Artificial*, trans. Henricus Salmuth (John Nicholson ..., and sold, 1715), 331: "*I am told, there was a clock presented by one of Cremona to Charles V which contained the whole Frame and Machine of the Heavens together with all the Stars of the Firmament and Signs also, which were wheeled about just as they are in their celestial Orbs; so that the Heavens seem to be brought down to the Earth. It cannot be denied, but that the Invention is rare and excellent, and worth our Observation*". From the original Latin: Guido Panciroli, *Rerum memorabilium iam olim*

All these accounts agree on the astonishing complexity of the clock and on its novelty. Further physical details are provided by the above-mentioned account made in 1602 by the officials in charge of evaluating the clocks in the service of the king of Spain. We read that the machine was:

> A big gilded brass clock, which displays ... all the heavenly motions, [it is] octagonal, made in the manner of a tower. It rests upon eight balls on its pedestal and upon it in each corner a pillar of the aforesaid metal with its bases and capitals and above it its frieze and cornice, and on top a flattened cupola and above that another in which the bell of the clock is, and over this part a sphere, and in each eight parts a wheel of mathematical things. It is half a vara [41.8 cm] in diameter, and the aforesaid Juanelo [Janello] made it. This clock was appraised by Jorge Estaurez [and] Jacome Diana, clockmakers, at two thousands ducats, in Madrid on May the second 1602.[37]

This description, which recalls both the polyhedric shape of Dondi's Astrarium (seven sides) and the clock painted by Jan Bruegel the Elder, also coincides with Zencoarus' report: "it is a gilded bronze metal-plate closed in a shape of an octagon".[38] Just the dimensions seem to be slightly wider in Zenocarus and other sources compared with several accounts: therefore, we can determine the clock's dimensions as being between 42 and 54 cm high and wide. Additionally, several descriptions stated that this planetary clock comprised between 1500 and 1800 cogwheels (but Mercator says 700), driven by springs and wound by a single key. Before mysteriously disappearing, the Microcosm was described for the last time in the year 1773 in another royal inventory:

---

deperditarum, & contrà recens atque ingeniose inventarum libri duo (Ambergae: Typis Fosterianis, 1599); Jesús Domínguez Bordona, "Federico Zúccaro en España," *Archivo español de arte y arqueología*, no. 7 (1927): 77-89.

37   García-Diego, *Juanelo Turriano, Charles V's Clockmaker*, 61-62, where García Diego transcribe only a 1,000 ducats, the inventory of 1603 gives 2,000.

38   Ibid., 141-142: it has been suggested that Torriani's clock may appear in a painting by Jan Brueghel the Elder (1568-1625) representing *The Sense of Hearing* in which the alarm clocks of Philipp II's collection are included among several musical instruments; but the attribution to Janello of the octagonal clock and the one enclosed in a case, both of which appear in the painting, though suggestive, is not proven. On the other hand, there are many contemporary accounts describing the Microcosm and the Crystalline that widely coincide with the two clocks in the painting.

A famous bronze clock round in shape which represents the Castle of Sant'Angelo with its towers and capitals in which are represented all the motions of the stars and of the heavens, fabricated by the famous Janelo Turriano of Cremona, for Our Lord Charles the Fifth, the admirable construction of which Ambrosio de Morales describes ... [here is a summary referring to the number of wheels, the time of planning, fabrication and inscriptions] ... Placed on an octagonal table of ebony with sheets of bronze, which represent various stories, supported by eight Solomonic pillars.[39]

The humanist and university professor Giovanni Musonio (d. 1561), a fellow countryman of Torriani, said in an encomiastic poem on Torriani's clock that the device provoked awe in Germany.[40] Indeed, Emperor Charles v considered Torriani's Microcosm a unique achievement and the first of its kind, as we can read in the imperial diploma that granted the clockmaker a lifelong pension.[41]

---

39   Garcia-Diego doubted the veracity of this last description of the clock dated 1773. In his eyes, the reference to Castel Sant'Angelo in Rome did not correspond to the traditional accounts of the clock. I think this is a problem of codes: first of all Felipe de Castro was a sculptor and a learned man with a great interest in the three noble arts: sculpture, painting and architecture, and the idea of describing an octagonal clock made by an Italian with the round tower of Sant'Angelo in Rome, though not geometrically entirely precise, fits the unusual and plethoric volume of the clock, and it seems to suit the cultural framework of de Castro. De Castro's attention was indeed attracted by things that the clockmakers of the previous survey did not consider relevant: for instance, beside the volume described with an architectonical frame of reference, de Castro considered it important to refer to the order of the small bronze columns, something that other witnesses considered irrelevant, a particular that even Jan Bruegel, if he represented Janello's clocks, did not record with sufficient accuracy. Perhaps, as García Diego suggested, somebody will one day be able to find it in some vault of the vast *Palacio de Oriente* of Madrid. Ibid., 62, 141-142; Claude Bédat, *La bibliothèque du sculpteur Felipe de Castro*, vol. 5 (Paris: E. de Boccard, 1969), 363-410.

40   *Giovanni Musonio Apollo Italicus* (1551). My translation: "*I can hardly understand, that marvellous Sky / that I shall call, as people name it, Cesar's Sky / Golden Pyramid, that with great a name / recently, under Ferrando's command, brought to the Austrian / that Janiculus who is very famous in the art of smithing, / Janiculus pride of Italy, and of the high Cremona, / Whose famous Tower he took his familiy name from, / Caesar and all of Germany admires it Gazes astonished, and rising falls and stays close to the artisan, / Through the squares and through the streets / With the finger points at him who passes by*".

41   'We Charles v, by the Grace of Divine Mercy, August Emperor of the Romans ... recognize and, by the tenor of the present letters, make manifest to those whom it may concern, that, considering the praiseworthy artistic and practical work which for us, for Our Empire and for the lieges of the Empire itself has been executed by Our dear Janellus de Turrianis, a

All other eyewitnesses of Torriani's clock affirmed the same. It seems that, besides the amazing complexity of the dials and its hundreds of components, the Microcosm had some other characteristic that made it a prototype worthy of imitation, as Capilupi suggested in his aforementioned manuscript. This novelty was probably so evident to contemporary spectators that it was not necessary to emphasise it and to describe it in detail, but it emerges if we put together its contemporary descriptions and those medieval and Renaissance planetary clocks made before and in the decades immediately after the Microcosm was put on display.

About the planetary clocks produced between 1350 and 1550, beside the Astrarium, we know of only two of them in detail: one made by Lorenzo della Volpaia (ca. 1446-1512) and an anonymous German one, kept in Paris in the library of Sainte Geneviève, henceforth referred to as the clock of Oronce Finé (1494-1555). While the planetary clock by Giovanni de Dondi and one by the Florentine Lorenzo della Volpaia went missing, they were reconstructed in the twentieth century thanks to the accurate information provided by the still existing manuscripts produced by the clocks' creators.[42] Oronce Finé's clock is the only known original preserved planetary clock older than the Microcosm. In the 1550s it belonged to Cardinal de Lorraine-Guise, probably the foremost sixteenth-century French ecclesiastic prince. The clock is allegedly of late-fifteenth-century German origin, and was renovated in 1553 by the physician

---

mathematician of Cremona and, very probably, the foremost among the inventors of clocks, in constructing for Us, with admirable technique and talent, an exceptional clock and – so far as is known- never seen anywhere else up to present time, which shows not only all the hours of Sun and of the Moon, but also all the other signs of the planets and the coming, going and the reflections of the celestial motion in a true, exact and visible order with consummate ability and to Our greatest satisfaction. We have conceded, appointed and consigned to Janellus himself, and by the tenor of the present We appoint and consign an annual pension of one hundred gold escudos from all the revenues and income of the seignory of Milan, both ordinary and extraordinary, to be paid by the hands of the General Treasurer, or the other Officers of Our State of Milan on whom the aforesaid matter depends or may depend in the future, for as many years as the life of Janellus himself may last, from now onwards, at the rate of a fourth part each three'. The printed document is kept in the State Archive of Cremona. The English translation is taken from: García-Diego, *Juanelo Turriano, Charles V's Clockmaker*, 75-7.

42   Built from the 1480s (and finished in the following century) on behalf of Lorenzo de Medici as a present to the King of Hungary and Bohemia Matthias Corvinus, but later purchased by the Republic of Florence as something too precious to be alienated. Ambrosio de Morales, praising Torriani's planetary clock, claimed it to be superior to della Volpaia's; Morales, *Las Antiguedades*, 91-94; Brusa, "L'orologio dei pianeti di Lorenzo della Volpaia".

Oronce Finé, "the restorer of mathematical studies in France".[43] Unlike Torriani, who built his clock from scratch, Oronce Finé restored Cardinal de Lorraine-Guise's machine by adding a part to the mechanism (that displaying the solar hours and the astrolabe-dial).

Perhaps this planetary clock was the one illustrated in a set of three drawings made for Cardinal Albrecht IV Hohenzollern (the very same bishop of Brandenburg who provoked Martin Luther's protest), who acquired the title of Archbishop-Elector of Mainz (1490-1545). Indeed, both clocks share the same morphology: they both have five faces with mathematical dials (instead of the Astrarium's seven and the Microcosm's eight). The great difference between Oronce Finé's clock and the illustrated one is in the style of the pedestal and especially the shape of the upper part: the drawn clock has a small table-clock pinnacle on the top, upon which stands a little sculpture. The clock from Paris has an egg-shaped cover with a large rotating celestial sphere partially emerging from its top. It has also been suggested that the subject of the illustrations made for Archbishop Albrecht IV might have been the clock, which was begun by Regiomontanus in Nuremberg (who visited Pavia's castle to study the Astrarium) and later completed in different times by the above-mentioned blacksmith Hans or Jacob Bulman, and by the famous watchmaker Peter Helein in cooperation with the astronomer Johann Werner (d. 1522).[44] Archbishop Albrecht IV seems to have purchased that very clock in 1529 for 180 gulden.[45] Since the archbishop had passed away in 1545, Charles V might have acquired this clock from his heirs. The celestial sphere could have been added under Charles V's command, as in the case of the Microcosm. Further studies on the clock and on the drawing may shed new light on this problem.

The history surrounding Oronce Finé's clock may cross that of Torriani's Microcosm on the battlefield of Metz. Cardinal Charles de Guise-Lorraine had been bishop of Metz since 1550 (though he had been administrating the bishopric even before that date) and was brother to François de Lorraine, Duke of Guise. The Duke of Guise was named Governor of Metz after the French king

---

43   This is the theory of the genealogy of this clock presented by Poulle: Denise Hillard and
       Emmanuel Poulle, *Oronce Fine et l'horloge planétaire de la Bibliothèque Sainte-Geneviève*
       (Genève: Librairie Droz, 1971), 318-22; Arthur Augustus Tilley, *Humanism under Francis I.*
       (London, 1900), 465.

44   Bedini and Maddison, *Mechanical Universe*, 26-27; Neudörffer, *Nachrichten von Künstlern
       und Werkleuten*, 65- 6; Dietrich Matthes, "From Methaphysics to Astrophysics: Clocks to
       Represent the Cosmos," in *Janello Torriani: A Renaissance Genius*, ed. Cristiano Zanetti
       (Cremona: Comune di Cremona, 2016), 127-30.

45   Henry C. King and John R. Millburn, *Geared to the Stars: The Evolution of Planetariums,
       Orreries, and Astronomical Clocks* (Toronto: University of Toronto Press, 1978), 65-67.

Henry II took the imperial city from Charles V, thanks to an alliance with the Lutheran princes of the kingdom of Germany. At the end of 1552, Emperor Charles V had tried to recover Metz with a siege that proved unsuccessful. It was a few months since Charles V had granted Janello Torriani his pension for the creation of his planetary clock, and the Emperor was outside the city walls of Metz when he received a letter from the clockmaker asking him to press the administration of Milan to pay his pension. The request, despite the Emperor being significantly otherwise occupied, was immediately satisfied: Charles V wrote to the Governor of Milan, ordering him to fulfil Torriani's requests, a quite clear sign of the Emperor's attachment to his marvellous new clock and to its creator. A few days later the Emperor was defeated. By January 1553, the Duke of Guise could celebrate his victory and Charles V had to leave his camp in such a rush that the Duke was able to plunder the imperial tents. He seized a series of precious wall-tapestries, personal belongings of the Emperor that he later gave as a gift to his brother the cardinal.[46] I wonder if the German planetary clock from Metz, restored by Finé, may have accompanied those tapestries. Indeed, when Petrus Ramus wrote in 1569 that he had seen Finé's clock, he claimed that it was part of the war booty from the *"germanico bello"*, perhaps the battle against the Imperials at Metz.[47]

After the Microcosm had appeared between the years 1549 and 1552 in Milan, Augsburg, Innsbruck, and from 1554 in Brussels, the production of planetary clocks began to flourish. Here, we take into account the only three extant planetary clocks made after the Microcosm (1550) and before Janello Torriani's death in 1585 that have been extensively studied. Between 1554 and ca. 1559, Philip Immser of Strasburg (alias Philippus Imserus, Imsser, Imser, Ymbser), a professor at the University of Tübingen, fashioned together with the clockmaker Gerhard Emmoser von Rainen, a planetary automaton for the Count Palatine Ottheinrich. The clock would end up in the hands of Emperor Ferdinand II (today part of the collections of the Technisches Museum in Vienna).[48] In the 1560s, Eberhard Baldewein (1525-1593) made two planetary clocks. These are on display in museums in Kassel (Astronomisch-Physikalisches Kabinett) and Dresden (Mathematisch-Physikalisches Salon). These two clocks were made

---

46   Guido Gerosa, *Carlo V: un sovrano per due mondi* (Milano: A. Mondadori, 1990), 349.

47   *"alteram apud Orontium mathematicum professorem regium germanico bello similiter diteptam"*: Ramus, *Scholarum mathematicarum*, 31.

48   King, *Geared to the Stars*, 68-72; Emmanuel Poulle, *Science et astrologie au XVI siècle: Oronce Fine et son horloge planétaire* (Paris: Bibliotheque Sainte-Geneviève, 1971), 15-16; Bruce Chandler and Clare Vincent, "To Finance a Clock: An Example of Patronage in the 16th Century," in *The Clockwork Universe*, ed. Klaus Maurice and Otto Mayr (Washington: Smithsonian Institution, 1980), 103-13.

for Prince William the Wise of Hesse (1532-1592) – who became William IV Landgrave of Hesse-Kassel in 1567 – a champion of the Reformation and one of the foremost patrons of astronomy.[49] The first clock was probably started in the second half of the 1550s and was completed in 1562. The second was crafted between 1563 and 1568, when William had it made for the Elector Augustus of Saxony.[50]

Were there any major technical differences between those two sets of clocks crafted ante 1550 and post 1552? The first thing we can observe is that the clocks by Giovanni de Dondi, Lorenzo della Volpaia, the so-called Oronce Finé, and Albrecht IV (if this was not the same one as Finé's), were all weight-driven devices. The same can be reasonably stated for those other lost clocks crafted before 1550. They had to be rather tall to let the weight unroll for a reasonable time and the weights had to be heavy enough to provide the necessary energy to drive such a large number of rather voluminous components. Immser's and Baldewein's planetary clocks were instead smaller and spring-driven. Torriani's Microcosm, according to the many descriptions, was only two feet high (less then 54 cm) and it was also spring-driven. As the anonymous secretary to the Venetian ambassador Antonio Tiepolo wrote, the Microcosm could be taken anywhere and it was wound with a single key. Morales tells us that there were 3 coilsprings inside it. With its hundreds of wheels, this clock was an impressive example of miniaturization. From this chronological standpoint, we might infer that the planetary clocks built after the Microcosm were indebted to it for their spring-driven motors. Moreovoer, it seems that Janello's planetary automaton did not have a *fusée* either: in his *De Subtilitate* Cardano writes that Janello

---

49   Gaulke, *Der Ptolemäus von Kassel*; Korey, *The Geometry of Power – the Power of Geometry*.

50   In the 1560s, the Duke of Urbino Guidobaldo II della Rovere commissioned Gio. Maria Barocci to make a splendid clock. This clock dates from 1570 and was meant as a present for Pope Pius V. It is a single-spring driven clock, and scholars debate whether it should be classified as a planetary or an astronomical clock, though the second definition seems to better suited to the device. It was stolen from the Vatican during the Napoleonic Wars and today it belongs to the Bernard collection of Paris. I have not had the chance yet to examine it, except from the picture published in: Silvio A. Bedini, "La dinastia Barocci: artigiani della scienza in Urbino, 1550-1650 = The Barocci dynasty : Urbino's artisans of science, 1550-1650," in *Scienza del Ducato di Urbino = The science of the dukedom of Urbino*, ed. Flavio Vetrano (Urbino: Accademia Raffaello, 2001), 7-98; see also Antonio Lenner, "La scuola di Urbino: gli orologi rinascimentali italiani, dai Barocci ai camerini," in *La misura del tempo: l'antico splendore dell'orologeria italiana dal XV al XVIII secolo*, ed. Giuseppe Brusa (Trento: Castello del Buonconsiglio. Monumenti e collezioni provinciali, 2005), 220-27; Morpurgo, *Dizionario degli orologiai italiani*, 1974; and Del Vecchio and Morpurgo, *Addenda al Dizionario*.

made this amazing clock "without weights and without cords"; therefore, he probably applied a *stackfreed* system to uniform the progressively diminishing power of the coilsprings.[51]

Needless to say, Torriani was not the inventor of the spring-driven clock with *stackfreed*,[52] which had been invented in the previous decades, but he was most probably the first to apply its mechanisms to such a complex device. As Cardano remarked in 1554, after he had attributed this invention to him in the 1550 edition, Janello was "a man of sharp ingenuity ... who had invented many things, or who had improved these ones invented by others".[53] But there is an absolute innovation we can attribute to Janello, his machine-tool that allowed for such a miniaturization and for a speeding-up of the construction. It was thanks to his special lathe, the first known gear-cutting machine, that he could miniaturize the components of the mechanism so that a set of springs could substitute for the extremely heavy weights necessary to move what used to be voluminous clocks. Therefore, the result was that the Microcosm was the first portable planetary table-clock. The theory outlined above can also help us to answer Silvio Leydi's question: why was the Crystalline not as famous as the Microcosm, if it was even more compact and three times more expensive to make? The answer is that though a remarkable and precious planetary clock, the Crystalline was not as innovative as the Microcosm. The display of this innovative instrument at the imperial court in the early 1550s must have

---

51   Cardano, *De subtitulate (1554)*, 1554, 454: after describing the invention of a clock without a cord, and following a description of Torriani's planetary clock and its differences with Archimedes' glass one, Cardano adds: "after all, Janello did not combine weights, neither cords, but built all things with iron, cogwheels, and admirable craft" and: "*quanqua[m] Ianellus neq[ue] pondera nec funes miscuerit, sed omnia ferro denticulisq[ue] miro construxit artificio*".

52   In another occasion, I interpreted this "clock without cord" driven by a "spring in a shape of a snail" as a coil-spring-driven clock, as opposed to a clock driven by weight with a cord. Giuseppe Brusa, on the contrary, had interpreted this "clock withour cord" as a stackfreed as opposed to another coil-spring clock, the fusee. I have to admit that if one takes into account the description of the "clock with a cord" that Cardano gives in the first book of his *De Subtilitate*, Brusa's interpretation seems to be more likely than my own. However, we have to observe that Cardano, in the description of the "clock without cord", does not include what is the most important part of the stackfreed, which is the stackfreed itself: a spring-loaded cam to even out the mainspring. In conclusion, what we can be sure of is that, according to Cardano, Janello Torriani's innovation was about the use of a coil-spring motor. Brusa, "Early Mechanical Horology in Italy," 512; Zanetti, "Microcosm".

53   Cardano, *De subtitulate (1554)*, 1554, 453: "*vir acris ingenii*" Ianellus Turrianus Cremonensis, *cuius etiam supra meminimus, vir acris ingenii multa talia aut excogitavit, aut ab aliis excogitata in melius traduxit*".

rustled a few branches in technologically fertile central Europe. But it was probably not until after the Peace of Passau, in August 1552, that Catholic and Protestant princes could conceive of spending their resources on a competitive ground other than that of the battlefield: the patronage of arts and science.

The high level of craftsmanship involved in clockmaking in the cities in the Kingdom of Germany provided the right technical environment for the reproduction of the Microcosm's innovations. The exhibition of such a clock at court had probably inspired the protestant princes of the Empire to emulate their lord the Emperor, champion of Catholic orthodoxy. Besides the King of Denmark, the members of the great houses of Hohenzollern, Wittelsbach, Wettin and Hesse were the German princes who patronised the creation of planetary clocks. Considering that Charles v's brother Ferdinand was, as King of Bohemia, the fourth lay grand elector, we see that nearly all of the greatest feudal lords of the Holy Roman Empire were involved with planetary horology, and, except for Cardinal Albrecht IV Hohenzollern (whose planetary clock was older than the Microcosm), all of their projects to create planetary clocks were started shortly after Torriani's Microcosm was displayed at court.[54] Count Palatine Ottheinrich of Neuburg-Palatine (who in 1556 also became Prince Elector of the Palatinate) had his residence some 50km from Augsburg. He was the first among those lords to commission (to Immser) a planetary clock in the style of the Microcosm. In fact, we can consider Immser's machine as the morphological link between Torriani's (spring-driven with a rotating sphere mounted on top) and the later German examples,[55] which all had a four-faces case and a rotating sphere on top. Immser's planetary clock has a four-faces case, and a small octagonal tower that contains a rotating sphere tops it. All surviving planetary clocks made between 1554 and 1585, are slightly larger than Torriani's creations, but seem to draw on the model of the Microcosm and the Crystalline.[56]

The creation of these impressive clockworks was close to the limits of human capabilities. Indeed, although they are always remembered as the

---

54   The prince-electors were three spiritual and four lay: the three spiritual were the Archbishop of Cologne, the Archbishop of Mainz, the Archbishop of Trier; the four lay electors were the Count Palatine of the Rhine (House of Wittelsbach), the Duke of Saxony (House of Wettin), the Margrave of Brandenburg (House of Hohenzollern) and the King of Bohemia (House of Habsburg).

55   Henry C. King had already noted in 1978 that "*Baldewein's clock at Kassel presents a combination of features seen on Finé's and Immser's clocks*": King, *Geared to the Stars*, 72.

56   For instance, Immser's clock measured 85cm high, 37.5cm wide, and Baldewein's (1561) was 90cm high, 37cm wide: Ibid., 69 and 72.

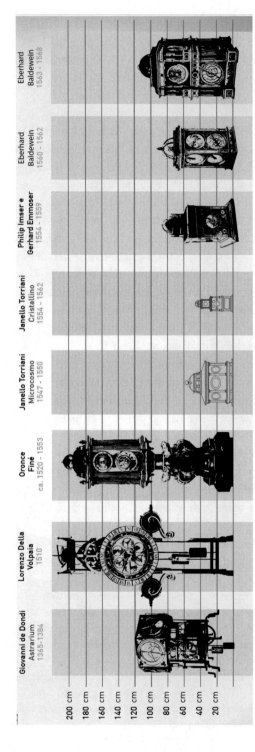

FIGURE 49   *The missing link in the evolution of planetary clocks.* One can see the spring-driven Microcosm and the Crystalline by Janello Torriani (hypothetical reconstructions by José A. García-Diego) representing a fundamental passage into miniaturisation. PANEL FROM THE EXHIBITION: JANELLO TORRIANI GENIO DEL RINASCIMENTO, SEPTEMBER 10, 2016-JANUARY 29, 2017. AUTHOR'S PHOTO.

product of one single master, the reality was that such pieces often reflected the efforts of a whole team, with work undertaken by multiple persons with different specialisations. In fact, the knowledge of both the relevant craftsmanship and theory required by such an endeavour was simply becoming too much for any one clockmaker. Philippe de Maisiers wrote in 1385 that Giovanni de Dondi:

> With his own hand forged the said clock, all of brass and copper, without assistance from any other person, and did nothing else for sixteen years, according to the information given to the writer of this book, who has had a great friendship with the said Master John [Giovanni de Dondi].[57]

We have seen how Torriani as well explained to Morales that during the twenty years he spent designing the clock, he had fallen gravely ill upon two occasions due to the fearsome intellectual efforts involved in planning his Microcosm. However, even if the great part of the job was performed by the clockmaker, Torriani, who was a master craftsman, could have never accomplished the Microcosm without the tutorship of a university-trained mathematical astrologer: the physician Giorgio Fondulo (d. 1545). Despite the fact that Torriani created all the mechanical components of his clock, Jacopo Nizzola da Trezzo worked on the rock-crystal components and in all probability also handled the gilded decorations for the Microcosm, unless Leone Leoni was responsible for these. The famous cartographer Mercator made the small painted globe that was inserted into Nizzola's celestial sphere.[58] The process of emulation was a difficult one and reveals how such complex sixteenth-century mechanical endeavours required the collaboration of people with differing specialisations, both theoretical and practical. The historians Henry C. King, Bruce Chandler

---

57   Bedini and Maddison, *Mechanical Universe*, 20.
58   From Mercator's letter dated 23 August 1554 to Melanchthon we learn about Mercator's audience with the Emperor Charles V in Brussels in May. "The Emperor asked him to paint a fist-sized ball with a world map, to be inserted into a crystal sky-ball. This celestial sphere should be placed on the top of a clock with mathematical dials that represented the movements of the 7 planets and the stars, which Janello had produced for the Emperor. On this occasion, Charles recalled the devices of Mercator, which were destroyed by fire at Innsbruck, and he asked for the best way to determine the meridian line. In contrast to Torriani, who suggested the Indian circle, Mercator explained his method of observation of the ascent and descent of the circumpolar star". My translation from: Ernst Zinner, *Deustche und niederländische astronomische Instrumente des 11.-18. Jahrhunderts, von Ernst Zinner* (München: C.H. Beck, 1956), 443.

and Claire Vincent have explored and reconstructed the painful process of the creation of Immser's mechanism.[59]

Immser's and Torriani's experiences show how courtly patronage promoted technological innovation based on both the highest craftsmanship and the best university mathematical theories. As shown by the collaboration between Torriani and Mercator, imperial patronage was able to gather the best practical and theoretical knowledge of the Empire at court and to make of it a place of scientific discussion and advancement. Renaissance planetary clocks are the material expression of this synergy. Even smaller courts such as Wilhelm of Hessen's could set up impressive teams of specialists: Eberhard Baldewein, who, according to Ramus, was, curiously enough, a tailor by training, made the two planetary clocks.[60] He managed to enter the court of the Landgrave of Hesse in his role as heating and lighting supervisor at Kassel Castle. During this appointment Baldewein achieved a certain familiarity with the son of the Landgrave, Wilhelm, the future patron of the astronomer Tycho Brahe (1546-1601) and of the famous clockmaker Jost Bürgi (1552-1632). After the election of Wilhelm in 1567, the common passion for astronomy that had fuelled their friendship won Baldewein the post of chief mechanic in the castle's workshop. But the cooperation with Wilhelm and the court astronomer Andreas Schöner

---

59   "Immser had been pupil to Johannes Stöffler at the University of Tübingen. Later he took his post there, teaching mathematics: Immser persuaded Elector Palatine Ott Heinrich, noted for his scholarly and artistic interests, to support the project, but he soon got into difficulties. He made matters worse by extending his original plans, and in 1556, after two years' work, had only the framework completed. In his report to the elector he requested payment of 1600 gulden, despite the terms of the original agreement, namely, payment of 700 gulden on delivery of the complete mechanism and an additional 100 gulden if it performed satisfactorily for a year. The elector refused to revise the contract and instead sent him the clockmaker Gerhard Emmoser von Rainen (d. 1584) of Heidelberg. This only made matters worse. Emmoser seems to have been capable of making and assembling the wheel-trains, but Immser resented his presence and preferred to work alone. At one stage his mistrust reached such a pitch that he dismantled the mechanism and locked the parts in a trunk. Eventually, at the end of 1557, the clock was finished, but the accuracy of the wheel-work left much to be desired, a fault which he laid at Emmoser's door. He begged the elector to accept the clock as it stood, for the project had assumed the proportions of a nightmare. This was done, but the astronomical indications proved so troublesome that the elector, confined to his bed through sickness, lost patience and demanded the return of the monies paid unless the clock performed correctly. He died in the following year, 1559, to be succeeded by elector palatine Frederick III". This passage comes from: King and Millburn, *Geared to the Stars*, 68-69. See also: Chandler and Vincent, "To Finance a Clock: An Example of Patronage in the 16th Century".

60   Moran, "German Prince-Practitioners". 253-374.

(1528-1590) had begun long before the title of landgrave was bestowed upon Wilhelm. Wilhelm was said to have constructed a machine useful for predicting planetary positions sometime before the 1560s. His machine was made of metal and moved by gear-work, and was based on Petrus Apianus' *Astronomicum Caesareum* (1540). Together with his astronomer, Wilhelm had also begun to attempt a correction of the Alphonsine Tables but, in order to do so, he needed precise scientific instruments.[61] Thus, he decided to build an accurate planetary clock, a project that involved Wilhelm himself, his astronomer Schöner, Baldewein, two clockmakers called Hans Bucher and Christoffel Müller, and a goldsmith, Hermann Diepel. The clock was successfully accomplished in 1562 and afterwards, the team worked on similar devices.

Thanks to the Microcosm we have been able to delve into a process of technological transfer, following the passage of knowledge from Renaissance Italy to Northern Europe and Spain. This case study has shown which agencies promoted the development of new technologies and these innovative technologies were transferred from one place to another. Thanks to Charles v's personal interests and his significant financial investments in them, Janello Torriani, a man of great talent, was given the chance to innovate planetary horology. The Emperor was also responsible for transporting the Microcosm to Germany, Brabant and Castile, and probably for the transportation of another planetary clock to Metz, and therefore to Paris (here I refer to Oronce Finé's clock). After the Microcosm's appearances in Milan, Innsbruck, Augsburg and Brussels, Immser, Baldewein and other German clockmakers applied spring-driven mechanisms to planetary clocks. In order to do so, they had to miniaturize the mechanisms and to create taskforces made of highly skilled craftsmen and university-trained mathematicians. Princely patronage, triggered by the desire to emulate their lord and innovate, provided the funding for such endeavours. The fact that the majority of these patrons were Lutherans probably gave them a further reason to compete with the Emperor, adding to a claim of a better knowledge of the metaphysical truth, a physical one.

The story of the Microcosm shows that Renaissance courts were very important *"luoghi di sociabilità"*, or "places of sociability" for innovative thought.[62]

---

61   Ibid., 257

62   Franceschi, "La bottega come spazio di sociabilità"; Planetary horology can also be considered an important "trading zone" where craftsmen, scholars and patrons met with communal enthusiasm: Long, "Trading Zones". In the 1970s and 1980s, Bruce T. Moran highlighted the role of courts "as institutional nodes of technical activity": Moran, "German Prince-Practitioners," 253; "Princes, Machines and the Valuation of Precision in the 16th Century".

The court of Charles V brought together skilled and learned people from different parts of the Empire and gave them a place to discuss exciting new projects,
despite their different linguistic, social, epistemological and religious backgrounds. Moreover, princely patronage was able to sustain the synergic work of
different experts on a specific project for several years by providing funding
that no individual could have ever raised alone, and has to be considered a
fundamental driving force in the development of miniaturized technologies
during the Renaissance.

From a long-term perspective, the story of the Microcosm reflects a major
technical trend of experimentation in sixteenth-century mechanics: beside
precision, some of the most relevant technological achievements in this period
were connected with the experimentation in macro- and micromechanics.[63]
Torriani's career provides the perfect representation of this trend, from the
gigantic device of Toledo to the crafting of tiny mechanisms set in planetary
clocks. The process of miniaturization was possible thanks to the designing of
a special machine-tool to cut small gears precisely and to other crafts, like
ground lenses, instruments necessary to magnify (though there is no evidence
of such use in Torriani's workshop) and later to gaze at the stars.[64]

---

63   Moran, "Princes, Machines and the Valuation of Precision in the 16th Century". 209-228.
64   See Tiemen Cocquyt, "Miniaturizzazione degli orologi e lenti d'ingrandimento," ed. Cris-
      tiano Zanetti, in *La Voce di Hora*, Atti della conferenza Janello Torriani, genio del Rinasci-
      mento (forthcoming 2017).

PART 3

Hydraulic Metamorphosis of a Clockmaker at the Court of the Never-Setting Sun

∴

CHAPTER 7

Mechanics: from Micro to Macro

Automata, Watches and Great Machines

We have seen how Padre Sigüenza wrote that, after the imperial abdication, Janello Torriani was accustomed to entertain Charles V with "*reloxes y otros ingenios*", which means clocks and other devices.[1] What other devices? This sibylline sentence keeps open the possibility for different interpretations. For instance, both Spanish and Italian literary sources convey an image of Janello Torriani as an ingenious automata maker. Contemporary and later reports tell stories that almost recall the atmosphere of fairy-tales, in which Janello was said to have invented a number of mechanical prodigies almost impossible to believe: according to these traditions the clockmaker had crafted a watch so tiny that it could be put in a ring, automata of different size and function, made in the shape of dancing dolls, soldiers that could fight as if on a battlefield, fighting dogs, or even flying birds.[2] The further we go in time from the sixteenth century, the greater the interest shown by historiography for these oddities. As in the children's game Chinese whispers, it seems that in Torriani's story news gets distorted in the passage from one teller to another: thus, Janello morphs from a court clockmaker to a kind of a wizard. In Italy there were rumours of perpetual motion machines and in Toledo, late sources from the following century describe how the clockmaker had made a wooden automaton that went every day to collect Janello's meal at the palace of the Archbishop to bring it back to his master. What kind of truth are these stories telling us?

One story, which started to circulate in the late seventeenth century, stated that Janello, on encountering the Emperor for the first time, had offered him a special present: a ring with a watch in the place of the jewel:

> Very small wheels in glorious though very minute rotations regulated its movements. By lightly pricking the finger [with a small hidden nail], it marked each of the hours, which were also marked on the dial by numbers, which, though certainly very tiny, were sufficiently visible. This marvellous little watch was greatly loved by the Emperor, both for its rare symmetry and for the perfect arrangement of its exquisite work. He could

---

1 See in this book the section: *A Clock Broth for the Emperor*.
2 See: Zanetti, *Janello Torriani, a Renaissance Genius*.

© KONINKLIJKE BRILL NV, LEIDEN, 2017 | DOI 10.1163/9789004320918_009

a

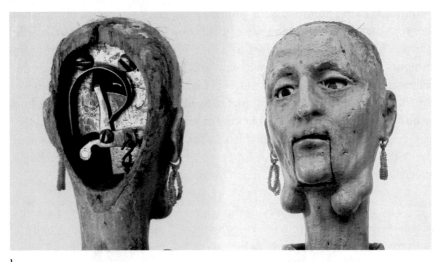

b

FIGURE 50a-b   *Janello Torriani* (?), Automaton and detail of head, *previously part of the Settala Museum (and later of the Veneranda Biblioteca Ambrosiana, Milan).* At this museum was also held the armillary sphere by Janello Torriani, as of now the only surviving object signed by the Cremonese master. This is the only extant automaton that resembles Janello's one described by Morales. PRIVATE COLLECTION. PHOTOGRAPH BY MARIO LIGUIGLI.

not cease praising it as a miracle of art, there being no jewel precious enough to equal it in value.[3]

This eighteenth century account circulating in Cremona drew upon earlier works that had spoken of a ring-watch made for Charles v, but without mentioning its creator.[4] Indeed, for the moment, we do not have any contemporary account attributing this invention to Janello. On the contrary, there are other sixteenth-century sources describing similar ring-watches: in 1550, Cardano, in the *editio princeps* of his *De Subtilitate*, wrote the same story about this ring given as a present not to the Emperor, but to the Duke of Urbino.[5] Someone attributed this ring-watch to Giovan Giorgio Capobianco, a famous goldsmith, engineer and clockmaker from Vicenza, and a contemporary of Torriani.[6]

---

3   Zaist, *Notizie istoriche depittori, scultori, ed architetti cremonesi* (my translation).

4   Most probably Giambattista Zaist (1700-1757) modeled his anecdote on the one provided by the lawyer and Muratorian historiographer Francesco Arisi (1657-1743). Arisi, *Cremona literata*, 1741, III. Francesco Arisi, in turn, took the anecdote from his brother Desiderio Arisi (who died in 1725) who, in his manuscript, mentions Fortunius Licetus (1577-1657): the Arisi brothers took for granted that the ring-watch mentioned by Licetus in Charles v's hands was Janello's, but Licetus did not mention the name of the craftsman: Fortunio Liceti, *De anulis antiquis, librum singularem in quo diligenter explicantur eorum nomina multa, prim[a]eua origo, materia multiplex, figurae complures, causa efficiens, fines, vsusve plurimi, diffenrentiae ... & contumulatio cum cadauere priscis temporibus* (Vtini: Typis Nicolai Schiratti, 1645), 35, 222. Licetus drew upon the works of Simone Maioli and Silvestro Pietrasanta, who both did not mention Torriani: Simeone Maiolo, *Dies caniculares seu colloquia tria, & viginti ...* (Romae: ex officina Ioan Angeli Ruffinelli. Typis Aloysij Zannetti, 1597), 1052, Colloquium XXIII; Silvestro Pietrasanta, *De symbolis heroicis, libri IX* (Antuerpiæ: Ex officina Plantiniana Balthasaris Moreti, 1634), bk. III, chap. III, 94-98.

5   Girolamo Cardano, *Hieronymi Cardani ... De subtilitate libri xxi* (Parisiis: Ex officina M. Fezendat & R. Granjon, 1550), 26v.: "*Principi Urbini, & hoc nostra etate contigit, dono datus est annulus, qui indice exciperetur, gemmam vero haberet, in qua horologium perfectum, quodque praeter lineam horas distinguentem uno ictu per singula horaria spacia gestantem admoneret.*"

6   Capobianco was said to have made for Guidobaldo II della Rovere (1514-1574), Duke of Urbino, "*a watch inside a portable ring, which had engraved in its top the 12 heavenly signs and a little figure in between, which was showing with their number, the hours of day and night, chiming them*": "*horologio dentro di un portatile anello, che haveva intagliati nella testa I dodici celesti segni, con una figurina fra mezzo, che signate mostrava per numero l'hore del giorno et notte pulsanti*": Giacomo Marzari, *La historia di Vicenza* (Vicenza: G. Greco, 1604); an older description is in Pietro Viola, *Petri Violae,... de Veteri novaque Romanorum temporum ratione libellus* (Venetiis: impressum apud N. de Bascarinis, 1546), 12v.; Hero of Alexandria, *De gli automati*; Morpurgo, *Dizionario degli orologiai italiani*, 1950; and Del Vecchio and Morpurgo, *Addenda al Dizionario degli orologiai italiani edizione 1974 di Enrico Morpurgo*; Lionello Puppi, "Capobianco, Giorgio," *Dizionario biografico degli italiani* (Roma: Istituto della Enciclopedia

Someone reported that Queen Elizabeth of England "wore one that not only told the time but [also] served as an alarm: a small prong came out and gently scratched her finger".[7] Louis Zapata de Chaves (1526-1595) reported enthusiastically that "Our Lord the King [Philip II] has a watch in a ring which marks the hours inside by picking the finger lightly."[8] In the biography of Bernardo Buontalenti, architect to the future Grand Duke Francesco I of Tuscany, written by Gherardo Silvani (1579-1675), one can read that Buontalenti had accompanied Prince Francesco de Medici to Spain in the years 1562-1563, and on that occasion he:

> Was so grateful to the King [Philip II] that he gave him as a present many miniatures and he made for him a small watch to carry around a finger, and thanks to this his master had more rapidly access to His Majesty; with no ease the King let him go.[9]

Bernardino Baldi from Urbino, famous mathematician and abbot of Guastalla, in his edition of Hero of Alexandria's book on automata, mentioned the very same ring-watch, this time attributing it to his fellow countryman Giovanni

italiana, 1975). It seems that Emperor Charles V was among Capobianco's notable customers,who also included Andrea Alciati, Cardinal Matteo Schiner, Duke Guidbaldo of Urbino, and Duke Federico of Mantua. It is said that Capobianco, who was convicted for homicide, was spared an unpleasant fate thanks only to the intervention of Charles V and the Duke of Urbino. He was exiled from Venice, and it seems that he was allowed to return only because he invented a machine for cleaning mud from the canals. It seems that Capobianco was also invited by Ferrante Gonzaga (one of Janello's patrons) to Milan for some fortification works. Other sources refer to the same ring-watch but this time made for Charles V: Maiolo, *Dies caniculares*; Pietrasanta, *De symbolis heroicis*, chap. III, 96; Gio. Felice Astolfi, *Della officina istorica* (Venezia: Turrini, 1642), bk. 2, 283.

7   Landes, *Revolution in Time*, 87.

8   Luis Zapata de Chaves, *Obra completa de Luis Zapata de Chaves (1526-1595)* (Badajoz: Institución Cultural "Pedro de Valencia," 1979); translation from: García-Diego, *Juanelo Turriano, Charles V's Clockmaker*, 93.

9   *"fu tanto grato al Re e gli donò molte sue miniature e gli fece uno oriolino da tenere in dito e per la sua virtù ebbe più presto udienza il padrone; con difficultà il Re lo lasciò andare"* (my translation): Vera Daddi Giovannozzi, "La vita di Bernardo Buontalenti scritta da Gherardo Silvani: appunti d'archivio," *Rivista d'arte*, 1932, 505-24.

Maria Barocci.[10] One Jacob Widmann from Antwerp was also said to have made a ring-watch for the Duke of Mantua in the 1570s.[11]

The Renaissance was an important time for miniaturized mechanics. Sometimes this technical turn was expressed through hyperbolic stories, such as the one of Regiomontanus' iron-fly. However, in the Cinquecento, ring-watches and other tiny timepieces inserted in different tools were belonging to the domain of reality, rather than fiction: they were luxury goods circulating among courts. Administrative sources can help us to corroborate the previous literary records: an inventory of the Gonzaga family, for instance, includes one such ring-watch some 30 years before the above-mentioned Jacob Widmann's one, the entry for which reads "*del Sultano*" i.e., of the Sultan.[12] This must be the same ring-watch that Pietro Aretino had attributed to Capobianco and that he names "*del Gran Turco*". Pietro Aretino refers to the object as if it was well known and he uses it to compare the mechanical competence of Capobianco in miniaturizing to his own ability in synthesis in the field of writing.[13] Aretino mentioned this ring-watch for the first time in a letter dated 1537. In order to visualize a similar device, we have to wait until 1561, when the French gold-smith and engraver Pierre II Woeiriot (1532-1599) published the *Livre d'anneaux d'orfèvrerie*, a book rich in illustrations of jewels. The book was also printed in Italian with the title *Libro d'anella d'orefici de l'inventione di Pietro Woerioto di Loreno*. Pierre Woeirot published this book after a trip to Italy where he may have seen a ring similar to this. The earlier still extant examples of this amazing craftsmanship seem to date to the late part of the century (1583-1585) and are attributed to the clockmaker Jacob Weiss from Nurnberg.[14]

The most important function of such watches was to dazzle rather than serve as practical timepieces. Master Pietro Guido, named dell'Orologio from Revere (state of Mantua), one of the most famous early watch-makers, had important customers such as the Marchioness of Mantua Isablella d'Este

10   "*Io non finisco di ammirare la diligenza di colui, che li richiuse in un castone di anello e fece sì che non solamente con l'indice, ma con la percossa ancora dividessero il tempo*": Hero of Alexandria, *De gli automati*, 8.

11   George Charles Williamson, *Stories of an Expert* (London: H. Jenkins Ltd., 1925), 191-192.

12   See the inventory published in: Daniela Ferrari, *Le collezioni Gonzaga: l'inventario dei beni del 1540-1542* (Cinisello Balsamo (Milano): Silvana, 2003).

13   Pietro Aretino, *Lettere*, ed. Paolo Procaccioli (Milano: Biblioteca Universale Rizzoli, 1991), lettera 108 al Fausto Longiano, Venezia XVII di Decembre MDXXXVII, 325-328.

14   One, whose movement is lost, is held in the collections of the Indianapolis Museum of Arts, whereas the other one, still intact, belongs to the collections of the *Münchener Residenz* in Bavaria.

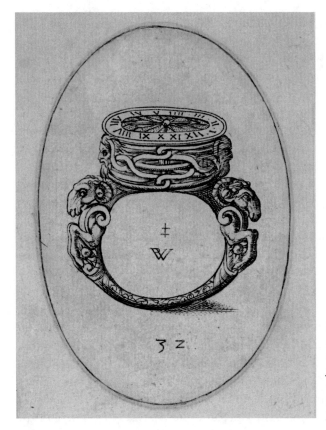

FIGURE 51
*Pierre Woeiriot
(engraving),* Ring-watch,
from his Livre d'Aneaux
of Orfevrerie, *1561,
published in the same
year.*

Gonzaga, her daughter the Duchess of Urbino, and members of the Venetian nobility. One of them, Pietro Bembo, writing about one of these watches, said: "because it is tiny and very well crafted, I was so grateful as one usually is when receiving a beautiful gem".[15] Indeed, these watches "were often a great deal closer to being jewels" than to modern timekeepers.[16] And, as previously men-

15   "...*per esser fino e molto ben fatto ne era gratissimo come se suole havere qualche volta una bella gemma* ..." Morpurgo, *Dizionario degli orologiai italiani,* 1950; and Del Vecchio and Morpurgo, *Addenda al Dizionario degli orologiai italiani edizione 1974 di Enrico Morpurgo,* 100.

16   John H. Leopold and Clare Vincent, "An Extravagant Jewel: The George Watch," *Metropolitan Museum Journal,* 2000, 137. And what jewel is more evocative than a ring as a universally recognized symbol of power? A ring can be made out of precious materials and may carry the emblem of authority. The customary representation of power of the patrons met the interest of Renaissance clockmakers to promote the preciousness of their miniaturizing skills on the stage offered by this small but suggestive object. Watches, becoming

tioned, these miniaturized mechanical jewels must be considered in relation to the technological development of lenses.[17]

Northern Italy, between Milan and Mantua, if not the fatherland of the watch, is among the first regions where portable clocks were invented.[18] The first portable watches are mentioned at the court of Milan around 1490 as "*horologini piccoli et portativi*" or small portable clocks. These were chiming timekeepers attached to ball costumes. In 1541 Lilio Gregorio Giraldi (1479-1552) wrote that, between 1513 and 1521, he often saw Pope Leo X carrying, when hunting or travelling, an eyeglass with a watch set in its handle, a timepiece probably crafted by Cherubino Parolaro Sforzani from the famous dynasty of

jewels, took the place of precious stones on the fingers of the most important rulers of the time. The transfer of value from a gem to a clock is symbolically significant: in our modern time we are used to watches that are mass-produced from industrial components. In the Renaissance those tiny parts were crafted and assembled one by one by hand, and such delicate processes needed a sound and specific education, rare talent and good eyes, and most probably magnifier lenses.

17 It is difficult to think of these achievements without mentioning the contributions provided by the glass-maker-masters of the Renaissance. Their role was not just basic to Galileo's telescope. Though written sources describing miniaturization masterpieces such as ring-watches do not mention the use of magnifiers, glassmaker workshops must have provided fine lenses for the purpose. Visual sources such as the above-mentioned engraving by Stradanus show the mature master clockmaker working with glasses perched on his nose. Vincent Ilardi, "Eyeglasses and Concave Lenses in Fifteenth-Century Florence and Milan: New Documents," *Renaissance Quarterly*, 1976, 341-60; Chiara Frugoni, *Medioevo sul naso: occhiali, bottoni e altre invenzioni medievali* (Roma: Laterza, 2001); Cocquyt, "Miniaturizzazione degli orologi e lenti d'ingrandimento."

18 Morpurgo, who is however often inaccurate, reports the tradition that sees Bernardo of Cremona, alias Caravaggio, as the first artisan who made watches, small spring-driven mechanisms to be carried on one's person, instead of weights-driven or spring-driven ones that had to stand in a single place when working. In the chapter entitled *Renaissance Scientific Instruments*, we have already seen that the first certain piece of news about a portable watch comes from Milan, and how Italian and German nationalists had tried to credit the birth of the watch to the lands south or north of the Alps, but without a final proof for one or the other hypothesis. Two recent exhibition in Nurnberg and in Cremona, and a conference in the latter city, have well explored this problem. Morpurgo, *Dizionario degli orologiai italiani*, 1950; and Del Vecchio and Morpurgo, *Addenda al Dizionario degli orologiai italiani edizione 1974 di Enrico Morpurgo*; Dohrn-van Rossum, *History of the Hour*, 118-123; Leopold and Vincent, "An Extravagant Jewel," 137; Eser, *Die älteste Taschenuhr der Welt?*; Zanetti, *Janello Torriani, a Renaissance Genius*; and Cristiano Zanetti, ed., "Atti della conferenza Janello Torriani, genio del Rinascimento," *Voce di Hora*, forthcoming 2017.

clockmakers.[19] Another Parolaro Sforzani, Girolamo, worked all his life in Mantua for the Gonzaga. He made eyeglasses with a small watch inside in 1529.[20] Paris and the German cities of Augsburg and Nurnberg were also important centres of mechanical miniaturization: King Francis I's daggers, made in 1518, displayed watches in their hilts, and Giovanni and Jacopo Marcoat from Augsburg (also recorded as Marquatti or Marquart) both employed by the Duke of Ferrara, in 1560 sold to Cardinal Luigi d'Este a harquebus with two flasks, one big and one small, with two watches set inside.[21] Another Italian city where the art of miniaturized watches flourished was Urbino, where during the 1580s, the above-mentioned Giovanni Maria Barocci made almond-clocks that were given as presents by the Duke of Urbino to the Duke of Este.[22] Thus, even if the story of the ring-watch does not apply specifically to Janello Torriani, the context in which he grew and worked was oriented towards the miniaturization of mechanics. We have already seen how Janello himself is acknowledged as the inventor of the first known gear-cutting machine, a most suitable device for accelerating manufacture and for precision-miniaturization. This gear-cutter lathe was contemporary with the flourishing machine-tool technologies of a city like Milan, adoptive fatherland of Leonardo da Vinci.[23]

---

19   The above-mentioned Pierre II Woeiriot engraved a book for this very Giraldi. Brusa, "Early Mechanical Horology in Italy," 510.

20   Francesco Malaguzzi Valeri, "I Parolari da Reggio e una medaglia di Pastorino da Siena," *Archivio storico dell'Arte*, no. I (1892): 36-37, fasc. I.

21   See the contributions by Günther Oestmann, Dietrich Matthes and Anthony Turner, in Zanetti, *Janello Torriani, a Renaissance Genius*, and "Atti della conferenza Janello Torriani, genio del Rinascimento." See also Auguste Wahlen, *Nouveau dictionnaire de la conversation; ou, Répertoire universel ... sur le plan du Conversation's lexicon ... Par une Société de Littérateurs, de Savants et d' Artistes ...* (Bruxelles: Librairie-Historique-Artistique, 1842). 134-136; Landes, *Revolution in Time*, 87 ; Campori, *Artisti degli Estensi*, 10.

22   Morpurgo, *Dizionario degli orologiai italiani*, 1950, 23.

23   Sixteenth century Milan is probably the most important centre for the development of these machines used in the flourishing industry of precious stone carving, rock crystal and ivory. Janello's friend Jacopo Nizzola da Trezzo, who made for him the Microcosm rock-crystal-sphere, the medal and other things, was said to have been the inventor of the first machine able to cut diamonds. Other Milanese craftsmen, such as the Maggiore and the Miseroni, set about inventing or improving tools involved in luxury production, in order to gain precision, new solutions and to save time. In Milan, there was a tangible memory of the experience of Leonardo da Vinci thanks to his manuscripts kept there by his pupil Francesco Melzi. Courts all around Europe hired these Milanese artisans who transferred their knowledge from Milan. See: Morigia, *La nobilta di Milano, diuisa in sei libri.*; Paola Venturelli, "La lavorazione di pietre dure e cristalli," in *Produzione e tecniche,*

More than with the ring-watch, the name of Janello Torriani is usually associated with automata. Several museums around the world had attributed to him sixteenth century specimens of this mechanical category. Anthropomorphic or zoomorphic automata, more exciting and evocative then other more complex mechanical devices, were needed to transmit the idea of ingenuity attributed to Torriani. Stories that could inspire awe were added to the dry records of Torriani's lost accomplishments. Already Antonio Campi, 30 years after Janello had moved to Brabant and Spain, wrote of his fellow countryman that: "like a new Archytas, he made, birds that besides moving their wings ... could sing as [if] they were alive, and this marvelled anyone".[24] Beyond these mechanical birds, the Roman Jesuit historiographer Famiano Strada (1572-1649) and later the Cremonese monk of the order of Saint Jerome Desiderio Arisi (ca.1659-1725) referred to a number of marvellous automata. The stories produced by these latter writers become richer in particulars that seemed to draw more on fantasy than reality: the Jesuit referred to Torriani entertaining Charles V at Yuste with little figures of soldiers jousting, together with flying birds that, once seen by the superior of the monastery, made him suspect witchcraft. Desiderio Arisi added that Janello, besides the things mentioned by Strada, created life-size warrior-automata, and mechanical barking dogs that could be petted, and that suddenly fought each other in a brawl until someone hit their tails with a stick. The little armed figures could ride on small horses, wounding each other, and the mechanical birds, after landing on one's hand as real birds do in search of food, used to fly away once more.[25] The existence in the sixteenth century of several records of mechanical flying birds suggest that there may well have been some experimentation in the creation of such automata.[26] However, the

---

ed. Philippe Braunstein and Luca Molà, vol. 3, Il Rinascimento italiano e l'Europa (Costabissara Vicenza: Fondazione Cassamarca : Angelo Colla Editore, 2007), 262-66.

24   "... haveva Lionello fabricto (à guisa d'un nuovo Archimede) uccegli, I quali non solo dibattevamo l'ali; ma cantavano anche, con meraviglia d'ognuno, come de vivi fossero stati" (my translation): Campi, Cremona fedelissima città, lv.

25   Famiano Strada, De bello Belgico decas prima (Romae: Scheus, 1648), bk. I, dec. 1; Desiderio Arisi, Accademia dei Pittori, Scultori, ed Architetti cremonesi altramente detta Galleria di Homini illustri, Biblioteca Statale di Cremona, fondo Civico, A.A:.2.16; Arisi, Cremona literata, 1741, III:338; Zaist, Notizie istoriche depittori, scultori, ed architetti cremonesi, vol 2, 150-156. Clockmaker Gottfried Hautsch from Nurenmberg (1634-1703) was said to have made a similar mechanical toy-army during the 17th century: Mario G. Losano, Storie di automi: dalla Grecia classica alla Belle Epoque (Torino: Einaudi, 1990), 77.

26   For instance Pietro Fanzago, the builder of the beautiful clock of Clusone (1583 and still in working order), or a member of his family, was said to have made a flying dove that could cover the distance of half a mile: Morpurgo, Dizionario degli orologiai italiani, 1950, 76.

level of control attributed to their flight makes of these automata nothing but a nice fairy-tale, as in the case of Regiomontanus who, besides the iron-fly, was also supposed to have made a mechanical eagle that, like the fly, could hover through the air and come back to its creator. Desiderio Arisi openly claimed that Janello had emulated him. Archytas provided the model for these most probably fictional stories. In the case of Janello, the classical model of Archytas rhetorically evoked by Campi had also misled later writers who thought, as in the case of the ancient Greek inventor, that Janello's mechanical birds must really have flown.[27] In fact, Antonio Campi does not talk about flying birds; he merely describes them moving their wings and singing. Such feats would have been perfectly consistent with the tradition of Medieval fountain-automata, like those in the Byzantine imperial palace of Constantinople or in the lavish Burgundian gardens of Hesdin, and testified by Giorgio Valla's manuscript and Cesare Cesariano's drawing.[28] As we have previously seen, Giorgio Fondulo, Janello's mentor, knew manuscripts by Hero and Pappus, the main sources for ancient automata models. Among the astonishing inventions that the Arisi brothers and Zaist attributed to their already mythical fellow countrymen was the chimera of any representation of Renaissance mechanics, the machine of perpetual motion: an impossible device that would produce enough energy to keep moving forever. Of course Torriani never made any instrument capable of producing perpetual motion for Philip II's desk, as these sources suggested.

Among all the legends, the most improbable and suggestive is the one already mentioned about the Wooden Man: *el Hombre de Palo*. In Toledo, since the seventeenth century, a central street that borders the northern side of the block of the Cathedral is called *Calle del Hombre de Palo*. Still today it is said that the street was named after a wooden automaton made by Torriani. It seems that the first to report this legend was the eighteenth century abbot Antonio Ponz.[29] Garcia Diego, Janello's Spanish biographer, tried to rationalize the story, attributing the name of the street to a wooden puppet used to collect money for a local religious institution. However, the legend about the wooden

---

27   Hero of Alexandria, *De gli automati*, 6.

28   Rovetta and Gatti Peter, *Cesare Cesariano e il classicismo di primo Cinquecento*, 42, fig. 20; Ceredi, *Tre discorsi*, 6; for what concerns Giorgio Valla, look: Gardenal, Landucci Ruffo, and Vasoli, *Giorgio Valla tra scienza e sapienza*; Anne van Buren, "Reality and Literary Romance in the Park of Hesdin," in *Medieval Gardens: [Dumbarton Oaks Colloquium on the History of Landscape Architecture, 9]*, ed. Elisabeth B. Macdougall (Washington D.C.: Dumbarton Oaks Research Library and Collection : Trustees for Harvard University, 1986), 115-34.

29   Antonio Ponz, *Viage de España: en que se da noticia de las cosas mas apreciables, y dignas de saberse, que hay en ella* (Madrid: Por D. Joachin Ibarra ..., 1776), 143-44.

automaton may also have had a humanist origin: there was an ancient Greek record, known by sixteenth century authors, which described wooden figures animated by internal devices that were a source of amusement to the peasants dancing among them.[30] The imposing physical presence in the urban space of Toledo of what remained of Torriani's water devices could easily have stimulated the minds of the locals to create or transform the memories of the praised clockmaker into something vivid and mysterious.

Despite this legendary tradition, Torriani had actually made automata in the service of his masters. Banquets and court activities often required mechanicals divertissements. An eyewitness, Ambrosio de Morales, gives us a certain description of an automaton made by Janello. It is remarkable that together with Antonio Campi, Morales was the only contemporary who openly mentioned such production.[31] If compared to the great number of texts written on Torriani's clocks and hydraulic devices during his own time, this side of his crafts appears to have been largely neglected, despite the great echo it later enjoyed. Likely, this was a consequence of the cultural value attributed to these works in the context of that period: a debate seems here to emerge between enthusiasts and denigrators of anthropomorphic (or zoomorphic) automata during the sixteenth century.

There is probably nothing more distant from the nineteenth-century crafted idea of Renaissance aesthetics than the automaton. The Renaissance automaton is one of the powerful keys that art historian Eugenio Battisti has used in his *L'antirinascimento* (1962) – meaning *The anti-Renaissance* – to disclose a far more complex representation of the stylistic panorama of the Cinquecento, and in the last decades, several historians had tried to reconsider the conceptual role of automata in their Early Modern context.[32] Bernardino

---

30   Hero of Alexandria, *De gli automati*, 4-6.

31   García-Diego, *Juanelo Turriano, Charles V's Clockmaker*, 102; Campi, *Cremona fedelissima città*, lv.

32   Eugenio Battisti, *L'antirinascimento, con una appendice di manoscritti inediti* (Milano: Feltrinelli, 1962). Here follows a non-exhaustive list of some of the most interesting contributions to the history of Renaissance Automata: Derek J. de Solla Price, "Automata and the Origins of Mechanism and Mechanistic Philosophy," *Technology and Culture* 5, no. 1 (1964): 9-23; Silvio A. Bedini, "The Role of the Automata in the History of Technology," *Technology and Culture* 5, no. 1 (1964): 24-42; Klaus Maurice and Otto Mayr, eds., *The Clockwork Universe: German Clocks and Automata, 1550-1650* (New York: Smithsonian Institution, 1980), Peter Dear, "A Mechanical Microcosm: Bodily Passions, Good Manners, and Cartesian Mechanism," in *Science Incarnate: Historical Embodiments of Natural Knowledge*, ed. Christopher Lawrence and Steven Shapin (Chicago: University of Chicago Press, 1998), 51-82; Alexander Marr, "Understanding Automata in the Late Renaissance," *Le*

Baldi, one of the sixteenth-century editors of Hero, located the appeal of the automaton in the marvel that a lifeless sculpture, supposedly still, can provoke if moving. This is probably why in anecdotal history an automaton is more evocative than a clock. However, technically speaking, a clock is an automaton too.[33] Automata, like watches or transportable clocks, were the offspring of the same technological advancements. From a formal and functional perspective an automaton is different from a clock, but from a substantial and theoretical point of view they belong in a unique category: a device that moves according to the program provided by the mathematical pattern impressed in its mechanism, and by a driving force connected to its mechanisms. Even Bernardino Baldi, an enthusiast apologist for automata, in his introduction to Hero's work, asserted that these devices depended on mechanics, as clocks and engines did.[34] Mechanics was to be considered a subdiscipline of mathematics, a *scientia media*, positioned between the two separated Aristotelian epistemic fields of mathematics and physics, i.e., the science of Nature.[35] Baldi tried also to rationalize the reason why spring-driven clocks did not excite him as automata did: he argued that, unlike in automata, clocks' movement was too slow to provoke awe in the beholder. Therefore, the attribution of automata to a separate category lay in this arcane atmosphere embracing anthropomorphic and zoomorphic automata, the source of whose motion was hidden as the heart is in a living being. It has been shown that medieval automata were

---

Journal de La Renaissance 2 (2004): 205-22; Alexander Marr, "'Gentille Curiosité': Wonder-Working and the Culture of Automata in the Late Renaissance," in *Curiosity and Wonder from the Renaissance to the Enlightenment*, ed. Robert John Weston Evans and Alexander Marr (Aldershot,: Ashgate, 2006), 149-70; Jessica Keating, The Machinations of German Court Culture, Early Modern Automata, doctoral dissertation, Northwest University 2010; Hanoch Ben-Yami, *Descartes' Philosophical Revolution: A Reassessment* (Basingstoke: Palgrave Macmillan, 2015).

33   Despite the fact that it does not imitate in its shape a living creature, and that it does not wander around as anthropomorphic and zoomorphic automata, it moves as they do by virtue of clockwork. Clocks seem to enter into one of the two categories created by Hero himself for automata. Indeed, according to Baldi's translation of Hero, there were *automata semoventi stabili* (stable self-moving automata) and *automata semoventi mobili* (mobile self-moving automata). A spring-driven clock seems to match the characteristics ascribed to the first of these two categories, whereas machines such as weight-driven clocks and mills (and consequently the Toledo device, we infer) were not considered automata stabili because the origin of their movement, a stream of water or the hanging weight, was external. Hero of Alexandria, *De gli automati*, 8-9.

34   Ibid., 4-8.

35   For instance see the epistemology of mathematical sciences during the 1540s in: Cardano, *De libris propriis* [*2004*], 148-149; Rose and Drake, "The Pseudo-aristotelian 'Questions of Mechanics' in Renaissance Culture," 83.

strongly evocative of fantastic literature, and strictly connected with the idea of magic.[36] Moreover, Aristotelian philosophy considered motion in individuals as being strictly dependent on soul, conferring an aura of philosophical ambiguity on automata. Hanoch Ben-Yami has shown how philosophers such as Descartes had made use of automata to unhinge this concept, constructing a new natural idea of purely mechanical movement independent from the idea of soul or supernatural forces.[37]

In 1539, the humanist Cristobal de Villalon, in his *La ingeniosa comparación entre lo antiguo y lo presente* (The ingenious comparison between Antiquity and Present Times), wrote that "nothing could be more admirable than men acquiring such skill that, by means of clocks, wooden images and statues can be made to walk around a table without anyone moving them as well as playing a guitar or kettledrum or other instruments".[38] Torriani was said by his acquaintance the royal historiographer Ambrosio de Morales to have made an identical dancing figure wandering around the table and playing a drum:

> Janello has also desired to renew for fun the ancient moving statues and that for this reason the Greeks called Automata. He made a more than a *tercia*-high lady, that positioned on a table, dances all around at the sound of a drum, that she plays herself, and she comes back to the place where she left ...[39]

As Morales, some sixteenth century humanists thought of automata as proofs of a recovery of Antiquity.[40] And perhaps Torriani's patrons had the words of Cristobal de Villalon in their ears when they commissioned to the Lombard

---

36 Elly R. Truitt, *Medieval Robots: Mechanisms, Magic, Nature, and Art*, University of Pennsylvania Press 2015.

37 Hanoch Ben-Yami, *Descartes' Philosophical Revolution*; and "L'ingresso degli automi nella filosofia al tempo di cartesio," ed. Cristiano Zanetti, *La Voce di Hora* Atti della conferenza Janello Torriani, genio del Rinascimento (forthcoming 2017).

38 Translation taken from: García-Diego, *Juanelo Turriano, Charles V's Clockmaker*, 101.

39 "*Tambien ha querido Ianelo por regozijo renovar las estatuas antiguas, que se movian, y por esso las llamavan los Griegos Automatas. Hizo una dama de mas de una tercia en alto, que posta sobre una mesa dança por toda ella al son de una tambor, que ella misma va tocando, y da sus bueltas, tornando a donde partio.": Y aunque es juguete y cosa de risa, todavia tiene mucho de aquel alto ingenio. Yo he dicho de las cosas deste raro y estremadamente insigne artifice: no perque piense aver acertado a declarar todo lo que ellos son, sino como deseoso de dar a entender alguna parte, y dexar aqui memoria de una cosa tan señalada, como en nuestros tiempos ha avido*" (my translation): Morales, *Antiguedades de las ciudades de España*, fols. 91-94.

40 Marr, "Gentille Curiosité," 149 and following.

clockmaker this automaton. The model was provided by mythical stories and by mechanical treatises: the greatest ancient Greek authors, from Homer to Plato and Aristotle, had told stories about automata. Baldi refers to a genealogy of the craft that goes back to Daedalus and his master Vulcan. This rhetorical strategy reflects once again the ennobling force of classical ancestry in the epistemological representation of Renaissance Christendom.[41] But humanists had more than stories about mysterious automata to draw upon. Ancient treatises on mechanics and automata were circulating in manuscripts long before their printed editions: the Pseudo-Aristotle, Hero of Alexandria, Pappo and Ctesibius were popular among scholars. Examples of these sixteenth century automata still exist in different museums. Like Torriani, Capobianco from Vicenza and Bulman from Nuremberg were each said to have made a dancing figure.[42] Capobianco was also acknowledged with having created a clock that marked the hours lighting candles, and with a silver vessel populated by moving sailors that stirred along the table thanks to hidden wheels, another model of automaton that left several specimens in modern collections.[43] If enthusiasts of Hero and ancient automation considered such devices praiseworthy, others disagreed. Capobianco's ship was a present made on behalf of the Venetian Republic to the sultan Suleiman the Magnificent. Pietro Aretino, in a letter dated 1537, ridicules this silver boat as an invention suitable to provoke the laughter of silly women.[44] From Morales' words transpires a similar judgement about automata, when, after describing Torriani's mechanical doll, he commented: "even if it is a little toy and something risible, it nevertheless reflects Torriani's high ingenuity."

Mechanical ingenuity and the final function of a device were perceived as two very different matters, not necessarily antithetical. Even Baldi, when tracing automata's noble ancient genealogy, could not help from observing that the Bible does not mention such mechanisms, perhaps because they were considered no more than jokes, he admitted.[45] It seems that even Francesco I de Medici too, who, at the time of Torriani's endeavours in Toledo, had created in Pratolino the most spectacular show of automata, was aware of their recreational function. In the great hall of his villa he had carved in stone that the

---

41   Hero of Alexandria, *De gli automati*, 4-8.

42   Neudörffer, *Nachrichten von Künstlern und Werkleuten*, 65-6.

43   Similar vessels can be seen in German, Austrian and French museums. See: Marina Belozerskaya, *Luxury Arts of the Renaissance* (Los Angeles: J. Paul Getty Museum, 2005); and the above-mentioned Jessica Keating.

44   Aretino, *Lettere*, lettera 108 al Fausto Longiano, Venezia XVII di Decembre MDXXXVII, 325-328; Morpurgo, *Dizionario degli orologiai italiani*, 1950, 582.

45   Hero of Alexandria, *De gli automati*, 4-8.

entire complex, with its springs, gardens and buildings, was made to cheer and to refresh his friends' spirit.[46] By the end of the century in Northern Europe automata retained their charm and their role in court-related sociability, especially in drinking-games,[47] whereas it appears that in Italy and Spain they lost part of their attraction, disappearing from several public clocks, as in Bologna, and not being mentioned in extremely rich and detailed inventories such as the one compiled at the death of Philip II.[48]

But what made Renaissance intellectuals consider automata silly and risible? Some scholars, following Baldi, think that Christian morality could not enumerate earthly amusements among memorable things. This was a mark of differentiation between Christianity and the ancient Hellenistic world that reckoned *ludus* as an important part of social life.[49] This reading seems to be not completely acceptable. Indeed, I think that the core of the question lay especially in two issues: on the on hand the pride of Renaissance intellectuals attempted to show that they were immune to the awe that automata provoked in ignorant people, and on the other, a Christian and a humanist had to feel unease when confronted with the lack of *utilitas* of those devices, i.e., their lack of virtues.

In their eyes automata were a mere exercise in skill. They had no respectable practical use or valuable function except to amuse children and uneducated people and to entertain courtiers as a joke or a sinful game. The problem of *utilitas* was an important issue in the humanist rhetoric of power: the employment of economic resources had to be justified through dignity or by such key-concepts. Moreover, a well taught scholar in mathematics could not show himself to be impressed by the simple mechanics hidden in automata, especially when they might just as well tackle mathematical puzzles such as Torriani's planetary clocks or his water-raising devices. A further negative characteristic of automata was their capacity to mislead simple minds to believe in wonders. Baldi's apology for automata tries to demonstrate that this kind of deception was as good as the lies a physician tells his patients to calm down their worried spirits. Baldi also defended the mathematical ingenuity behind

---

46   *"Fontibus, Vivariis Xyftis has Ædes Franc. Med. Magn. Dux Etruriae II. Exornavit Hilaritatique Et fui amicorumque fuorum Remifsioni animi dicavit Anno Dom. M. D. LXXV"*: Luigi Zangheri, *Pratolino: il giardino delle meraviglie*, seconda edizione con aggiunta di disegni e tavole (Firenze: Gonnelli, 1987), 105.

47   For instance see the *Trinkspiel* automata at the Kunsthistorisches Museum in Vienna, at the Metropolitan Museum of New York, or at the Yale University Art Gallery.

48   Sánchez Cantón, *Inventarios reales*, MCMLVI-MCMLIX.

49   Losano, *Storie di automi*, 5-6.

them (which was not necromancy, he specified),[50] and compares the pleasure derived from them to the honest pleasure of music. Baldi, despite his mathematical arguments, cannot hide his fascination with the power of marvel. He quite openly faces the core of the moral question behind this problem: amusement is a human necessity. A pastime is an instrument of happiness. Even the most severe moral philosopher has asserted the necessity of distraction, instituting fests, games, banquets, jousters and plays. The problem was therefore in the measure: the amusement provided by automata, argued Baldi, was to be considered honourable as a practice of recreation, but not as a pure pleasure having an end merely in itself.

It is difficult to believe that this argument could have won over automata's detractors. However, in Aristotle's first book on politics, there was a point of possible contact between the two factions: the philosopher considered slaves as living instruments. Aristotle argued that one did not need any slave if looms, saws and other instruments were able to vibrate on their own as those of Daedalus did. Aristotle's call for useful automata had probably a strong impact on the minds of sixteenth century mechanicians. Torriani's water-raising device has to be set in this field of research: a science of mechanical automation that was able to substitute for muscular force. Another example that unifies in Torriani's career the category of utility and his skill in mechanical miniaturization is reported once again by Morales, who wrote:

> Moreover Janello invented an iron mill so small that it could be hidden in a sleeve; and grinds wheat *celemines*, more than two per day [9 Kg ca.] moving by itself, and without anyone to do it. And [this mill] had another great function, that separated the flour from the dregs, so that the good one falls into a sack, and the rest in another. [This invention] could be very useful for an army, in a siege, and to those who sail, given that it moves without anyone doing it.[51]

The portability and the fact that this device was able to move on its own recall a spring-driven device, like Janello's planetary automata. Such devices were

---

50   Hero of Alexandria, *De gli automati*, 4-6.

51   "*Demas de todo esto ha inventado Ianelo un molino de hierro tan pequeño que se puede llevar en la manga: y muele mas de dos celemines de trigo el dia, moviendose el a si mismo, y sin que nadie lo trayga. Y tiene otro grandissimo primor, que derrama la harina cernida, assi qui ella cae por si bien apurada en un saco, y el salvado en otro. Puede ser de mucho provecho para un exercito, para un cerco, y para los que navegan: pues se mueve el mismo sin que nadie lo menee*" (my translation): Morales, *Antiguedades de las ciudades de España*, fols. 91-94.

not fictional: seven small iron-mills, probably like this one, were still kept among Janello's instruments in the royal collections at the beginning of the following century.[52] Perhaps he had invented this portable and automatic mill remembering that far winter, 1511-1512, when a cruel cold had iced for many days the river Po, preventing the mills from grinding the meal to prepare the necessary bread to feed the people of Cremona. Certainly these mills were the fruit of an ingenious experimentation driven by a noble scope, suitable to be kept in the royal collections, until rust had probably eaten them out. But, as said before, there is no trace of automata in the detailed inventories of the things that belonged to Charles v and to Philip ii when they died. If Janello made automata at court, perhaps in occasion of some banquet, theater performance or festivity, these were given away without too many concerns and were not kept in the royal possession.

Models and the Problems of Scale

If Janello Torriani won popularity thanks to his Microcosm, an impressive miniaturization of a complex mechanism, he later achieved fame thanks to his hydraulic "mega-machine" engaging the field of macro-mechanics, and proving himself capable of transferring mechanical properties from a model to its larger reproduction on a gigantic scale: the Toledo Device was a 300 metre long snake made out of timber, metal and bricks, capable of elevating water on a slope of a good 100 metres.[53] It was something never seen before. Not even the

---

52   Sánchez Cantón, *Inventarios reales,* MCMLVI-MCMLIX, item 4731.

53   The term *megamachine* was created by Lewis Mumford to describe those organized bodies of people orchestrated to achieve an engineering aim, such as the draining of the Pontine marches in Mussolini's Italy or the construction of German highways in the same period. Sawday has used this term in order to describe great Renaissance engineering projects that gave birth to"*structures of enduring beauty and complexity*" such as the movement of the obelisk of Saint Peter by Domenico Fontana. In my view, enterprises that required the organized work of people by the thousands did not differ so drastically from the construction sites of the great cathedrals, fortifications and canals of previous centuries. Instead, I would argue that during the Renaissance something much more innovative was attempted on a gigantic scale: if the *megamachine* described by Mumford and Sawday belonged to all urban societies since antiquity, Torriani's Toledo Device was more an innovation of the experimental technological milieu of Renaissance culture. No automatic device had ever before been built to such a scale; Jonathan Sawday, *Engines of the Imagination: Renaissance Culture and the Rise of the Machine* (London; New York: Routledge, 2007), 58.

large hydraulic structures of ancient Rome – such as the complex mill of Barbegal, near Arles, which linked a complex canalisation system to a number of separate mills – or the hidden *Stangenkunsten* of sixteenth century German mines could compete with this cyclopean machine invading the urban centre of one of the most representative cities of the first "global empire".

The model Janello Torriani made for this water-raising device looked convincing.[54] But once the real device was created, the amount of elevated water was quite different from the figure calculated on the model. Fortunately enough for Janello, the quantity of water was far greater than that expressed in the contract.[55] Furthermore Janello was able to make the device elevate more or less water by speeding up or slowing down the movement of its mechanism by the means of interchanging components; when he was asked by the royal officials to explain how he was doing it, he laconically answered that there was not a precise explanation, the secret was in Torriani's own touch: "*en su mano*".[56]

We cannot say whether Janello tried to maintain the secret of this successful outcome in order to protect his knowledge, or if he really did not know the reason behind the incongruent mechanical results between the model and device executed in its real scale. As we have seen, Janello gave similar justifica-

---

54 It was customary for constructors on important building sites to present to the committee assigning the work small-scale models of the buildings and the most complex pieces of machinery required to erect them. In Venice, the physician and hydraulic engineer Giuseppe Ceredi described the state-building in which many of these mechanical models were housed as presenting a curious sight to the visitor: "*Many models of never built machines are to be seen in the secret rooms of the Ufficio deProvveditori di Commune in Venetia where everyone who believes to have found a new device goes in the wish of receiving an inventor's privilege*". These models were kept in glass cases of Murano or in copper or tin boxes (in Germany): Ceredi, *Tre discorsi*, 11, 19; some years later Torriani would deliver one of his own models to this location. In 1568, the Senate of the Republic granted him a patent for invention.

55 In a letter to the King (dated 1575), Janello complains about the fact he was not paid for the Toledo device. Janello reminds the King that, before the contract was signed, he had shown a model to His Majesty and to the representatives of the city of Toledo. With the help of this model, Janello had explained to them the functioning of the device. Janello also wrote to the King that the real machine was even more efficient than its model: indeed, the water-whel of the prototype had to turn 83 times before the 4 pipes emptied into the upper pond, whereas the real device needed to turn only 40 times in order to pump up the quantity of water required by the contract and even more: Cervera Vera, *Documentos biográficos*, doc. 44.

56 "*La causa de aver tanta diferençia desta medida la que se embió a vuestra mercedes lo que el mismo Juanelo nos confesó allí, que es estar en su mano haçer que el Yngenio camine más aprisa o no y así en esto non puede haver precisa justificaçión*": Ibid., doc. 123.

tions to both Zenocarus and Morales when they asked about the most difficult accomplishments of the Microcosm: Ambrosio de Morales, reporting a dialogue he had with Torriani, observes that the master claimed he had problems organizing the motion of Mercury, the movement of the eighth sphere and the irregular hours of the Moon. This must have been a very difficult task to accomplish. Indeed, once carried out, it was a perpetual cause for praise and admiration, as the motion of the Moon had been in the cases of the fourteenth century clock in the abbey of Chiaravalle of Milan (sketched by Leonardo) and in the clock of the German abbey of Bad Doberan. Torriani found a solution too, but according to what he told Morales, it was not a theoretically proper one: he indeed claimed that where arithmetic did not arrive, his craft did.[57] It is difficult to interpret these sibylline words from a technical perspective. Moreover, when Zenocarus asked about the complete revolution of the eighth sphere, Torriani answered that only God could answer such a question, because the calculus involving exponents two and three was too complex. Janello also added that because of the enormous size of this divine number and because of the briefness of the lives of those who observed it, this calculus was impossible.

An interesting parallel to Janello's story appears in the case of the above-mentioned Bartolomeo Campi da Pesaro (ca. 1520-1573), a famous and much-admired mechanician who died as engineer in the service of the Duke of Alba at the siege of Haarlem. Despite the fact he had a good theoretical knowledge of mechanics that allowed him to work on a small scale, sometimes he was unable to increase the magnitude of his mechanical works. Bernardino Baldi describes an automaton Bartolomeo Campi had made for his master: it was a silver turtle that walked the table moving its head, tail and feet. Once it reached the centre of the table its carapace opened up, showing a box with toothpicks inside. When Bartolomeo Campi tried to apply his theoretical knowledge of mechanics to the task of recovering a sunken Venetian galleon, he failed miserably.[58] The confidence that one might devise a mechanical system for raising ships from the seabed, was based on the tradition that attributed a similar invention to Archimedes. Moreover, the most famous apothegm attributed to the mathematician of Syracuse was "give me a lever long enough and a point to stand upon, and I will move the world", a message earlier transmitted by Aristotle's *Mechanics* and circulated in Italy thanks mainly to Cardinal Bessarion in the fifteenth century.[59] According to this text, all the principles of

---

57 Morales, *Antiguedades de las ciudades de España*, fols. 91-94.

58 Hero of Alexandria, *De gli automati*, 12v.-13r.

59 Aristoteles, *Meccanica*, 178 and following. Rose and Drake, "The Pseudo-aristotelian 'Questions of Mechanics' in Renaissance Culture," 65-104.

the lever are set in the principles of the circle.[60] It should be noted that clock-making and engineering were the main professional fields of application of such theories. The *Mechanics*, describing the paradoxes of the circle, explains how the points on a moving radius of the circle, like the pointer of a watch, are moving with different speeds. The further from the centre, which stands immo-bile, the faster a point has to move.[61] The meaning here expressed well represents the direction of investigation that sixteenth century mechanicians had to follow in order to overcome the problem of maintaining the same effects with different scales. The scientific community of this period, like the practical world of engineers, was aware of the need for a better tool for calcu-lating forces applied in different proportions. The questions here implicitly raised, as the *Mechanics* stated, belonged to the field of mathematics (as con-cerned the method) and of natural philosophy – for the empirical location in which practical experience demonstrated the existence of a problem.[62] The new science of Physics has its roots in this experimental period, often unsuc-cessful in its outcomes, but for this very reason plentiful in terms of practical examples calling for comprehension and solutions.

The 1560s were a very vivacious moment for hydraulic engineering: in this very decade Torriani accomplished the construction of the most impressive device of his century; however, beside him, there were many other people competing in order to find new economical mechanical solutions to raise water. If we compare the experience of some of them with Torriani's, we will see how relevant and widely acknowledged the problem of scale was. From a document dated 1564, coming from a controversy about the granting of a privi-lege for invention in Milan, we learn that in order to discredit the other, one of the two parties asked the authority to judge the water-raising devices only once they were built in full scale, instead of taking into account only their models. Indeed, an official of the *Magistrato Straordinario* of the duchy of Milan wrote to the governor of the state that:

> After considering sufficient the utility of the hydraulic machine, it re-mains to be determined its feasibility: they [the two inventors associ-ated, the painter Carlo Urbino[63] and the sculptor Gio. Antonio Buzzi]

---

60   Ibid., 166-167.

61   Ibid., 166-169.

62   Ibid., 166-167.

63   It is here important to note that the so-called *Codex Huygens*, largely inspired by Leon-ardo da Vinci's manuscripts, is attributed to Carlo Urbino. This demonstrates how six-teenth century Milanese mechanics, as paintings, was indebted to Leonardo: Sergio

say that the realisation of such a machine cannot be judged until constructing the very device as it should be in large scale, real and not virtual, but if required, they will provide, as they never refuse to do, another demonstration with drawings and a small model. However it will not be possible to clearly demonstrate in this way the feasibility of the machine, because small prototypes and models do not always telling the true about their effects, but there are several examples of models that despite promising marvellous effects, once turned into a large scale have shown themselves fallacious and useless. This is well known to anyone who has some knowledge in those practical issues.[64]

Marinelli, "The Author of the Codex Huygens," *Journal of the Warburg and Courtauld Institutes* 44 (1981): 214-20.

64   ASMi, Acque Parte Antica, cartella 7, Acque e Mulini: *"1564 adi 17 de decembre in terciis in venerdì. Dinanti al Mag.co magr.o straordinario di Milano. D: Carlo Urbino pittore et Gio Antonio Buzio scultore compaiono/cittati per questa hora ad isantia de D. Antonio Stoche et/ Ludovicho Cavalazo a vedere presentar alcune lettere de lo/Eccelso Consiglio segretto et dicono con riserva sempre d'ogni/ lor ragione che dette lettere son surrepte concedute asupl/ catione che non espone quello che narrar doveva, inpetrate/non per altro che per impedire la espeditione del previlegio che dal predetto consiglio, egli ne sono a termine di optenere et/in herendo sempre alle proposte che in questo negotio han/fatto sotto giongono che la relatione che il predetto Mag.o/ha da fare a sua ecc.a per esecutione di dette lettere ha da/contenere dui Capi l'uno a portare alla certezza della inv/ntione, l'altro alla utilitate di quelle come in esse lettere si leg/ge, questa utilità si mostra apertamente nelle proposte de detti comparenti pero che essi promettono modo espediente et/bello da solevare in altezza de braza cinque et più sino/alla quantità di quaranta onze d'acqua et adaquarre de/beni vicini et di fare che con l'opera d'uno homo solo/si nalzerà tanta che sene farà uno acquedutto di doi onze, et se bene li aversari oferiscono di voler fare/uno simile effetto et agiongendo di più di voler far gira/r rote operar peste lavorar folle et cose si fatte non/per questo la loro oferta deve essere stimata megliore/di quella de li comparenti perché se essi non hanno specifi/cato questo tanto che di più li adversari hanno detto che cio ha/nno fatto per che alor parer cosa soverchia venire a si fat/ti particolarità sapendosi ancora apresso de gliomeni/ di mediocre inteligentia che sempre che lacqua arà alza/ta secondo la oblatione fatta dalli comparenti ella si/potrà poi fare decorrere non tanto avoltar rote operar/peste et lavorar folle ma ancora in uso de alter ser/vitù/quasi senza numero onde per quello che tocca alla uti/lità la cosa è chiara a favore de comparenti restaci/la parte de la certezza et qui dicono che una siffatta/cosa no' pole essere giudicata se no' dallo artificio posto/efectualmente in opera grande, vera et no' finta, però/ che se bene si volesse venire, il che no' rifiutano mai di/fare, alt(r)a mostra con disegno et modelletto dar certezza/non di meno no' si potrà infallantamente cognoscere, et la/ragione è quest ache li esemplari et li modelli non/sempre dicono la verità nelli efetti loro ma ce ne so/ no de quelle come tal hora si è creduto, che danno mostra/di cose maravigliose li quali vedutte in forma più/grande si scno scoperte bugiardi et di niun valore, et/questo è nottisimo a chi ha lume di pratiche si fatte/et questa ragione dimandono li ditti comparenti che/con le*

Carlo Urbino and Giovanni Antonio Buzzi's point proved to be the right weapon to obtain the privilege for invention disputed by Antonio Stocchi and Ludovico Cavalazzo, who claimed the right of precedence to the privilege. Indeed, the authorities ordered the two parties to build their devices within three months. Only Carlo Urbino and Giovanni Antonio Buzzi were able to produce a functioning device within the deadline. This demonstrated that the other two competitors most probably were not able to transfer the mechanical properties of their model to a larger scale. This kind of problem was not a novelty in Milan: already in the year 1548, Bartolomeo Brambilla, later quoted in the first book of Cardano's *De Subtilitate* in relation to this pump, had obtained an inventor privilege for it. Writing a supplication to Ferrante Gonzaga, the engineer made sure that the Governor could come to his garden and observe his machine at work. Indeed, Brambilla adds:

> In this way one can know the truth, instead of being a matter of dreams and imagination, like in the case of some people who after having been granted with a privilege, they were not able to do anything of what they had dreamed of doing.[65]

These extraordinary testimonies were corroborated a couple of years later by the previously mentioned Giuseppe Ceredi, physician to the Duke of Parma and Piacenza Ottavio Farnese. Ceredi provides us with some considerations worthy of note about models and experiments in the field of sixteenth century hydraulic engineering. In 1567, in his book about the Archimedean screw, which we have already quoted, he wrote:

---

alter cose il Mg.co Mag.to voglia referire a sua ecc.za dalla quale spereno di optenere il previl-legio che /le hanno suplicato non ostante le caluni dalli ave/rsari ateso che se la mentione che essi dicono si troverà/in opera al iuditio del predetto Mag.to nova et di meglior/bontà di quella delle comparenti essi si contentano che/tal previllegio non preiudichi alli adversary ne ad/alcuno altero che se le inventioni de luna de laltra/parte sono egualmente bonela ragion vole che li/comparenti siano preferiti per che primi sono statti/ a proporre onde si hanno aquistato la ragione di talpre/cedenza oltra che la nobeltà delle loro professioni han/no da movere lo animo de hogni principe et de ogni/altra persona a favorire un sifatto vertuoso pensiero/più in uno pittore et in scultore che in altri di prof/esione vagabonda instabile et ambigua, et qui produco/no et esibiscono tutte quelle ragioni che fanno per loro/in questo negotio et dimandano che sia data ripulsa/a detti aversioni con condonatione delle spese et come /di sopra detto" (my translation).

65   ASMi, Autografi 92, Bartolomeo Brambilla (my translation).

I could make an almost infinite number of small and big models ... It is well known by those scientists,[66] who at least once have attempted the process, that so great is the number of those theories that at once one has to keep in mind in order to accomplish some new and important effect, that it is nearly always impossible to order them all together well, and direct them to an ordered work, if not just after a good deal of mistakes. Experience then recognizes those errors at the time, and reason corrects them.[67]

In this book, while describing the water-lifting devices used in Italy in the 1560s, Ceredi quoted this very Carlo Urbino, mentioning his "yet unpublished but very well projected device for water elevation".[68] We can infer that the painter Carlo Urbino from Crema and Giovanni Antonio Buzzi, after being triumphant in 1565 in the above-mentioned controversy over the water-lifting device, had not yet received the privilege, or perhaps they had not yet circulated their invention outside of Milan, if Ceredi is not here referring to another invention.

In the second part of Ceredi's book, the author adds that it is because of the wrong calculation of the multiplication of the force, of the relation between the motor and the weight, and between the weight and the speed, that one can easily obtain the correct result "in almost all small models", but "in the real work, the same authors are then misled, because they do not have at hand the reasons of proportions".[69] Together with Torriani's alleged stochastic method applied to mechanical problems (devoid, as it may seem, of sufficient and strict mathematical calculation), these words help us to understand how

---

66   Here the term "scientist" (*scientiato*) has a different semantic meaning from our moder-
      none. In sixteenth century context it corresponded more closely to the notion of wise and
      learned man, humanistic curriculum included, than a modern scientist, although already
      in this period the importance of mathematical science seems to develop a special relation
      to this category.

67   "*ho potuto fabbricare quasi infiniti modelli piccoli, & grandi ... Che si sà bene da quei scien-
      ziati, che pure una volta si son dati all'operazione, che si numeroso, & grande è il mucchio di
      quelle affermationi, le quali tutte a un tratto bisogna avere nella fantasia per far nascere
      qualche nuovo, & importante effecto, che quasi sempre è impossibile assettarle bene insieme,
      et indirizzarle sicuramente all'opera ordinata; se non dopo molti errori in varii tempi
      dall'esperienza riconosciuti, & di modo corretti con la ragione*" (my translation): Ceredi, *Tre
      discorsi*, 7; for what concerns the meaning of "experience" see: Dear, "The Meanings of
      Experience."

68   Ceredi, *Tre discorsi*, 19.

69   Ibid., 47-48

Renaissance "men of science" were heuristically experimenting in mechanics. They would have been no different from those brilliant architects of gothic cathedrals if their cultural milieu had not become so avid a consumer of ancient theoretical works. The systematic collection, editing, printing and diffusion of ancient texts throughout the last half of the previous century provided the sixteenth century *homo experimentalis* (i.e., experimental man) with a vast range of theories ready to be tested and re-shaped. Renaissance men dealing with engineering were conscious of improving their field: already in the previous century Filippo Brunelleschi believed in technical progress, and in fact it is not possible to look at his endeavours without thinking about a soul eager for new engineering solutions.[70] If we look at the letter of self-promotion written in 1482 by Leonardo da Vinci to the Ludovico il Moro seeking an appointment in Milan, we may recognize a tradition in the same attitude towards practical demonstration as the base of sure knowledge:

> And if any one of the above-named things seem to anyone to be impossible or not feasible, I am most ready to make the experiment in your park, or in whatever place may please your Excellency, to whom I commend myself with the utmost humility.[71]

To these Renaissance engineers, intuition and experience were as important as theory, in line with the teachings of Vitruvius and the *Mechanics* of the *Corpus Aristotelicum* and the reminders issued by Tartaglia.[72] Empirical experience gave to these practitioners of "*scientiae mediae*" a general grasp on things, and allowed them to find extemporal solutions. The demonstration of one's capacity to reach a solution made of his demonstration a true statement. This is an interesting point concerning how historians of science looked in different ways at practical experiment as the base of scientific knowledge.[73]

However, in Janello's accomplishment of his Toledo Device, more efficient than the precious explanatory model, we find the clue of something perhaps even more surprising than pure practical skill to adjust things. According to

---

70   Anthony Grafton, *Leon Battista Alberti: Master Builder of the Italian Renaissance* (New York: Hill and Wang, 2000), 94.

71   Translation in John T. Paoletti and Gary M. Radke, *Art in Renaissance Italy* (London: Laurence King Publishing, 2011), 371.

72   Niccolò Tartaglia, *Euclide Megarense acutissimo philosopho solo introduttore delle scientie mathematice* (In Venetia: appresso gli heredi di Troian Nauo alla libraria dal Lione, 1586), 3r.-3v.

73   James McLachlan, "Experimenting in the History of Science," *Isis* 89, no. 1 (March 1998): 90-92.

Ambrosio de Morales, the true genius of Torriani was most expressed not in the originality of the design and the ergonomics of these bronze vessels, which he called "tubes", but above all in the harmonization (by calculus) of the motion of such a large number of containers:

> There are many particularities of great marvel in this, but two baffle more than the others. The first consists in the tempering of all the movements with this measure and proportion, that each thing moves in harmony at the command of only one gear, moved by the waters of the river ... and if all the tubes had had the same weight, it would seem to not have been as much of a marvel in the admiring of this concerto of movements. But one being empty, as we said, and the other full, controlling such a big uniformity of the movement one together with the other, is something that surpasses any comprehension, even after having seen it, moreover having been invented and put into practice. More than this, if all the movement was continuous it would not have been such a marvel; but being [the movement] so different, this marvels and confounds the understanding without being able to comprehend, nor take one step in the astounding invention. Seeing as without ever stopping the wooden structure and being tied to it the tubes and recipients of bronze, and moving to the rhythm of that when they conjoin to dump and receive the water, there they arrest and immobilise, as if they were stopped, for the time that it takes to empty one and fill the other, not ceasing in the meantime the movement of the wooden part. And finished the giving and receiving, the tubes return to their movement, as if nothing had ever stopped them. This would not have been possible without the art of the proportions [being] very different ... from what is usually taught in arithmetic ... This was not able to be done [transmit smoothness and speed to a machine that big and heavy] without big calculations of proportion in the balance of the movement; and the finding of them with the ingenious is something rare and never heard of, and the putting of it into execution with such precision was an even bigger marvel.

When the same Janello explained to Morales the theoretical basis of the machine, the historiographer grew confused, unable to follow the complexity:

> I then understood something of this when Janello showed me for the first time the model of the aqueduct to see how in the small components of wood there were implicated of the sum of the arithmetic numbers so large that I still cannot understand them. Seeing this I said to him: "Sir

FIGURE 52 Portrait of Ambrosio de Morales, *Enrique Florez*, Viage de
Ambrosio de Morales por orden del rey D. Phelipe II a los reynos
de León y Galicia, y Principado de Asturias, para conocer las
reliquias de Santos, *Madrid, 1765.*

Janello, this system of proportions is different from that of which we
know".

Considering Torriani's prior residence in Milan during the same years in which
Cardano was teaching mathematics at the Scuole Piattine, when the physician

published his *Ars Magna sive de Regulis Algebricis* (1545), we must ask whether in defining the proportions of his mechanical marvels Janello might have used the algebraic calculations with cubic and quartic equations as Tartaglia, Lodovico Ferrari, and Cardano himself had learned to resolve in the 1530s and 1540s. As previously seen, from what he told Zenocarus, Janello was aware of the concept of equations of the second and third power. In any case, what unites the Microcosm and the Device of Toledo is Janello's incredible mastery in the transmission of controlled movement from a motor to a large number of mechanical components, independently from their dimensions. As he had with the Microcosm and the Crystalline, in the Toledo Device Janello demonstrated himself as the most worthy disciple of Archimedes and his "long enough" lever.

From Clockmaking to Hydraulic Engineering

Our clockmaker trajectory allows us to discuss now another important issue reflecting Renaissance society, regarding the professional categorization of practical mathematical careers. Torriani's greatest accomplishment, the one that gave him eternal fame, was his water-lifting device of Toledo, an engineering masterpiece. The vast majority of sixteenth century documents dealing with Janello refer to him as a "master clockmaker"; some refer to him as mathematician, some other as architect of clocks, and only in four cases (of over more than a hundred), in two administrative documents from the court of Milan (1540s), in the letter to the doge of Venice (1567), and in a notary act from Toledo, is Janello addressed as engineer.[74] According to Morales, the Lombard clockmaker had already started thinking of a technical solution for supplying Toledo with water since the time he was living in Milan. According to the Spanish historiographer, who reported some discussions he had had with the clockmaker, Janello heard for the first time about Toledo's lack of water from the Governor of Milan Alfonso d'Avalos Marquis del Vasto, who happened to be of ancient Toledan stock. Indeed, on visiting his ancestral city the Marquis described it as an excellent and noble place that, however, lacked water.

---

74   For the definition of mathematician, see in Part 1: *Fashioning the Aura of the Genius*. About the definition of engineer, see: "*M.ro Jomello (sic) Torriano ingegnero* Leydi, "Un cremonese del Cinquecento," 133. In the plea addressed to the Doge of Venice (1567, 28th of June, ASVe, Senato, Terra, filza 50, doc. 2) one can read: "*Volendo adunque il predetto ingegniero prime per bene universale, e dapoi per suo particolare*". In the AHPT, a document reads: "*Sepan quantos desta carta de poder ... como yo Juanelo / Torriano yngen[i]ero de su Mag[stad]*": Juan Sotelo prot 1637, fol. 1104 v., 14th 1567.

Torriani confessed to Morales that he had already started to think about how to solve the problem at that time. Nevertheless, he had to suspend his search for a solution because he had to put all his energies into the construction of the planetary clock.[75] The reasons behind the King's appointment of Torriani to this task are unknown. As previously mentioned, as a royal clockmaker, Janello had to perform by contract all sorts of tasks connected with his profession.[76] Evidently, building a gigantic water-machine called on his professional competences.

After 1558, the year of the death of his old master Charles v, Torriani took care of the clocks, which had been handed over to him immediately after the Emperor passed away. In the year 1562 Philip ii doubled his wage.[77] Already in March of the following year, Janello bought a plot in Madrid to build a house.[78] His home was erected in the quarter of Lavapiés, in a street that since that time has borne the name of *Calle de Juanelo*.[79] The doubling of Janello's wage kept him at the Castilian court when other European princes could have attracted him to different places (as Pope Pius iv tried to do). In the contract the King specified what Janello had to do: "You will have to serve us and you will serve in

---

75   Alfonso d'Avalos had been governor of Milan from 1538 to 1546. In this last year, he had to go to Madrid in order to defend himself from an accusation of power abuse. He died the same year after returning to Italy. If Torriani's story reported by Morales is true, it must have been in 1546 that the Marquis del Vasto told him about Toledo, unless the imperial captain had previously been to Spain. When in April 1565 Janello signed the contract with the King and with the city of Toledo to build the water-supply system, among the three representatives of the City Council was one Alonso de Ávalos, *"jurado y comisario de la çiudad"*. Was this just a coincidence, or had he somehow used the authority derived from the relation he once had with the deceased marquis Alfonso d'Avalos to win consensus and support at Toledo? Is the presence of this Alonso de Ávalos among the three Toledan signatories a sign of a temporarily succesful strategy applied by Janello to win the favour of a reluctant City Council? Gustavo De Caro, "Avalos, Alfonso d'," *Dizionario biografico degli Italiani* (Roma: Istituto della Enciclopedia italiana, 1962); Federico Chabod, "Usi e abusi nell'amministrazione dello stato di Milano a mezzo il '500," in *Studi storici in onore di Giocchino Volpe* (Firenze: Sansoni, 1958), 95-194; Garcia-Diego maintained that Morales' version was not reliable.

76   16th of July 1562, Madrid: Cervera Vera, *Documentos biográficos*, doc. 5.

77   In the official inventory of Charles' properties in Yuste, Torriani (who is one of the witnesses as well) was ordered by Philip ii to take care of the clocks he had made for his Majesty. Cadenas y Vicent, *Hacienda de Carlos v al fallecer en Yuste*, 51.

78   *"nos ayáis de servir y sirváis en hazer los reloges y otras cossas de vuestra professión"*: 13th of March 1562, Madrid: Cervera Vera, *Documentos biográficos*, doc. 4.

79   Ibid., 1st of January 1590: *"en la calle que llaman de Joanelo"* doc. 136.

making clocks and other things belonging to your profession".[80] As we have mentioned, when in 1579 Janello presented his mathematical treaty on the reform of the calendar to Philip II, he stated: "Despite the fact this is not my main profession".[81] Indeed, officially he was paid as clockmaker royal. What did it mean to be a clockmaker in that time? What did Janello's profession encompass? How could a clockmaker work as an engineer?

Once he entered Philip's court, Torriani filled several roles requiring a knowledge of practical mathematics, from designer and smelter of bells for El Escorial to royal surveyor, from automata maker to writer of manuals for his planetary clocks, from creator of mathematical instruments to inventor of ingenious machines.[82] According to Juan de Herrera, Torriani worked on methods to determine longitude, and from other sources we know that he even took part in the royal project of systematic observation of the lunar eclipse of 1577 simultaneously involving scholars in Spain and Mexico under the direction of the cosmographer Lopez de Velazco.[83] In the following years, Philip II responded to the papal call for a mathematical model to actuate the reform of the calendar, and called on his clockmaker together with the universities of Salamanca and Alcalà to aid him in this task. Under royal patronage, Torriani contributed to Spanish engineering in miniaturized mills and in the domain of water-lifting technologies, a knowhow that Vida and Cardano, already in the 1550, attributed to him when the clockmaker was said to have made use of the principle of the Ctesibian pump. Janello Torriani also had competences in the art of creating canals and levelling waters: in January 1571, he was chief of the royal committee that judged the hydraulic works for the construction of the

80   Ibid., 16th of July 1562, Madrid, doc. 5.

81   "*e quantunque questa non sia principale mia professione*": 19th of June 1579, Janello's letter
       and treatise on the reduction of the calendar sent to the King: Turriano, *Breve discurso*, 73.

82   During the siege of August-September 1526, Cremona was constantly bombarded, pro-
       voking damage for the incredible sum of 800,000 ducats. During these months all men
       between 15 and 50 (gentlemen included) had to reinforce bastions and curtains. Further-
       more, all the city bells were seized and cast into ten large pieces of artillery. The generals
       of the League were astonished at the efficient works of fortification carried out by the
       defenders under the direction of the Spanish captain Urias. The bells cast into cannons
       had later to be replaced. It is possible that a young Janello picked up some special knowl-
       edge on this occasion, as we infer from his later expertise in bell-casting for El Escorial. If
       he was in Cremona at that time, it seems clear that he had to join his fellow countrymen
       in compulsory works for the defence, under the threat of violence and plunder. See the
       first chapter: *Cremona, the Italian Wars and the Desire for a better Life.*

83   Vicente Maroto, "Juan de Herrera," 81; Esteve Secall, "Aspectos histórico-gráficos de una
       observación a escala intercontinental."

Colmenar de Oreja's canal by Francesco Sitoni.[84] Furthermore, it seems that in the last years of his life, he was studying a system to make the river Tagus navigable from central Spain down to Lisbon.[85] Janello was not an isolated case; like many other Italian engineers he was transferring knowledge he had picked up in his fatherland, as a guild-master coming from an area renown for its assets in hydraulic engineering.[86]

Torriani was not an exception in moving from clock-making to engineering. He participated in a vast group of highly skilled craftsmen who customarily dealt with hydraulic problems, weaponry and clock-making. Some of the foremost among sixteenth century clockmakers, such as Torriani, Louis de Foix, Capobianco, Jost Bürgi – probably the most famous German-speaking clock-maker, who designed water-pumps at the court of Emperor Rudolph II – and the mathematician Orence Finé embodied this alleged "multidisciplinary pro-

---

84 Cervera Vera, *Documentos biográficos*, doc. 24.

85 The court historiographer Esteban de Garibay reports this piece of news: "*Este insigne varon, antes de ver acabada esta navigacion murió*" : Llaguno y Amirola, *Noticias de los arquitectos y arquitectura de España*, 250. This data has not yet been clarified. I did not have the chance to check Garibay's manuscript. See: *Tesoros de la Real Academia de la Historia* (Madrid: Real Academia de la Historia, 2001), 334; however, as García Tapia has shown, Giovan Battista Antonelli was in charge of this task. If he had any assignment in the project, Janello may have been, like in another documented case (see below), just an adviser. García Tapia dedicated some pages to the problem of fluvial navigation in Castile, noting that since the beginning of Philip II's reign, there have been projects to render some rivers navigable. Nicolás García Tapia, "La ingeniería," in *Historia de la ciencia y de la técnica en la Corona de Castilla*, ed. José María Lopez Piñero, vol. III, Siglos XVI y XVII (León: Junta de Castilla y León, Consejería de Educación y Cultura, 2002), 445-48. Spain, from this point of view, was importing Italian and Flemish know-how;: however, García Tapia seems to read these events in a post-nineteenth century ideological manner. Indeed, an attempt is made to downplay the importance of foreign engineers in Spanish history, promoting an idea of technological autarchy. Ciriacono interprets such a theory as having a nationalistic bias. I think that the weight of the so-called *Leienda Negra* has provoked a desire for compensation in Spanish academia, which may lead to the kind of position stigmatised by Ciriacono. Salvatore Ciriacono, "Trasmissione tecnologica e sistemi idraulici," in *Il Rinascimento italiano e l'Europa*, ed. Philippe Braunstein and Luca Molà, vol. 3, Produzione e tecniche (Treviso: Fondazione Cassamarca : Angelo Colla Editore, 2007), 449-50.

86 Cesare S. Maffioli, "Hydraulics in the Late Renaissance 1550-1625: Mathematicians' Involvement in Hydraulic Engineering and the Mathematical Architects," in *Engineering and Engineers: Proceedings of the xxth International Congress of History of Science, (Liège, 20-26 July 1997)*, ed. Michael Claran Duffy, vol. 60, De Diversis Artibus, XVII (Tumhout: Brepols, 2002), 67-75.

fessionalism", as Brunelleschi did a hundred years earlier:[87] According to Antonio di Tuccio Manetti, in the fifteenth century, Filippo Brunelleschi applied his knowledge in clockmaking to the study of the erection of the massive buildings of ancient Rome: in this way he was able to understand how the Ancients were "lifting, moving and pulling" colossal columns and enormous blocks. This background was later relevant when Filippo applied his *ingenio* to the construction of Santa Maria del Fiore's dome,[88] a pattern repeated by Janello in Toledo. Both Renaissance engineering and clockmaking were professions devoted to a careful measured programming of movement. Clockmakers, beside their practical know-how in blacksmithing and other metallurgic technologies, had enough mathematical knowledge to engage with scientific instrument-making and organ-making.[89] Trigger devices for crossbows and guns, locks, clockworks, mills and other geared machines were based on the same rotary mathematical principles involving set controlled concatenated motion. If we define Renaissance engineering as the art of finding precise solu-

---

87   It seems that in the second half of the fifteenth century, the clockmaker Bartolomeo Manfredi dell'Orologio, beside clocks, constructed mills. Claude Grenet-Delisle, *Louis de Foix: horloger, ingénieur, architecte de quatre rois* (Bordeaux: Fédération historique du Sud-Ouest, 1998) ; King and Millburn, *Geared to the Stars*, 78 ; Gorla and Signorini, *Orologio astronomico astrologico di Mantova*, 17-18; Puppi, "Capobianco, Giorgio."

88   Antonio Manetti, *Vita di Filippo Brunelleschi*, ed. Carla Chiara Perrone, Minima 34 (Roma: Salerno Editrice, 1992), 65: "*andossene a Roma (...) E veduto le gran cose e dificili che erano intra esse, che pure si vedevano fatte, non gli venne meno pensiero d'intendere e modi che coloro avevano tenuti e con che strumenti. Ed essendosi dilettato nel passato e fatto alcuno oriuolo e destatoio, dove sono varie e diverse generazioni di mole e da varie e multitudine d'ingegni multiplicate, che tutte o la maggiore parte aveva vedute, gli dettono grandissimo aiuto al potere immaginare diverse machine e da portare e da levare e da tirare, secondo le opportunità ch'egli aveva veduto che erano state di bisogno*"; Vasari, *Le vite de più eccellenti pittori, scultori e architettori*: "*Laonde, avendo preso pratica con certe persone studiose, cominciò a enrarli fantasia nelle cose detempi e demoti, depesi e delle ruote, come si possono far girare e d ache si muovono, e così lavorò di sua mano alcuni oriuoli bonissimi e bellissimi.*" Both passages reported in Fondelli, "*Oriuoli mechanici*" di Filippo di ser Brunellesco Lippi, 5.

89   Johannes Mutina de Organis, ducal engineer in Milan in 1352 and organ-maker: he built the public clocks for Monza and Genoa: Dohrn-van Rossum, *History of the Hour*, 130; Giovanni Fontana of Venice, organ-maker and builder of fountains, from which profession he obtained his name; Henry Arnault of Zwolle wrote the first tract about keybord-making. White, "Medical Astrologers," 309-10; Similar competence had Antonio Bovelli, the papal plumber at Avignon, and clockmaker to the Aragonese king at Perpignan Beeson, *Perpignan 1356;* on the background of Renaissance clockmakers see also: Cipolla, *Le macchine del tempo*, 24-5.

tions to practical problems thanks to mathematical design, one could say that clockmaking was a branch of engineering.

Besides clockmakers, other professions involved with *disegno* dealt on a daily basis with hydraulic engineering: in Toledo the painter Juan de Coten had worked on the water-lifting project.[90] In the same decade Torriani's first device was built in Toledo, an antiquarian, a physician, a jurist and a city-magistrate were involved in problems and projects concerning hydraulic engineering in Rome.[91] In Germany, the famous painter Matthias Grünewald (ca. 1475-1525), also worked as a hydraulic engineer in Mainz and Halle,[92] and the painter Albrecht Altdorfer (1480-1538) was elected public architect and engineer of the city of Regensburg.[93] In 1567, the physician Giuseppe Ceredi published his book about the Archimedean screw for water elevation, accompanied by privileges for inventions granted to him by different states. The list could include still many more names. Indeed, in many places there was an office for engineers but not a specific theoretical or practical *ratio studiorum* that could define a clear education for it: people coming from different backgrounds like Taccola (notary and sculptor), Brunelleschi (goldsmith), Leon Battista Alberti (jurist), Pierre Belle (carpenter) Giuseppe Ceredi (physician), Aristotile Fioravanti

---

90   In 1564 Torriani and a certain Juan de Coten tried together to work out some solutions for the hydraulic problem of Toledo. By April 1565, after showing a model to the King and to the representatives of the City of Toledo explaining the functioning of the device and convincing his audience, Janello signed the contract for the construction of a device that had to supply the square close to the *Alcázar* with water. This document has to be dated 1574 (the new device already started and the new agreement with Toledo had not yet been made) and not 1575 as in: Cervera Vera, *Documentos biográficos*, doc. 44.

91   Pamela Long in her article promises that other studies will be made on urban hydraulic engineering at the time of Pius IV and Pius V. Indeed, Toledo provides a useful comparison to Rome in the 1560s. First of all, important similarities should be individuated in the different backgrounds of most of the people involved with hydraulic engineering. In the case of Rome, there were at least four professions involved in the same project: Andrea Bacci (physician), Antonio Trevisi (military engineer), Luca Peto (jurist and magistrate), and Pirro Ligorio (antiquarian-archaeologist and architect): Long, "Hydraulic Engineering and the Study of Antiquity," 1114; however, Long does not mention the clockmaker Gio. Bartolomeo Gritti who, together with Giacomo della Porta, completed Antonio Trevisi's works on the Aqua Virgo: Antonino Bertolotti, *Artisti lombardi a Roma nei secoli XV, XVI e XVII: studi e ricerche negli archivi romani* (Milano: Ulrico Hoepli, 1881), 65; see also: Giovanni Beltrani, *Leonardo Bufalini e la sua pianta topografica di Roma: (Estratto della Rivista Europea-Rivista Internazionale.)* (Firenze: Tipogr. della Gazzetta d'Italia, 1880).

92   Martin F. Schloss, "Grünewald and the Chicago Portrait," *The Art Journal*, 1963, 10.

93   Harold Joachim, "About a Landscape by Altdorfer," *The Art Institute of Chicago Quarterly* 49, no. 3 (1955): 51.

(goldsmith), Leonardo da Vinci (painter), Janello Torriani and Luis de Foix (clockmakers), to name a few, worked as engineers.

Prior to Torriani's lifetime, Renaissance engineering did not have a precise institutional or practical curriculum; but when he died, Milan and Florence had institutionalised the curricula to become engineers. For this reason, Torriani's trajectory is very interesting to address such a complex process of transformation of a profession so important for any narrative involving modernity. As in the case of the technical office of public clock keeper, the engineer had a long tradition in terms of function. In Spain, the office of royal engineer was granted from the second half of the fifteenth century, but in the medieval city-states and *signorie* of Italy such an office was established long before to run the construction and management of commercial, monumental and military infrastructures.[94]

The occupational designation of engineer (*ingeniarius/ingeniator*) appeared during the high Middle Ages and became more frequent during the twelfth century. This term could be exchanged with the learned terms of *architectus* or *machinator*. In Northern Italy, since the thirteenth century, there had been communal officials called *engineers* dealing with public constructions and water-management. In this context, *magister aquarum* or *extimator civitatis* were also synonyms for *ingenierus*. Engineers constructed mills and the necessary machines to build cathedrals, fortifications and canals. In Europe, outside of Italy, one could find mainly monks and friars performing this function, whereas in the peninsula, these engineers were also laymen. During the fourteenth century, with the extension of the Visconti lordship over a good part of

---

94   In the year 1480, we have for the first time in Spain an *"ingeniero de los reyes"* with the name of Abrahan de los Escudos. Ibid., 147. As concerns hydraulic engineering in Italy, Squatriti collected interesting data about eight medieval fountain-makers from Central Italy. Three of them were brethren, three were called "master" (in the case of Siena a stone-mason) and the other two had no apposition. Perugia's commune searched for trained craftsmen for years, seeking them in other towns or monasteries. This shows that at this time it was hard to find people with knowledge in hydraulic engineering: these engineers were often monks, who besides their knowledge in the quadrivium, were probably the only ones, beside Cathedral libraries, in possession of ancient treatises in their monastic libraries. Another category employed in hydraulic projects was the stone-mason: his practical knowledge was necessary to build aqueducts and stone-fountains. From other documentary material, the craftsmen involved the most with such projects were blacksmiths, most probably because of their competence in fashioning pipelines, siphons and hydraulic mills. Paolo Squatriti and Roberta Magnusson, "The Technologies of Water in Medieval Italy," in *Working with Water in Medieval Europe: Technology and Resource-Use*, ed. Paolo Squatriti (Leiden: Brill, 2000), 244-51.

Northern Italy, we find the creation of the office of *"ingeniarii et architecti"* in any city subject to the Viper – the dragon-snake of the Visconti's coat of arms – where two public engineers were constantly kept in charge with a higher salary than masters, masons and carpenters. At the end of the Quattrocento one can eventually find in the juridical-administrative language of the duchy of Milan the titles of *architect, engineer, agrimensor* and *water-leveller*. These were considered to be synonyms, though engineer was the most popular one among illiterates.[95] Engineer – it is prevalently considered to derive etymologically from *ingenium*, i. e., talent or wits – could be defined as a person who finds new mechanical solutions or a new design to solve practical problems.[96] This need to find some new way ties the Medieval and Renaissance engineer to the concept of "invention". In order to perform such duties, the engineer, besides his practical knowledge of materials, had to be acquainted with measurement, therefore with mathematics.

Besides the practical efficiency of engineers to solve practical problems, institutional power could use technical officials as political tools. Observing how conflicts related to technical problems were solved in the fifteenth and sixteenth centuries, it becomes clear that one strategy for institutionalised power was to create offices that could deflect opposition from its subjects. Violence could be substituted with a more efficient end economical method based on persuasion. When during the fifteenth and sixteenth century water management created conflicts, mathematical argumentation was the weapon often used by officials to enforce decisions from above. Mathematicians had a special weapon of persuasion-based demonstration thanks to measure. For instance, a report that the famous engineer Aristotile of Bologna was required to write by the Duke of Milan brings to our attention interesting news about the activities of such kind of officials, those engineers involved with hydraulic

---

95   Valeria Poli, *Architetti, ingegneri, periti agrimensori: le professioni tecniche a Piacenza tra XIII e XIX secolo* (Piacenza: Banca di Piacenza, 2002), 22-23: "*Architectos seu Agrimensores et Libellatores aquarum, qui omnes vulgo Ingegneri appellantur*". This document dates 22nd December 1497 (Parma), but it was already issued during the rule of Gian Galeazzo Maria Sforza (1476-1494); Since the time of Gian Galeazzo Visconti, the construction of the huge cathedral had developed a system of canals to transport blocks of stone into the city. Military needs were still the first issue in terms of ducal spending, and here too engineers had a central role. The great importance of such duties created a very close personal link between these officials and the person of the duke: "*acteurs et agents de la politique ducale des grands travauxs*": Patrick Boucheron, *Le pouvoir de bâtir: urbanisme et politique édilitaire à Milan (XIVe-XVe siècles)* ([Rome]: Ecole française de Rome, 1998), 245-379.

96   Hélène Vérin, *La gloire des ingénieurs: l'intelligence technique du XVIe au XVIIIe siècle* (Paris: Albin Michel, 1993).

technologies during the fifteenth century in the very land where Janello Torriani grew up.

During the year 1459, the city of Cremona was authorized by its master the Duke of Milan to open a new canal in order to bring more water from the river Oglio (the frontier between the duchy of Milan and the land of Brescia, at that time belonging to the Republic of Venice) down to the city's *Naviglio Civico* (i.e., the Civic Canal) which, as already mentioned, brought water to Cremona. The new canal was meant to be open two kilometres northwards of Soncino, a subject town of Cremona; but the people of Soncino were afraid that the new waterworks might cause dangerous floods. Therefore, they stole the papers for the project and the mathematical instruments of the engineers and sent to Milan ambassadors to ask the Duke to stop the hydraulic project. The Sforza then sent his ducal engineer Aristotile of Bologna to Cremona and Soncino where he checked the hydrological situation and explained "scientifically" to the people of Soncino that there was no danger. Consequently, the people of Soncino agreed that the project could go ahead. This story shows that engineers were public officials who made use of scientific instruments and projects drawn on paper and they were commonly recognised as reliable *men of science*,[97]

The military employment of engineers made of these technicians a professional group closely connected with nobility. For instance, during the first half of the fifteenth century, Renaissance petty tyrants such as Sigismondo Pandolfo Malatesta engaged with military experimentation and patronised Valturio's work, the popular work on sieging and warfare engineering. This kind of knowledge showed itself to be particularly suitable to the seignorial courts and to the *condottieri* who aspired to be seen as ancient Roman princes. If any Renaissance *signore* had his new Apelles and Phidias at court, he needed a new Archimedes as well. The engineer was a wheel of the state-engine as he was a cog for the propaganda machine: their "inventions" gave both real and symbolic power to their lords, who had to flaunt the best of any luxury production, even in this field. Stable employment at court uprooted those men from their social web in the city; at court, guild obligations vanished, giving the master a malleability, which the prince could employ in various fields. The academy, a new institution that brought together different branches of practical and theoretical knowledge platonically united under the sign of mathematics, was created in this environment.

---

97   Luca Beltrami, *Aristotele da Bologna al servizio del duca di Milano* MCCCLVIII-MCCCCLXIV: *documenti inediti pubblicati* (Milano: A. Colombo & A. Cordani, 1888), 33.

Mario Biagioli has observed that mathematicians' activity on the battle-fields "gave them the chance to ennoble themselves and their discipline by partaking in the high social status of the *milites*".[98] The *milites* were the knights, i.e., feudal nobility. The big money at stake for military engineering and the feudal tradition of noble monopoly on warfare probably made of the military engineer an appropriate label for the gentleman. In the sixteenth century even the profession of civil engineer, as we might call it today, was boasted of by members of the gentry, as in the case of Giovanni Francesco Sitoni from Milan. On the other hand, as we have seen, non-noble engineers felt also entitled to climb the social ladder. As in the case of Torriani, the humanist classical tradi-tion made of the professional definition of "*architectus*" a suitable dress for any modern engineer who wanted to be elevated to a higher social status.[99] Medals, paintings and other encomiastic cultural products, celebrating Janello, testify an important and early step in the process of ennoblement of these technical professions. Janello's marble bust was even placed in the Toledo Device while the clockmaker was still alive.

Janello's forerunner was Brunelleschi, who was the first engineer to be cel-ebrated with a marble portrait, though posthumously. In 1446, in the Florentine Cathedral, his features were reproduced upon a cenotaph, with the following explanation:

> How outstanding has been Filippo in the art of Dedalus, it is demon-strated by the wonderful dome of this very famous temple, and by the many machines invented by him with divine ingenuity. And for his excel-lent qualities of his soul, and his singular virtues, his body worthy of merit has been here buried ...[100]

---

98   Biagioli, "The Social Status of Italian Mathematicians, 1450-1600," 46.

99   See the Imperial privilege, Trezzo's medal and the last lines of Alessandro Lamo's acclam-atory description of Janello: "*Now King Philip held dear this Architect as had his father Charles*". Lamo, *Sogno non meno piacevole, che morale,* 60.

100   "*Quantum Philippus architectus arte dedalea aluerit cum huius celeberrimi templi mira tes-tudo tum plures machinae divino ingenio ab eo adinventae documento esse possunt. Qua-propter ob eximias sui animi dotes singulares que virtutes ...*" (my translation). There are a few known exemples of medieval artists who produced self-portraits in manuscripts, sculpted altars or stain-glass windows. However, Brunelleschi's cenotaph inside the pub-lic space of the Cathedral of Florence celebrating his mechanical abilities appears to be the only real precursor for Janello's portraits. Castelnuovo, *Artifex bonus: il mondo dell'artista medievale.*

FIGURE 53
Pompeo Leoni (?), Bust of
Janello Torriani, *Carrara
marble, ca. 1560.* PHOTO
JEAN LAURENT. 1860-1886,
COURTESY OF THE
FUNDACIÓN JUANELO
TURRIANO. CURRENTLY
HELD IN THE MUSEO DE
SANTA CRUZ TOLEDO.

It is astonishing to find in this brief cenotaph four of the most characteristic
rhetorical elements used to celebrate Torriani's great achievements more than
a hundred years later: we find in 1552 the definition of "architect" in the imperial diploma, in 1610 Luis de Góngora in *Las firmezas de Isabela* called him
"Daedalus from Cremona", the virtue is named in his medal, and we have seen
how Vida and others have written of his divine ingenuity.[101] The apotheosis of
the engineer is well represented by the processes of emancipation from the
guild system: Filippo Brunelleschi refused to pay dues to his guild, and Janello,
as many other Vitruvian artisans involved with engineering and other practical
mathematical professions, moved from the guild system of the cities to the
court.[102]

---

101   *"El Tajo, que hecho Icaro, a Juanelo, Dédalo cremonés, le pidió alas, Y temiendo después al Sol*
       el Tajo, Tiende sus alas por allí debajon" in: Sánchez Mayendía, "El artificio de Juanelo en la
       literatura española," 81.

102   Grafton, *Leon Battista Alberti,* 76 and note 10, where two works (Carroll W. Westfall, "Paint-
       ing and the Liberal Arts : Alberti's View," *Journal of the History of Ideas* 30, no. 4 (1969):

We have already seen the classical models behind this apotheosis. It remains for us to discuss the process that created a curriculum for engineers and that raised them among the elite of urban professions. The oldest record relating to a guild of engineers comes from Milan and goes back to the beginning of the century, when we find a *Universitas Ingenierorum et agrimensorum Mediolani* (the *Guild of Engineers and Surveyors of Milan*).[103] Around the year 1564, this *Universitas* was transformed into the *Collegio degli Architetti ed Ingegneri di Milano* (the *Association of Architects and Engineers of Milan*). A document from this *Universitas* dated 1505 explains that in order to be admitted into this association, one had to be able:

> To measure both on a paper and on the field for contracts of sale or rental, to measure and level the water, to install openings for waterworks, to be able to measure and estimate houses, to measure ditches, canals, drains and karst springs, and to be knowledgeable about rivers, torrents and navigable canals, to be able to perform all types of delivery and redelivery [probably in relation to contracts of lease], and to be quite able to draw in architecture.[104]

Each one among the professions involved with engineering previously listed (painting, sculpture, clockmaking or an academic background in the *qua-*

487-506, and Martin Warnke, *Der Hofkünstler: zur Vorgeschichte des modernen Künstlers* (Köln: DuMont, 1985), 3) focussed *"on the courtly context in which such claims to freedom from guild restriction arose"*. Pamela Long has noted that Brunelleschi's refusal to pay his fee to the guild of masons was perhaps depending on the fact that the status of that *Arte* was inferior to the one he previously belonged with: Pamela O. Long, *Openness, Secrecy, Authorship, Technical Arts and the Culture of Knowledge from Antiquity to the Renaissance* (Baltimore: Johns Hopkins University Press, 2001), 96-7.

103   Giovanni Liva, "Il Collegio degli ingegneri architetti e agrimensori di Milano," in *Il Collegio degli Ingegneri e Architetti di Milano*, ed. Giorgio Bigatti and Maria Canella (Milano: Franco Angeli, 2008), 10.

104   Ibid.,*"sapere misurare in dissegno, et in campagna, in vendita, et in affitto, misurare, et livelare acque et piantare bocchetti et saper misurare e stimar case, et saper misurar fossi, roggie, scolatori, et fontanili, st saper le raggioni dei Fiumi, torenti, e Navigli, et saper fare di tutte le sorti di Consegne e riconsegne, et sapere alquanto disegnare di architettura"* (my translation). As some documents in the State Archive of Milan demonstrate, in order to become engineers of the ducal chamber, it was however necessary to direct a supplication to the governor. Further investigations in this archive may produce more evidence about Torriani's alleged nomination as an engineer in the 1540s: ASMi, Autografi: Ingegneri e architetti, cartella 86, year 1556: *Supplica di Francesco Sitoni ingegnere e agrimensore per essere investito dell'ufficio di ingegnere della Ducal Camera.*

drivium) could perform these tasks. Applied geometry (together with good drawing skills) was the greatest common divisor of all these professions. The passage from *Universitas* to *Collegio*, besides its alleged juridical quality,[105] also shows that engineers and surveyors aspired to a higher social status: indeed, the traditional *collegi* grouped together highly respected professions which were built on a liberal education such as jurists (judges and lawyers), notaries and physicians. Moreover, the passage for the definition of "engineers and surveyors" to the one of "architects and engineers", shows a clear humanist agenda of ennoblement.

After the association was transformed into a *collegio*, new rules better defined the educational pattern of the engineer and not just his competencies: in order to become a member of this guild, one had to spend four years as apprentice to a master (who could therefore boast such a professional title in architecture or engineering), one had to pay a fee, and one had to pass an exam. This exam had to test both knowledge and respectability.[106] The working papers of the single engineers and architects would from this point onward belong to the *collegio*. Furthermore, anyone found performing these professions without the license granted by the *collegio* had to pay a fine of a 100 *soldi*.[107] This meant the end of engineering as an open professional field and the beginning of a process of exclusion controlled by a close number of practitioners recognised by the state. Henceforth, only a specific kind of education would be able to legitimately produce engineers in Milan. On the other hand, in the very same period in Florence we face a very different trend: Grand Duke Cosimo I created in 1563 the *Accademia del Disegno*, an institution that was consecrating the old epistemic system which unified all professionals involved with *disegno*, from the painter to the sculptor, from the architect to the engineer. This model was then co-opted in the territories of the *Patrimonum Sancti Petri*: similar academies were indeed founded in Perugia (1573), Bologna (1582), and Rome (1593). This institutional shift in Milan and Florence conveyed the "revolutionary" idea that it was possible to train ingenious people, and that ingenuity was no longer a divine gift, but the outcome of a curricular training. Was the theoretical background of this process originating from Trent's theology? Was the season of intellectual freedom of the Renaissance artist-engineer over?

---

105 Ibid., According to Giovanni Liva, this *Universitas* was not recognised as a guild – i.e., an institutionalised art – but just as a private corporate body.

106 We have seen the case of Cardano, who for a long time was unable to enter the *Collegio* of physicians in Milan because he had been born an illegitimate son, though his father had soon recognised him as a legitimate son.

107 Ibid., 10-11.

CHAPTER 8

Janello in Spain as a Royal Hydraulic Engineer (1563-1585)

Hydraulic Engineering in the Habsburg Empire

Renaissance water-lifting technologies dealt mainly with three problems: land reclamation, irrigation and urban water-supply systems. The privileges for invention obtained by Janello for his water-lifting devices offered to solve both problems in a decade that was crucial for this kind of technology: it has been noted that it was especially after the Peace of Cateau-Cambrésis (1559) that a more systematic public and private effort in reclaiming swampy lands was made in order to increase crop production, triggering a period of economic growth.[1] Though the special size and technical difficulty of the Toledo enterprise has monopolized contemporary literature, especially Spanish literature during its golden age, one can observe that the whole of sixteenth century Western Europe saw the creation of numerous water-supply systems and the circulation of many hydraulic engineers like Torriani, though the latter undoubtedly represents the cutting edge of this technology in his time.

When in 1567 Giuseppe Ceredi dedicated his book *Tre discorsi sopra il modo d'alzar acque da' luoghi bassi* (i.e., *Three propositions about how to elevate water from low spots*) to the ducal heir of Parma and Piacenza Alessandro Farnese, nephew to Philip II, he claimed that the young prince would find some utility in this treatise about the Archimedean screw (the screw-pump used to elevate water). Indeed, Ceredi had heard the rumour that when young Farnese was at the Spanish court he had been very much interested in mathematical issues such as cosmography and the art of fortification. Ceredi added that it was common to all nobles of this court to share this very same interest, as we have previously seen discussing Charles V's interests in mathematics and clocks.[2] Giuseppe Ceredi wrote this book with the intent of selling his model of the Archimedean screw. This book represents a very precious document for our contextualization of Torriani's experience in Milan and especially in Toledo.

---

1 Franco Cazzola, "Le bonifiche cinquecentesche nella valle del Po: governare le acque, creare nuova terra," in *Arte e scienza delle acque nel Rinascimento*, ed. Alessandra Fiocca, Daniela Lamberini, and Cesare Maffioli (Venezia: Marsilio, 2003), 29.

2 Ceredi, *Tre discorsi.*

Ceredi was born in Piacenza, a few kilometres away from Torriani's home-town. The Italian wars had separated the political destiny of the twin cities founded by the Romans to control the two shores of the River Po at the time of Hannibal's invasion of Cisalpine Gaul. Now, the ambitious political activity of the Farnese had alienated Piacenza from the Duchy of Milan, to which it had belonged, with Cremona, since the beginning of the fifteenth century. The cultural tradition of these two cities was therefore the same, stretching between the academic centre of Pavia, and the powerful patronage of the families traditionally connected with the Visconti and Sforza rule. The name traditionally taken as the personification of Renaissance hydraulic engineering's excellence in the Duchy of Milan is that of Leonardo da Vinci; however, Leonardo's service as technician at the Milanese court is not to be considered unique, at least as concerns hydraulics:[3] many other engineers were involved with water-management during the late Middle Ages and the Early Modern period.[4] The Castilian crown that would come to possess the Duchy of Milan after the end of the House of Sforza would later draw on this rich tradition of importing numerous Milanese engineers such as Janello around the Spanish empire. Even the Republic of Venice, queen of the seas and one of the main political actors in Renaissance water-management, imported many of its best hydraulic engineers from the newly acquired lands of Lombardy, such as Brescia, Bergamo and Crema.[5] Renaissance Tuscan tradition also played a

---

3   Already at the beginning of the nineteenth century, Ammonetti highlighted the need to down-play the influence of Leonardo on Lombard hydraulics: Carlo Ammoretti, *Memorie storiche sulla vita e sugli scritti di Leonardo da Vinci* (Milano: Alfieri & Lacroix, 1804).

4   The Duke of Milan had many specialists employed as *"ingegneri della ducal camera"*. Before the arrival of the famous Leonardo da Vinci, Aristotile from Bologna, Bertola from Novate and Aguzio from Cremona operated with great proficiency in the field creating some outstanding *navigli*, or canals, in Milan, Cremona and Parma. Furthermore, Dalla Valle and Missaglia were authoritative hydraulic engineers contemporary with Leonardo: Beltrami, *Aristotele da Bologna al servizio del duca di Milano*.

5   Other territories of the Republic also produced famous Renaissance engineers, such as Fra Giocondo of Verona. Alessandra Fiocca has stressed the importance of scientific institutions such as the Accademia Olimpica of Vicenza and the offices involved with water-control as the base for the fortune of Venetian hydraulics. The Accademia Olimpica was founded in 1555 and it presented two particularities: its curriculum focused mainly on mathematical sciences *"that are the true ornament of a noble and virtuous soul"* and it was opened to people from any social background (craftsmen included). Among the promoters of this cultural institution was one Silvio Belli. A contemporary of Palladio and Torriani, Belli had here a relevant role and held a position as *"lector of the Sphere and other mathematical things"*. Belli was later appointed in Venice with the important office of *proto*: his duty consisted of controlling the balance of the hydrology of the lagoon and of the rivers that fed it. Patrice Beck, "Le techniciens de l'eau à

major role in the development of early hydraulic engineering, and had a great influence on the Milanese environment.[6] However, we must stress that even if Northern Italy was the biggest European centre for water-technology application in the fifteenth and sixteenth centuries, the engineers involved were mainly, but not only, local.[7]

Besides land reclamation and water control, on the other side of water-lifting technology we have supply systems for cities and for princely residences. The creation of new water supply systems created better conditions for urban

Dijon," in *Le technicien dans la cité en Europe Occidentale 1250-1650*, ed. Mathieu Arnoux, Collection de l'École Française de Rome 325 (Roma: École Française de Rome, 2004), 109-43; Salvatore Ciriacono, *Building on Water: Venice, Holland, and the Construction of the European Landscape in Early Modern Times* (New York: Berghahn Books, 2006), 103 and following; Alessandra Fiocca, "Silvio Belli ingegnere: empiria e matematica nella cultura tecnica del Rinascimento," in *Acque e terre di confine. Mantova, Modena, Ferrara e la bonifica di Burana: studi nel centenario dell'apertura della Botte napoleonica*, ed. Daniele Biancardi and Franco Cazzola (Ferrara: Editrice Cartografica, 2000), 25.

6   Florence is well-known as the fatherland of the most famous Renaissance engineers. Siena was another prolific host of engineers involved with hydraulics: the best known among them being Mariano di Iacopo, called Taccola (1382-1458), who illustrated the ancient texts of mechanics in his two manuscripts: *De ingeniis* (1419-1450) and *De machinis* (1430-1449); Francesco di Giorgio Martini (1439-1502), who wrote four technical books including a *Trattato di architettura,* which was studied by Leonardo da Vinci; Vannocchio Biringuccio (1480-1537); and Baldassarre Peruzzi (1481-1536). Francesco di Giorgio can be considered innovative for introducing a strict rendering of the proportions and distances existing among the elements of the machinery represented in his manuscripts and for attempting consistent classification of the devices. Paratext such as visual representations were to reach a level of quality equal to that of written text: Paolo Galluzzi, *Gli ingegneri del Rinascimento da Brunelleschi a Leonardo da Vinci* (Firenze: Giunti, 1996),11-18; Mario Taddei, Edoardo Zanon, and Castello sforzesco, eds., *Leonardo, l'acqua e il Rinascimento* (Milano: Federico Motta, 2004).

7   Between 1559 and 1580, the Este actuated the reclamation of 30,000 hectares of swampy land thanks to the excavation of 300 kilometres of ditches. Ciriacono emphasises that the Venetian Republic wrestled from the swampy waters around 70,000 hectares of land, as much as the reclaimed polders at the end of the sixteenth century in Holland, the main centre of exportation for water-technologies in the following centuries. See: Cazzola, "Bonifiche cinquecentesche nella valle del Po"; Ciriacono, "Trasmissione tecnologica e sistemi idraulici". Some scholarship anticipates the establishment of an intensive Lombard irrigation system back to the eleventh century: Squatriti and Magnusson, "The Technologies of Water in Medieval Italy," 2; Potito D'Arcangelo, "Acque e destinazioni colturali nel Cremonese dal XIII al XV secolo," in *Storia di Cremona: il Quattrocento: Cremona nel Ducato di Milano (1395-1535)*, ed. Giorgio Chittolini (Bergamo: Bolis, 2008), 148; Ciriacono, *Building on Water,* 24.

life and increased citizens' number and their reputation.[8] For instance, during the first decade after the peace of Cateau-Cambrésis, as one can still see today, Florence and Bologna were embellished with monumental fountains fed with siphon-aqueducts. Rome, which housed the most important courts-network of Europe, played a major role in this process: this crowd of competitive popes,

---

8   The political choice to set a centre of power in a specific place has often encouraged the creation of more efficient water-supply systems. Naples, which was already supplied with some fountains by King Ferdinand of Aragon in the late fifteenth century, had at that time 100,000 inhabitants. Some scholars propose a slightly lower number. In the second half of the sixteenth century, thanks to the restoration of an ancient Roman aqueduct, the urban population of Naples, in the turn of a few decades, doubled to a figure of around 280,000, becoming the largest Christian city of the Mediterranean. Another similar example is given by the growth of Madrid during the sixteenth century, when the King made the town the centre of his system of Castilian royal residences. At the beginning of the century the population of Madrid numbered 10,000. A hundred years later it numbered 90,000. Scholarship has highlighted that during this century of vertiginous growth, Madrid's main administrative concerns were directed to the practical problems of water and dumping. This process of growth is not always controllable by the very same authority that triggers it. For instance, Philip II planned to rationally enlarge his capital, but the King appeared to have lost hold on it. It has been observed that the urban development of Madrid ended in a haphazard way. The *Real junta de obras y bosques*, instituted by Charles V in 1544 and controlled by Prince Philip, was a board made for this purpose. Also in Naples, in the last two decades of the sixteenth century, the political authority tried to stop the abnormal growth of the city, but without success. The number of inhabitants increased still more rapidly in the beginning of the following century to 450,000, when other ancient aqueducts were then restored: Braudel, *Civiltà e imperi del Mediterraneo nell' età di Filippo II*, 365-67; Giuseppe M. Montuono, "L'approvvigionamento idrico della città di Napoli: l'acquedotto del Serino e il Formale Reale in un manoscritto della Biblioteca Nazionale di Madrid," in *Atti del 2° Convegno nazionale di Storia dell'Ingegneria (Napoli, 7-9 aprile 2008)*, ed. Salvatore D'Agostino (Napoli: Cuzzolin, 2008), 1038; Giuseppe Galasso, "Aspetti della megalopoli napoletana nei primi secoli dell' età moderna," in *Mégapoles méditerranéennes géographie urbaine rétrospective: actes du colloque, Rome, 8-11 mai 1996*, ed. Claude Nicolet et al. (Paris: Maisonneuve et Larose, 2000), 565; Luis Miguel Enciso Recio, "Nápoles en tiempos de Felipe II: historiografía reciente," in *Madrid, Felipe II y las ciudades de la monarquía*, ed. Enrique Martínez Ruiz, vol. I, Poder y dinero (Madrid: Actas, 2000), 27-72; Escobar, "Francisco de Sotomayor and Nascent Urbanism in Sixteenth-Century Madrid," 358 and 363; Juan Carlos Zofío Llorente, "Trabajo y socialización: los aprendices en Madrid durante la segunda mitad del siglo XVI," in *Madrid, Felipe II y las ciudades de la monarquía*, ed. Enrique Martínez Ruiz, vol. II, Capitalismo y economía (Madrid: Actas, 2000), 521; Virginia Tovar Martín, "Lo urbano y lo suburbano: la capital y los sitios reales," in *Madrid, Felipe II y las ciudades de la monarquía*, ed. Enrique Martínez Ruiz, vol. II, Capitalismo y economía (Madrid: Actas, 2000), 205; Claudia Conforti, "Acque, condotti, fontane e fronde: le provvisioni per la delizia nella Villa Medicea di Castello," in *Il teatro delle acque*, ed. Attilio Petruccioli and D. Jones (Roma: Edizioni dell'Elefante, 1992), 79.

cardinals, local nobles and ambassadors, together with the impressive models provided by ancient imperial water-works and medieval fountains, and with the unpredictable Tiber (which brought sudden floods), made the Eternal City a central place for the development of water-technologies.[9] The text *De aquae ductu urbis Romae*, written by Sextus Julius Frontinus[10] in the first century A. D., rediscovered by the humanist Poggio Bracciolini in 1425, inspired in Rome a renaissance of the ancient water-supply systems, and was a prelude to the magnificent theatre of fountains that gave the city its lavish appearance in coming centuries.[11] Many fifteenth and early sixteenth century popes attempted a *renovatio urbis* (a renewal of the City) that sought to recast Rome as a universal Christian capital.[12] Pius IV, Pius V, Gregory XIII and Sixtus V, the four popes living during the reign of Philip II, supported imposing hydraulic works.[13] Torriani, as previously seen, was in contact with three of them, and the first two were probably interested in his hydraulic knowledge.

While introducing the problem of urban water supply, Ceredi quoted three paradigmatic examples of cities in need of his rediscovered Archimedean screw, likely those he considered to be the most renowned cases during the 1560s: Rome, Ferrara and Toledo.[14] It is not surprising that the third place men-

---

9   Despite the damage of the classical water-supply system, Rome had also a medieval hydraulic tradition: the most famous fountain of the high Middle Ages was probably the bronze *Pigna* (pine cone) in S. Peter's atrium in Rome. The rest of the *Patrimonium Sancti Petri* and its surroundings provided as well many examples of this early rebirth Squatriti and Magnusson, "The Technologies of Water in Medieval Italy"; Paolo Buonora, "Cartografia e idraulica del Tevere: secoli XVI–XVII," in *Arte e scienza delle acque nel Rinascimento*, ed. Alessandra Fiocca, Daniela Lamberini, and Cesare Maffioli (Venezia: Marsilio, 2003), 169–93.

10 Sextus Julius Frontinus, *De aquae ductu urbis Romae*, ed. Fanny Del Chicca (Roma: Herder, 2004).

11 Simon Schama, *Landscape and Memory* (New York: A.A. Knopf, 1995), 286.

12 The patronage of the pontiffs in supplying the city with water was embedded in both Christian and antiquarian rhetoric, from Moses to Caesar. The popes saw themselves not just as the successors of Peter but also as new Roman emperors: for the first time a new Rome attempted to be no less than her glorious memory, and the restoration of the city's ancient aqueduct was among the main tasks of these popes. Long, "Hydraulic Engineering and the Study of Antiquity."

13 *Felipe II: los ingenios y las máquinas : ingeniería y obras públicas en la época de Felipe II* ([Madrid]: Sociedad Estatal para la Conmemoración de los Centenarios de Felipe II y Carlos V, 1998), Introduction.

14 Ferrara, with its 35,000 inhabitants, was the seat of one among the most brilliant courts of Renaissance Europe. Nevertheless, at the end of the sixteenth century, Ferrara was set to lose its political relevance as the city was detached from the Este Duchy. Ceredi, writing in

tioned by Ceredi was Toledo, the stage of our story:[15] not all cities were placed in plains just outside of hills rich in springs like Rome, Bologna, Florence or Naples. Urban centres with different geographic situations needed mechanical solutions, such as water-lifting devices, in order to be supplied with water, and Toledo was the most famous case in Christendom.

In his book, Ceredi divided water-raising machines existing in Italy at the midst of the sixteenth century in three categories. He ascribed to the first group those machines that raised water: "because it is not possible to have a body without a space and a space without a body", such as suction or piston pumps. According to Ceredi and the science of his time, Nature abhorred a vacuum. The machines belonging to the second group were activated "with the science of moving weights", i.e., through counterweights, and especially scoop wheels that use the weight and force of moving water. To the third group Ceredi ascribed those machines that worked "because of a craft that facilitates the motion of fluid bodies", a category created ad hoc in order to include in it his Archimedean screw:[16] Ceredi's agenda was to advertise the superiority of his

the 1560s, had a strong interest in Ferrara (neighbouring power to the Farnese of the Duchy of Parma and Piacenza, at whose court he was employed), which might have made him rich if applying on a large scale his screw for watering, land reclamation and water-supplying. However, instead of supplying the city with water, Alfonso II preferred to employ an engineer such as Aleotti in the project of hydraulic automata for his urban villa of Castellina. Lino Marini, "Lo Stato estense," in *Storia d'Italia*, ed. Giuseppe Galasso, vol. 17, I Ducati padani, Trento e Trieste (Torino: UTET, 1979), 3-59; Patrizio Barbieri, "Ancora sulla 'Fontana dell'organo' di Tivoli e altri Automata sonori degli Este (1576-1619)," *L'organo: rivista di cultura organaria e organistica* 37 (2004): 187-221; Giambattista Aleotti, *Della scienza et dell'arte del ben regolare le acque di Gio. Battista Aleotti detto l'Argenta architetto del Papa, et del publico ne la città di Ferrara*, ed. Massimo Rossi (Modena: F.C. Panini, 2000), 15-16.

15   The problems Toledo faced in terms of getting a proper water supply were widely known and discussed by engineers for a century. For instance, a manuscript by Francesco di Giorgio Martini (1439-1501) contained an illustration representing a hydraulic wheel moving a pump supplying a fortified town with water. On the city-walls reads "*Toledo*". Galluzzi, *Gli ingegneri del Rinascimento da Brunelleschi a Leonardo da Vinci*, 28, fig. 21.

16   The of technology Marcus Popplow has calculated that a good quarter of all machines illustrated in Renaissance books were water-lifting devices. The taxonomy of water-lifting devices can change substantially according to the function one focuses on. Popplow, interested in technological complexity, individuates two types of Renaissance water-lifting devices: simple ones that worked thanks to muscular force (such as raising a bucket from a well by means of a winch), and complex ones that raised water continuously. The techniques implicated in this second type of device were scoop wheels, Archimedean screws, chains of buckets and force pumps, all mechanical devices already known in Antiquity. Cardinal Bessarion, who brought to Italy all basic ancient Greek texts for the

Archimedean screw and to discredit other competitors with their water-lifting devices. The list he drew is a precious catalogue to contextualize Torriani's inventions in Milan and Toledo.

To the first group Ceredi ascribed a number of devices, such as the machine of San Giorgio Maggiore in Venice, the monumental complex partially built by Palladio, and two Milanese examples very relevant to us: the machine of a certain "Milanese Philosopher" and one made at the church of San Pietro in Gessa. Ceredi wrote that as "futilely had done many wonderful minds", the mysterious Milanese philosopher had sought perpetual motion.[17] This unknown Milanese might be identified with that Brambilla mentioned – unlike Torriani – in all three editions of Cardano's De Subtilitate. Both Brambilla and Torriani may be behind the pump of the church of San Pietro in Gessa of Milan, in Ceredi's eyes similar to the previous one, but not that efficient. Ceredi adds that Ferrante Gonzaga ordered to build a similar one that was only different in what he called "the balance" moving the system, whose principle is described in the Mechanics of the Aristotelian corpus. If Ferrante ordered as governor (1546-1554) the construction of this second pump, this must have happened before the year 1554, when Torriani was still in Milan, and Cardano was still mentioning his name in the editions dedicated to the very governor. However, it is more probable to attribute this machine to Brambilla. Instead, there is another one from Ceredi's list that we may connect to Torriani: invented by Ctesibius, and transmitted by Vitruvius, and therefore here called the Vitruvian pump, which was so famous that, according to Ceredi, it made an appropriate topic for nobility. Ceredi added that, at his time, one could find many reconstructions of the Vitruvian pump in Milan, Venice, Genoa and Naples, and he stated that, except for his Archimedean screw, this was the most efficient type of pump.[18] This machine, added Ceredi, was also to be used in the project for the fountain of the garden of Parma, ordered by Duke Ottavio. The Flemish

---

so-called Mathematical Renaissance, found an already thriving hydraulic mechanical tradition in the peninsula: see Alex G. Keller, "A Byzantine Admirer of 'Western' Progress: Cardinal Bessarion ," Cambridge Historical Journal 11 (1955) 343-348. Marcus Popplow, "Hydraulic Engines in Renaissance Privileges for Inventions and "theatres of Machines"," in Arte e scienza delle acque nel Rinascimento, ed. Andrea Fiocca, Daniela Lamberini, and Cesare Maffioli (Venezia: Marsilio, 2003), 73-84.

17 Ceredi, Tre discorsi, 20.

18 He also added that the problem concerning these pumps was related to the engines required to power them. In fact they had to be extremely powerful. Consequently, because of the great mechanical stress provoked by the rapidity of the movement, these pumps were easily ruined. The mechanical stress, said the physician, was too strong even when instead of a reed one had a metal ball.

masters he had mentioned in another passage of his book were probably in charge of it. Ceredi added to this first group also a mill activating what seems to be a single piston pump extant at the Ancona harbour, and a chain pump inspired by Pliny the Younger to a certain Mario Pellegrino:[19] this latter was a machine particularly useful for drawing water at construction sites in swampy areas, and for armies, Ceredi commented. Then, he cited a vertical pump commonly used in the lowest deck of ships and in galleys described by Valturio and Flavius Vegetius and finally the bellows also described by Valturio in his 10th book of *De re militari*, at that time in use in three or four places of Milan. This bellows was very efficient and able to elevate a small quantity of water to any height, Ceredi wrote.

To the second group the Farnesie physician ascribed those machines activated by the very same weight of water. According to Ceredi, they were more durable than the ones described in the first group. The majority of hydraulic wheels listed came from the Venetian Republic, and Palladio had personally showed to Ceredi one of his inventions, which appeared to him to be the most efficient in this category.[20] To the second group belonged also a buckets-chain wrapped around two cylinders, which he claims to be a very expensive one, even advising against its use. However, he added, if one wanted to attempt building it, despite his discouragement, the engineer would have to avoid the mistake made previously by an anonymous "Cremonese" who designed an excessively large diameter for the external cylinder carrying the buckets in relation to the internal one transmitting the motion. It is a pity that Ceredi failed to quote the name of this engineer from Cremona. If this were to be identified with Janello Torriani, the most famous Cremonese of the time deal-

---

19   Perhaps related to Pellegrino de Pellegini Tibaldi.

20   This list included the following: (1) a Veronese wheel built on the river Adige. In order to fabricate such a wheel, one needs a river with a stable stream and fixed shores; (2) a Vitruvian tympanum – a modification of a scoop wheel. At Ceredi's time, there were such wheels placed in Lucia Fucina (seven miles away from Venice). The smallest one took the emerging water from the fields and poured it into the lagoon. The bigger one elevated water from the river Brenta, and it reversed it over the lagoon's embankment, where boats collected it and transported it to Venice's water-tanks. (3) Palladio's *timpano* or scoop wheel. Palladio himself showed Ceredi a project for a very efficient machine that had also been lauded by Marcantonio Barbaro, brother of Daniele (editor of a vernacular edition of Vitruvius, a copy of which Janello possessed in his library), to whom, Ceredi observes, because of his real merit, noble Venetians submitted almost all mathematical works. This Palladian invention is the most advantageous after the Archimedean screw, Ceredi says. Caccialupi, from the duchy of Parma and Piacenza, showed once a model built after Alfarabio's description that is similar to Palladio's but less perfect.

ing with the subject, we may have witnessed a very interesting attempt to discredit a competitor within the system of European courts. The type of machine resembles the one made by Torriani in the lower level of the Toledo device and the royal clockmaker was at the same time seeking privileges for invention throughout Italy in the very year of Ceredi's publication.[21] Finally Ceredi mentions two other devices, one made by the already mentioned Carlo Urbino from Crema, who was active in Milan, and the device made by a fellow countryman of his: Bossio from Piacenza, who had crafted a mysterious S-shaped noria, or bucket pump. In a cunning move of self-promotion, Ceredi ascribed to the third and last group his Archimedean screw only. The fact that its motion was supposed to go along with the natural movement of water, he argued, made it theoretically more durable and less difficult to activate.

It is interesting to note that Carlo Urbino, to whom is attributed the *Codex Huygens*, cooperated with Bernardino Campi, the friend of Torriani, and his patrons were people like Cardinal Taverna, in whose service Alessandro Lamo was employed as a secretary, and later Vespasiano Gonzaga, Duke of Sabbioneta.[22] This testifies that practitioners of hydraulic engineering like Brambilla, Ceredi, Urbino and Torriani were all gravitating around a group of patrons that would have easily provided them with the necessary links to know about each other's works and thus enhanced their competitive spirits. It is also remarkable to observe how hydraulic engineering was in the 1560s a field in which craftsmen, nobles, and humanists came together to discuss issues published in ancient texts and modern books, and Ceredi's book tells us how Milan and Venice were places *par excellence* where this process was especially taking place. Intellectuals such as Barbaro, noble patrons such as Ferrante Gonzaga, and Vitruvian craftsmen such as Palladio contributed to the field, drawing examples from Classical writers.[23]

---

21   Perhaps, but this is just a supposition, the functional discrepancy between Torriani's device and its model had reached Ceredi's ears. Only two years after the publication of Ceredi's work, Torriani's device would be officially tested, showing that it elevated much more water than calculated according to the demonstrations made with his model, a very convenient mistake for the buyer, after all. At this early stage, when Ceredi was writing his book (probably in the years 1566-1567), the favourable outcome was not yet evident. This mysterious Cremonese could have also been Janello's nephew Bernardino Torriani, who was lacking his uncle's skills.

22   See the chapter about Janello Torriani's networks.

23   Such as Vitruvius, Frontinus, the two Plinis and Vegetius, or modern authors of technical books such as Vergerio and Valturio.

Water technology had also a long tradition in Spain, starting with the legacy of Roman civilisation.[24] For what concerns the material Classical tradition in Spain, the presence of Roman aqueducts in the two religious areas of the peninsula, was relevant. The example of these ancient structures must have been an important source of technological inspiration.[25] Muslim rulers of Spain had cultivated this inheritance and sometimes had implemented it by importing know-how from elsewhere.[26] However, despite a few remains of such systems, no substantial records of them have survived. Christian settlers seem to have changed them completely, except for those enclaves (*alquerias*) of Valencia.[27]

---

24   Ignacio González Tascón and Isabel Velázquez Soriano, *Ingeniería romana en Hispania: historia y técnicas constructivas* ([España]: Fundación Juanelo Turriano, 2000); Paolo Squatriti, ed., *Working with Water in Medieval Europe: Technology and Resource-Use* (Leiden: Brill, 2000), 130.

25   New medieval aqueducts were built in Christian Oviedo and Santiago de Compostela (respectively ninth and eleventh century) and in Muslim Seville (twelfth century). The materials employed by Alfonso II at Oviedo's aqueduct were pottery-pipes running through stone structures or timber ones reinforced with iron and lead. Rudimentary aqueducts such as this were improved during the fifteenth century, when capacity was increased, or when ancient Roman aqueducts were restored, as in the impressive case of Segovia. García Tapia, "La ingeniería," 441.

26   Large-scale urban spaces are a feature of Islamic civilisation. Life in huge cities such as Bagdad, Cairo and Cordoba would have been impossible without water supplies and deep-rooted agriculture. It has been decisively argued that Islamic civilisation was not innovative in this field: methods of irrigation that had existed in Greco-Roman Antiquity in Mesopotamia and Egypt were adopted by the administrators of Arab caliphates and sultanates. However, this ancient legacy had an impressive diffusion in the Islamic world, far superior to Europe. Hill observes that in the vast Islamic world of the Middle Ages and early Modern Times, it was not possible to distinguish an irrigation system from that of a town water supply, because they were the very same thing. As in the case of ancient Rome and China, the water-systems were so big they needed the patronage of a wide and strong state. D.R. Hill, "Engineering," ed. Rushdī Rāshid and Régis Morelon, *Encyclopedia of the History of Arabic Science* (London; New York: Routledge, 1996), 751-755; Ciriacono, *Building on Water*, 26; Yaḥyá ibn Muḥammad Ibn al-'Awwām, *Le livre de l'agriculture d'Ibn-al-Awam (kitab-al-felahah)* ..., ed. Jean Jacques Clément-Mullet (Paris: A. Franck, 1864); Thomas F. Glick and Helena Kirchner, "Hydraulic Systems and Technologies of Islamic Spain: History and Archaeology," in *Working with Water in Medieval Europe: Technology and Resource-Use*, ed. Paolo Squatriti (Leiden: Brill, 2000), 267-329.

27   This may have brought historians, in search of something refined and complicated, to underestimate the importance of Islamic water technology in Spain. The simplicity of a productive but silent countryside had left little archaeological or written evidence. "*Technology alone is not the crucial factor in organizing an agro-ecosystem: knowledge, experience and the social organization of work are*": Ibid., 323 and 329. After the *Reconquista*, a

Already at that time, political and religious reasons made of the Islamic legacy in Spain a delicate and controversial issue. For instance, Gabriel Alonso de Herrera, considered one of the foremost Spanish agronomists of the sixteenth century, deliberately omitted the *Mudéjar* tradition from his works in favour of the Classical one.[28] After the *Reconquista*, local Muslim craftsmen carried out important works for their Christian masters and for their communities. Moreover, conversion played an important role in the conservation of local traditions despite political shifts: entire households involved with a workshop and its specific technical knowledge often chose to convert to remain.[29] A

---

shift in agricultural trends caused the shift from an Iberian crop-culture to a livestock-centred one, leading to the disappearance of the traditional Moorish irrigation systems. The Venetian ambassador Andrea Navagero, who was an amateur botanist and a refined humanist, helps us to understand what was happening in the newly acquired and Christianized Kingdom of Granada. Here Muslims were forced to change their lives or to leave the country; many had left their houses and possessions, and in the 1520s the Inquisition was in the process of moving into the city. Navagero documents a countryside that until a few decades previously had been cultivated by Muslim peasants, who had turned it into a garden. Every house was a part of this efficient and expansive watering system. By Navagero's time these dwellings webre abandoned and falling apart, leaving the countryside in a state of abandon. Cammy Brothers, "The Renaissance Reception of the Alhambra: The Letters of Andrea Navagero and the Palace of Charles V," *Muqarnas* 11, no. 1 (1994): 79-102.

28   By *Mudéjar* we mean those Spanish Muslims that remained after the Christian conquest. Later, when all Musilms were forced to choose between expulsion or conversion, those *Mudéjar* who did not leave became Christians and were now called *Moriscos*. Moorish contributions to Spanish culture have somethimes been neglected. An article about engineering in Renaissance Spain appears in the third volume of the "*Historia de la ciencia y de la técnica en la corona de Castilla*" but deals only with Christian technology. Nicholás Garcia-Tapia does the same, as partially has even done David C. Goodman. Gabriel Alonso de Herrera, *Libro di Agricoltura utilissimo, tratto da diuersi auttori ... Dalla Spagnuola nell'Italiana lingua traportato (per Mambrino da Fabriano).*, trans. Mambrino Roseo (Venetia: Michel Tramezzino, 1557); García Tapia, "La ingeniería," 435-65; Goodman, *Power and Penury*. Other historians such as Ciriacono and Brothers, on the contrary, are convinced supporters of the importance of a Moorish hydraulic inheritance within Mediterranean Renaissance Europe: Ciriacono, *Building on Water,* 26; Brothers, "The Renaissance Reception of the Alhambra."

29   The case of Seville is exemplary: when in 1502 a royal decree ordered the expulsion of all the *Mudéjares*, the local Muslim Sevillian family of pipe-masters converted to Christianity, changing its name to Hernández. This shift of identity provided this family with the chance of maintaining its privileged position, keeping its service to the royal *Alcázar*. Manuel Fernández Chaves, *Los Caños de Carmona y el abastecimiento de agua en la Sevilla moderna* (Sevilla: Emasesa Metropolitana, 2011), 43.

specific form of practical knowledge could therefore remain in a specific place despite dramatic political changes, thanks to individual strategies. Recent archaeological excavations have shown that even Janello, building his hydraulic machine in Toledo, integrated the local technological tradition, perhaps a *Mudéjar* one, within his innovative design: he erected a great waterwheel that lifted water from the river Tagus and he used an old mill to power the Toledo Device. Both wheels were set within a masonry structure that had existed since at least the thirteenth century.[30]

Besides Spain and Italy, the Habsburg Empire embraced other areas with a remarkable hydraulic tradition: the Kingdom of Germany and the other possessions of the Holy Roman Empire were among them. Ancient Roman aqueducts and the hydraulic tradition coming from monasteries were an inspiring source for the flourishing medieval towns of the Empire.[31] The production

---

30   Several *Mudéjar Norias* (or great waterwheels) are documented as having existed by the river Tagus at Toledo. One of them, active during the twelfth century, seems to have had the considerable diameter of a good 90 cubits, which consisted of around 40 meters: See: Juan Manuel Rojas Rodríguez-Malo and Alejandro Vicente Navarro, "The Contribution of Archaeology to Understanding Janello's Device," in *Janello Torriani, a Renaissance Genius*, ed. Cristiano Zanetti (Cremona: Comune di Cremona, 2016), 173-74; Glick and Kirchner, "Hydraulic Systems and Technologies of Islamic Spain," 308; Hill, "Engineering," 776.

31   It has been observed that during the Middle Ages, in the Kingdom of Germany, there was a difference between newly-founded towns and older ones: new towns had planned pipe-systems and channels for water-supplying and fire-prevention, whereas older towns were more reluctant to introduce such systems. Among the most significant cases of water supplying systems were Frieburg in Breisgau and Goslar in Saxony. Freiburg had a pipe-system taking water into the town across a river through aqueduct-bridges. Already in the year 1333 is mentioned an official in charge of the maintenance of the town-water-systems: an early date for such a development. The problem of deterioration of the materials employed in pipe-systems sometimes led to experimentation. For instance, in 1501, the master potter Ulirich of Saulgau made more then 5 kilometres of clay-pipes in order to replace the wooden conduits-system. Unfortunately, the pipes were not watertight. In the Saxon city of Goslar a small aqueduct was already serving the *Kaiserpfalz* since the eleventh century, and a public fountain, which still exists, was created at the very beginning of the thirteenth century, probably the oldest in Germany. The city water-system consisted of open channels and wooden pipes. Another city with an outstanding water-supplying system was Nuremberg. At the beginning of the nineteenth century it still had 8,500 kilometres of wooden pipes, plus 3,700 kilometres of lead ones. Extremely wealthy and powerful between the Middle Ages and the Renaissance, Nuremburg employed a number of public officials to take care of the waterworks. The most famous hydraulic feature of Nuremberg is *Schöner Brunnen* (reconstructed at the beginning of the twentieth century). This outstanding fountain was built by Heinrich Beiheim between 1385 and 1396. The fountain used the impressive quantity of a good 100 litres per minute: Klaus

of beer, one of the main aliments of the northern European diet, needed vast quantities of cold water. Already at the end of the thirteenth century, Lübeck had endowed itself with a water-elevating device. This consisted of a bucket-water-wheel and was given the significant name of *Brauwasserkunst*, i. e., Brew-water-fountain. During the second half of the fifteenth century, the bucket-wheel seems to have been replaced by a pumping-station constructed by a master builder: Hinrich Helmstede. In 1532 master builder Claus Moller from Hannover built a water-tower in Brunswick with four water-powered pumps (a structure he had already built for the Abbot of Lünemburg). The water was pumped into a pond placed at the height of 16 metres, a level that was higher than any other private building of the city. Any dwelling or workshop could be connected via pipes to the basin in order to receive water through gravity-pressure. In 1525 when Ägidien-Wasserkunst was built, Brunswick could be adorned with artificial fountains activated by water-pumps. Hannover had a bucket-wheel in 1532, but this device could work properly only when the river level was high enough to transmit the mechanical power of the stream over the wheel. Therefore, by 1535, a new device, a six-pumps-station, was built. Even Breslau/Wroclaw on the Oder had an artificial fountain during the Middle Ages. This was replaced in 1538 with considerable stonework by the river, where a 15 metres high wheel elevated 500 litres per minute. The foundations of this massive building rested on 12,300 oak poles. A second artificial fountain was added to the first one between 1529 and 1539. One of the most representative water-elevation devices was without doubt the wheel of Bremen, which, along with the Cathedral and the statue of Roland, was the pride of the city. This waterwheel was built in the last decade of the fourteenth century, and numerous documents about its construction and management have survived. There was an association that took care of the device, from its construction to the maintenance. In the end of the eighteenth century, the wheel was still supplying 450 taps. In another part of Germany, we find an interesting well-preserved sample of Renaissance water-supply structure: Bautzen today still maintains its 47-metre-high sixteenth century water-tower. This *Wasserkunst* was built in 1558 under the direction of the council's master builder Wenzel Röhrscheidt, after the medieval mechanical elevating system got destroyed. By 1523, Paderborn as well had its water-elevation-system: it was an undershot water-wheel activating a plunger pump. The water was stored into a 19 metres high pond.

Grewe, "Water Technology in Medieval Germany," in *Working with Water in Medieval Europe: Technology and Resource-Use*, ed. Paolo Squatriti (Leiden: Brill, 2000), 129-59.

To describe one of the most remarkable water-lifting systems of the Empire, that of Augsburg – a city visited by Torriani during the 1550s – we can count on a special witness: the French philosopher Michel de Montaigne. Since 1430, Augsburg had a medieval device that elevated water. The new water-supply system, the Water-Tower, was built in the 1540s. In 1580, Michel de Montaigne, on his journey to Italy, visited Augsburg and described this device as a 25 metres high tower equipped, it seems, with Archimedean screws activated by wheels powered by a stream of water that was carried on an aqueduct. This construction spanned a river and at the same time passed under a bridge. By the means of pipes, the Water-Tower fed many fountains and private houses.[32] This system allowed the Fugger, the most famous German banking family, to display sophisticated water-games.

A few days earlier, Montaigne was in Konstanz, where by the shore of the river Rhine he saw a huge covered building: it was a construction site for a pumping-station, ready to house a dozen wooden-wheels to elevate water to the height of a building's first floor. Then, two iron-wheels were supposed to elevate the water further up to the level of around 15 metres. From here, through an artificial canal, the water would feed numerous mills within the town. The master in charge of this device was also preparing a mechanical system in order to raise or lower the wheels according to the river's level, something that Janello had also provided for in his Toledo Device. The payment asked by this master engineer, according to Montaigne, was 5,700 golden florins, besides the wine. Though it is not clear how many years this remarkable wage was supposed to cover, Konstanz and the other cities of the Empire constituted a good market for hydraulic engineers. For instance, we know that in the year 1545 a certain Hans Hedler obtained a privilege for inventions from Charles v: his water-lifting devices were in use in different parts of Austria and Bavaria.[33] Not far from Konstanz, Montaigne could also see watermills powered by a stream of water accelerated by an artificial canal inclined in a special way.[34]

Still within the borders of the Empire was the cultural region of the Low Countries, set at the estuary of the river Rhine, a mosaic of feudal entities (nowadays mainly Netherlands and Belgium's Flanders and Brabant) which during this time was considered one of the most advanced European centres

---

32   Montaigne reports the prices for such a service: 10 florins per year or 200 florins *one time*.

33   Popplow, "Hydraulic Engines in Renaissance Privileges for Inventions and "theatres of Machines"," 81.

34   Michel Eyquem de Montaigne, *Viaggio in Italia*, ed. Giovanni Greco and Ettore Camesasca (Milano: Rizzoli, 2008), 142.

for large-scale water-technologies. Just before Charles V's abdication, between 1554-1556 Janello Torriani lived in Brussels, the capital of the Duchy of Brabant, for long periods at the imperial court. The medieval hydraulic history of the Low Countries is connected to the two actions of land drainage and flood control already started by the monks during the high Middle Ages.[35] Since the sixteenth century, Dutch hydraulic engineers were ubiquitous across Europe: from Germany to France, and from England to Poland and Italy as well.[36] According to Ceredi, Flemish engineers had built different water-raising machines that of course he claimed to be less efficient than his Archimedean *coclea*: a certain

> Michele of Liegi describes in a very elegant poetry (exceptionally impressive for a Barbarian!) the machine used in Flanders in order to obtain drinkable water. However this machine as well needs a strong motor for a small quantity of water elevated: insufficient for irrigation ... A similar machine existed on the Rhine, where the river powered it. This machine was described by a very excellent author in a print in Louvain.[37]

Ceredi also refers that his lord, the Duke Ottavio Farnese, who had "said that Flemish are the best in these issues like the ones working at the machines of the fountain of his palace in Parma".[38]

---

35   Around the eleventh century waterlogged peat lands and lowlands began to get drained on a large scale by means of ditch digging. Already in the fourteenth century, wind-power was used to drain water from the ditches dug into the swampy areas of the region. In Alkmaar, in the North of Holland, Floris van Alcmade built in 1408 the first windmill of the *wipmolen* type (a mill with rotating head that could catch the best of the wind even when this was changing direction). William H. TeBrake, "Hydraulic Engineering in the Netherlands during the Middle Ages," in *Working with Water in Medieval Europe: Technology and Resource-Use*, ed. Paolo Squatriti (Leiden: Brill, 2000), 101-27.

36   In the seventeenth century, Holland finally overtook Italy as the leading exporter of hydraulic engineers. Ciriacono, *Building on Water*, 196.

37   The ethnonym "Flemish" applied in Southern Europe to both today's Flemishand Dutch as well as other ethnic groups at the time under the duke of Burgundy (my translation).

38   Ceredi, *Tre discorsi*, 3-19 (my translation). The *tjasker* (an Archimedean screw connected with a windmill) was patented by the end of the sixteenth century, a couple of years after the arrival in Flanders of Alessandro Farnese as governor, to whom Ceredi's book about the *coclea* was dedicated. Did the Farnese export Ceredi's book to the Netherlands? Ciriacono claims that the Archimedean screw connected with windmills was already used there by the middle of the century. I could not find the source for such a claim. It seems to me that for the moment it remains an open question whether Ceredi's book had any influence on Dutch hydraulic engineering.

Outside of the vast Habsburg empire, the British islands and the kingdom of France presented a similar but less innovative landscape in hydraulic engineering.[39] In conclusion, we can say that Renaissance Europe, and especially many lands under Habsbourg control, constituted a fertile ground for water-supplying enterprises. Networks made of monasteries and other religious institutions, universities, and large-scale monarchies such as the Habsburg's, made of Latin Christendom a unified cultural area for the application, improvement and circulation of water technologies. Lay and ecclesiastical princes and wealthy cities were the main patrons of such projects. Janello Torriani's waterworks in Toledo were not an isolated enterprise.

Toledo: a Paradigmatic Stage for Renaissance Water-Technology

When medieval and Renaissance Spanish towns did not have functioning aqueducts or mechanical water-supply systems to feed their fountains, the *aguador* was employed. The *aguador* was a man who, often with the help of a donkey, carried water to the places where he could sell it. The water was poured into uniformly sized vases filled directly in the river.[40] This was the main system for

---

39   Until 1620, when the Arcueil aqueduct was constructed, Paris was supplied only by the meagre spring in Belleville. The pipe system, made out of lead and terracotta, bypassed a canal and entered the city. The heat of the summer, the frost of the winter, and the too numerous pipe-connections to the residences of the nobility (whose status required an increasing availability of water) impoverished the flow of the small spring to the effect that the numerous population had to search for water elsewhere. Of course wells were the traditional source for water, but Paris had a water-table extremely polluted by heavy urbanisation. Therefore, the favourite source for drinkable water remained the river Seine, like the river Tagus in Toledo or the river Tiber in Rome. In 1606 Jean Lintlaer built a pump system under Pont Neuf. These water-wheels provided water for the royal palaces of Louvre and Le Tuileries. Because of the abundance of rain, water-elevation-systems appeared in England very late. This happened in highly urbanised centres like London, probably in order to avoid polluted water from the city-wells. In 1582, a certain Peter Metris built a wheel-powered pump that elevated water from the river Thames up to the level of the top of a bell-tower. Only in 1613 was London's New River canal dug. It was a 30-km-long man-made waterway created to supply the capital of England with freshwater. Richard Holt, "Medieval England's Water-Related Technologies," in *Working with Water in Medieval Europe: Technology and Resource-Use*, ed. Paolo Squatriti (Leiden: Brill, 2000), 51-100.

40   José Ortega Valcárcel, "El microcosmo humanizado," in *Historia de la ciencia y de la técnica en la Corona de Castilla*, ed. Luis García Ballester, 1, Edad Media (Valladolid: Junta de Castilla y León, Consejería de Educación y Cultura, 2002), 349-50.

providing water to a city like Toledo, which was situated on a spring-less hill, and where, beside the *aguadores*, the only sources of water were deep wells dug into the rock, ice-houses and rainwater water-tanks.[41] Cervantes made famous the *aguadores* of Toledo with the novel *La Ilustre Fregona*: their function remained of basic importance for Toledo before and even after the time of Torriani.[42]

Toledo lies on the top of a hill closed on three sides by the sinuous curve of the river Tagus. Formidable fortifications once enclosed its skyline. Toledo's central position in the peninsula made it one of the favourite places of Charles V and his ancestors. It was a populous urban centre, the biggest in the Meseta. However, the scarcity of springs and wells in its environs placed the water-supply problem firmly on the agenda of the city administration. Even Saint Teresa of Ávila lamented the lack of water when she stayed in the city.[43] And it was also possible to witness, as in 1576, an official of the city government, the *alférez mayor de Toledo*, requesting the nomination of commissaries charged with convincing the priors of the city's monasteries to pray against the lack of water, which was "affecting the quality of the bread".[44]

Torriani managed to build his first efficient device between 1565 and 1569. However, this was not the first time a reconstruction of the water supply system to Toledo had been considered. In the city there was still memory of the ancient Roman aqueduct bringing water from far away springs, and perhaps also of the medieval hydraulic wheel elevating water from the river up to an aqueduct-bridge.[45] At the time of Torriani only the *aguadores* remained to

---

41   Alfredo Aracil, *Juego y artificio: autómatas y otras ficciones en la cultura del Renacimiento a la ilustración* (Madrid: Cátedra, 1998); Santiago Cantera Montenegro, "Los usos del agua en las cartujas de la corona de Castilla, en la transición del Medievo al Renacimiento," in *El medio natural en la España medieval: actas del I Congreso sobre Ecohistoria e Historia Medieval*, ed. Julián Clemente Ramos (Cáceres: Universidad de Extremadura, 2001), 257-75.

42   A decree of the *Ayuntamiento's* session ordering the restoration of the path parallel to the water device, descending to the river, mentions "*el camino que baxa al rrio por donde se provee de agua a esta çiudad*": 1st of December 1574: Cervera Vera, *Documentos biográficos*, doc. 42.

43   A description of the city of Toledo from the beginning of the sixteenth century written by the Italian Marineo Siculo states that there were around 4000 private wells. The contradictory report might have been the result of an exaggeration. It is also possible that these wells were not very efficient. Montemayor, *Tolède entre fortune et déclin (1530-1640)*, 17; Moreno Nieto and Moreno Santiago, *Juanelo y su artificio*, 59.

44   21st of March 1576, Toledo: Cervera Vera, *Documentos biográficos*, doc. 73.

45   This information is taken from the *Geographie d'Eldrisi*, a medieval arabic work, translated into French for the first time in the nineteenth century: Antonio Martín Gamero,

take the water up to the top of the hill where Toledo's most important build-ings were located. Records from the end of the fifteenth century show that at that time there was already a plan to build a machine to bring water into the city. In the year 1485, the commander of the Order of San Juan expressed his wish to convey water from the river Tagus to Zocodover square, only 200 metres from the *Alcázar*.[46] The restoration of the Roman aqueduct would have been too expensive for Toledo, which, unlike Rome or Segovia, did not have ancient aqueducts still in good condition. Naples, as Toledo, could not afford the resto-ration of its ancient Roman aqueduct. In the middle of the sixteenth century, the cost for the restoration of a 43 kilometres long Roman aqueduct was esti-mated at the enormous sum of 185,000 ducats. The high cost prevented the realisation of the project.[47] Toledo's Roman aqueduct was a similar length (38 km).[48] Local administrations had scarce resources, and in order to complete such projects, they often had to raise taxes. This brought new conflicts with local groups of power.[49] The cost for a mechanical device, which could elevate water from the Tagus appeared to be far cheaper than such a project of restoration.

Toledo offers a good example of the importation of Renaissance hydraulic technology from different European regions to Spain, and in this specific con-text, the King, members of his court and of the *Ayuntamiento* or City Council were the motor of attraction.[50] The strategy of the Habsburgs was to employ

Aguas potables de Toledo, ed. Gabriel Mora del Pozo, Clásicos toledanos, v (Toledo: Insti-tuto Provincial de Investigaciones y Estudios Toledanos, 1997).

46   Moreno Nieto and Moreno Santiago, *Juanelo y su artificio*, 53-54.

47   Montuono, "Approvvigionamento idrico della città di Napoli," 1036.

48   Montemayor, *Tolède entre fortune et déclin (1530-1640)*, 58.

49   For example, in the year 1536, the City Council of Ciudad Rodrigo (Castile and Leon) added new taxes to the consumption of wine in order to raise funds to restore the aque-duct. The clergy, thanks to an old privilege, challenged this disposition and refused to pay. A further insistence of the local administration led to the excommunication of the mem-bers of the Council. García Tapia, "La ingeniería," 441.

50   Castilian cities were ruled by a *corregidor*, that is, a co-governor named by the King and by a Council called *Ayuntamiento*. See the on-line version of: Esperanza Pedraza Ruiz, "Cor-regidores toledanos," *Toletum* 8 (1977): 153-75. During the sixteenth century, monasteries, missions, universities, city councils, guilds, workshops, princes and their courts, acade-mies, public libraries, merchants and armies were responsible for the circulation of tech-nical knowledge, transferred by men with their practical expertise, by books (manuscripts and printed ones), by scientific drawings, often inserted in books as para-texts (by scien-tific drawings I refer to technical and descriptive images based on the technique of math-ematical perspective and naturalistic representation), and by scientific objects. The circulation of technological knowledge could also follow the pathways of religious and

their vassals from Spain, Germany, Burgundy and Italy in the administration of the empire. These vassals brought their servants along with them, and among the entourages there were also engineers. In the early 1520s, after the repression of the *Communeros*, Emperor Charles V rewarded his *camerero mayor*, or upper chamberlain Heinrich III (1486-1538) Count of Nassau-Dillenburg, Lord of Breda, Lek, etc.[51] with the title of Marquess of Cenete, which he acquired through marriage.[52] Two years later, in 1526, according to a contemporary record, Nassau-Cenete brought to Toledo some servants, fellow-countrymen of his from the region of the lower Rhine (it is not clear whether from what is today Germany or from the Low Countries), who built a device for elevating water up to the point of Zocodover square, the centre of commerce and social life in Toledo.[53] The same story, but dated this time to the year 1528, is reported by royal historiographer and Torriani's acquaintance Esteban de Garibay y Zamalloa in his voluminous book about the history of the Kingdoms of Spain.[54] Garibay reported how just one servant of the Nassau had set a system of mills by the river Tagus in order to provide a machine with the necessary power to

---

curative pilgrimages, curiosity, trade, marriages, colonial administration, slavery, flights from persecutions and war.

51   This powerful family's most famous scion is the future Netherlands' *stadtholder* and leader of the Dutch uprising, Wilhelm the Silent of Orange-Nassau, son to the Marquess of Cenete's brother. The house of Nassau was divided between numerous branches mainly pivoting around the lower Rhine. This area of the Empire owns for the most part its fertility and richness to the abundance of fresh water. The vast net of rivers provided this region (especially the districts of today's Netherlands) with relevant strategic advantages as it served as an important means of both communication and defence, as well as an inexhaustible source of fresh water for agriculture, urban life, commerce and industry. In fact, the flow of these large rivers offered an easy source of energy. Because of this, as previously said, the area was one of the most advanced regions of Christendom for hydraulic-technology development. It is therefore not a surprise that Nassau's patronage could first bring to Castille a know-how originated from such a geographical context. A further investigation into the structure of the household of Heinrich III of Nassau-Breda might help in discovering the identity of these hydraulic mechanics. I am planning to visit the *Arxiu del Palau Requesés* in Sant Cugat del Vallès (Barcelona), where the archives of the "Marquesado de Cenete" are kept.

52   Mencía de Mendoza, daughter of Maria de Fonseca, daughter of Alonso Fonseca de Toledo and Maria de Toledo: <http://www.grandesp.org.uk/historia/gzas/cenete.htm>.

53   From the receiving report book of the Monastery of the *Conceptión Francisca*, quoted by Sisto Ramon Parro, *Toledo en la mano* (Toledo: S. Lopez Fando, 1857).

54   Esteban de Garibay y Zamalloa (Mondragon 1533-Madrid 1600), a friend of Janello Torriani, was an important Basque historiographer of the time. He ceased to write his history of Spain in 1566. Garibay y Zamalloa, *XL libros*.

elevate water up to the *Alcázar*.[55] This machine consisted of mallets (most likely pistons), which beat the water so violently that the liquid "*a puro impetu*" rose up through the pipes. In all probability, this was a system of forcing-pumps, activated by the stream of the Tagus. This machine must have had a series of pumping stations positioned at different levels between the river and the top of the hill where the city lay, because water cannot physically rise when elevated by these kinds of pumps for more than some 10 metres.[56] Moreover, the pressure needed to raise the water was too high for the metal pipes through which the water was pumped. Garibay writes that the creator of this device asked then to be provided with a better steel from Mondragon, in order to cast new pipes in this more resistant material. Despite this measure, the device did not work. The lower part of it was even destroyed by a river flood. This was probably the same area where Torriani would build his devices, close to the bridge of Alcántara, where the river is cleaner because not yet polluted by the dumping of the city's rubbish.

In the year 1553 Charles' son Philip, while still a prince, gave authorisation to a certain Sebastián Navarro[57] (perhaps a subject of the crown of Navarra) to attempt a similar enterprise. A few years later, during the early 1560s, when Philip was now King, new efforts were directed to the solution of this technical problem: a plate without date, held in the archive of Simancas, shows a naive sketch of a water-supply system for Toledo: a chain of towers, probably shells for pumping stations, connect the Tagus to the Royal Castle: the *Alcázar*.[58] This drawing demonstrates a renewed interest in the project in these years: indeed, in 1561 there was a new attempt to build a water-raising device.[59] The King, who needed water for his *Alcázar*, urged a project for a new device to be designed. The City Council of Toledo seemed to be extremely cautious about the plan. And it had good reason to be so, as we shall see. In this agreement preceding Torriani's, the time for the creation of the machine was set as three years. The scepticism of the local administration was based on both technical

55   In the book by Luis Moreno Nieto and Ángel Moreno Santiago the authors quoted the description by Francisco de Pisa, who published his *Description de la Imperial Ciudad de Toledo* in 1605. This passage is a literal transcription of what Esteban de Garibay had written some 40 years before (although his work was published only in 1571). Meanwhile, Janello was building his first device. Ibid.

56   Shapin, *Scientific Revolution*, 38-39.

57   Moreno Nieto and Moreno Santiago, *Juanelo y su artificio*, 54.

58   Nicolás García Tapia, "Los ingenieros y sus modalidades," in *Historia de la ciencia y de la técnica en la Corona de Castilla*, ed. José María Lopez Piñero, vol. III, Siglos XVI y XVII (León: Junta de Castilla y León, Consejería de Educación y Cultura, 2002), 157.

59   Cervera Vera, *Documentos biográficos*, doc. 2: 13th and 17th October 1561, Toledo.

problems and the elevated cost. However, the King showed his determination to realize the project claiming that, after the construction of the first device (if it was going to work properly) he would provide his own money for the payment of the construction of a second one in order to match the quantity of water determined in the first contract with the city. According to the agreement, the water had to be used first for the *Alcázar's* needs. What remained would be for Toledo. Yet the King added in his own hand that he could get more water for his castle, should the need arise. Here lies the root of a conflict, which Torriani would have to face.

Unfortunately, according to technical calculations, the royal technicians judged the pipes far too big: the pressure of such a great flow of water would have cracked them, and the project had to be changed. The vertical sections of the previous pipe and of the new proposed one were sketched together with their measures.[60] The reduction of the gauge was to decrease the hypothetical quantity of supplied water too, requiring the construction of a second device. As a result of the important changes to the original agreement, the *Ayuntamiento* of Toledo judged that a ratification was needed: the representatives of the city stressed that the costs for the construction for both the maintenance of the water device and the salaries of the engineers were far too high to be paid as decided in the first contract.[61] Despite such concerns, in the following years a number of scattered and cryptic documents testify that the enterprise was proceeding.[62] A new attempt was then made by two northern European mas-

---

60   Not reported in Cervéra-Véra's transcript. Maybe "quantity and measure" (*cantidad y medida*) were to be inferred from the relation between the geometrical proportions of the vertical sections sketched in the document.

61   Unfortunately the names of these engineers are not mentioned in the document.

62   A note three lines long, probably dating from the year 1562 (this is what supposes Cervèra Vèra), mentions that a certain Luis de Vega, perhaps the architect of the royal palace of Valladolid, was working on the problem of the *Alcázar* water supply, and "*if he will manage it is a big deal*" (*Luis de Vega según me ha dicho tiene concertado su nivel y dize que cree dará nivelado en toda esta semana lo del agua que ya ha de venir al alcázar, que si alcancase seria gran cossa*: Cervera Vera, *Documentos biográficos*, doc. 3). A royal officer with some authority in the works' organisation most probably wrote the note. However, it is not clear, from Cervéra-Véra's transcript, whether the *Alcázar* here quoted is the one of Toledo. Another note dated May 1563 was written from some royal informers to the King about the water-supply problem: "*One thing to raise water in Toledo*" ("*una cosa de subir el agua a Toledo*"). In the note one can read that a certain Robles had exposed this issue to the King, who decided to deal with the problem a second time (May 1563, Posada de Algora, *Ibid.*, doc. 6).

ters: the painter Joan de Coten and *maestre* Jorge from Flanders.[63] However, the endeavour must have failed since on 15th of September 1564 Philip II ordered a new project. As previously seen, in the year 1563 Janello had just received from the King an inventor privilege for a water-lifting device. The new royal order was sent from Madrid by means of a special royal decree: with this document the King ordered the payment of 300 ducats to the French clock-maker Louis de Foix a *criado* (i.e., a servant at court), for some models of devices for the water supply in Toledo.[64]

As discussed above, the 1560s were the starting point for a great boom in hydraulic engineering. The peace treaty of Cateau-Cambrésis in 1559 and the great hydraulic projects of leading patrons such as the Pope and the King of Spain created a lively market in which engineers competed to sell their ideas. Nobles and officials were most of the time the brokers who made it possible for fellow-countrymen technicians to access this market.[65] Many European princes attracted the best craftsmen of Christendom, with, however, a strong preference for those coming from areas politically connected to their own. Thus it is not surprising to find in Toledo, in addition to Spaniards, Flemish, Germans and Italians, who were all natural subjects of King Philip II, even a Frenchman such as Louis de Foix, who was most probably part of the entourage of Elisabeth of Valois, the new French Queen of Spain.

The failure of all previous attempts gave Janello Torriani his chance to win new fame and (in his dreams) great wealth, yet the danger he faced was also great. The agreement to which Torriani subscribed in 1565 in Toledo presented a high level of risk for the clockmaker. The contract stated that Janello had to build his device without any economic help either from the King or from the City of Toledo. If the device failed to meet the prerequisites mentioned in the contract, Janello would not be rewarded, and all the expenses would have been his own loss. On the other hand, in case of a successful realisation of the device

---

63   Juan de Coten is registered as a court painter for the year 1561: José Martínez Millán and Santiago Fernández Conti, *La monarquía de Felipe II: la casa del rey*, vol. II (Madrid: Fundación Mapfre Tavera, 2005), 550.

64   Cervera Vera, *Documentos biográficos*, doc. 11.

65   In Italy, beside Italian engineers, we find Northern European ones. Alongside the already mentioned Flemish water-masters working for the Duke of Parma, we have in Lucca the Flemish William Raed de Bolduc, who made a project for vast land-reclamation in the year 1577. In the villa D'Este the Neapolitan Pirro Ligorio (ca. 1510-1583) and the French fountain-makers Claude Venard and his uncle Luc le Clerc created spectacular waterworks. The Frenchmen built the famous hydraulic organ automaton between 1567-1568. Montaigne, *Viaggio in Italia*, 402 and 496. Barbieri, "Ancora sulla 'Fontana dell'organo' di Tivoli e altri Automata sonori degli Este (1576-1619)."

(to be tested by official inspectors), the King and the city council would pay one time in equal parts to Torriani the sum of 8,000 ducats.[66] After that, the City of Toledo had to pay 1,900 ducats annually to Janello and his heirs for at least the next 30 years.[67] In case of a successful outcome of the enterprise, Janello had to take care of the device, providing for its maintenance and repairs at his own expense. Torriani was committed to provide a constant flow of water, and in case of breakage he had only six days to repair the device. Expropriations and damages to third parties during the construction had instead to be paid by the King,[68] who was supposed to use only one seventh of the water, with the remaining six parts for the city. Any remaining water not used by the King or the city could be sold to a third party. In the contract, the King committed himself to help the city with the necessary permissions and help fulfil all the contractual agreements.[69] From a technical point of view we can summarize the requirements in the agreement as follows:

1 – the water had to reach the square outside the castle and had to be delivered some around 9 metres from the *Alcázar* itself;
2 – the water had to arrive through four pipelines;
3 – each pipe had to pour 16 pounds of water;
4 – the pipes had to pour water alternatively one per time, so that the delivery of water would be regular;
5 – Torriani, once paid, had to take care of the problem of urban distribution of the water throughout the main streets and their secondary branches.
6 – in order to better serve the (i. e. the *Alcázar* Royal Castle) with the necessary water and to direct the rest to any other part of Toledo, the Community committed itself to pay for the construction of a pond to collect the water in the square in front of the *Alcázar*. This cistern had to measure 16 feet in height (around 4 meters and 20 cm).[70]

---

66   This amount of money, to be paid 15 days after the successful test, had to be considered comprehensive of all the expenses incurred by Janello during the construction. Janello was then obliged to present a list of the expenses and to swear to their veracity, and he was also supposed to furnish the device with brick-walls at his own expense, unless the King and the City would delay the payment of the 8000 ducats.

67   16th of April 1565, Madrid: Cervera Vera, *Documentos biográficos*, doc. 13.

68   This action was probably easier if made in the name of the crown than in the name of the community itself, whose members were going to suffer from the expropriation itself.

69   Ibid.,16th of April 1565, Madrid, doc. 13.

70   Ibid. For a partial Italian and English translations of the document, see: Zanetti, *Janello*

In a letter to the King, written some years later, Janello would recall the three main stages of the construction of the first device: on the 23rd of March 1565 the contract was signed by Janello himself, the King and the representatives of the City Council of Toledo; on the 23rd of January 1569 the construction was ended according to the agreement; and on the 13th of June 1569 the *corregidor* and other members of the *Ayuntamiento* visited the device and recognised that everything was working in a satisfactory way.[71] Janello's invention of a system of oscillating towers was an accomplishment of a feat that had proven impossible for others. With a letter dated to the 1st of October 1570, Philip II made Don Diego de Zuñiga,[72] *corregidor* of Toledo, responsible for the care of the device, and he communicated to him that he had also ordered Janello to add a new pipeline to the device and to project the construction of two similar mechanisms. The efficiency of the first one must have convinced everybody if such a work of replication was put into action.[73] However, despite the successful enterprise, the city of Toledo did not honour the contract. Why?

Torriani, genio del Rinascimento and its English version, *Janello Torriani, a Renaissance Genius.*

71   Ibid. 12th of December 1573, Madrid: doc. 32.

72   Ibid. In the document: Don Diego de Çuñiga

73   Ibid. 1st of October 1570, El Escorial: doc. 20.

CHAPTER 9

The First Global Empire Produced the First Giant Water-Machine

Qui Bono? Janello Caught between the Devil and the Deep Blue Sea

Water is, was, and will always be a source of conflict. And so it was in Toledo, where the scarcity of this resource made it even more valuable. Scholarship has long emphasised the link between Torriani's water-elevation device and this city; but this endeavour has to be understood more as a contribution to Philip II's program of self-celebration than as a service for the community of Toledo.[1] In addition to navigation, the river Tagus, from which the water was taken, was also used for irrigation, to power mills, to supply drinking water, for waste disposal – for nearly all activities, from husbandry to a housewife's daily chores.. Any variation to the delicate balance customarily established would be a source of conflict. The Crown, with its gargantuan projects such as Janello's device and attempts to make the Tagus more navigable threatened this balance. Only recently have some clues, which emerge from the administrative documents of the period, revealed the problematic context in which the construction project had taken place. These vaguely mentioned troubles are to be considered the prelude to a long *querelle*, which would haunt Janello until the day he died.

Since 1561, King Philip II had preferred to hold his court in Madrid instead of Toledo. If not a proper capital, Madrid became one of the main political centres

---

1 Since Cervera-Vera published Torriani's documents about the construction of the Toledo Device in 1991, historians have a very well documented Renaissance technological endeavour to work on. So far, Daniel Damler has been the only one to have looked at this collected material addressing questions about the institutional conflict behind Janello's Toledan experience. Damler has noted how the King took all the water of the devices and how Toledo was forced to enter in the business. However, Damler claims that the City of Toledo was reluctant to back this project because of a general opposition of the City towards an enterprise perceived more as a pointless amusement than a useful machine, and because of the influence exercised by the local *aguadores*. Damler does not take into account that the city of Toledo tried to hire independent engineers for the construction of a water-supply system alternative to the royal ones (see below). For what concerns Damler's second point, it is difficult to believe that the 40,000 daily litres of water that the Device could elevate – though repairs and other problems make this figure a too optimistic one – would have jeopardized the income of the *aguadores* in a city of 60,000 inhabitants. Damler, "Modern Wonder."

of Spain. As seat of the court of the most powerful kingdom of Christendom, Madrid needed a suitable display of power: a crown of monumental royal residences was set around it. El Escorial, El Pardo, Aranjuez and the Royal *Alcázares* of Madrid and Toledo, just to mention the most visible, were cast in the new shape of a "modern" display of power, which was made of a mix of Habsburg princely traditions and the powerful rhetoric of the Renaissance. Among these *Sitios Reales*, Toledo offers the possibility to observe from a close distance how Philip's agenda of power-representation could be carried out within a large and ancient urban space. Two political powers, the King with the court and his officials on the one side, and the city administration with its representatives on the other, entered into a conflict that reaffirmed the general hierarchy on a local level. It is interesting to observe how, in this conflict, the head of the city was the Governor, or *Corregidor*, who was however nominated by the King. In this ambiguous system of power, it seems that the King used Torriani as an instrument to gain advantages at the city's expense. Once Janello was not rewarded as stated in the contract, he found himself too weak to impose any *ultimatum* on the city or the crown. Nevertheless, he was also too close to the King to be completely annihilated; though he was used as a scapegoat, it seems that his status of royal servant made of him an untouchable.

In the letter from the year 1570 in which the King communicated the order to double the first of Janello's machines and to construct two new devices for raising water – something that remained just a project – he also ordered the Governor of Toledo to resolve a number of problems. Indeed, Janello was said to have complained about the fact that there were people taking water unlawfully from the river, through the means of breaking the canal of alimentation of the hydraulic wheel of his mechanism. In this way they were damaging the first device, which consequently could not move as fast as it was supposed to. The King then ordered the *Corregidor* to solve the problem and also to forbid dumping around the Device.[2] On the other hand, in the same letter, Philip apologised to the same governor of Toledo for the inconveniences and difficulties that occurred to the city in relation to the construction of the first Device: the two new ones were meant to repair the damage, supplying Toledo with

---

2   1st of October 1570, El Escorial: Cervera Vera, *Documentos biográficos*, doc 20. These two concerns have to be related to a practice that constantly threatened the town of Toledo: often rich families tried to divert water from the course of the river upstream in order to water their possessions. This lowered the water level, which was bad for the mills. The other problem was related to water-pollution, when people dumped rubbish upstream of the town. Water contamination was of course a cause of disease: Julian Montemayor, *Tolède entre fortune et déclin (1530-1640)* (Limoges: PULIM, 1996), 59.

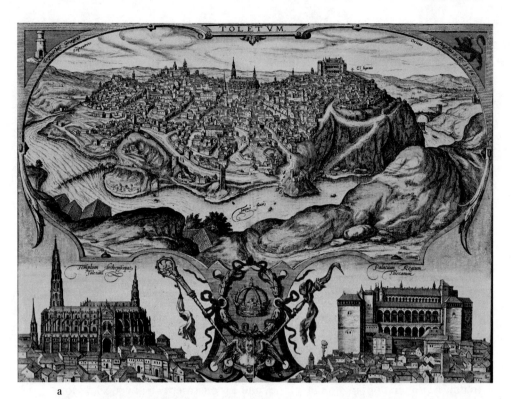

a

b

FIGURE 54a-b   View of Toledo and detail of Janello's Device: el Ingenio, *Georg Braun,*
Civitates Orbis Terrarum, 1598. COURTESY OF THE BIBLIOTECA DE LA
FUNDACIÓN JUANELO TURRIANO, MADRID.

water, he said.[3] In spite of not being openly stressed, it is likely that the problem was indeed the King's hoarding of the entire amount of water delivered by the first Device. Indeed, the agreement stated that only one seventh of the entire amount of water had to go to the King, while the rest was designated for the city. It seems that Philip II took all the water in blatant violation of the contract, as the worrying line added by the King's hand had probably predicted to a careful reader. This occurrence was the origin of a long-lasting litigation between Janello and the city of Toledo. Janello, as seen before, committed himself to financing the construction of the first Device. In order to do that, he had to ask for loans. At the end of the construction he had borrowed 8,000 ducats. After four years the interest Janello had to pay on this amount of money amounted to 7,000 ducats, for a total loss of 15,000 ducats.[4]

A royal official, in a document written around the year 1573, said that Janello had been begging the King to pay for the first device and to spur the community of Toledo to finally pay his salary since the day he brought water to the square of the *Alcázar* for the first time four years earlier.[5] Despite the fact that

---

3   1st of October 1570, El Escorial: Cervera Vera, *Documentos biográficos*, doc. 20.

4   Ibid., doc. 18 (1569, 4th of January, Toledo): on this date Torriani signed for a loan of 1,580,013 *maravedis* (a little more than 4,200 ducats) with Ippolito Afetate, Jacobo de Bardi, and Xpoval Riba (in Italian: Ippolito Affaitati, Giacomo de Bardi, and Cristoforo Riva). Ippolito Affaitati was a member of the powerful Affaitati family from Cremona. This common territorial background of Janello and the Affaitati, if not accidental, raises a question: did Philip II's financing policy push the groups of power present at court into supporting their former protégées now in service of the king? Were the Borromeos-Gonzagas-Medicis di Marignano who silently asked to stand surety for Torriani by the bankers Affaitati? Or was Ippolito Affaitati, under the sword of Damocles of the next Spanish bankruptcy, indirectly threatened to lend money to Janello without hoping to see it returned as fast as it should have? See the Affaitati bankers and merchants in Europe: Jean Denucé, *Inventaire des Affaitadi, banquiers italiens à Anvers de l'année 1568* (Anvers: Éditions de "Sikkel," 1934); and María Emelina Martín Acosta, "Las remesas de Indias y la política imperial," in *Dinero, moneda y crédito en la monarquía hispánica : actas del Simposio internacional Dinero, moneda y credito. De la Monarquía hispánica a la integración monetaria europea : Madrid, 4-7 de mayo de 1999,* ed. Antonio M. Bernal (Madrid: Marcial Pons : Fundación ICO, 2000), 413-4.

5   Ibid., Janello was also said to be willing to do with the best of intentions what had to be done for the Devices because it was for the service of the King. Moreover, he committed himself to producing documentation (*recaudo*) about his work and industry. Torriani – says the document – did not ask for anything more than what the King wished to give him after seeing the benefit he received from the clockmaker's work. In the end of this letter, Janello is described as having pleaded with the King to pay his creditors before the interest would grow to the extent that he would undoubtedly be ruined: Cervera Vera, *Documentos biográficos*, doc. 17. The first evidence of works in the second device are mentioned in the last paragraph of doc. 44, which has to be dated to the year 1574.

Janello had honoured the contract executing his device as required by the pact, he did not see any money. Nevertheless, he went on working. Every day he lost even more because he had to repair the building continuously and he had also to pay the wages of three officials who were daily running the Device. However, in order to maintain his word, he protested, he had dealt with the problems and kept on working. Moreover, the clockmaker had built, with his own money, all the additional facilities that the non-payment ostensibly allowed him to leave unfinished. In order to avoid imprisonment Torriani had to contract debts with loan sharks, signing *mohatra*-contracts so that he was soon ruined.[6] Janello called vehemently for justice.

Torriani's economic situation was becoming worse and worse during the wrestling match between the King and Toledo, which obstinately refused to pay the clockmaker. In 1572 the King granted Janello's salary in advance "in order to do a good deed to the above-mentioned Janello".[7] This was however very little help. Therefore, after four years of waiting for a solution, Janello was finally tempted to start a legal action: on the 10th of November 1573, an Italian relation of Torriani, a certain Giovan Antonio Fasolo, requested a transfer of three documents related to the contract of the first Toledo device.[8] A month later, Torriani wrote directly to the King lamenting the situation and asking for an intervention in order to repair all the torts he had to suffer. Janello concluded this very detailed supplication with the wish that the issue could be closed before his death: indeed, he claimed, because of all these afflictions, he might have easily died soon.[9] The letter seems to have had an effect: on the very same day, with a special royal decree, the King ultimately ordered the *Ayuntamiento* (City Council) and the *Corregidor* of Toledo to envoy to him a

---

6   The *mohatra* is a fake buying and selling contract which shields a usury loan: Joaquín Escriche, *Diccionario razonado de legislación y jurisprudencia*, ed. Juan María Biec, León Galindo y de Vera, and José Vicente y Caravantes (Madrid: Impr. de E. Cuesta, 1876). This document has to be dated 1574 (the new device was just started but the new agreement with Toledo not made yet) and not 1575 as in: Cervera Vera, *Documentos biográficos*, doc 44.

7   Ibid. "*por le hazer buena obra al dicho Juanelo*". The salary is given by a royal official to Jacopo da Trezzo on behalf of Janello Torriani. The amount of money is 55,000 maravedis (the first third of the annual salary granted by the King): 5th of January 1572, Madrid: doc. 27 and doc. 28.

8   As previously seen, this Giovan Antonio Fasolo is said to be a nephew of Janello. It is curious to see that this character appears in the documents related to Torriani in the same year a homonymous painter from Vicenza, partner of Veronese and Palladio, died there. Ibid. 20th of March 1575: doc. 50 and doc. 29 dated 10th of November 1573, in Madrid.

9   Ibid., "... *se acabe este negocio antes que me muera porque según ando afligido, podría ello ben ser*": 12th of December 1573, Madrid: doc. 32.

person with the necessary power to discuss and find a solution to the contract with Torriani.[10] It was in the following months that the document we have already discussed above was written by somebody close to Janello (perhaps that Fasollo who asked for the transfer of official documents) in order to convince the King to exert royal pressure on the *Ayuntamiento*, asking Philip to force it to pay the 4,000 ducats or the expenses, plus the promised annual salary of 1,900 ducats since the end of the work. Janello had also made clear that the King as well had to pay him 12,000 *ducats* for the damage suffered because of Philip's breech of the contract. The sense of frustration must have been great for Janello.[11] The evident injustice was most probably a reason of uneasiness for him and for his family, friends and working entourage. As we have seen, Janello was often remembered as someone not very talkative, but remarkably caustic in the moments he decided to make manifest his contempt, especially when he believed he had suffered an injustice despite having accomplished something outstanding, as in this very case.

How could such a delay be possible? First of all, we should recall that the Spanish Crown was at the time strongly committed to other major enterprises, more expensive and of a greater priority than Janello's job in Toledo: the war fronts in Flanders and in the Mediterranean, the colossal enterprise of El Escorial, just to mention a few among the expenses of the Spanish crown, attracted a great deal of its liquidity, without mentioning the debts with Italian and German bankers and the inflation triggered by the rivers of silver coming from the New World. All these problems had thrown Philip II's economy into a chronic instability, and a new bankruptcy seemed imminent in 1575.[12] The city of Toledo had important and expensive endeavours to carry out too, as it lamented during this controversy: in fact, the *Ayuntamiento* had written to King Philip that Toledo had no treasure nor wealth left, having restored the city-wallsand sustained the costs of the lawsuits and for other not better defined reasons.[13]

However, during the year 1574 something had changed: the King eventually paid to the clockmaker 15,000 ducats of gold, covering the cost of the device plus four years interest on the money borrowed by Torriani, amounting to

---

10   Ibid., 12th of December 1573, Madrid: doc. 31.

11   14th of December 1574:. Janello is very ill and cannot get up from his bed.

12   José I. Fortea Pérez, "Fiscalidad real y politica urbana en el reinado de Felipe II," in *Haciendas forales y Hacienda Real : homenaje a Miguel Artola y Felipe Ruiz Martín: encuentro de Historia Económica Regional*, ed. Emiliano Fernández de Pinedo (Bilbao: Universidad del País Vasco, 1990), 63-79; Braudel, *Civiltà e imperi del Mediterraneo nell'età di Filippo II*, 546-50.

13   18th of January 1575, Toledo: Cervera Vera, *Documentos biográficos*, doc. 48.

7,000 ducats. The same could not be said of the city of Toledo, which despite the clockmaker's many solicitations remained insolvent. By September 1574, the royal secretary Martin de Gaztelu wrote a letter to the Governor of Toledo urging the conclusion of the controversy related to the contract with Torriani.[14] On the 27th of September 1574, the issue of Janello's contract was the order of the day of the first session of convocation of the *Ayuntamiento* of Toledo.[15] The result was the emission of a decree issued on the 1st of October 1574 in which two commissaries were nominated to analyse the issue.[16] Five days later another decree of the same assembly showed the decision to communicate to the King the nomination of these two commissaries. The tone used by the representatives to address Janello appears to have been scathing: "keeping more in account His Majesty's will than the arguments of Janello's pretensions".[17] Another decree was then issued by the *Ayuntamiento's* session of the 24th of October 1574 when it was decided to communicate to the King the choice to put an end to the controversy about the payment of Torriani's device by the time of the next gathering.[18] During this assembly it was indeed decided to make clear to the King that Torriani's device had not brought any good (*ningund fruto*) to the city of Toledo, as all the water being supplied was being consumed by the royal *Alcázar*.[19] A few days later Luis Gaitán de Ayala was officially granted with the authority to appear before the King in Madrid, in order to hand him the approved letter. In the *Ayuntamiento's* words, the blame was entirely on Torriani:

> It is well known that Janello did not accomplish what he offered, neither to this city results or will result in the future any good.[20]

All the next steps made by the government of Toledo during the following three months (from November 1574 till February 1575) led to the same request: dispensation from the payment to Torriani. The *Ayuntamiento* claimed that the city had received nothing but inconvenience from Janello's device. Toledo, it was said, had fair reasons not to pay Janello, as de Ayala would personally

---

14   Ibid. September 1574, Madrid: doc. 34.
15   Ibid. 27th of September 1574, Toledo: doc. 35.
16   Ibid. 1st of October 1574, Toledo: doc. 36.
17   Ibid., 6th of October 1574, Toledo: doc. 37.
18   Ibid., 24th of October 1574, Toledo: doc. 38.
19   Ibid., 24th of October 1574, Toledo: doc. 39.
20   29th of October 1574, Toledo: "...*es notorio Juanelo no ha cumplido con lo que ofresció ni a esta çiudad le a rresultado ni puede rresultar nyngund benefiçio ni aprobechamiento...*": doc. 40.

explain to the King. The city of Toledo also lamented repeatedly the support, which Philip II had given to Torriani. In addition, the *Ayuntamiento* expressed the hope that the King would understand the reasons that its representative Luis Gaitán de Ayala was going to express personally to him, in the last of several interviews.[21] Unfortunately, all the documents referring to these interviews lack more detailed data.

As previously seen in detail, on the 15th of February 1575, Janello sent a new letter to Philip II. The clockmaker probably was expecting Toledo's decision not to satisfy his requests. Therefore, he performed an act of pressing on the King, appealing to Philip's responsibility in the situation. He communicated to the monarch his grave poverty due to the device that the King himself had ordered him to build. Torriani had lost in this enterprise all his capital and his friends' money too. In conclusion, after six years in which Janello had fairly accomplished what by contract he was supposed to do, the city of Toledo had not yet paid, but on the contrary, it was now even claiming that it had no need of water. Janello kept repeating that the device was functioning and it was properly made; after all Janello reminded Philip that the King himself, while visiting the Device several times, had expressed satisfaction through "his own royal mouth". Therefore, in Torriani's eyes, nobody could have maintained that he had not accomplished all the things stipulated. This letter tried to highlight the evidence of injustice he had suffered, not just from a moral point of view, but also from a legal perspective: indeed "the inequity of the treatment I have suffered is well-known by all learned men [in law] – *letrados* – and two of them have even proposed to defend me in exchange for a third of the money I have to receive and which I demand": "*que me se deue y pretendo*". The appeal ends with a reminder of the advanced age, poverty and tiredness of the clockmaker, who begs the King to conclude the controversy before the clockmaker's death. On the cover of the letter one can read a note by two royal officials: "It seems to licenciado Fuenmayor and to don Yñigo that 4000 or 5000 ducats have to be paid on Toledo's behalf ... [because] the justice he requests is obvious".[22]

---

21   Ibid., 11th (and 20th) of November 1574, Toledo; 10th of January, 1575; 15th of January 1575; 17th of January 1575, Toledo; 18th of January 1575, Toledo: documents 41-48. As a further justification for its request of payment-exoneration, the City Council used the already mentioned fact that the treasure of Toledo was empty because of too many expenses in the maintenance of the infrastructures.

22   See the traslation of this letter in the section: *From Commoner to Courtier*. About these two royal officials commenting in favour of Janello's call for justice, one can see that in the collection of laws and privileges about the guild of the shepherds, published in 1609, there is a royal decree dated 1579. At the foot the page, among several royal officials, are mentioned the names of two homonymous characters: "*El Licenciado Fuenmayor ... El Doctor*

Thus, a solution was reached during the following March. Juan Antonio Fasolo, the juridical delegate of Janello, Luis Gaitán de Ayala, the juridical delegate of the City Council of Toledo, and the *liçençiado* Joan Díez de Fuenmayor as delegate of "His Majesty's Council and Chamber, and in His name, and on His mandate"[23] drew up a new contract to replace the old one written on the 16th of April 1565.[24] With the project of a second device, a new agreement that might satisfy all the contenders was sought: first of all Janello, in order to shortcut the lawsuit, gave up his pretensions over the former contract with Toledo. The King, meanwhile, remitted to Janello 8,004,769 *maravedís*, equivalent almost to the enormous sum of 22,000 ducats of gold from his treasure to cover Torriani's expenses and debts for the device.[25] As previously seen, since Janello was a resident servant of the court, he indeed had the assurance of having all his working expenses paid by the King who, even if with a good delay, had covered his costs.[26] Any further debts accumulated after this money was delivered would instead have to be paid by Janello himself.

In the new contract the clockmaker and his heirs committed themselves to provide 400 *cargas* of water a day (day and night) *ad infinitum*. It was by contract Janello's duty to run and supervise the Device. The costs of maintenance, additional works,[27] and repair had to be paid by his majesty. The excess water not used by the King for his *Alcázar* was then granted to Janello who was

---

don *Yñigo de Cardenas Zapata"*. It would appear reasonable to believe them to be the same people appearing on the cover of Torriani's letter: indeed, they were working in the same decade as fiscal royal officials: *Libro de las leyes, privilegios, y provisiones reales del Honrado Concejo de la Mesta general, y cabaña Real destos Reynos: confirmados, y mandados guardar por Su Magestad* (Madrid: en casa de Iuan de la Cuesta, 1609), 112.

23   *"Consejo y Cámera de su Magestad, y en su nombre e por su mandado".*

24   Ibid., 29[h] of October 1574, Toledo: document certifying that Luis Gaitán de Ayala is the juridical delegate of the city of Toledo; 14th of December 1574: Janello's document of nomination of Joan Antonio Fassole, his grandson, as his juridical delegate. Janello at this point is very ill and bedridden; 20th of March 1575, Madrid: new contract between the King, the city of Toledo and Janello.

25   It is not clear when the King paid this money. As we have seen it must have happened, at least for 15,000 ducats, around the year 1574. Ibid. 20th of March 1575 *"Ocho quentos y quatro mill y seteçientos y sesenta y nueve maravedís"* doc. 50.

26   We have already mentioned that according to his contract of employment in 1562, Janello was paid 400 ducats a year, plus the pension and the further works he was going to make for his majesty's service: Madrid, 16th of July 1562: Cervera Vera, *Documentos biográficos*, doc. 5.

27   Janello has to build – with his majesty's money – a pool or a cistern in order to collect the water.

allowed to sell it until the accomplishment of a second device that was to be attached to the first one, as it was its doubling. The deadline for its construction was fixed at five years. The King would then grant the second device to Janello and his heirs, so that they could sell the water, but only over a royal licence. It is important here to note that Philip II substitutes the yearly income of 1,900 ducats the city of Toledo had been supposed to pay to Torriani since 1569 with the alleged income Janello would have made from the sale of the water the second device was going to elevate. The salary of 1,900 ducats was most probably calculated in relation to the selling cost of the yearly amount of water that was supposed to be elevated according to the first contract.[28] But this was the theory: in practice, counting all the days the machines did not work, the price of the water rose very high. The Flemish gentleman Jehan L'Hermite, who visited the devices some decades later, albeit being highly impressed by the ingenuity of the mechanism, considered it too expensive for the benefit it produced:

> Since the time it was made, the device was never used on Sundays and holydays, neither by night, because it needs somebody to govern it, and it has probably been only in use for the times His Majesty took his residence here. The machine cannot be used in winter when it ices, nor when the level of the river is too high, so that this machine is useless most of

---

[28] According to Hamilton, a *carga* of water was sold in sixteenth century Toledo at the price of 0.9 maravedis per *cantaro*. The amount of water that the first device was supposed to elevate remains unclear: in the contract it was said that each of the four pipelines had to lift 16 pounds of water (2,880 litres in total), but unfortunately we do not know in what span of time. What we know is that because of some changes, Janello made his first machine elevate more water than required in the agreement. We are better informed regarding the second device: each of the four pipelines had to elevate 100 cargas per day. One carga was 40 *cántaros* of four *azumbres*. Each *cántaro* was about 8 litres. The amount of water elevated daily was supposed to be 12,800 litres, that is 1,600 *cántaros*. If we apply the price of 0.9 marav. per cántaro, we will have a yearly equivalent cost of 1401 ducats. This means that theoretically Janello was supposed to sell the water from his device at 1.35 times the current price. Even with the second device, thanks to some technical tricks (such as the inversion of mechanical components of the machine), Janello was able to elevate 2,160 *cantaros* a day instead of the 1,600 of the contract (17,463 litres of water per day instead of 12,800). In this case the proportion of the price Janello sold one *cántaro* was very close to the one of the market: according to this calculation we can see that the hypothetical money earned with the sale of the water was 1,892 ducats per year, instead of the 1900 that Toledo had promised to pay; Earl J. Hamilton, *American Treasure and the Price Revolution in Spain, 1501-1650* (New York: Octagon Books, 1965), 171; Cervera Vera, *Documentos biográficos*, docs. 29, 122, 123.

the time, indeed it is of little use and very expensive, and in the end there was never a *cántaro* of water for which the King has not paid more than a *real* [a silver coin]. Because of that, in this machine, it has to be praised only the rarity of the invention which allows the elevation of water from the river upwards to an altitude of 300 feet.[29]

This means that the King was paying per *cántaro* more than 37 times the current price (1 *real* = 34 *maravedis*). This reflects what Juan de Herrera had also observed: one year after Janello passed away, and following the above-mentioned survey he made on mandate of the royal secretary Juan de Ibarra, he wrote that the two devices were complementary: they could not work constantly. Indeed, they needed to be alternated. On that occasion Herrera gauged that each of the two machines could elevate ninety *cantaros* of water per hour, which would have amounted to a total for both structures of 33,480 litres of water per day. Herrera estimated the daily average to be 26,040 litres of water, reached effortlessly, and the total value of the two devices to be 27,000 gold ducats (20,000 for the new and 7,000 for the old one).[30] This equated to the value of 1,134,000 pounds of wheat. For the construction of the second device, whose cost had been estimated at between 8,000 and 10,000 ducats, the King committed himself to provide around 2,000 ducats annually.[31]

What did Toledo have to do, according to this new contract? Janello and his heirs would have the land occupied by the two devices in perpetual possession, and they would not have to pay any fee to the city of Toledo until they were able to maintain the two functioning devices. At the same time, as noted above, Janello had to give up his claim to the money the community had to pay to him according to the former contract: 1,900 per year. Toledo would then pay one time 6,000 ducats instead of the 4,000 of the first contract. The act of imposition of Philip over the city of Toledo's will is clear from the fact that the representative of the city stressed in the new contract that "the city pays this money not for having received any good, but just to please the King". This was in effect a compromise: Toledo agreed. The question arises: did Toledo have any choice? Could the local government have refused to sign the contract? Philip II reminds that Toledo was ordered by him to sign the contract of 1565.[32] In the end, the city paid this amount of money without receiving any benefit,

---

29   Lhermite, *El pasatiempos de Jeham Lhermite*, 535-6 (my translation).
30   6th of March 1586, Madrid doc. 129: Cervera Vera, *Documentos biográficos*.
31   Cervera Vera, *Documentos biográficos*, doc. 53.
32   Philip II, writing from El Pardo a Special royal decree on the 28th of March 1575, ordered the city of Toledo to fulfil this new contract: Cervera Vera, *Documentos biográficos*, doc. 52:

except for the immaterial prize of fame following the success of the mechanical endeavour.

Janello got some satisfaction from the new contract, but even in his case the gain was not what he was expecting to achieve, especially having done all he was supposed to do. Of the 15,400 ducats he was supposed to get from Toledo for the past 6 years (11,400 as yearly wage, plus 4,000 one time at the delivery of the first device), he was granted only 6,000. Moreover, he lost the perpetual wage of 1,900 ducats a year, assured to him and his heirs by the former contract. Nevertheless, the royal concession and the new contract were in theory an almost satisfactory solution. However, even if the second device were to work continuously in order to lift the amount of water sellable for the sum of 1,900 ducats, the fact that Philip II was going to be the only real costumer (Torriani needed a royal licence to sell the water!) and at the same time the patron and master of Torriani, opened a tricky perspective. The King, constantly harassed by financial problems, was not to be considered a trustworthy client. The personal dependency of Torriani on Philip as his servant put him in the position of not being able to negotiate. He could only beg. And this is what he would continue to do even from his tomb after the end of his days when a posthumous letter of supplication by Torriani would be sent to his majesty after the clockmaker's death. If being a servant of the King had a price, it was this one.

In spite of the fact that the King granted Toledo with the extraordinary power to collect the 6,000 from any possible source, in order to have Torriani paid without delay, the *Ayuntamiento* of the 18th of April 1575 decided to give Janello a deposit of just 2,000 ducats. The city excused itself for delaying the payment of the remaining 4,000 (without paying any interest!) for a year, due to some economic difficulties.[33] Janello, perhaps too tired, surprisingly agreed to wait in order to meet Toledo's needs.[34] But his desire to please the needs of Toledo did not bring a more relaxed climate in the construction place. The tensions, emerging from the documents, between the royal officials and the city administration are still evident. It seems we are here facing a small-scale retaliation of the local government against the royal servants, or perhaps we are observing a fragment of the daily conflicts engendered by the friction between two contingent powers playing on the very same stage. Philip II pushed Toledo into an expensive business, which the city could easily have avoided. The King could not be the direct target of any attack: too dangerous. But what exactly

---

"... haviendo tomado la çiudad de Toledo, por nuestra orden y mandado ... çierto asiento con Joanello Turriano , nuestro relogero ...."

33   Ibid., 18th of April 1575, Toledo: doc. 60.
34   Ibid., 20th of April 1575, Toledo: doc. 61.

FIGURE 55   *Berge* (*fec.*), Regiae Toletane conspectus, a parte Pontis, Tago injecti; Vista del Palacio de Toledo, *engraving, ca. 1650, Madrid. By*
the 1650s, one could still admire the imposing external shell of the Toledo Device, though the machine inside was already
dismantled. COURTESY OF THE BIBLIOTECA DE LA FUNDACIÓN JUANELO TURRIANO, MADRID.

happened? On the 8th of July 1575 the royal officials of the construction site of the castle of Toledo wrote a letter to the King's secretary lamenting a situation that had become unbearable. The officials wrote to be aware of the fact they were annoying secretary Gaztelu with these kinds of problems. That is why they had preferred to remain silent until that moment about the reluctance and negligence of the *Corregidor* of Toledo as concerned the service of the King, and they could no longer remain silent about the problem: indeed, although they were promised a cart of quicklime a day for the construction of the device, nothing had been delivered except for "many bad words"! In this letter, Janello is said to be very annoyed (*tan mohíno*) and nothing of the building can be made to proceed. The officers also asked to be invested with the authority of patrolling the construction site by night, because there were many attempts of sabotage against the device. Furthermore, they ask Gaztelu not to give the *Corregidor* the authority to solve these problems because "*es cosa de burla*", meaning that it would have been a joke. An example of the *Corregidor's* negligence is given with reference to the fact that, in spite of many requests, he had never repaired the paths around the construction site, so that five mules had died carrying stones. In conclusion, the *Corregidor* was not helping at all.[35] The King intervened during the following months ordering the *Corregidor* and all components of the government of Toledo to provide all that was necessary for the construction of the *Alcázar* and Janello's device: everything was supposed to be paid according to the right price, including the labour force and the officers needed to carry on the necessary work. Moreover, the *Corregidor* had to repair the paths around the device and to provide the transportation of the purchased materials to the construction sites. This special royal decree ends with the King ordering that anybody who might attempt to encumber the proceeding of the construction of the device will have to pay a fine of 50,000 *maravedis* (roughly 133 ducats) to the *Cámara Real,* the Royal Chamber.[36]

The *Corregidor*, formally the King's representative on a local level, seemed here to act in manifest defiance of royal interests, and in harmony with the local powers. Torriani's story might then serve to produce a more articulate representation of the Spanish political chain of power, according to which real conflicts are to be placed also within the royal administrative structure and not just between local interests against those of the Crown. Could a *Corregidor* have sought an alliance with the local nobility of Toledo in order to reinforce his position in the rivalry with other royal officials such as those who moved in and around the King's residence? Or were King and *Corregidor* just pretending

---

35   Ibid. 8th of July 1575, *Alcázar* of Toledo: doc. 64.
36   Ibid. 25th of August 1575, Madrid: doc. 66.

to arm-wrestle in order to confuse local powers? Indeed, in the end of this story, the King obtained all he wanted: he had the two devices for his castle, he maintained the *status quo* in the city, and he appeared as a slow but good protective father to his officials involved in his building sites.

However, it is possible that the moral framework of the time pushed the *Corregidor* of Toledo to sympathize with the local aristocracy against a project that looked like a useless loss of resources for the community. In sixteenth century Europe, city administrators and princes were concerned with the problem of water supply; but in Toledo this excuse was used to justify the construction of an aqueduct devoted to sumptuary purposes, clashing against the model of the "common good", an argument based on ancient authority and Christian religion.[37] In Castile, a work published at the end of the sixteenth century by a member of the Royal Council, Jerónimo Castillo de Bovadilla (1547-1605), provides a precious insight into the cultural and political discourses behind the experience of Torriani in Toledo. Castillo de Boavdilla, after a career as a royal official, having also gained a certain experience as *Corregidor*, published in 1597 the *Policy for governors and feudal lords in war and peace time*.[38] This was a voluminous guide-book for the profession of the *corregidoria*. In the chapter about cities and water, the humanist style and contents set out an ideological base for public works concerning urban water supply. Chiefly, ancient authors from Plato to Aristotle, but also some modern ones, are precisely quoted with their bibliographical references.[39] These moral writers gave the arguments and the authority to Castillo de Bovadilla's assertions on the public utility of water-supply-systems.[40] Paragraph 53 reads:

---

37   "*the reference to the common welfare is also an old formula that legitimated public invest-
       ments against potential objections from special interests, and justified the right to pass the
       costs on the general public*": Dohrn-van Rossum, *History of the Hour*, 143. Dohrn-Van Ros-
       sum takes this assertion from: Martin Warnke, *Bau und Überbau: Soziologie der mittelal-
       terlichen Architektur nach den Schriftquellen* (Frankfurt am Main: Syndikat, 1976), 78-92.

38   Jerónimo Castillo de Bovadilla, *Politica para corregidores y señores de vassallos, en tiempo
       de paz, y de guerra, y para juezes eclesiasticos y seglares y de sacas, aduanas, y de residen-
       cias, y sus oficiales: y para regidores, y abogados, y del valor de los corregimientos, y govier-
       nos realengos, y de las ordenes*, vol. 1 (En Amberes: En casa de Juan Bautista Verdussen,
       impressor y mercador de libros, 1704); I thank professor Yun for this bibliographical infor-
       mation.

39   Because of the interest in Castillo de Bovadilla's general discourse, when quoting his text,
       I decided not to reproduce the numerous footnotes that one finds in the original.

40   The rhetoric of public good and utility and their traditions have been investigated by
       Skinner and Dolza: Skinner, *Visions of Politics*, 25-6, 47-8, 56, 611, 72, 99, 372, 379-80, 382;
       Dolza, *Storia della tecnologia*, 30-5; Marcus Tullius Cicero, *De Officiis* (London: A.L. Hum-

Providing abundance of water, among all duties of the Republic, is one of the most necessary for cities and villages: thus Plato in his laws entrusting the city-councillors with the supplying of the most abundant and clearest springs, not just for the citizens' sake, but also to adorn the city. Aristotle suggested the same, saying that the first care of the governor is to provide the city with a great abundance of water. And he says that the Ancients, when water was missing, were used to find some remedy, i. e., the use of cisterns that were collecting winter rainwater. These cisterns had great capacity in order to supply the population as much in peace as in war, when it was not possible to exit the fortifications. Plato too had suggested building large ... rainwater reservoirs and cisterns in the countryside for the irrigations of the estates, big and notable as one can find in the city of Almansa. Thus, our governor has to mostly care that in this city there is an abundance of water, building beautiful and healthy fountains, and he has to position in different places reservoirs for the livestock and for washing places, because they are useful not just for their drinking and for the daily washing and for private use, but for extinguishing urban fires. Because of a lack of these reservoirs and wells, the city of Oviedo was burned down several times.[41]

In the following passage (paragraph 54) Jerónimo Castillo de Bovadilla defines the duties in urban water-management, attributing to the governor the main role as inspector:

> With this care the governor has to check the pipes and streaming of water, and personally inspect them. The plumber has often to do the same, as the jurist Venuleyo says, this is more important than repairing the roads. And in the same way, one has to avoid any seizure of the water, turmoil, or fraud against the law, in relation to the irrigation systems or in relation to domestic supply. For this problem, the Romans created special judges that they called curators of the waters, according to imperial and civil right, and to the things that Julius Frontinus and others interestingly wrote about.

---

phreys, 1902), bk. I, chap. XVI: the first action of the government of the Republic has to be the common good.

41   All translations of this text are mine.

The next paragraph (55) reads:

> If there is any requirement, the governor has to take care as well of build-
> ing public baths and restore the old ones in order to cure certain diseases
> as one can find in many places: because as Hippocrates, Plato, and Paulus
> Egineta said, a proper use of baths is more healthy than the cures of phy-
> sicians, and I can say this out of experience. According to Galen, baths are
> singularly good cures for literate men. Bishop Simancas in his *Republica*
> stressed very much this point to the governors, and ordered the restora-
> tion of the baths close to Ledesma. The ruins of Antoninus' or Diocletian's
> baths and thermal plans that one can see now a-day in Rome, where one
> found many columns and buttresses of many different coloured marbles,
> with many rooms and places ordered according to the different uses with
> great care, shows how the Romans took care of the baths and the great
> expense and heed they put, without limit and measure, in their construc-
> tion. Indeed the Ancients were accustomed to wash themselves almost
> every day, and they were promoting sweating to maintain and preserve
> health, as wrote in particular the French Gulielmo Choul, whose work
> translated master Perez del Castillo. They also had for recreation and
> amusement other baths, like the ones the Moors use. King don Alfonso
> ordered to dismantle the baths he got in Toledo, because he did not want
> his people to get effeminate.

After this excursus on the utility and, if coming from Moorish tradition, on
dangers of baths, a very interesting paragraph (56) expresses some harsh criticism of situations that resemble Torriani's enterprise in Toledo:

> Because of the things said in this matter, it would be good for the gover-
> nor, according to them, to try to do public works, and he should not spend
> the money of the Republic in making inventions and new things, as some
> governors think, because they believe this would do good to their fame.

And the negative similitude is once again found in Levantine tradition,
opposed to the positive legacy of ancient Rome:

> Thus it has not to be for conceit, nor without proper consideration, as the
> Egyptians, who were used to building things more vane and foolish than
> the Romans, after King Minos, in order to show his power, made water
> rise from the river Nile on a mountain for a hundred stadia (that for us is
> 12,5 leagues). And Meris ordered to create a lake of such a capacity that in

order to fill it up it required the entire water of the river Nile for a whole month.[42]

Beside Bovadilla's exaltation of the ancient philosophical, medical and civil tradition of the Greco-Roman World over the effeminate Muslim and vain ancient Egyptian ones, it is interesting to see the criticism towards innovations and inventions promoted by the desire for glory – is this a reference to what happened in Toledo with Torriani's enterprise? Could the story of the vain Egyptian king who brought the water of the river Nile to a mountain have been used to delicately criticize what had happened in Toledo in the previous decades, where an astonishing invention failed to bring a single drop of water to the city, but instead took it exclusively to the Royal Castle? The only good gained by the city of Toledo was an increased fame because of the exceptionality of Torriani's Device, something that Castillo de Bovadilla would have probably defined as "Egyptian" in its vanity.

Humanist tradition gave a moral framework to these expensive enterprises, convincing governments to invest in public water-supply systems, and making of monumental public and private fountains, a necessary facet of respectability. This classical model was well-established long before Bovadilla's *Politica*, as is evident from the work of the controversial antiquarian Giovanni Tarcagnota from Gaeta (ca.1508- before 1566):

> Above all, what a city most requires are fountains of fresh, flowing water because, beyond the practical benefits that these offer (which is huge and incomparable), they render beautiful, pleasant and lively that city in which they are found.[43]

A previously mentioned anecdote about the knight of partial Jewish ancestry shows how such lineage or Moorish one could play a role in the complex scenario of Spanish nobility.[44] One wonders if Torriani's anecdote reflects some underground rivalries played between Toledo and the court by new and old Christians. For the moment this remains an open question. A remarkable thing

---

42   Castillo de Bovadilla, *Politica para corregidores*, 53-56.

43   Giovanni Tarcagnota, *La città di Napoli dopo la rivoluzione urbanistica di Pedro di Toledo*, ed. Franco Strazzullo (Roma: Gabriele e Mariateresa Benincasa, 1988), here quoted and translated in English as in Edelstein, "Acqua Viva e Corrente." Edelstein, "Acqua Viva e Corrente"; Gennaro Tallini, "Giovanni Tarcagnota: bibliografia," *Cinquecento plurale. Bibliografie*, 2012 <http://www.nuovorinascimento.org/cinquecento/tarcagnota.pdf>.

44   See the section: *From Commoner to Courtier.*

emerging from this anecdote is the fact that the amount of water delivered by Torriani's device, though larger than the one required by the contract, was still not satisfactory: Philip tried to have two more devices built there, but only one more was eventually constructed. Perhaps, it was for this reason that the King ordered his minister Cardinal Granvelle, in the year 1584, to search in Italy for another hydraulic engineer who could dry mines and swampy lands, and elevate water in Toledo: thus, Cardinal Granvelle wrote to cardinal Deza, who was in Rome, to find such an engineer. Deza found one in the service of a Cardinal Medici who had built an impressive water-supply system for the vineyards of his patron. Deza tried to have this engineer, a certain Valente Valenti from Brescia, a neighbouring city of Cremona, to move to Toledo, though upon agreement with old Janello, as required by Granvelle. But the negotiation ran aground.[45]

During the long-lasting controversy, the City Council of Toledo, with less tact than Cardinal Granvelle, had tried to promote a number of possible competitors for Torriani, looking for alternative solutions to its thirst of water. This testifies that Toledo did not pay Torriani because of lack of interest in the water-supply project, but because the King had taken all the water available. In 1575, a committee from the local Toledan government was analysing the project of a certain Bernal Esfalar, who had promised to supply the city with water through a new machine: this was the time when a second agreement had been signed with Janello and the King to double the Toledo Device![46] This is a clear proof that the local government of Toledo was not satisfied with the new contract the King had enforced upon the City Council. After the above-mentioned lamentations coming from the royal officials of the *Alcázar* of Toledo that the local government was not complying the contract, the King had to order the *Corregidor o Juez de residençia* of the city to provide Turriano with an *alcuazil*, a kind of a police officer, for patrolling the device and to recruit the necessary officials and labour-force for its construction site, as used to be the case before the city of Toledo had fired the *alcuazil* that Torriani had previously employed for running and guarding the device's construction site.[47] But once again things did not proceed as smoothly as one would expect after the expedition of a royal decree. It is evident that the City Council tried to hit the royal body in its less sacred (non aristocratic) and less dangerous segments, the ones daily

---

45   Eloy Benito Ruano, "Un competidor de Juanelo Turriano (y otro proyecto de 'Artificio' para Toledo)," in *Studia Historica e Philologica in honorem M. Batllori*, ed. Jorge de Esteban (Roma: Instituto Español de Cultura, 1984), 83-88.

46   15th of July 1575: Cervera Vera, *Documentos biográficos*, doc. 65.

47   19th of November 1575, El Pardo: Cervera Vera, *Documentos biográficos*, doc. 70.

exposed to its anger: Janello and the other royal officials. In this way, nobody belonging to the King's entourage was personally offended, but the result was that the entire project was put at risk. This looks like a strategy of passive resistance: after a good three months, the *Corregidor* had still not provided Torriani with any *alcuazil*. Therefore, the royal officials of the *Alcázar* had to send another letter of lamentation to the King's secretary, asking for it. They also complained about the *Corregidor's* negligence in keeping order around the *Alcázar* and in proximity of the device, where all the garbage was dumped and where vagabonds and thieves were circulating and throwing things on worked stones and glass windows, breaking them.[48] The King answered his servants' request eight days later, ordering with a special royal decree the *Corregidor* Joan Gutiérrez Tello to address these problems, arresting and carefully punishing anybody breaking the law on this site.[49] By the following year, in 1576, things had to get eventually better if in a letter to the King's secretary – as already mentioned – the officials of the *Alcázar* of Toledo said that:

> Janello is carrying on so fast the construction of the device that there is an immediate need of the money scheduled for the year. Otherwise Janello will be so sad not to get it that we will have to appeal to all our diligence in order not to get him very upset (*que él no se disguste*).[50]

Around 1578, a further petition from the *Ayiuntamiento* to the King was made in order to receive a royal licence for the construction of another device by a certain Capitan Cristóbal de Çuaco or Zuaco. The new device, never built, was supposed to provide both the city and the *Alcázar* with water without damaging Janello's device.[51] As in the above-mentioned case of the Grand Duke of Francesco de Medici and Torriani, was the King here protecting his own servant, conditioning the local market and consequently the technological environment of his subjects? However, in 1584, perhaps when it was becoming clear that Janello was soon going to die, it was the King himself who triggered the search for a new engineer. In the end, by the year 1580, Janello finished successfully his second device. However, he never recovered completely from the economic damage occurred by the delay in payment after the construction of the first device, if still in 1584 he was reminding the King of his debts due to the city of Toledo's past behaviour. The following year, after

48   Ibid., 15th of February 1576, *Alcázar* of Toledo.

49   Ibid., 23rd of February 1576, Madrid.

50   Ibid., 30th of May 1576.

51   Ibid. See. doc. 85 and doc. 65.

Janello's death, two trustworthy witnesses from the very court, Juan de Herrera and Garibay, confirmed that Janello and his family were broke, and Philip II would provide for Janello's family only after his servant had passed away.

In the end, the result was that neither Torriani nor the community of Toledo got what they were expecting from the construction of the ingenious Device. Both contenders blamed each other and (except for some reproaching complaints by Janello to the King), they did not attack Philip directly. The King, on the other hand, had all the water he wanted for his *Alcázar* and the property and possession of the doubled Device. Janello did not keep the second machine, as was his right according to the contract signed in 1575, but he decided to deliver it into the hands of the King, who, however, was his only possible customer.[52] The King, in exchange, kept Torriani's heirs in charge of the device and of the clocks, and provided for Janello's only daughter and universal heir. This story seems to follow a precise pattern, as Cardinal Federico Borromeo once wrote:

> A very important cardinal used to say, I do not know whether seriously or joking, that if one day the many courtiers of his household were going to be united, it would have been a great trouble for him.[53]

The story of the Toledo Device has probably to be seen in two perspectives: on the one hand it shows how Renaissance princely patronage created technological innovation. On the other hand, it illustrates how the King of Spain financed such enterprises without paying for their costs in their entirety. The legacy of Torriani's machines for water elevation has to be found in the future great royal water-supply systems of Christendom. Torriani had demonstrated how it was possible to build enormous machines under royal patronage. The machine for the royal palaces of Paris, the device of Valladolid made by Zubiaurre for the recreation of the King of for the Duke of Lerma, and the colossal device of Marly, built in the next century to provide Versailles with water, had their root in the Toledo Device.[54] It remains an important question to ask: why did such

---

52   Ibid. 12th of November 1584 (?): doc. 119.

53   "... *un Cardinale assai grande, il quale, no sò se per beffe, o daddovero, soleva dire, che male per lui sarebbe stato, se i suoi cortigiani, dequali egli ne havea in sua casa un gran numero, si fossero uniti insieme*" (my translation): Borromeo, *Gratia de' principi*, 155.

54   Nicolás García Tapia and Javier Rivera Blanco, "Juan Bautista de Toledo, Jerónimo Gil y Juan Herrera: autores de la 'Mar de Ontígola,'" *Boletín del Seminario de Estudios de Arte y Arqueología* 51 (1985): 319-44. Nicolás García Tapia, "El ingenio de Zubiaurre para elevar el agua del rio Pisuerga a la huerta y palacio del Duque de Lerma," *Boletín del Seminario de Estudios de Arte y Arqueología*, 1984, 299-324; Nicolás García Tapia and Javier Rivera

Perspectiva de la Ciudad de Toledo vista de la otra parte de la Puente del camino de Sevilla

FIGURE 56   *Carlos y Fernando de Grunenbergh*, Perspectiva de la ciudad de Toledo, *in* Memorial sobre rendir navegable el Manzanares, *1668*. On the left: Janello's Toledo Device stretching from the river Tagus up to the Royal Castle. COURTESY OF THE BIBLIOTECA DE LA FUNDACIÓN JUANELO TURRIANO, MADRID.

a hydraulic marvel, so widely celebrated by contemporaries, leave no trace of its mechanisms?

Invention and the Practice of Secrecy[55]

Christmas 1579 was approaching in Madrid, and the Apostolic Nuncio Filippo Sega was restlessly waiting for the written opinion by the King's clockmaker on a very important issue that would have significant effects on all Papal Ecumene and beyond.[56] The Nuncio had already received the written opinions about the reform of the Julian Calendar that the King had ordered from the famous universities of Salamanca and Alcalá. Janello's opinion, according to the clockmaker, was in the process of being printed. Only Janello's instruments were missing. Eventually the craved instruments arrived. They were personally carried by the octogenarian clockmaker:

> The day before yesterday, master Janello arrived in Madrid together with all the instruments [he made for the reform of the calendar]. I beg Your Majesty to give him audience as soon as possible, and to give him those orders, which You will find the most suitable to the situation. Indeed he refuses to deliver his instruments to anybody that is not Your Majesty, and this is time that this business finds an end.[57]

Janello, who was almost 80-years old, recovering from a grave illness and heavily indebted, decided to undertake an 80 kilometres winter-trip from Toledo to Madrid in order to deliver his creation personally into the hands of his master,

---

Blanco, "La presa de Ontígola y Felipe II," *Revista de Obras Públicas* 132, no. 3236 (1985): 477-90.

55   Knowledge, the problems of authorship and its relation to the practices of secrecy and openness have been widely debated among historians of technology. For instance see: Long, *Openness, Secrecy, Authorship, Technical Arts and the Culture of Knowledge from Antiquity to the Renaissance*.

56   The Apostolic Nuncio Filippo Sega was Gregory XIII's Bolognese fellow-countryman. Between 1577 and 1581 he held this diplomatic office in Madrid, and since 1578 he was invested with the bishopric of Piacenza, the twin-city of Cremona.

57   "*Hieri l'altro giunse Maestro Gianello in Madrid con tutti li suoi instrumenti. Supplico Vostra Maestà di farlo chiamare a se quanto prima et dargli quelle ordine che le parerà conveniente perchè egli non vuole consegnare questi suoi instrumenti senon alla Maestà Vostra et hormai è tempo che si venghia fine di questo negotio*" (my translation): in Turriano, *Breve discurso*, 25.

the King. Why did he go to such trouble? We have already noted the stubborn temper of Janello Torriani when Ferrante Gonzaga and the very same Nuncio Filippo Sega had acknowledged that he did things only when he felt like.[58] However, what seems to emerge here is what one may call a strategy involved with the practice of invention. Debts were strangling Janello and his house-hold to the extent that any possible form of extra income must have looked like manna from the Heavens. Furthermore, meeting the King in person might have reminded Philip of the tragic situation into which he had pushed Torriani and his family. Perhaps a personal supplication might have opened the unyield-ing purse of the King. Janello likely feared two things: firstly, the King receiving Janello's works through an emissary might have postponed his compensation. By delivering personally his works to Philip II, Janello thought he could prob-ably enforce the contractual obligation of payment on delivery. Janello's physical arrival at court should be read as a willingness to personally negotiate with the King for remuneration. Secondly Janello might have feared that some-body else could have stolen the credit for his mathematical calculating instruments before his master had officially acknowledged their authorship. Once the prince had recognised someone's work as worthy, it was no longer possible for anyone to challenge its authorship and the remuneration it deserved. It was indeed the prince who granted inventor privileges: neither the technical office testing the machine nor the scholar publishing one's invention (like Cardano did with different inventions by Janello in *De subtilitate*) had the juridical authority to make of the invention a protected saleable item con-nected to the name of its author. It was the prince or the supreme legislative body of a Republic who recognised authorship and granted the legal right to exploit an invention for commercial profit in a certain jurisdictional area. The invention has a value on the market while in its whole, or in part, it is kept secret. Once it is unveiled it loses this value, unless the unveiling is officially made in front of the power that can acknowledge its authorship. The 22nd of December, probably the very day Janello delivered his instruments into Philip's hands, the King granted him with the office of "*alcayde de la carcel de la villa de Ocaña*", the post of director of a prison belonging to the military order of *Santiago*, together with its salary and probably with all the bribes that such a position of power could extort from prisoners. Both Janello and the King knew that this was a vendible item at court. Indeed, Janello sold this office as soon as he could for 800 golden ducats.[59] Probably Janello was expecting something else from the King, something that could have put an end to his economic trou-

58   See Part 2, Chap. 5.1.

59   Cervera Vera, *Documentos biográficos*, doc. 100 and doc. 102.

bles, but Philip had no cash for him. The only remuneration he could offer was this post made vacant by the recent death of the former *alcayde*.

By the 22nd of February 1580, abbot Biçeno wrote from Rome to King Philip II that the written opinions had all arrived, but not Janello's instruments, which were still in the Nuncio's hands:

> His Holiness answered me that he had very much enjoyed the books [written by the Universities of Salamanca and Alcalá, and by Janello], but that the instruments of Master Janello had not yet arrived: the Nuncio has written that he had not yet found a comfortable way to send them, and he would do it soon.[60]

In spite of all the haste previously manifested, the Nuncio kept hold of the instruments for some months, instead of sending them along with the treatises to Rome! Why was this expedition delayed? It seems this happened because of security reasons. Janello had visited the Nuncio in the last two years informing him about his calculating instruments. Nuncio Sega wrote following one of these occasions to Rome that Janello's treatise and instruments were to be considered the best solution to the problem of the reduction of the calendar: in my eyes ... one cannot desire a better solution to this reduction".[61] Using the calculations of certain volvelles and mathematical tables, whose drawings are still kept in the Biblioteca Apostolica Vaticana (Figs 57, 58, 59 and 60), the solution proposed by Janello would have allowed the continual use of old liturgical books and prayers calibrated on the ancient system, as would have been preferred by King Philip II.[62] With the introduction of bissextile years, Janello's proposed solution would have progressively but definitively reduced the delay over a long span of time.

From another letter he had written to the Bishop of Como at Rome, it seems that the Nuncio was stalling, waiting to send the paper-instruments, merely because he needed to have them brought by a secure emissary:

---

60   *"Su Santitad me respondio que havia holgado mucho con los libros, y que los instrumentos del Maerstre Juanelo no havian aun llegado por que el Nunçio escrivia que no havia tenido comodidad de embiallos que lo haria presto"*: Ibid., doc. 97.

61   *"a mi juicio ... no se puede desear mejor modo para esta reducción"*: Ángel Fernández Collado, "Juanelo Turriano y la aportación española a la reforma del calendario gregoriano," *Toletum: boletín de la Real Academia de Bellas Artes y Ciencias Históricas de Toledo*, no. 23 (1989): 154.

62   Zanetti, *Janello Torriani, a Renaissance Genius*, 54-56.

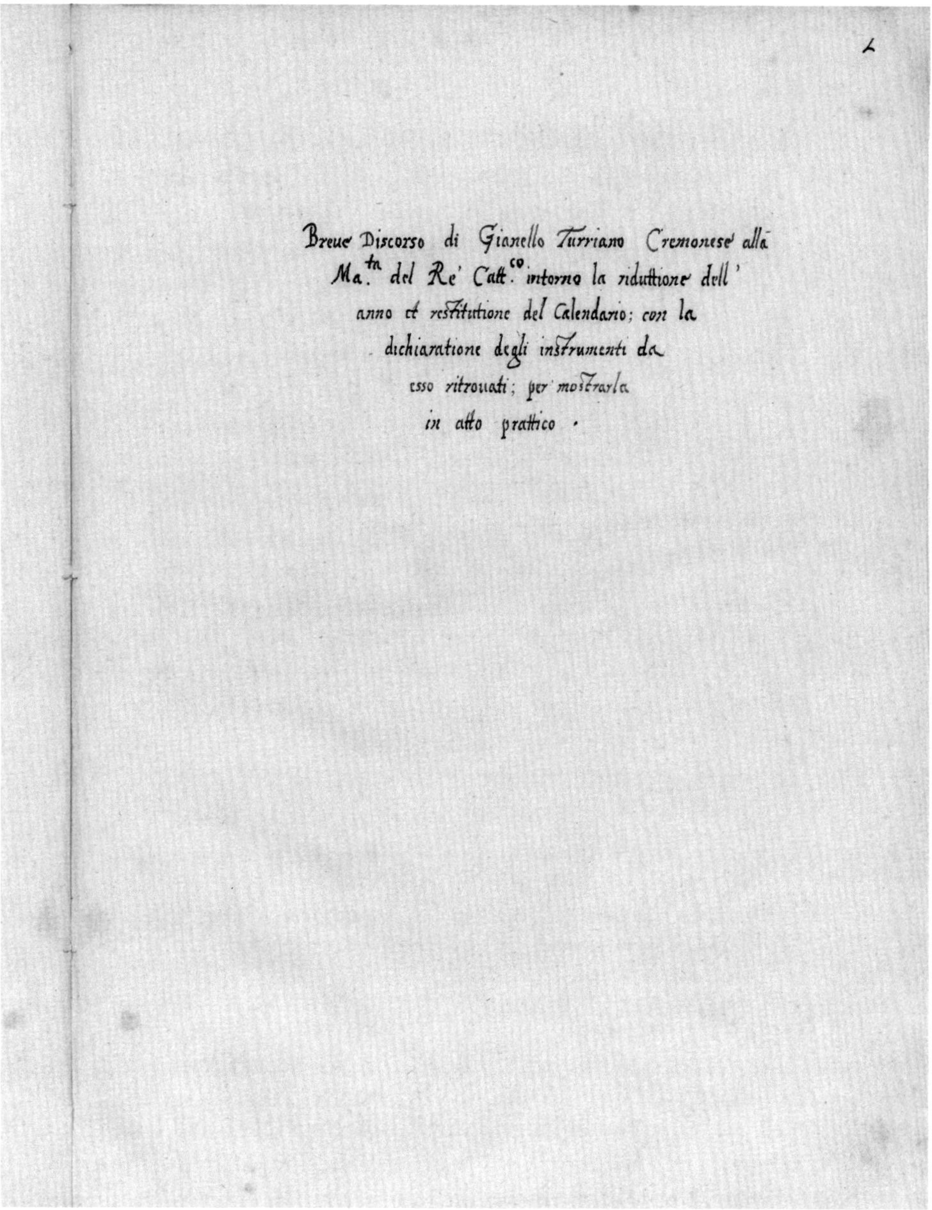

Breue Discorso di Gianello Turriano Cremonese alla
Ma.^{ta} del Re' Catt.^{co} intorno la riduttione dell'
anno et restitutione del Calendario; con la
dichiaratione degli instrumenti da
esso ritrouati; per mostrarla
in atto prattico ·

FIGURE 57   *Janello Torriani,* Breve discorso di Gianello Turriano Cremonese alla Ma[es]ta del
Re Catt[oli]co intorno la riduttione dell'anno et restitutione del Calendario ...,
1579, front page. COURTESY OF THE BIBLIOTECA APOSTOLICA VATICANA, LAT
7050.

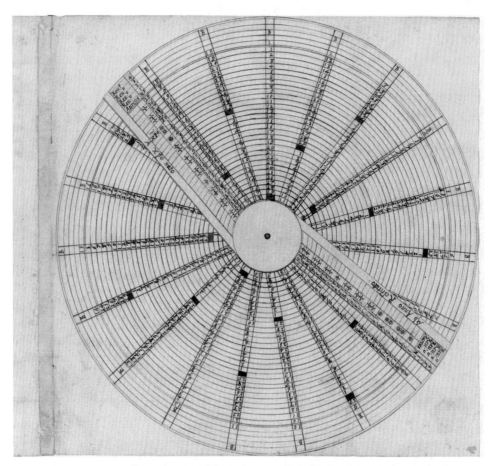

FIGURE 58   *Janello Torriani, one of the mathematical volvelles from his* Breve discorso *on the
reduction of the calendar, 1579.* COURTESY OF THE BIBLIOTECA APOSTOLICA
VATICANA, LAT 7050.

I could not send them in any other way because there is necessity, as You
shall see, to send them together with someone that could take special
care of them. I beg Your Very Illustrious [Lordship] to forgive me for the
delay and I beg You to inform me of his arrival in order to reassure my
soul.[63]

63   *"No he podido enviarlos de otra manera por tener necesidad, como verà, de ser enviados con
una persona que tuviese particular cuidado de ellos. Suplico a V.Iima perdone el retraso y me
envie noticia de su llegada para apaciguar mi ánimo"* (my translation): Ibid., 158.

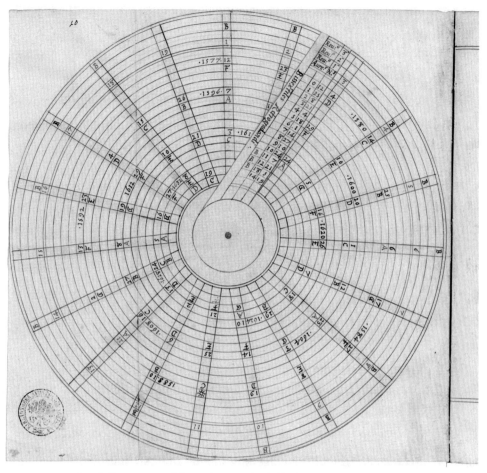

FIGURE 59   *Janello Torriani, one of the mathematical volvelles from his* Breve discorso *on the reduction of the calendar, 1579.* COURTESY OF THE BIBLIOTECA APOSTOLICA VATICANA, LAT 7050.

Janello's instruments eventually arrived in Rome by June 1580 in a good condition. Perhaps the Nuncio had also worried about the weather conditions, since Janello's instruments were very delicate: they consisted of cardboard volvelles with different rotating components for calculations, and of some other obscure instruments "to be kept in a box because they needed to be well-protected". It was the Count of Montebello who was recognised as sufficiently reliable to take care of the delicate mission.[64]

---

64   Ibid.

Tabula Epactarū A Xp̄o Nato.

Circul		B		B		B		B		B	
Circul	B		B		B		B		B		
Circul	B		B		B		B		B		
Circul		B		B		B		B		B	

Reuo lucio Pri°	Aureus / Anno	1	2	3	4	5	6	7	8	9	10	11	12	13	14	15	16	17	18	19	Num° Anno finis	Reuo lucio
1	Xp̄o	7	18	29	10	21	2	13	24	5	16	27	8	19	☼	11	22	3	14	25	303	1
2	304	8	19	☼	11	22	3	14	25	6	17	28	9	20	1	12	23	4	15	26	607	2
3	608	9	20	1	12	23	4	15	26	7	18	29	10	21	2	13	24	5	16	27	911	3
4	912	10	21	2	13	24	5	16	27	8	19	☼	11	22	3	14	25	6	17	28	1215	4
5	1216	11	22	3	14	25	6	17	28	9	20	1	12	23	4	15	26	7	18	29	1519	5
6	1520	12	23	4	15	26	7	18	29	10	21	2	13	24	5	16	27	8	19	○	1823	6
7	1824	13	24	5	16	27	8	19	○	11	22	3	14	25	6	17	28	9	20	1	2127	7
8	2128	14	25	6	17	28	9	20	1	12	23	4	15	26	7	18	29	10	21	2	2431	8
9	2432	15	26	7	18	29	10	21	2	13	24	5	16	27	8	19	○	11	22	3	2735	9
10	2736	16	27	8	19	○	11	22	3	14	25	6	17	28	9	20	1	12	23	4	3039	10
11	3040	17	28	9	20	1	12	23	4	15	26	7	18	29	10	21	2	13	24	5	3343	11
12	3344	18	29	10	21	2	13	24	5	16	27	8	19	○	11	22	3	14	25	6	3647	12
13	3648	19	○	11	22	3	14	25	6	17	28	9	20	1	12	23	4	15	26	7	3951	13
14	3952	20	1	12	23	4	15	26	7	18	29	10	21	2	13	24	5	16	27	8	4255	14
15	4256	21	2	13	24	5	16	27	8	19	○	11	22	3	14	25	6	17	28	9	4559	15
16	4560	22	3	14	25	6	17	28	9	20	1	12	23	4	15	26	7	18	29	10	4863	16
17	4864	23	4	15	26	7	18	29	10	21	2	13	24	5	16	27	8	19	○	11	5167	17
18	5168	24	5	16	27	8	19	○	11	22	3	14	25	6	17	28	9	20	1	12	5471	18
19	5472	25	6	17	28	9	20	1	12	23	4	15	26	7	18	29	10	21	2	13	5775	19
20	5776	26	7	18	29	10	21	2	13	24	5	16	27	8	19	○	11	22	3	14	6079	20
21	6080	27	8	19	○	11	22	3	14	25	6	17	28	9	20	1	12	23	4	15	6383	21
22	6384	28	9	20	1	12	23	4	15	26	7	18	29	10	21	2	13	24	5	16	6687	22
23	6688	29	10	21	2	13	24	5	16	27	8	19	○	11	22	3	14	25	6	17	6991	23
24	6992		11	22	3	14	25	6	17	28	9	20	1	12	23	4	15	26	7	18	7295	24
25	7296	1	12	23	4	15	26	7	18	29	10	21	2	13	24	5	16	27	8	19	7599	25
26	7600	2	13	24	5	16	27	8	19	○	11	22	3	14	25	6	17	28	9	20	7903	26
27	7904	3	14	25	6	17	28	9	20	1	12	23	4	15	26	7	18	29	10	21	8207	27
28	8208	4	15	26	7	18	29	10	21	2	13	24	5	16	27	8	19	○	11	22	8511	28
29	8512	5	16	27	8	19	○	11	22	3	14	25	6	17	28	9	20	1	12	23	8815	29
30	8816	6	17	28	9	20	1	12	23	4	15	26	7	18	29	10	21	2	13	24	9119	30

FIGURE 60   *Janello Torriani, one of the mathematical tables from his* Breve discorso *on the reduction of the calendar, 1579.* COURTESY OF THE BIBLIOTECA APOSTOLICA VATICANA, LAT 7050.

The day after Christmas 1579, Janello played the other card he still had in his hands in order to squeeze the best he could from this last endeavour: he gave a letter to the Nuncio to be forwarded to the Pope. In this missive Janello excused himself for the delay, which he attributed to his weak health, and he supplicated to be granted a privilege for invention if his instruments were to be printed. In this case Janello promised the Pontiff he would send him some further explanations (*"raggioni"*) about the calculating system. It is possible that Janello kept secret some parts of his computation in order to maintain some negotiating power. The letter followed with another supplication: if instruments and treatise were not going to be published, the clockmaker asked the Pope for a pension or some other kind of remuneration for his efforts, because he was old and poor.[65] Gregory XIII wrote to the Nuncio soon after ordering him to assure Janello that after receiving treatise and instruments in Rome he would send him an answer. This answer was positive and was sent from Rome the day before Christmas 1581, even if Janello did not gain much from this privilege: at the end, the papal commission for reform of the calendar preferred a solution less definitive and precise than Janello's one, but more immediate, reducing the delay of ten days in one instalment in October of 1582 with the publication of the papal bull *Inter Gravissimas.*[66]

An even more stunning strategy of secrecy was played around the Toledo Device. The greatest Spanish writers of the so called "Golden Age", from Cervantes to Lope de Vega, from Quevedo to Calderón de la Barca, mentioned Torriani's Device. Many of these writers emphasised that this machine was a wonder, a marvel worthy to be seen, even just from the outside:[67] indeed, since

---

65   "... *se [Tavole e compendio] saranno approvate, et che non si giudichino indegne di luce, supplico humilmente V. S.tà a farmi gratia che s'imprimano; et di concedere a me questa impressioneper quel tempo che la S.tà V.ra sarà servita, che io in tal caso non mancarò per più giustificazione mia, et sodisfattione de gli altri di aggiongervi le raggioni, sopra le quali sono fondate*": Turriano, *Breve discorso,* 54.

66   The papal administration invited the Nuncio to ask Philip II for a privilege of 10 years in favour of M. Antonio Giglio (brother to the deceased Luigi Giglio), who was in charge of the official publication of the new calendar. According to the request, nobody but Giglio should have been entitled to print or sell the new Calendar and *Martirologium*. Antonio Giglio was keeping in Rome the privilege granted to Torriani. Unfortunately for the clockmaker, Philip II refused to grant any privilege for the Calendar saying that it was his right to print it in his kingdoms free from any obligation. After some months of negotiation the King found a compromise with Antonio Giglio so that he could print Calendar and *Martirologium* in some specific Spanish places: Fernández Collado, *Gregorio XIII y Felipe II en la Nunciatura de Felipe Sega (1577-1581),* 248-249.

67   Moreno Nieto and Moreno Santiago, *Juanelo y su artificio,* 39-52; Sánchez Mayendía, "El artificio de Juanelo en la literatura española," n. 103. For a list of these texts see in the

it was enclosed in a shell made out of bricks and wood, its real mechanism remained hidden to the observer from the outside. Only respectable guests of the Crown, and people willing to pay a fee, were taken inside the building. Even then, they could not have a full view of the machine that stretched some 300 metres up the hill, following a non-linear pathway.[68]

One of the most important descriptions which helps us to better comprehend the mechanics of the Device is the account given by the priest and historiographer Manuel Severim de Faria in 1604, which includes the only known explanatory sketch of its mechanics (Fig. 61).[69] In 1598, after an epidemic of the plague in Portugal, Manuel Severim participated in a pilgrimage to Guadalupe, and passed through Toledo, where he noticed the many *aljibes* (cisterns), where:

> One gathers the water ... to drink it fresh during the summer, for the extreme heat in this city and for there not being other sources of water in this city if not for the Tagus river. And there are cisterns like this in private homes, that can contain two *cargas* of water, and all the water that we drink is transported [by the aguadores, the transporters of water] except that of the *Alcázar*, [whose water is supplied by] that admirable work – so famous and celebrated in the whole of Spain and in almost the whole world – with the quintessential name of the Ingenious Device of Toledo [*Ingenio de Toledo*]. [It] brings water to the highest parts of the royal palace, which is on an elevated peak above the river, and the first part of the Device is direct and vertical, and then it climbs on the steep slope with its profiles one straighter and the other more hanging, completely covered on the outside by wood and French bricks so as nothing would be visible from the outside. Only by the continuity of this construction can it be seen that inside there is the Device. The motor of this Device is the same river, which pushing the wheels moves its bronze tubes by five or six

---

bibliography the section entitled: "*Spanish printed literary texts from the Siglo de Oro mentioning Janello Torriani and his Toledo Device.*"

68   One of the reason for the Toledo Device's successful reception can be found in the *Mechanics* of the *Corpus Aristotelicum*: "Among the events that occur in harmony with nature, the ones whose cause is unknown amaze us. Among the events that are instead happening against nature, the ones that are realized with craft for humanity's sake are the ones that astonish us". "*Meravigliano, tra gli eventi che accadono in armonia con la natura, quelli dei quali la causa è ignota; tra gli eventi che accadono invece contro natura, destano stupore quelli che sono realizzati con arte per utilità dell'uomo*": taken from: Helbing, "La scienza della meccanica nel Cinquecento," 573-76.

69   See image.

FIGURE 61   *Baltasar de Faria Severim, Sketch of Janello's Toledo Device, in* Peregrinação de
Baltasar de Faria Severim, Chantre de Evora, ao Mosterio de Guadalupe, *1604.*
This sketch is the only known representation of the mechanisms present in
Janello's Device of Toledo. COURTESY OF THE BIBLIOTECA NACIONAL DE
PORTUGAL, LISBON.

spans of length [around 125 cm, a span being 20.9 cm] which on one side
they have an open and square case one span in length and more than half
in height in which it receives the water. And the tube that is folded in this
is three fingers and of a squared shape [5.25cm, measuring a Castilian
finger as 1.75cm] and open on top, and finished diminishing with a round
section with a hole in the centre [the diameter] of a coin. These tubes are
placed in the manner of scissors and rotate the one above the other for
the head, the way that one's head corresponds to the sharp side of the
other. And there are many of them one on top of the other in a way that
the first to take the water with the larger part [its head] when it moves up
with its sharp part loads the water in the head of the upper one that stays
at the summit, as seen in these two figures. The first, A, represents when,
in the first movement, the first tube takes water with the large extremity,
and the second, B, represents how much this first tube raises and pours
the water in the large extremity of the second tube. When it returns to the
lower position as in figure A so that the first tube takes again as much
water as before, already the second tube, which had previously received

water when it was in position B, passes it now to the third tube, and it is to be noted that, because there will be the need for a sufficient time to let the water flow across the tube, there is a half gear cut in a way that it passes a post, that they call forcing, through the crescent, and they cannot move. The brilliance of this system of tubes is so great that the pole passes through this half-gear and falls into the hole that there is in it [stopping the motion for the necessary time;] therefore these pipes are raised and positioned in another way. To ensure that they lower their points all together and also to raise the other side in unison, they have been vertically linked together with thin iron bars, as shown by the lines C D. This same invention is the one that moves ahead when the Device does not rise so much vertically and it approaches the hill. The only difference is that the tubes are here longer and the iron bars are of pieces joined one to the other. This Device does not always go up evenly, in some parts it climbs 21 or 20 spans (4-5m) in level: between the different levels there are some reservoirs for the water where the following tube is going to draw it [to lift it].

It seems that not even the members of the Spanish Court had access to the whole picture, in the form of drawings illustrating the entire Device. For instance, the Flemish gentleman Jehan L'Hermite, who was in charge of the royal clocks in Madrid (among them Torriani's planetary ones), visiting Toledo by the end of the century, wrote full of resignation:

> I was never able to obtain the copy of the project or a reliable representation of this machine in order to satisfy my desire to provide my readers with its vivid image, and though I believe that human intellect can grasp what it is about, it is difficult to understand the apparatus, the workmanship and invention [*artificio, industria, invencion*] of this device without seeing a representation of it. Thus, I have tried very hard to get a reliable image, but until now, I have not succeeded, and all the many requests I have forwarded with the royal architect Juan Babtisto Monnegro [the famous architect and sculptor Juan Bautista Monegro], president of the above-mentioned palace [the *Alcázar*], being the person in charge of its administration, were useless. The royal architect Juan Babtisto Monnegro has never missed to persuade me in any occasion to desist from my plan, because of the difficulties and troubles [*molestias*] that my desire would have provoked, and one can easily see how little use I could do of the few words about this matter he sent to me with a short note, in which he wrote that in order to properly understand and worthily explain the

mystery of this device, in the first place one would need a book filled with all its different illustrations, and furthermore, to give a more vivid understanding of it, it would also require the construction of a series of wooden models, because – he wrote to me – there is nobody in the world able to understand what this device is about in just one single illustration. And even if one had properly understood the whole of this work or machine, nobody will be found able to get any benefit or to imitate it, because of its great expense and cost ... Because of that, in this machine, it has to be praised only the rarity of the invention which allows the elevation of water from the river upwards to an altitude of 300 feet. In the end, because this device is kept like an entirely royal work [*como una obra enteramente real*] and very suitable of admiration, and because an individual cannot aspire to imitate it, I decided to give up trying to obtain a reliable representation of it, and because of that I encourage the spirit of the curious which has desire to see it, to find it by himself in the place where its lies. This is really very worthy and ... I hold it as one of the most rare and admirable things that one can see only a few times in all the days of his life.[70]

The stubborn refusal to give any information about the machine was justified by the fact that the Device was too difficult to understand and too expensive to reproduce. However, the official did not openly say that the design of the device was to be kept secret. This foggy behaviour hides probably a sum of different agendas. Even if Plato had disapproved of the popularization of mathematical knowledge, we are here a long way from moral humanist plans. Other issues were at stake: first of all, the prestige deriving from the uniqueness of the machine required secrecy. Secretive practices were not just in the interest of the inventor, as Brunelleschi once explained to Mariano Taccola, and had also shown with the projects for the Florentine Dome.[71] The patron, once a new creation was delivered in his hands, had interest to maintain it undisclosed as well, as in the case of the contract for the Toledo Device, which protected both Torriani's intellectual property (we may say anachronistically) and King Philip's monopoly on such an astonishing machine. The contemporary lack of specific documentation about it can probably be explained in this way.

But if it appears reasonable that the mechanism of the Device were kept secret to the outside world by the court officials in charge of it, how can one

---

70   Lhermite, *El pasatiempos de Jeham Lhermite*, 535-36 (my translation).

71   Long, *Openness, Secrecy, Authorship*, 98.

explain that the same happened even inside this very environment? It has puzzled more than a historian that not even in the manuscript called *Los veinte y un libros de los ingenios y máquinas de Joanelo*, no real drawing of the Toledo Device survives. This book was a kind of technological encyclopaedia of Spanish engineering made under royal patronage around the last quarter of the sixteenth century, and it was for a long time attributed to Torriani and now to Juan de Lastanosa. One of the arguments used as evidence of the rejection of Janello's authorship of *Los veinte y un libros* was the absence of any description or illustration of the most important of all Spanish sixteenth century machines. But if one considers that this technical book was supposed to be used as a manual at court, in the service of royal engineers and by anybody with technical curiosity, we may understand the reason for this absence, especially if Janello was its author or curator: it was a protectionist strategy.[72] The technical model of this water-raising system was lost with the destruction of the device and with the dispersion of the secret paper that Janello wrote and that, after the clockmaker's death, the King transferred to other places where probably they were lost in a fire.[73]

There were also some models of the Toledo Device sent to Mantua, Rome, Venice and Cremona in order to obtain privileges for invention and, in the case of Cremona, to insure the glorious memory of Torriani. This introduces a further problem about the King, his engineer, and the secrecy of the technical details of his device: was the King aware of Janello's pursuit of privileges in Italy? Most probably he was: the mechanical principles invented by Janello for the Toledo Device were indeed published by these models; what was not accessible to everyone was Janello's know-how in turning the principle expressed by these models into such a large scale functioning machine. The Toledo Device, as the Microcosm and the Crystalline, was a unicum. We have seen how a royal official had urged Philip II to maintain Janello at court, because without him, his amazing planetary automata were to be considered worthless, because no one would have been able to run them properly.

---

72   This manuscript was indeed an important medium for the dissemination of models and ideas. And even if the book was the work of some competitor, it should not surprise us that the Toledo Device was not there: why should Janello and his heirs have ever given up their monopoly?

73   When Janello died, the King asked for "... *los ynstrumentos y otras cossas del dicho Juanello que se avran de tomar para nuestro servicio* ...": these consisted of six chests full of papers, books and iron instruments that were brought to Madrid. They may have been lost during the fire of the *Alcàzar* of Madrid, or in the one of El Escorial. Cervera Vera, *Documentos biográficos*, doc. 125, 126 and 132.

The true genius of Torriani, according to Ambrosio de Morales, the other best witness of the mechanism of Janello's Device, was most expressed not in the originality of the design and the ergonomics of these bronze vessels, which he called "tubes", but above all in the harmonization of the motion of such a large number of containers whose extremities were:

> Accommodated with a strange design, to pour and receive without losing one drop. Said like this, it would seem a small thing but in seeing it, it surely surprises; because you see how it was necessary to have that [specific] design, not any another, and this is extremely new ... there are many particularities of great marvel in this, but two baffle more than the others. The first consists in the tempering of all the movements with this measure and proportion, that each thing moves in harmony at the command of only one gear, moved by the waters of the river ... and if all the tubes had had the same weight, it would seem to not have been as much of a marvel in the admiring of this *concerto* of movements. But on being empty, as we said, and the other full, controlling such a big uniformity of the movement one together with the other, is something that surpasses any comprehension, even after having seen it, moreover having been invented and put into practice. More than this, if all the movement was continuous it would not have been such a marvel; but being [the movement] so different, this marvels and confounds the understanding without being able to comprehend, nor take one step in the astounding invention. Seeing as without ever stopping the wooden structure and being tied to it the tubes and recipients of bronze, and moving to the rhythm of that, when they conjoin to dump and receive the water, there they arrest and immobilise, as if they were stopped, for the time that it takes to empty one and fill the other, not ceasing in the meantime the movement of the wooden part. And [having] finished the giving and receiving, the tubes return to their movement, as if nothing had ever stopped them. This would not have been possible without the art of the proportions very different and astounding from that which is usually taught in arithmetic ... This was not able to be done [transmit smoothness and speed to a machine that big and heavy] without big calculations of proportion in the balance of the movement; and the finding of them with the ingenious is something rare and never heard of, and the putting of it into execution with such precision was an even bigger marvel.

As previously mentioned, when the same Janello explained to Morales the theoretical basis of the machine, the historiographer grew confused, unable to follow the complexity of "a system of proportions ... different from that of which we know", perhaps a legacy of the Milanese algebraic tradition. Any explanatory drawing – with or without algebraic caluclations – needed to be kept secret, and perhaps to be hidden under a narrative of *sprezzatura*, as previously seen in the case when Janello used to explain the improvement of his machines as effects of his hand, as if it were magic.

Janello's concerns with secrecy seem to have been based on a real necessity. We have seen how Cardano in 1560 had operated a *damnatio memoriae* on Janello's prestige as an inventor, and how Agostino Ramelli in 1584 did not give credit to the royal clockmaker for the machines n. xcv and n. xcvi (Figs 62, 63) of his famous book (see illustrations). Moreover, difficulties regarding communication could easily result in the transfer of one man's credit to some ambitious trickster: at the beginning of the seventeenth century, Pedro Salazar de Mendoza wrote that there was a person in France who had taken credit for the construction of the Toledo Device:

A certain Luis de Fox [Louis de Foix], a Parisian architect, master-builder of San Lorenzo el Escorial and author of Toledo's aqueduct ... We are going now to see who was Luis de Fox ... He was a servant to Janello Torriani native of Cremona in Lombardy, sole inventor and author of the famous Toledo Device. This Petty-Luis was serving him playing the bellows at the forge that he was keeping by this great machine. In San Lorenzo El Real there was no architect named Luis de Fox. The principal ones were Juan Bautista de Toledo and Juan de Herrera. There was however an "*albañil*" [a mason] called Luis, who was said to be French. This could be and for this reason he probably called himself "architect". "*Masón*" is called the "*albañil*" in France, thus like that he could have called himself, and not architect.[74]

---

74   "*Preguntando el Tuano cómo ... cómo vinieron a su noticia estas patrañas, dice se las contó un Luis de Fox, arquitecto natural de París, maestro de las obras de San Lorenzo el Real de El Escorial y autor del acueducto de Toledo ... Veremos ahora quién fue Luis de Fox ... Fue un criado de Juanelo Turriano, natural de Cremona, en Lombardía, único fundador y autor del insigne artificio de Toledo. Servíale Luisillo de sonarle los fuelles de la fragua que tenía para esta gran máquina. En San Lorenzo el Real no ha habido un arquitecto que se llamase Luis de Fox. Los principales fueron Juan Bautista de Toledo y Juan de Herrera. Un albañil hubo que se llamó maese Luis, que dicen era francés. Este pudo ser y por esto llamarse "architector". Masón se llama el albañil en Francia, y así se pudiera llamar y no arquitecto*" (my translation): Pedro Salazar de Mendoza, 1618, en Sánchez Mayendía, "El artificio de Juanelo en la literatura española," 85.

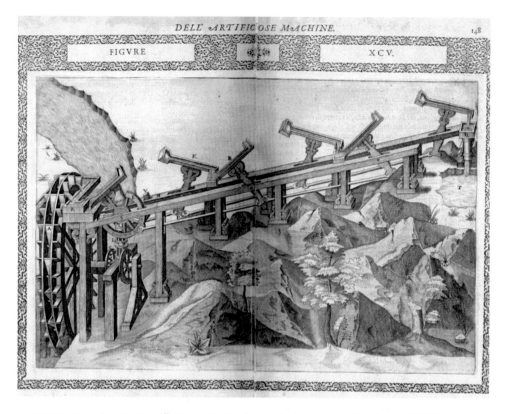

FIGURE 62   *Agostino Ramelli,* Figure xcv, Le diverse et artificiose machine, *Paris, 1588.* This
image presents a hydraulic machine probably inspired by those invented by
Janello Torriani.

In reality, Louis the Foix was a clockmaker and he proved himself to be a good
architect as well. Louis' boastfulness testifies to the fact that sixteenth century
distances were efficient barriers against the circulation of certified news. A
clockmaker like Louis the Foix, once back in his native France, could take
credit for many famous Spanish buildings, increasing his fame and consequently being put in charge of an important work of architecture such as the
light-tower of Bordeaux.[75] In spite of puffing himself up like a balloonfish,
Louis de Foix had indeed been involved to some extent with the water-supply
problem of Toledo. Indeed, from a special royal decree dated 15th of September
1564, we know that he was a royal servant and that he was paid 300 ducats for
models and devices he had made for His Majesty. Louis de Foix could have

75   Grenet-Delisle, *Louis de Foix.*

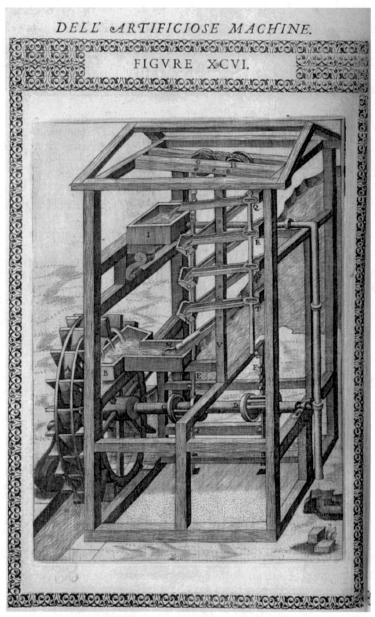

DELL ARTIFICIOSE MACHINE.

FIGVRE XCVI.

FIGURE 63   *Agostino Ramelli,* Figure XCVI, Le diverse et artificiose machine,
              Paris, 1588. This second image presents another hydraulic machine
              probably inspired by those invented by Janello Torriani.

made such models under Janello's direction or in competition with him. From the laconic sources in our possession, we cannot say more. The payment of the 300 ducats was supposed to be divided in three payments of 100 ducats, to be paid in September, at Christmas and in April 1565,[76] the very same month Janello had signed his contract with the city of Toledo and with the King for the construction of the famous water-work. Could Janello have stolen Louis' project? The lack of any doubt to Janello's authorship in the contemporary abundant documentation, plus his successive successful direction of the works of construction, and Louis' fantastic claims even over the project of San Lorenzo del Escorial, makes one reject this possibility.

---

76   1564, 15th of September, Mardid: Cervera Vera, *Documentos biográficos*, doc. 11.

Conclusions

Janello Torriani, the greatest inventor and constructor of machines of the Renaissance, died in poverty at the advanced age of 85. If his purse was empty, his family was still under royal protection, and his fame would survive for centuries, leaving a relevant corpus of documents that allows anyone who is interested in the role of artisans in the history of science and technology to look into the dynamics that are at the base of that phenomenon known as the Scientific Revolution, one of the pillars of the idea of Modernity. Torriani's trajectory, crossing the first three quarters of the sixteenth century, provides us with a useful tool to address questions on how Renaissance techno-scientific knowledge was created, how it was used, how practitioners of mechanical arts gained visibility and glory, why technological innovation developed in certain fields, and how innovation and invention were practiced and circulated. Vitruvian artisans like Torriani are central to a cultural shift that brought what today we call technology, at the core of political action, in terms of invention and innovation; in this period, warfare, trade and administration went through a process of deep transformation that demonstrated the centrality of the rational practices of measurement and design. This book is an attempt to look into three problems that are crucial to explain how these successful artisans were educated, how they had access to power, and how they interacted with these powerful patrons to perform as professional inventors or technological innovators.

Torriani was a commoner of modest background, son of a small businessman who had to provide for the cost of his practical education at the workshop as a clockmaker in the guild of blacksmiths. Like a good part of the urban male population of late Medieval Italian cities, he was also trained in writing vernacular and in elementary mathematics. Probably even before going to school, Janello showed some special talent in mathematics. For this reason, a university-trained physician found it worthy to encourage this talent, transmitting to him astronomical knowledge that was usually the preserve of the university. Physician Giorgio Fondulo played a role of cultural mediator translating Latin knowledge into the vernacular for Janello. To become a successful clockmaker and engineer, Janello Torriani drew upon the best mathematical theoretical knowledge of his time. Renaissance physicians, because of the necessity of sidereal calculations for their profession, were most closely involved with the study of the *quadrivium* at a university level. Humanist ideology created the conditions for such a tutorship. The result was a mixed education that drew on refined technical skills and sophisticated university knowledge, contradicting the representation of a society in which theory and practice stood completely separated. The workshop of the "superior craftsman" was a place of sociability where curious and ambitious artisans met university-trained literati to exchange knowledge in the sign of Vitruvius – an architect at the time of the Emperor Augustus – and of the model produced by his *De Architectura*. Superior craftsmen like Janello Torriani demonstrated the enormous ideological impact of this work on the whole of Renaissance Europe. In his tenth and final book, Vitruvius speaks at length of machines, but it is in his first one that he presents what may be defined as the Hellenistic paradigm of knowledge endorsed by these ingenious artisans "with the compass in the eye", to use Michelangelo's terms. It is remarkable, that of all books of Janello Torriani dispersed library, at the moment we know only of two, and they are both editions of Vitruvius' *De Architectura*. Taking this book as a model, Janello constructed a hydraulic machine and a ballista. Janello, like Brunelleschi, Aristotile of Bologna, Leonardo da Vinci, and many others implemented this Vitruvian ideal, demonstrating how powerful was the application of mathematics to control and consequently understand Nature.

Since the Middle Ages, in a cultural climate of growing awareness of the importance of Archimedes, Vitruvius and other great engineers of Antiquity, communal governments and petty tyrants created specific technical offices for specialized mathematical craftsmen. Engineers, architects, surveyors and public-clock keepers were appointed long before any curricula were created for such practical mathematical professions. During the sixteenth century the first guilds and court academies were created to unify and institutionalize the educational pattern of these artisans. Janello still embodies the free model of clockmaker and engineer, where the capacity to invent, despite the professional background, was a basic asset in the quest for public posts, especially at court. The court offered different opportunities to these Vitruvian craftsmen: power, dignity, wealth, personal security and freedom from guild obligations. The above-mentioned models from Roman and Greek antiquity created a new space of respectability for these inventive artisans.

Torriani did not enter the imperial entourage just because he was exceptionally gifted and skilled. He entered court because, besides being exceptionally talented, he interacted with a chain of doormen who held the keys to access different steps on the way to the vertex of the pyramid, where practical mathematics was considered to be a suitable field of investment for princely patronage. These doormen or mediators had an interest in selecting and promoting skilled craftsmen who could solve problems in specific fields particularly relevant to the prince. The Habsburg dominions linked specific territorial practitioners of crafts and science who moved along the channels provided by feudal nobility. Observing the long career of Janello Torriani, one can see emerging a consistent network of power: in his case it was a

Ghibelline (or philo-Spanish) group based geographically around Torriani's city. Presenting the best craftsmen of one's land to the Emperor allowed one to gain prestige in the eyes of the ruler, especially if this art was reflecting the prince's personal interests. A Vitruvian craftsman like Janello Torriani found at court a fruitful garden to cultivate in the sign of practical mathematics: formally a blacksmith, under princely patronage he partook in different activities such clockmaking, automata-making, hydraulic engineering, architecture, astronomical observation, surveying, musical instruments-making, the invention of combination-locks, of mechanical instruments, and in the writing of technical-treatises. Torriani's experience shows that there was no contradiction among these endeavours, especially at court where guild obligations did not exist, and where the personal interest of sixteenth century princes was a paramount driver in technological innovation in different fields. Charles v's love for clocks and Philip ii's program of monumental power-representation imposed on Torriani the field in which he had to employ his inventive energy. Torriani did not create new clocks, water-lifting devices and calculating mathematical instruments of his own accord, but because the Emperor, the King and the Pope precisely asked him to do that.

Janello, Jacopo da Trezzo and Leone Leoni were used as instruments of credit by the governors of Milan and by their local allies, in a game played to increase and maintain the monarch's favour. Celebrating Janello Torriani in the sixteenth century meant celebrating his patrons. It is noteworthy that excluding Miguel de Cervantes and Luis de Góngora, all the other most representative authors of the Spanish *Siglo de Oro* literature who mentioned Torriani's creation were linked to the court and consequently to the memory of Torriani and of his glorious patrons: Charles v and Philip ii. Ambrosio de Morales, Esteban de Garibay, José de Sigüenza and Sebastián de Covarrubias were all employed at the court of Philip i; Francisco de Quevedo, Calderon de la Barca and Lope de Vega had instead studied at the *Colegio Imperial* of Madrid, the Castilian royal academy. The clockmaker made use of the cultural codes to fashion a new identity for himself. He changed his family name into one that sounded more aristocratic, and he proudly enjoyed the status of imperial and royal servant who moulded his professional identity into the more respected modes of architect and mathematician. Humanist culture provided Torriani with models he endorsed in order to claim for himself and for his household a better place in society: according to classical ethos, *virtus* could substitute for an obscure origin and turn a vile craftsman into a new Archimedes or a modern Daedalus worthy to be celebrated with the brush, the pen, the chisel and bronze.

Courtly patronage also facilitated the circulation of knowledge. After Torriani's innovative planetary clock – the Microcosm, also known as Caesar's

Sky – appeared in Milan, Innsbruck, Augsburg and Brussels, Immser, Baldewein and other clockmakers also applied spring-driven mechanisms to planetary automata. Janello's story shows us how princely patronage drew knowledge from different parts of the empire to meet at court, as it happened in Toledo and in Brussels, when Charles V opened a floor of scientific exchange for Gerhard Mercator and Janello Torriani. Torriani's innovations in clockmaking and hydraulic engineering were directed, as elsewhere in that time, to precision and efficiency in the micro and macro scales. Janello played a remarkable role in this regard: he produced the first mega-device of history, and the first known gear-cutting machine-tool that allowed for an increased standardisation, miniaturization, and for a quicker production of mechanical components.

Princely patronage in hydraulic engineering may have triggered a speeding up of the market for inventions in this field: great hydraulic enterprises such as the one in Toledo deployed vast resources, setting a standard for European elites. Craftsmen able to design water-lifting devices understood that, beside warfare technology, this was a market that offered great opportunities. Torriani himself sought for privileges for invention, and thanks to the brokerage of members of his patronage network, he was able to obtain them in several states; but not in Florence. This specific case is very interesting in terms of illustrating how the market for invention worked in this period: indeed, it depended as much on personal loyalties as on the quality of technological creation. The Grand Duke of Florence refused to grant Torriani a privilege for invention because he wanted to protect engineers from his own household working on a similar project. Moreover, Torriani's career shows how power and inventor had different ways of funding technological innovation that could be more or less favourable to the artisan. Janello Torriani's constant search for economic resources was not a matter of greed. In fact, he was the head of a large household made of relatives and employees, and his constant lack of money put all these people in jeopardy. Janello's trajectory shows that behind the name of a Renaissance Vitruvian artisan there was an enterprise involving numerous people.

As concerns the practice of secrecy and openness, Janello offers some interesting insights. His amazing planetary automata and the Toledo Device were unique secret marvels. The Emperor and the King of Spain were the only rulers who could boast such technological *monstra* as tangible signs of their prestige, and, together with their Cremonese Archimedes, they maintained the technical anatomy of these machines secret. They were so effective in that that we do not have technical images of Janello's device, a situation that paradoxically provoked a nearly total oblivion for the most successful creator of machines of that time.

Bibliography

Manuscript Sources

Alieri, Bernardino. "Epistolario tra Giorgio Fondulo e Paolo Trizio," n.d. Bibliothéque Nationale de France, Parigi.

Arisi, Desiderio. "Accademia de' Pittori, Scultori, ed Architetti Cremonesi altramente detta Galleria di Homini illustri," end XVII-beginning XVIII century. Deposito Libreria civica, A.A.2.16. Biblioteca Statale e Libreria Civica di Cremona.

Borromeo, Federico. "De fabricatis olim Typis orbium Caelestium Libellus," 1628. Cod. G 9 inf, n. 4. Biblioteca Ambrosiana.

Borromeo, Federico. "De Principium gratia," 1625. G 310 inf, n. 3. Biblioteca Ambrosiana.

Capilupi, Camillo. "Maestro Gianello Cremonese a Carlo V," n.d. Ms. Vittorio Emanuele 1062.

Capilupi, Camillo. "Quesito Elegantísimo Di Maestro Gianello a Carlo V Imperatore," n.d. Biblioteca Nazionale di Roma.

Da Vinci, Leonardo. "Cod. Madrid I," n.d. Biblioteca Nacional de España.

Da Vinci, Leonardo. "Ms. B," n.d. Institut de France, Paris.

Divizioli, Giovanni Francesco. "Memorialle Jovanes Francisci Divittioli Ferariis Cremonensis. 1575 – La Fabbricazione Dell'astrolabio," 1575. Biblioteca Statale e Libreria Civica di Cremona.

Kyeser, Conrad."Bellifortis," n.d. Niedersächsische Staats – u. Universitätsbibliothek – Göttingen.

Lhermite, Jehan. "Le Passetemps," n.d. Bibliothèque royale de Belgique.

Pietragrassa, Bartolomeo. "Annotazioni diverse Spettanti alla Fondaz[io].ne della Regia Città di Pavia," 1636. Manoscritti Ticinesi, n. 113. Biblioteca Universitaria.

Seuse, Heinrich. "Ms. IV, 111," n.d. Bibliothèque royale de Belgique.

"Statuta paratici ferrariorum 1474-1590," n.d. B. B. 1. 7. / 17.

Tritio, Paolo. "De Astrolabio," n.d. Biblioteca Ambrosiana.

Turriano, Juanelo. "Breve discorso di Gianello Turriano Cremonese alla Ma[es]ta del Re Catt[oli]co intorno la riduttione dell'anno et restitutione del Calendario; con la dichiaratione degli instrumenti da esso ritrovati; per mostrarla in atto prattico, copertina del trattato che accompagnava gli strumenti e tabelle matematici," 1579. Biblioteca Vaticana.

Turriano, Juanelo. "Del planispherio," n.d. Manuscrito cod. no 2054. Biblioteca de la Universidad de Salamanca.

Printed Sources

Abeler, Jürgen. *In Sachen Peter Henlein*. Wuppertal: Selbstverlag, 1980.

Acidini Luchinat, Cristina, Vincenzo Cazzato, Francesco Gurrieri, and Alessandro Vezzosi. *La Fonte delle fonti: iconologia degli artifizi d'acqua*. Firenze: Alinea, 1985.

Ackerman, James S. "'Ars Sine Scientia Nihil Est': Gothic Theory of Architecture at the Cathedral of Milan." *The Art Bulletin* 31, no. 2 (1949): 84-111.

Addomine, Marisa. "Cenni di storia dell'orologeria da torre." In *Orologi da torre: MAT, Museo arte tempo di Clusone*, edited by Marisa Addomine and Daniele Pons. Milano: Skira, 2008.

Aegidii, Guillermus. *Liber desideratus super celestium motuum indagatione sine calculo*. Edited by Bonino Bonini. Cremona: Carolus de Darleriis, 1494.

Alberti, Leon Battista. *Autobiografia e altre opere latine*. Edited by Loredana Chines and Andrea Severi. Milano: Rizzoli, 2012.

Alberti, Leon Battista. *I libri della famiglia*. Edited by Ruggiero Romano and Alberto Tenenti. Torino: G. Einaudi, 1994.

Aleotti, Giambattista. *Della scienza et dell'arte del ben regolare le acque di Gio. Battista Aleotti detto l'Argenta architetto del Papa, et del publico ne la città di Ferrara*. Edited by Massimo Rossi. Modena: F.C. Panini, 2000.

Almansi Sabbioneta, Carla, ed. *L'Università dei Mercanti e le corporazioni d'arte a Cremona dal medioevo all'età moderna: mostra iconografica e documentaria; catalogo; sala contrattazioni 3-15 giugno 1982*. Cremona: Linograf, 1982.

Almansi Sabbioneta, Carla, ed. "L'arte ferrarecia a Cremona tra i secoli XV e XVIII." In *Il Ferro nell'arte: documenti e immagini*. Cremona: Camera di commercio, industria, artigianato e agricoltura, 1989.

Ammoretti, Carlo. *Memorie storiche sulla vita e sugli scritti di Leonardo da Vinci*. Milano: Alfieri & Lacroix, 1804.

Andretta, Elisa. "Dedicare libri di medicina: Medici e potenti nella Roma del XVI secolo." In *Rome et la science moderne: entre Renaissance et Lumières*, edited by Antonella Romano, 207-55. Collection de l'École française de Rome, 403. [Rome]: École française de Rome, 2008.

Antolín, P. Guillermo. "La Librería de Felipe II (Datos Para Su Reconstitución)." *La Ciudad de Dios* 116 (1919): 36-49.

Aracil, Alfredo. *Juego y artificio: autómatas y otras ficciones en la cultura del Renacimiento a la ilustración*. Madrid: Cátedra, 1998.

Arcangeli, Letizia. "La città nelle guerre d'Italia (1494-1535)." In *Storia di Cremona*, edited by Giorgio Chittolini, 6, il Quattrocento: Cremona del Ducato di Milano: 1395-1535:40-63. Bergamo: Bolis, 2008.

Aretino, Pietro. *Lettere*. Edited by Paolo Procaccioli. Milano: Biblioteca Universale Rizzoli, 1991.

Arisi, Francesco. *Cremona literata*. Vol. II. Cremona: typis A. Pazzoni & P. Montii, 1706.

Arisi, Francesco. *Cremona literata*. Vol. I. Parmae: Pauli Montii, 1706.

Arisi, Francesco. *Cremona literata*. Vol. III. Cremonae: Apud Petrum Ricchini, 1741.

Aristotele. *Analitici secondi*. Edited by Mario Mignucci. Vol. Organon IV. Roma: Laterza, 2007.

Aristotele. *Meccanica*. Edited by Maria Fernanda Ferrini. Milano: Bompiani, 2010.

Asch, Ronald G, and Adolf M Birke. *Princes, Patronage, and the Nobility: The Court at the Beginning of the Modern Age, ca. 1450-1650*. [London, England]; Oxford [England]; New York: German Historical Institute London ; Oxford University Press, 1991.

Astolfi, Gio. Felice. *Della officina istorica*. Venezia: Turrini, 1642.

Atkinson, Catherine. "Inventing Inventors in Renaissance Europe: Polydore Vergil's De Inventoribus Rerum." Mohr Siebeck, 2007.

Azzolini, Lidia. *Palazzi Del Quattrocento a Cremona*. Cremona: Editrice Turris, 1994.

Azzolini, Monica. *The Duke and the Stars: Astrology and Politics in Renaissance Milan*. Cambridge, Mass.: Harvard University Press, 2013.

Babelon, Jean. *Jacopo da Trezzo et la construction de l'Escurial: essai sur les arts à la cour de Philippe II, 1519- 1589*. Paris: E. de Boccard, 1922.

Bacon, Roger. *The "Opus Majus" of Roger Bacon: Ed., with Introduction and Analytical Table Volume 2*. Edited by Henry Bridges. Nabu Press, 2010.

Badoer, Federico. *Notices of the Emperor Charles the Fifth, in 1555 and 1556 Selected from the Despatches of Federigo Badoer, Ambassador from the Republic of Venice to the Court of Bruxelles*. Edited by William Stirling Maxwell. London: Printed for the Philobiblon Soc., 1856.

Baldi, Bernardino. *Cronica de matematici, overo Epitome dell'istoria delle vite loro*. Urbino: Angelo Ant. Monticelli, 1707 <https://archive.org/details/Cronicadematema00Bald>.

Barbieri, Patrizio. "Ancora sulla 'Fontana dell'organo' di Tivoli e altri Automata sonori degli Este (1576-1619)." *L'organo: rivista di cultura organaria e organistica* 37 (2004): 187-221.

Barbisotti, Rita. "Gli inizi della stampa a Cremona (1473-1500)." In *Storia di Cremona : il Quattrocento: Cremona del Ducato di Milano (1395-1535)*, edited by Giorgio Chittolini. Bergamo: Bolis ed., 2009.

Barbisotti, Rita."Janello 'Torresani', alcuni documenti cremonesi e il 'baptismum' del Battistero." *Bollettino Storico Cremonese*, Nuova Serie, 7 (2001): 255-68.

Baron, Hans. *The Crisis of the Early Italian Renaissance : Civic Humanism and Republican Liberty in an Age of Classicism and Tyranny*. 2 vols. Princeton, N.J.: Princeton University Press, 1955.

Baron, Hans. "The Limits of the Notion of 'Renaissance Individualism': Burckhardt after a Century." In *In Search of Florentine Civic Humanism: Essays on the Transition from Medieval to Modern Thought*, [1960, Then modified and enlarged in 1973]., II:155-81. Princeton, N.J.: Princeton University Press, 1988.

Barrera-Osorio, Antonio. *Experiencing Nature: The Spanish American Empire and Theearly Scientific Revolution*. Austin, TX: University of Texas Press, 2006.

Battelli, Giulio. "Il libro universitario." In *Civiltà comunale: libro, scrittura, documento: atti del Convegno, Genova, 8-11 novembre 1988*, 2:281-313. Atti della Società ligure di storia patria, XXIX. Genova: Società ligure di storia patria – Associazione italiana dei paleografi e diplomatisti – Istituto di civiltà classica cristiana medievale: Università di Genova, 1989.

Battisti, Eugenio. *L'antirinascimento, con una appendice di manoscritti inediti*. Milano: Feltrinelli, 1962.

Beck, Patrice. "Le techniciens de l'eau à Dijon." In *Le technicien dans la cité en Europe Occidentale 1250-1650*, edited by Mathieu Arnoux, 109-43. Collection de l'École Française de Rome 325. Roma: École Française de Rome, 2004.

Bédat, Claude. *La bibliothèque du sculpteur Felipe de Castro*. Vol. 5. Paris: E. de Boccard, 1969.

Bedini, Silvio A. "Falconi, Renaissance Astrologer and Astronomical Clock and Instrument Maker." *Nuncius* 19, no. 1 (2004): 31-76.

Bedini, Silvio A. "La dinastia Barocci: artigiani della scienza in Urbino, 1550-1650 = The Barocci dynasty : Urbino's artisans of science, 1550-1650." In *Scienza del Ducato di Urbino = The science of the dukedom of Urbino*, edited by Flavio Vetrano, 7-98. Urbino: Accademia Raffaello, 2001.

Bedini, Silvio A. "L'Orologio notturno: un'invenzione italiana del XVII secolo." In *La misura del tempo: l'antico splendore dell'orologeria italiana dal XV al XVIII secolo*, edited by Giuseppe Brusa, 189-219. Trento: Castello del Buonconsiglio. Monumenti e collezioni provinciali, 2005.

Bedini, Silvio A. "The Role of the Automata in the History of Technology." *Technology and Culture* 5, no. 1 (1964): 24-42.

Bedini, Silvio A., and Francis Maddison. *Mechanical Universe: The Astrarium of Giovanni De' Dondi*. Philadelphia: American Philosophical Society, 1966.

Beeson, Cyril Frederik Cherrington. *Perpignan 1356: The Making of a Tower Clock and Bell for the King's Castle*. London: Antiquarian Horological Society, 1982.

Bellabarba, Marco. *Seriolanti e arzenisti: governo delle acque e agricoltura a Cremona fra Cinque e Seicento*. Cremona: Biblioteca statale e libreria civica di Cremona : Distribuzione, Libreria del convegno, 1986.

Belozerskaya, Marina. *Luxury Arts of the Renaissance*. Los Angeles: J. Paul Getty Museum, 2005.

Beltrani, Giovanni. *Leonardo Bufalini e la sua pianta topografica di Roma: (Estratto della Rivista Europea-Rivista Internazionale.)*. Firenze: Tipogr. della Gazzetta d'Italia, 1880.

Beltrami, Luca. *Aristotele da Bologna al servizio del duca di Milano MCCCLVIII-MCCCCLXIV: documenti inediti pubblicati*. Milano: A. Colombo & A. Cordani, 1888.

Ben-Yami, Hanoch. *Descartes' Philosophical Revolution: A Reassessment*. Basingstoke: Palgrave Macmillan, 2015.

Ben-Yami, Hanoch. "L'ingresso degli automi nella filosofia al tempo di cartesio." Edited by Cristiano Zanetti. *La Voce di Hora* Atti della conferenza Janello Torriani, genio del Rinascimento (forthcoming 2017).

Benisovich, Michel N. "The Drawings of Stradanus (Jan van Der Straeten) in the Cooper Union Museum for the Arts of Decoration, N.Y." *The Art Bulletin / Ed. John Shapley [U.a.]*. 38, no. 4 (December 1956): 249-51.

Benito Ruano, Eloy. "Un competidor de Juanelo Turriano (y otro proyecto de 'Artificio' para Toledo)." In *Studia Historica e Philologica in honorem M. Batllori*, edited by Jorge de Esteban, 83-88. Roma: Instituto Español de Cultura, 1984.

Bennett, Jim. "Gli strumenti astronomici prima del Seicento." In *Galileo : immagini dell'universo dall'antichità al telescopio*, edited by Paolo Galluzzi, 219-25. Firenze: Giunti, 2009.

Benzoni, Gino. "Francesco I de' Medici, granduca di Toscana." *Dizionario biografico degli italiani*. Roma: Istituto della Enciclopedia italiana, 1997.

Berghahn, Volker R, and Simone Lässig. *Biography between Structure and Agency: Central European Lives in International Historiography*. New York: Berghahn Books, 2008.

Berkel, Klaas van. "'Cornelius Meijer Inventor et Fecit': On the Representation of Science in Late Seventeenth – Century Rome." In *Merchants & Marvels: Commerce, Science, and Art in Early Modern Europe*, edited by Pamela H Smith and Paula Findlen, 277-94, 2002.

Berra, Giacomo. "L'Arcimboldo 'c'huom forma d'ogni cosa': capricci pittorici, elogi letterari e scherzi poetici nella Milano di fine Cinquecento." In *Arcimboldo / Palazzo Reale*, edited by Sylvia Ferino-Padgen. Milano, 2011.

Bertinelli Spotti, Carla, Maria Teresa Mantovani, and Giovanna Ferrara Bondioni. *Cremona: momenti di storia cittadina*. Cremona: Turris editrice, 1996.

Bertolotti, Antonino. *Artisti lombardi a Roma nei secoli XV, XVI e XVII: studi e ricerche negli archivi romani*. Milano: Ulrico Hoepli, 1881.

Biadi, Luigi. *Notizie sulle antiche fabbriche di Firenze non terminate e sulle variazioni alle quali i piu ragguardevoli edifizj sono andati soggetti: operetta*. Firenze: Stamperia Bonducciana, 1824.

Biagioli, Mario. *Galileo, Courtier: The Practice of Science in the Culture of Absolutism*. Chicago: University of Chicago Press, 1993.

Biagioli, Mario. *Galileo's Instruments of Credit: Telescopes, Images, Secrecy*. Chicago: University of Chicago Press, 2006.

Biagioli, Mario. "The Social Status of Italian Mathematicians, 1450-1600." *History of Science* 27 (1989): 41-95.

Biral, Alessandro, and Paolo Morachiello. *Immagini dell'ingegnere tra Quattro e Settecento: filosofo, soldato, politecnico*. Edited by Antonio Manno. Milano, Italy: F. Angeli, 1985.

Black, Antony. *Guilds and Civil Society in European Political Thought from the Twelfth Century to the Present*. London; New York: Methuen, 1984.

Bleichmar, Daniela, ed. *Science in the Spanish and Portuguese Empires, 1500-1800*. Stanford, CA: Stanford University Press, 2009.

Bödeker, Hans Erich, ed. *Biographie schreiben*. Göttingen: Wallstein, 2003.

Boethius, Anicius Manlius Torquatus Severinus. *La consolazione della filosofia*. Edited and translated by Ovidio Dallera. Milano: Rizzoli, 1977.

Bonetti, Carlo. "La libreria dello storico e pittore Antonio Campi." *Cremona* IV, no. I (1932): 5-11.

Bonetti, Carlo. "L'assedio di Cremona (Agosto-Settembre 1526)." *Rivista Militare Italiana*, 1916.

Bonetti, Carlo. *Memorie: la fabbrica della Cattedrale: laica od ecclesiastica?* Cremona, 1936.

Bordigallo, Domenico. *Urbis Cremonae syti designum*. Edited by Emanuela Zanesi. Cremona: Associazione ex Alunni del Liceo-ginnasio "Daniele Manin," 2011.

Borgogni, Gherardo. *La fonte del diporto* ... Bergamo: Per Comin Ventura, 1598.

Borromeo, Federico. *Il libro intitolato La gratia de' principi di Federico Borromeo* ... In Milano: [s.n.], 1632.

Boskovits, Miklós. "Giotto di Bondone." *Dizionario biografico degli italiani*. Roma: Istituto della Enciclopedia italiana, 2001.

Boucheron, Patrick. "L'artista imprenditore." In *Produzione e tecniche*, edited by Philippe Braunstein and Luca Molà, volume terzo:417-38. Il Rinascimento italiano e l'Europa. Treviso: Fondazione Cassamarca : Angelo Colla Editore, 2007.

Boucheron, Patrick. *Le pouvoir de bâtir: urbanisme et politique édilitaire à Milan (XIVe-XVe siècles)*. [Rome]: Ecole française de Rome, 1998.

Braudel, Fernand. *Civiltà e imperi del Mediterraneo nell' età di Filippo II*. Vol. 1. Torino: Einaudi, 2002.

Breventano, Stefano. *Istoria della antichita nobilta, et delle cose notabili della citta di Pauia, raccolta da m. Stefano Breuentano cittadino pavese*. In Pauia: appresso Hieronimo Bartholi, 1570.

Brothers, Cammy. "The Renaissance Reception of the Alhambra: The Letters of Andrea Navagero and the Palace of Charles v." *Muqarnas* 11, no. 1 (1994): 79-102.

Bruni, Leonardo. "Funeral Oration for Nanni Strozzi, 1427." In *Major Problems in the History of the Italian Renaissance*, edited by Benjamin G. Kohl and Alison Andrews Smith. Lexington, Mass.: D.C. Heath and Co., 1995.

Brusa, Giuseppe. "Early Mechanical Horology in Italy." *Antiquarian Horology* 18, no. 5 (1990): 485-513.

Brusa, Giuseppe. "I primi orologi da persona in Italia: nuovi indizi e nuove eccellenti testimonianze." *Voce di Hora* 3 (1997): 3-20.

Brusa, Giuseppe. "L'orologio dei pianeti di Lorenzo della Volpaia." *Nuncius* 9 (1994): 645-69.

Bruschi, Arnaldo. *Filippo Brunelleschi*. Milano: Electa, 2006.

Buchanan, Milton A. "Short Stories and Anecdotes in Spanish Plays." *The Modern Language Review* 4, no. 2 (1909): 178-84.

Bugati, Gasparo. *Historia uniuersale*. In Vinetia: Appresso Gabriel Giolito de Ferrarii, 1570.

Bugati, Gasparo. *Historia uniuersale*. In Vinetia: Appresso Gabriel Giolito de Ferrarii, 1571.

Buning, Marius, and Cristiano Zanetti. "Discovering Inventions: A Short History of Inventor's Privileges." In *Janello Torriani, a Renaissance Genius*, 59-60. Cremona: Comune di Cremona, 2016.

Buonora, Paolo. "Cartografia e idraulica del Tevere: secoli XVI-XVII." In *Arte e scienza delle acque nel Rinascimento*, edited by Alessandra Fiocca, Daniela Lamberini, and Cesare Maffioli, 169-93. Venezia: Marsilio, 2003.

Burckhardt, Jacob. *The Civilisation of the Renaissance in Italy*. Salt Lake: Project Gutenberg, 2000.

Burckhardt, Jacob. *The Civilization of the Renaissance in Italy*. Edited by Peter Burke and Peter Murray. Translated by Samuel G.C. Middlemore. London: Penguin Books, 1990.

Burckhardt, Jacob. "The Development of the Individual." In *The Civilization of the Renaissance in Italy*, 82-103. Mineola, NY: Dover Publlications, 2010.

Buren, Anne van. "Reality and Literary Romance in the Park of Hesdin." In *Medieval Gardens: [Dumbarton Oaks Colloquium on the History of Landscape Architecture, 9]*, edited by Elisabeth B. Macdougall, 115-34. Washington D.C.: Dumbarton Oaks Research Library and Collection : Trustees for Harvard University, 1986.

Burke, Peter. *The Historical Anthropology of Early Modern Italy: Essays on Perception and Communication*. Cambridge [Cambridgeshire]; New York: Cambridge University Press, 1987.

Cadenas y Vicent, Vicente de. *Carlos de Habsburgo en Yuste, 3-II-1557-21-IX-1558*. Madrid: Hidalguia, 2000.

Cadenas y Vicent, Vicente de. *Hacienda de Carlos V al fallecer en Yuste*. Madrid: Hidalguia, 1985.

Caine, Barbara. *Biography and History*. Basingstoke; New York: Palgrave Macmillan, 2010.

Cámara Muñoz, Alicia. "Immagini della Orano e della Mazalquivir di Vespasiano Gonzaga in un manoscritto inedito di Leonardo Turriano." *Civiltà mantovana* 3 (2010): 6-35.

Cámara Muñoz, Alicia, Rafael Moreira, Marino Viganò, and Daniel Crespo Delgado. "Leonardo Turriano al servicio de la Corona de Castilla." In *Leonardo Turriano: ingeniero del rey*. Madrid: Fundación Juanelo Turriano, 2010.

Camerani Marri, Giulia, ed. *Statuti delle arti dei Corazzai, dei Chiavaioli, Ferraioli e Calderai, e dei Fabbri di Firenze, 1321-1344, con appendice dei marchi di fabbrica dei fabbri, dal 1369*. Firenze: Olschki, 1957.

Camerota, Filippo. "L'eredità di Euclide: la tradizione dell'abaco." In *I Medici e le scienze: strumenti e macchine nelle collezioni granducali*, edited by Filippo Camerota and Mara Miniati, 23-30. Firenze: Giunti, 2008.

Camerota, Filippo, and Mara Miniati, eds. *I Medici e le scienze: strumenti e macchine nelle collezioni granducali*. Firenze: Giunti : Firenze musei, 2008.

Campagna Cicala, Francesca. "Fondulo, Giovan Paolo." *Dizionario biografico degli Italiani*. Roma: Istituto della Enciclopedia italiana, 1997.

Campano da Novara. *Equatorium planetarum*. Edited by William R. Shea and Tiziana Bascelli. Translated by A. Bullo. 2 vols. Padova: Conselve, 2007.

Campi, Antonio. *Cremona fedelissima città et nobilissima colonia de Romani rappresentata in disegno col suo contado: et illustrata d'una breue historia delle cose piu notabili appartenenti ad essa et de i ritratti naturali de duchi et duchesse di Milano e compendio delle lor vite*. In Cremona: In casa dell'istesso auttore, 1585.

Campi, Antonio. *Cremona, fedelissima città et nobilissima colonia de' Romani: rappresentata in disegno col suo contato, et illustrata d'una breve historia delle cose più notabili appartenenti ad essa, et dei ritratti naturali de' duchi et duchesse di Milano, e compendio delle lor vite*. Milano: Bidelli, 1645.

Campo y Francés, Angel del. *Semblanza iconográfica de Juanelo Turriano*. Madrid: Fundación Juanelo Turriano, 1997.

Campori, Giuseppe. *Artisti degli Estensi: orologieri, architetti ed ingegneri : con documenti inediti ed indici*. Modena: Vincenzi, 1882.

Campori, Giuseppe. *Lettere artistiche inedite*. Modena: Erede Soliani, 1866.

Cantera Montenegro, Santiago. "Los usos del agua en las cartujas de la corona de Castilla, en la transición del Medievo al Renacimiento." In *El medio natural en la España medieval: actas del I Congreso sobre Ecohistoria e Historia Medieval*, edited by Julián Clemente Ramos, 257-75. Cáceres: Universidad de Extremadura, 2001.

Cañizares-Esguerra, Jorge. "Iberian Science in the Renaissance: Ignored How Much Longer?" *Perspectives on Science* 12, no. 1 (Spring 2004).

Cañizares-Esguerra, Jorge. *Puritan Conquistadors: Iberianizing the Atlantic, 1550-1700*. Stanford, CA: Stanford University Press, 2006.

Capilupi, Giulio. *Fabrica et uso di alcuni stromenti horari universali*. Roma: Giliotti, 1590.

Capra, Carlo, and Claudio Donati, eds. *Milano nella storia dell'età moderna*. Milano, Italy: Franco Angeli, 1997.

Carabias Torres, Ana María, ed. *Las Relaciones entre Portugal y Castilla en la época de los descubrimientos y la expansión colonial*. Salamanca, España: Ediciones Universidad de Salamanca, Sociedad V Centenario del Tratado de Tordesillas, 1994.

Cardano, Girolamo. *De libris propriis: the editions of 1544, 1550, 1557, 1562, with supplementary material*. Edited by Ian Maclean. Milano: FrancoAngeli, 2004.

Cardano, Girolamo. *De subtilitate*. Edited by Elio Nenci. Vol. 1, Libri I-VII. Milano: F. Angeli, 2004.

Cardano, Girolamo. *De subtilitate libri XXI: ab authore plusquam mille locis illustrati non-nullis etiam cum additionibus ; addita insuper Apologia adversus calumniatorem, qua vis horum librorum aperitur.* Basileae: Ex Officina Petrina, 1560.

Cardano, Girolamo. *Hieronymi Cardani ... De subtilitate libri xxi.* Parisiis: Ex officina M. Fezendat & R. Granjon, 1550.

Cardano, Girolamo. *Hieronymi Cardani,... De Sapientia libri quinque. Ejusdem de consolatione libri tres... Ejusdem de libris propriis liber unus...* Norimbergae: apud J. Petreium, 1544.

Cardano, Girolamo. *Hieronymi Cardani De svbtilitate libri XXI: nvnc demum recogniti atq perfecti.* Basileæ: per Lvdovicvm Lvcivm, 1554.

Cardano, Girolamo. *Hieronymi Cardani De svbtilitate libri XXII.* Basileæ: per Lvdovicvm Lvcivm, 1554.

Cardano, Girolamo. *Hieronymi Cardani Medici Mediolanensis De svbtilitate Libri XXI.* Norimbergae: Apud Ioh. Petreium, 1550.

Cardano, Girolamo. *Hieronymi Cardani Mediolanensis, De propria vita liber ...* Edited by Gabriel Naudé. Amstelaedami: Apud Joannem Ravesteinium, 1654.

Cardano, Girolamo. *Hieronymi Cardani Mediolanesis Medici Liber de libris propriis.* Lugduni: Apud Gulielmum Rovillium, 1557.

Carlsmith, Christopher. *A Renaissance Education: Schooling in Bergamo and the Venetian Republic, 1500-1650.* Toronto: University of Toronto Press, 2010.

Carocci, Sandro. *Il nepotismo nel Medioevo: papi, cardinali e famiglie nobili.* Roma: Viella, 1999.

Castelnuovo, Enrico, ed. *Artifex bonus: il mondo dell'artista medievale.* Roma: Laterza, 2004.

Castiglione, Baldassare. *Il Libro Del Cortigiano.* Edited by Giulio Preti. Torino: Einaudi, 1965.

Castillo de Bovadilla, Jerónimo. *Politica para corregidores y señores de vassallos, en tiempo de paz, y de guerra, y para juezes eclesiasticos y seglares y de sacas, aduanas, y de residencias, y sus oficiales: y para regidores, y abogados, y del valor de los corregimientos, y goviernos realengos, y de las ordenes.* Vol. I. En Amberes: En casa de Juan Bautista Verdussen, impressor y mercador de libros, 1704.

Cavitelli, Lodovico. *Annales: quibus res ubique gestas memorabiles a patriae suae origine usque ad annum salutis 1583 breviter ille complenus est.* Cremonae: Draconius, 1588.

Cazzola, Franco. "Le bonifiche cinquecentesche nella valle del Po: governare le acque, creare nuova terra." In *Arte e scienza delle acque nel Rinascimento*, edited by Alessandra Fiocca, Daniela Lamberini, and Cesare Maffioli, 15-36. Venezia: Marsilio, 2003.

Ceredi, Giuseppe. *Tre discorsi sopra il modo d'alzar acque da' luoghi bassi [...].* Edited by M. Favia Del Core. Parma: Appresso Seth Viotti, 1567.

Cervera Vera, Luis. *Documentos biográficos de Juanelo Turriani.* Madrid: Fundación Juanelo Turriano, 1996.

Cesariano, Cesare. *Volgarizzamento dei libri IX (capitoli 7 e 8) e X di Vitruvio, De architectura, secondo il manoscritto 9-2790 Seccion de Cortes della Real Academia de la Historia Madrid.* Edited by Barbara Agosti. Pisa: Scuola normale superiore, 1996.

Cesariano, Cesare, and Marco Vitruvius Pollio. *Di Lucio Vitruuio Pollione De architectura libri dece traducti de Latino in vulgare affigurati: commentati & con mirando ordine insigniti.* Impressa nel amoena & delecteuole citate de Como: P[er] Magistro Gotardo da Po[n]te, 1521.

Ceserani, Remo. "Besozzi, Antonio Giorgio." *Dizionario biografico degli italiani.* Roma: Istituto della Enciclopedia italiana, 1967.

Chabod, Federico. "Usi e abusi nell'amministrazione dello stato di Milano a mezzo il '500." In *Studi storici in onore di Giocchino Volpe,* 95-194. Firenze: Sansoni, 1958.

Chandler, Bruce, and Clare Vincent. "To Finance a Clock: An Example of Patronage in the 16th Century." In *The Clockwork Universe,* edited by Klaus Maurice and Otto Mayr, 103-13. Washington: Smithsonian Institution, 1980.

Cimilotti, Ercole. "Lezioni tenute presso l'Accademia degli Inquieti, in casa di Muzio Sforza, marchese di Caravaggio," n.d. Y 93 sup.

Cicero, Marcus Tullius. *De Officiis.* London: A.L. Humphreys, 1902.

Cipolla, Carlo M. *Le macchine del tempo: l'orologio e la società.* Bologna: Il mulino, 1981.

Cipolla, Carlo M. *Vele e cannoni.* Bologna: Mulino, 1983.

Ciriacono, Salvatore. *Building on Water: Venice, Holland, and the Construction of the European Landscape in Early Modern Times.* New York: Berghahn Books, 2006.

Ciriacono, Salvatore. "Trasmissione tecnologica e sistemi idraulici." In *Il Rinascimento italiano e l'Europa,* edited by Philippe Braunstein and Luca Molà, 3, Produzione e tecniche:439-56. Treviso: Fondazione Cassamarca : Angelo Colla Editore, 2007.

Clagett, Marshall. *Archimedes in the Middle Ages.* Vol. 1, The Latin tradition. Madison: The University of Wisconsin Press, 1964.

Clagett, Marshall. *Archimedes in the Middle Ages.* Vol. 2, The translations from the Greek by William of Moerbeke. Philadelphia: The American philosophical society, 1976.

Clagett, Marshall. *Archimedes in the Middle Ages.* Vol. 3, The fate of the medieval Archimedes. Philadelphia: The American philosophical society, 1978.

Clagett, Marshall. *Archimedes in the Middle Ages.* Philadelphia: The American Philosophical Society, 1978.

Clagett, Marshall. *Archimedes in the Middle Ages.* Vol. 4, A supplement on the medieval Latin traditions of conic sections. Philadelphia: The American philosophical society, 1980.

Clagett, Marshall. *Archimedes in the Middle Ages.* Vol. 5, Quasi-Archimedean geometry in the thirteenth century. Philadelphia: The American philosophical society, 1984.

Clericuzio, Antonio. *La macchina del mondo: teorie e pratiche scientifiche dal Rinascimento a Newton.* Roma: Carocci, 2005.

Cocquyt, Tiemen. "Miniaturizzazione degli orologi e lenti d'ingrandimento." Edited by Cristiano Zanetti. *La Voce di Hora* Atti della conferenza Janello Torriani, genio del Rinascimento (forthcoming 2017).

Coll, Jaume, and Ivana Iotta, eds. *Realismo y espiritualidad: Campi, Anguissola, Caravaggio y otros artistas cremoneses y españoles en los siglos XVI-XVIII*. Alaquàs: Ayuntamiento de Alaquàs, 2007.

Conforti, Claudia. "Acque, condotti, fontane e fronde: le provvisioni per la delizia nella Villa Medicea di Castello." In *Il teatro delle acque*, edited by Attilio Petruccioli and D. Jones, 76-89. Roma: Edizioni dell'Elefante, 1992.

Cook, Harold J. "The Cutting Edge of a Revolution?: Medicine and Natural History near the Shores of the North Sea." In *Renaissance and Revolution: Humanists, Scholars, Craftsmen and Natural Philosophers in Early Modern Europe*, edited by J.V. Field and Frank A.J.L. James, 45-62. Cambridge: Cambridge University Press, 1993.

Coppel Areizaga, Rosario. "Carlos V y el Furor." In *Los Leoni (1509-1608): escultores del Renacimiento italiano al sefvicio de la corte de España*. Madrid: Museo del Prado, 1994.

Cormack, Lesley B., Steven A. Walton, and John Andrew Schuster, eds. *Mathematical Practitioners and the Transformation of Natural Knowledge in Early Modern Europe*. Studies in History and Philosophy of Science 45. Cham: Springer, 2017.

Cortesi, Mariarosa. "Libri memoria e cultura a Cremona nell'età dell'Umanesimo." In *Storia di Cremona*, edited by Giorgio Chittolini, 6, il Quattrocento: Cremona del Ducato di Milano: 1395-1535:202-27. Bergamo: Bolis, 2009.

Covarrubias y Orozco, Sebastián de. *Tesoro de la lengua castellana o española*. Madrid: Luis Sanchez, 1611.

Covarrubias y Orozco, Sebastián de. *Tesoro de la lengua castellana o española según la impresión de 1611, con las adiciones de Benito Remigio Noydens publicadas en la de 1674*. Barcelona: S.A. Horta, 1943.

Crespo Delgado, Daniel. "Juanelo Turriano: Genius and Fame." In *Renaissance Engineers*, edited by Alicia Cámara Muñoz and Bernardo Revuelta Pol, English edition 2016., 9-24. Juanelo Turriano Lectures in the History of Engineering. Madrid: Fundación Juanelo Turriano, 2016.

Crespo Delgado, Daniel. "Juanelo Turriano: Janello Torriani in Spanish Literature." In *Janello Torriani, a Renaissance Genius*, edited by Cristiano Zanetti. Cremona: Comune di Cremona, 2016.

Crivello, Fabrizio. "Tuotilo : l'artista in età carolingia." In *Artifex bonus : il mondo dell'artista medievale*, edited by Enrico Castelnuovo, 26-34. Roma: Laterza, 2004.

"Cronaca di Cremona dall'anno 1494 al 1525." *Bibliotheca historica Italica*, 1876, 189-276.

Cruceius, Gasparo Annibal. "Epigramma in Ianelli Turriani Cremonensis horologium." In *Carmina poetarum nobilium Io. Pauli Vbaldini studio conquisita*, by Gio. Pietro Ubaldini, 12-13. Mediolani: apud Antonium Antonianum, 1563.

Da Pozzo, Giovanni. *Storia letteraria d'Italia*. Vol. 1, Il Cinquecento. Padova: Piccin Nuova Libraria, 2007.

Daddi Giovannozzi, Vera. "La vita di Bernardo Buontalenti scritta da Gherardo Silvani: appunti d'archivio." *Rivista d'arte*, 1932, 505-24.

Dal Monte, Guidobaldo. *Guidiubaldi e marchionibus Montis Mechanicorum liber*. Pisauri: apud Hieronymum Concordiam, 1577.

Damler, Daniel. "The Modern Wonder and Its Enemies: Courtly Innovations in the Spanish Renaissance." In *Philosophies of Tecnology: Francis Bacon and His Contemporaries*, edited by Claus Zittel, 11:429-55. Intersections 11. Leiden: Brill, 2008.

Dante Alighieri. "Purgatorio." In *La divina commedia*, edited by Natalino Sapegno. Milano: Ricciardi, 1957.

D'Arcangelo, Potito. "Acque e destinazioni colturali nel Cremonese dal XIII al XV secolo." In *Storia di Cremona: il Quattrocento: Cremona nel Ducato di Milano (1395-1535)*, edited by Giorgio Chittolini, 148-61. Bergamo: Bolis, 2008.

D'Arco, Carlo. "Famiglie mantovane e mille scrittori mantovani," n.d. B.V. Archivio di Stato di Mantova.

Daston, Lorraine, and Katharine Park. "Introduction: The Age of the New." In *The Cambridge History of Science*, edited by Lorraine Daston and Katharine Park, 3. The Early Modern Science:5-6. Cambridge: Cambridge University Press, 2006.

Daston, Lorraine, and H. Otto Sibum. "Introduction: Scientific Personae and Their Histories." *Science in Context* 16, no. 1 (2003): 1-8.

De Caro, Gustavo. "Avalos, Alfonso d'." *Dizionario biografico degli Italiani*. Roma: Istituto della Enciclopedia italiana, 1962.

De Caro, Gustavo. "Capilupi, Camillo." *Dizionario biografico degli Italiani*. Roma: Istituto della Enciclopedia italiana, 1975.

De Caro, Gustavo. "Capilupi, Ippolito." *Dizionario biografico degli Italiani*. Roma: Istituto della Enciclopedia italiana, 1975.

Dear, Peter. "The Meanings of Experience." In *The Cambridge History of Science*, edited by Lorraine Daston and Katharine Park, 3:106-31. Cambridge: Cambridge University Press, 2006.

Dear, Peter. "A Mechanical Microcosm: Bodily Passions, Good Manners, and Cartesian Mechanism." In *Science Incarnate : Historical Embodiments of Natural Knowledge*, edited by Christopher Lawrence and Steven Shapin, 51-82. Chicago: University of Chicago Press, 1998.

Dee, John. *The Mathematicall Praeface to the Elements of Geometrie of Euclid of Megara (1570)*. New York: Science History Publications, 1975.

Del Vecchio, Gian Carlo, and Enrico Morpurgo. *Addenda al Dizionario degli orologiai italiani edizione 1974 di Enrico Morpurgo*. Milano: Tipografia Nava, 1989.

Delbourgo, James, and Nicholas. Dew. *Science and Empire in the Atlantic World*. New York, NY: Routledge, 2008.

Dell'Anna, Giuseppe. *Dies critici: la teoria della ciclicità delle patologie nel 14. secolo*. 2 vols. Galatina (Lecce): M. Congedo, 1999.

Denucé, Jean. *Inventaire des Affaitadi, banquiers italiens à Anvers de l'année 1568*. Anvers: Éditions de "Sikkel," 1934.

Destrez, Jean. *La Pecia dans les manuscrits universitaires du XIIIe et du XIVe siècle*. Paris: Éd. Vautrain, 1935.

Di Renzo Villata, Gigliola. "Grassi, Pietro." *Dizionario Biografico Degli Italiani*. Roma: Istituto della Enciclopedia italiana, 2002 <http://www.treccani.it/enciclopedia/pietro-grassi_(Dizionario_Biografico)/>.

Di Teodoro, Francesco Paolo. "Pacioli, Luca." *Dizionario biografico degli italiani*. Roma: Istituto della Enciclopedia italiana, 2014.

Di Tullio, Matteo. "L'Estimo di Carlo V (1543-1599) e il Perticato del 1558: per un riesame delle riforme fiscali nello Stato di Milano del secondo cinquecento." *Società e Storia* 131 (2011): 1-35.

Dohrn-van Rossum, Gerhard. *History of the Hour: Clocks and Modern Temporal Orders*. Chicago: University of Chicago Press, 1996.

Dolce, Lodovico. *Vita dell'invittissimo e gloriosissimo Imperatore Carlo Quinto*. Vinegia: Appresso Gabriel Giolito de Ferrarii, 1561.

Dolza, Luisa. *Storia della tecnologia*. Bologna: Il mulino, 2008.

Domínguez Bordona, Jesús. "Federico Zúccaro en España." *Archivo español de arte y arqueología*, no. 7 (1927): 77-89.

Drei, Giovanni. "La politica di Pio IV e del cardinale Ercole Gonzaga." *Archivio della R. Società Romana di Storia Patria* XL (1917).

Edelstein, Bruce L. "'Acqua Viva e Corrente': Private Display and Public Distribution of Fresh Water at the Neapolitan Villa of Poggioreale as a Hydraulic Model for Sixteenth-Century Medici Gardens." In *Artistic Exchange and Cultural Translation in the Italian Renaissance City*, edited by Stephen J. Campbell and Stephen J. Milner, 187-220. Cambridge: Cambridge University Press, 2004.

Elias, Norbert. *The Court Society*. New York: Pantheon Books, 1983.

Elliott, John Huxtable. *The Old World and the New 1492-1650*. NY: Cambridge Univ., 1998.

Enciso Recio, Luis Miguel. "Nápoles en tiempos de Felipe II: historiografía reciente." In *Madrid, Felipe II y las ciudades de la monarquía*, edited by Enrique Martínez Ruiz, 1, Poder y dinero:27-72. Madrid: Actas, 2000.

Epstein, Stephan R. "Trasferimento di conoscenza tecnologica e innovazione in Europa (1200-1800)." *Studi storici. Rivista trimestrale dell'Istituto Gramsci* 50, no. 3 (2009): 717-46.

Epstein, Stephan R., and Maarten Roy Prak. *Guilds, Innovation, and the European Economy, 1400-1800*. Cambridge; New York: Cambridge University Press, 2008.

Epstein, Steven A. *Wage Labor and Guilds in Medieval Europe*. Chapel Hill: University of North Carolina Press, 1991.

Erasmus, Desiderius. *Collected Works of Erasmus.* Edited by Dominic Baker-Smith, 1997.

Erasmus, Desiderius. *The Praise of Folly.* Translated by John Wilson. Rockville, MD: Arc Manor LLC, 2008.

Erichson, Klaus E. *Juanelo, der Schmied von Toledo Roman.* Norderstedt: Books on Demand, 2011.

Erizzo, Sebastiano. *Discorso di M. Sebastiano Erizzo sopra le medaglie de gli antichi Con la dichiaratione delle monete consulari & delle medaglie de gli imperadori romani. Nella qual si contiene vna piena & varia cognitione del'istoria di quei tempi.* In Vinegia: Appresso Gio. Varisco & Paganino Paganini, 1559.

Ernst, Germana. "Giuntini, Francesco." *Dizionario biografico degli Italiani.* Roma: Istituto della Enciclopedia italiana, 2001.

Escobar, Jesús. "Francisco de Sotomayor and Nascent Urbanism in Sixteenth-Century Madrid." *The Sixteenth Century Journal* 35, no. 2 (2004): 357-82.

Escriche, Joaquín. *Diccionario razonado de legislación y jurisprudencia.* Edited by Juan María Biec, León Galindo y de Vera, and José Vicente y Caravantes. Madrid: Impr. de E. Cuesta, 1876.

Eser, Thomas. *Die älteste Taschenuhr der Welt?: Der Henlein-Uhrenstreit.* Nürnberg: Germanischen Nationalmuseums, 2014.

Eser, Thomas. "Die Henlein-Ausstellung Im Germanischen Nationalmuseum: Rückblick, Ausblick, Neue Funde." *Deutsche Gesellschaft Für Chronometrie: Jahresschrift* 54 (2015): 23-34.

Esteban Piñeiro, Mariano, and Maria Isabel Vicente Maroto. "La Casa de la Contratación y la Academia Real Matemática." In *Historia de la ciencia y de la técnica en la Corona de Castilla,* edited by José María Lopez Piñero, III, Siglos XVI y XVII:35-52. Valladolid: Junta de Castilla y León, Consejería de Educación y Cultura, 2002.

Esteve Secall, Carlos Enrique. "Aspectos histórico – gráficos de una observación a escala intercontinental: Las Instrucciones del Cosmógrafo Lopez de Velasco." Zaragoza, 2004 <http://www.egrafica.unizar.es/ingegraf/pdf/Comunicacion17110.pdf>.

Fane, Lawrence. "The Invented World of Mariano Taccola: Revisiting a Once-Famous Artist-Engineer of 15th-Century Italy." *Leonardo : Journal of the International Society for the Arts, Sciences and Technology,* 36, no. 2 (2003): 135-43.

Fara, Amelio. *Bernardo Buontalenti.* Milano: Electa, 1996.

Farina, Rachele. *Dizionario biografico delle donne lombarde: 568-1968.* Milano: Baldini & Castoldi, 1995.

Favaro, Antonio. "Nuove ricerche sul matematico Leonardo cremonese." *Bibliotheca matematica,* III, 1905, 326-41.

Felipe II: los ingenios y las máquinas : ingeniería y obras públicas en la época de Felipe II. [Madrid]: Sociedad Estatal para la Conmemoración de los Centenarios de Felipe II y Carlos V, 1998.

Fenti, Germano. *La zecca di Cremona e le sue monete: dalle origini nel 1155 fino al termine dell'attività*. [Cremona]: Linograf, 2001.

Ferguson, Wallace K. *The Renaissance*. New York: H. Holt and Co., 1940.

Fernández Álvarez, Manuel. *Corpus documental de Carlos V (1539-1548)*. Vol. II. Salamanca: [Ed. Universidad], 1975.

Fernández Álvarez, Manuel. *La España del emperador Carlos V: (1500-1558; 1517-1556)*. 3a ed. Historia de España (Espasa Calpe, S.A.) 20. Madrid: Espasa-Calpe, 1982.

Fernández Chaves, Manuel. *Los Caños de Carmona y el abastecimiento de agua en la Sevilla moderna*. Sevilla: Emasesa Metropolitana, 2011.

Fernández Collado, Ángel. *Gregorio XIII y Felipe II en la Nunciatura de Felipe Sega (1577-1581): aspectos políticos, jurisdiccional y de reforma*. Toledo: Estudio Teológico de San Ildefonso, Seminario Conciliar, 1991.

Fernández Collado, Ángel. "Juanelo Turriano y la aportación española a la reforma del calendario gregoriano." *Toletum: boletín de la Real Academia de Bellas Artes y Ciencias Históricas de Toledo*, no. 23 (1989): 151-59.

Ferrari, Daniela. *Le collezioni Gonzaga: l'inventario dei beni del 1540-1542*. Cinisello Balsamo (Milano): Silvana, 2003.

Ferrero, Jesús. *Juanelo O El Hombre Nuevo*. Punto de Lectura. Madrid: Alfaguara, 2000.

Fiocca, Alessandra. "Silvio Belli ingegnere: empiria e matematica nella cultura tecnica del Rinascimento." In *Acque e terre di confine. Mantova, Modena, Ferrara e la bonifica di Burana: studi nel centenario dell'apertura della Botte napoleonica*, edited by Daniele Biancardi and Franco Cazzola, 15-50. Ferrara: Editrice Cartografica, 2000.

Fiorelli, Piero. "Azzone." *Dizionario biografico degli Italiani*. Roma: Istituto della Enciclopedia italiana, 1962.

Fléchon, Dominique. *L'orologiaio: mestiere d'arte*. Milano: Il saggiatore, 1999.

Fondelli, Mario. *Gli "Oriuoli mechanici" di Filippo di ser Brunellesco Lippi: documenti e notizie inedite sull'arte dell'orologeria a Firenze: l'orologio dipinto da Paolo Uccello nel Duomo fiorentino : nuovi studi e precisazioni per la sua lettura*. Edited by Umberto Baldini. Quaderni della critica d'arte. Firenze: Le lettere, 2000.

Fortea Pérez, José I. "Fiscalidad real y politica urbana en el reinado de Felipe II." In *Haciendas forales y Hacienda Real : homenaje a Miguel Artola y Felipe Ruiz Martín: encuentro de Historia Económica Regional*, edited by Emiliano Fernández de Pinedo, 63-79. Bilbao: Universidad del País Vasco, 1990.

Franceschi, Franco. "La bottega come spazio di sociabilità." In *Arti Fiorentine: la grande storia dell'artigianato*, edited by Gloria Fossi and Franco Franceschi, 2, Il Quattrocento:65-83. Firenze: Giunti, 1999.

Freedberg, David. *The Power of Images: Studies in the History and Theory of Response*. Chicago: The University of Chicago Press, 1989.

Freedman, Luba. *Titian's Portraits through Aretino's Lens*. University Park, Pa.: Pennsylvania State University Press, 1995.

Frontinus, Sextus Julius. *De aquae ductu urbis Romae.* Edited by Fanny Del Chicca. Roma: Herder, 2004.

Frugoni, Chiara. *Medioevo sul naso: occhiali, bottoni e altre invenzioni medievali.* Roma: Laterza, 2001.

Gachard, Louis-Prosper. *Relations des ambassadeurs Vénitiens sur Charles-Quint et Philippe II.* Bruxelles: M. Hayez, imprimeur, 1855.

Gachard, Louis-Prosper. *Retraite et mort de Charles-Quint au monastère de Yuste.* Bruxelles: C. Muquardt, 1854.

Gaetano Blancalana, Gilimón. *Disertaciones Y Opúsculos Sobre Toledo.* Toledo: Celya, 2011.

Galasso, Giuseppe. "Aspetti della megalopoli napoletana nei primi secoli dell' età moderna." In *Mégapoles méditerranéennes géographie urbaine rétrospective: actes du colloque, Rome, 8-11 mai 1996,* edited by Claude Nicolet, Robert Ilbert, Jean-Charles Depaule, and École française de Rome, 565-74. Paris: Maisonneuve et Larose, 2000.

Galeati, Giuseppe. *Il Torrazzo di Cremona.* Cremona: Emilio Bergonzi, 1928.

Galluzzi, Paolo. "Dall'artigiano all'artista-ingegnere: Filippo Brunelleschi uomo di confine." In *Arti Fiorentine. La grande storia dell'artigianato,* edited by Franco Cardini and Riccardo Spinelli, 1, Il Medioevo:285-94, 1998.

Galluzzi, Paolo. *Gli ingegneri del Rinascimento da Brunelleschi a Leonardo da Vinci.* Firenze: Giunti, 1996.

Galluzzi, Paolo, Edoardo Speranza, Alessandro Roccati, Giovanni Di Pasquale, Giorgio Strano, Eugenio Lo Sardo, Annamaria Ciarallo, et al. *Galileo: immagini dell'universo dall'antichità al telescopio.* Giunti arte mostre musei. Firenze: Giunti, 2009.

Gamba, Enrico, and Vico Montebelli, eds. *Le scienze a Urbino nel tardo Rinascimento.* Urbino: QuattroVenti : Distribuzione P.D.E., 1988.

García Rey, Verardo. "Temas de arte: Juanelo Turriano ; matemático y relojero." *Arte español,* 1929, 524-26.

García Tapia, Nicolás. "El ingenio de Zubiaurre para elevar el agua del rio Pisuerga a la huerta y palacio del Duque de Lerma." *Boletín del Seminario de Estudios de Arte y Arqueología,* 1984, 299-324.

García Tapia, Nicolás. "La ingeniería." In *Historia de la ciencia y de la técnica en la Corona de Castilla,* edited by José María Lopez Piñero, III, Siglos XVI y XVII:437-66. León: Junta de Castilla y León, Consejería de Educación y Cultura, 2002.

García Tapia, Nicolás. "Los ingenieros y sus modalidades." In *Historia de la ciencia y de la técnica en la Corona de Castilla,* edited by José María Lopez Piñero, III, Siglos XVI y XVII:147-60. León: Junta de Castilla y León, Consejería de Educación y Cultura, 2002.

García Tapia, Nicolás. *Pedro Juan de Lastanosa: : el autor aragonés de Los veintiún libros de los ingenios.* Huesca: Instituto de Estudios Altoaragoneses, 1990.

García Tapia, Nicolás. "Y sin embargo es Lastanosa." *Técnica industrial,* no. 203 (1991): 54-61.

García Tapia, Nicolás, and Javier Rivera Blanco. "Juan Bautista de Toledo, Jerónimo Gil y Juan Herrera: autores de la 'Mar de Ontígola.'" *Boletín del Seminario de Estudios de Arte y Arqueología* 51 (1985): 319-44.

García Tapia, Nicolás, and Javier Rivera Blanco. "La presa de Ontígola y Felipe II." *Revista de Obras Públicas* 132, no. 3236 (1985): 477-90.

García Tapia, Nicolás, and Jesús Carrillo Castillo. *Tecnología e imperio: ingenios y leyendas del Siglo de Oro : Turriano, Lastanosa, Herrera, Ayanz.* Madrid: Nivola, 2002.

García Tapia, Nicolás, and María Isabel Vicente Maroto. "Las escuelas de artillería y otras instituciones técnicas." In *Historia de la ciencia y de la técnica en la Corona de Castilla*, edited by José María López Piñero, 73-82. III, Siglos XVI y XVII. León: Junta de Castilla y León, Consejería de Educación y Cultura, 2002.

García-Diego, José Antonio. *Juanelo Turriano, Charles V's Clockmaker: The Man and His Legend.* Madrid: Castalia, 1986.

Gardenal, Gianna, Patrizia Landucci Ruffo, and Cesare Vasoli. *Giorgio Valla tra scienza e sapienza.* Firenze: L.S. Olschki, 1981.

Gargan, Luciano. "«Extimatus per bidellum generalem studii Papiensis». Per una storia del libro universitario a Pavia nel Tre e Quattrocento." In *Per Cesare Bozzetti: studi di letteratura e filologia italiana*, edited by Simone Albonico. Milano: Fondazione Arnoldo e Alberto Mondadori, 1996.

Gargan, Luciano. "Le note conduxit: libri di maestri e studenti nelle Università italiane del Tre e Quattrocento." In *Manuels, programmes de cours et techniques d'enseignement dans les universités médiévales : actes du Colloque international de Louvain-la-Neuve, 9-11 septembre 1993*, edited by Jacqueline Hamesse, 385-400. Louvain-La-Neuve: Institut d'Études Médiévales de l'Université Catholique de Louvain, 1994.

Garibay y Zamalloa, Esteban de. *Los XL libros d'el compendio historial de las chronicas y vniuersal Historia de todos los reynos de España.* Anueres: por Christophoro Plantino, 1571.

Garibay y Zamalloa, Esteban de. "Memorias." In *Memorial histórico español: colección de documentos, opúsculos y antigüedades que publica la Real Academia de la Historia*, edited by Pascual de Gayangos, Vol. VII. Madrid: Imprenta de José Rodríguez, 1854.

Garin, Eugenio. "Aristotelismo veneto e scienza moderna." In *Umanisti, artisti, scienziati: studi sul rinascimento italiano*, 205-28. Roma: Riuniti, 1989.

Garin, Eugenio. *Lo zodiaco della vita: la polemica sull'astrologia dal Trecento al Cinquecento.* Roma; Bari: Laterza, 1976.

Garin, Eugenio. *Medioevo e rinascimento: studi e ricerche.* 2a ed. Biblioteca di cultura moderna 506. Bari: Laterza, 1961.

Garin, Eugenio. *Umanisti, artisti, scienziati: studi sul rinascimento italiano.* Roma: Riuniti, 1989.

Garzoni, Tommaso. *La piazza universale di tutte le professioni del mondo.* Venetia: Ad instantia di Roberto Meglietti, 1605.

Gaulke, Karsten, ed. *Der Ptolemäus von Kassel: Langraf Wilhelm IV. von Hessen-Kassel und die Astronomie.* Vol. 38. Kataloge der Museumslandschaft Hessen Kassel. Kassel: Museumslandschaft Hessen Kassel, 2007.

Gazzini, Marina. "Confraternite/Corporazioni: i volti molteplici della schola medievale." In *Corpi, "fraternità", mestieri nella storia della società europea,* edited by Danilo Zardin, Trento (Italy : Province), Università degli studi di Trento, and Dipartimento di economia. Roma: Bulzoni, 1998.

Gerosa, Guido. *Carlo V: un sovrano per due mondi.* Milano: A. Mondadori, 1990.

Ghisetti Giavarina, Adriano. "Fioravanti Aristotele (Fieravanti)." *Dizionario biografico degli Italiani.* Roma: Istituto della Enciclopedia italiana, 1997.

Giazzi, Emilio. "Fragmenta codicum: la biblioteca e lo 'scriptorium'presso la Cattedrale di Cremona: sulle tracce di una biblioteca dispersa." In *Cremona: una cattedrale, una città : la cattedrale di Cremona al centro della vita culturale, politica ed economica, dal Medio Evo all'Età Moderna,* edited by Giancarlo Andenna, 92-131. Milano: Silvana, 2007.

Giazzi, Emilio. "Frammenti di codice a Cremona: testimonianze per una storia della cultura cittadina." In *Cremona: una cattedrale, una città : la cattedrale di Cremona al centro della vita culturale, politica ed economica, dal Medio Evo all'Età Moderna,* edited by Giancarlo Andenna, 22-49. Milano: Silvana, 2007.

Giazzi, Emilio. "Letteratura specialistica e biblioteche professionali a Cremona tra Medioevo ed età Moderna." In *I professionisti a Cremona: eventi e figure di una storia centenaria,* edited by Valeria Leoni and Matteo Morandi, 15-23. Cremona: Libreria del Convegno, 2011.

Giglioni, Guido. "La divinazione: motivi filosofici e aspetti sociali." In *Le Scienze,* edited by Antonio Clericuzio and Germana Ernst. Il Rinascimento italiano e l'Europa 5. Costabissara (Vicenza): Angelo Colla Editore, 2008.

Giussani, Achille. *L'ospitalità ai Principi nel palazzo Trecchi.* Seconda edizione. Cremona: Ind. Grafica Editoriale Pizzorini, 1992.

Glick, Thomas F., and Helena Kirchner. "Hydraulic Systems and Technologies of Islamic Spain: History and Archaeology." In *Working with Water in Medieval Europe: Technology and Resource-Use,* edited by Paolo Squatriti, 267-329. Leiden: Brill, 2000.

Góngora y Argote, Luis de. *Teatro completo: Las firmezas de Isabela ; El doctor Carlino ; Comedia venatoria.* Edited by Laura Dolfi. Madrid: Cátedra, 2016.

González, Francisco Antonio, and Juan Tejada y Ramiro, eds. *Colección de cánones de la Iglesia Española.* Madrid: Imprenta de Don José María Alonso, 1849.

Gonzáles-Palacios, Alvar. *Il mobile in Liguria.* Genova: Sagep, 1996.

González Tascón, Ignacio, and Isabel Velázquez Soriano. *Ingeniería romana en Hispania: historia y técnicas constructivas.* [España]: Fundación Juanelo Turriano, 2000.

González Vega, Adela, and Ana Ma Díez Gil, eds. *Títulos y privilegios de Milán: siglos XVI-XVII.* Valladolid: Archivo General de Simancas, 1991.

Goodman, David C. "Philip II's Patronage of Science and Engineering." *British Journal for the History of Science* 16 (1983): 49-66.

Goodman, David C. *Power and Penury: Government, Technology, and Science in Philip II's Spain*. Cambridge: Cambridge University Press, 1988.

Gorla, Alberto, and Rodolfo Signorini. *L'orologio astronomico astrologico di Mantova: le ore medie e solari, lo zodiaco, le fasi lunari, le ore dei pianeti, i caratteri umani, le attività giornaliere e le previsioni astrali*. Edited by Rosa Manara Gorla. Mantova: A. Gorla, 1992.

Goselini, Giuliano. *Vita di don Ferrando Gonzaga principe di Molfetta,*. Collezione di ottimi scrittori italiani in supplemento ai classici milanesi 16. Pisa: Presso N. Capurro co'caratteri di F. Didot, 1821.

Grafton, Anthony. *Leon Battista Alberti: Master Builder of the Italian Renaissance*. New York: Hill and Wang, 2000.

Grandi, Angelo. *Descrizione dello stato fisico, politico, statistico, storico, biografico della provincia e diocesi di Cremona*. Vol. 1. Cremona: Ed. Turris, 1856.

Grant, Edward. *The Foundations of Modern Science in the Middle Ages: Their Religious, Institutional, and Intellectual Contexts*. Cambridge: Cambridge University Press, 1996.

Grasselli, Giuseppe. *Abecedario biografico dei pittori, scultori ed architetti cremonesi*. Milano: Co' Torchj D'Omobono Manini, 1827.

Greenblatt, Stephen. *Renaissance Self-Fashioning: From More to Shakespeare*. Chicago: University of Chicago Press, 1980.

Grendler, Paul F. *Schooling in Renaissance Italy: Literacy and Learning, 1300-1600*. Baltimore: Johns Hopkins University Press, 1989.

Grendler, Paul F. *The Universities of the Italian Renaissance*. Baltimore: Johns Hopkins University Press, 2002.

Grenet-Delisle, Claude. *Louis de Foix: horloger, ingénieur, architecte de quatre rois*. Bordeaux: Fédération historique du Sud-Ouest, 1998.

Grewe, Klaus. "Water Technology in Medieval Germany." In *Working with Water in Medieval Europe: Technology and Resource-Use*, edited by Paolo Squatriti, 129-59. Leiden: Brill, 2000.

Groiss, Eva. "The Augsburg Clockmakers' Guild." In *The Clockwork Universe: German Clocks and Automata, 1550-1650*, edited by Klaus Maurice and Otto Mayr, 57-86. Washington; New York: Smithsonian Institution ; N. Watson Academic Publications : National Museum of History and Technology : Bayerisches Nationalmuseum, 1980.

Gualandi, Maria Letizia. "Roma resurgens: fervore edilizio, trasformazioni urbanistiche e realizzazioni monumentali da Martino V Colonna a Paolo V Borghese." In *Roma del Rinascimento*, edited by Stefano Andretta and Antonio Pinelli, 123-60. Roma del Rinascimento 3. Roma: GLF editori Laterza, 2001.

Gualazzini, Ugo, and Gino Sollazzi, eds. *Statuta et ordinamenta comunis Cremonae facta et compilata currente anno Domini MCCCXXXIX*. Milano: A. Giuffrè, 1952.

Guarinus Veronensis. *Carmina differentialia*. Cremona: Rafainus Ungaronus & Caesar Parmensis, 1494.

Guazzoni, Valerio. "Pittura Come Poesia: Il Grande Secolo Dell'arte Cremonese." In *Storia Di Cremona*, edited by Giorgio Politi, 4, L'età degli Asburgo di Spagna: 1535-1707:350-415. Bergamo: Bolis, 2006.

Guenzi, Alberto, Paola Massa, and Fausto Piola Caselli, eds. *Guilds, Markets, and Work Regulations in Italy, 16th-19th Centuries*. Aldershot, Hampshire, Great Britain; Brookfield, Vt., USA: Ashgate, 1998.

Haag, Sabine. "A Signed and Dated Ivory Goblet by Marcus Heiden." *The J. Paul Getty Museum Journal*, 1997, 45-59.

Hall, A. Rupert. "The Scholar and the Craftsman in the Scientific Revolution." In *Critical Problems in the History of Science*, edited by Institute for the History of Science and Marshall Clagett. Madison: University of Wisconsin Press, 1969.

Hall, A. Rupert. *The Scientific Revolution, 1500-1800 : The Formation of the Modern Scientific Attitude*. London: Longmans, Green, 1954.

Hamilton, Earl J. *American Treasure and the Price Revolution in Spain, 1501-1650*. New York: Octagon Books, 1965.

Hamilton, Nigel. *Biography: A Brief History*. Cambridge, Mass.: Harvard University Press, 2007.

Helbing, Mario. "La scienza della meccanica nel Cinquecento." In *Il Rinacimento italiano e l'Europa*, edited by Antonio Clericuzio and Germana Ernst, 5, Le scienze:573-92. Costabissara (Vicenza): Angelo Colla Editore, 2008.

Hero of Alexandria. *Di Herone Alessandrino De gli automati, ouero, Machine se mouenti, libri due*. Translated by Bernardino Baldi. In Venetia: Appresso Girolamo Porro, 1589.

Hero of Alexandria. *Gli artifitiosi et curiosi moti spiritali di Herrone. Aggiontoui dal medesimo Quattro Theoremi ... Et il modo con che si fà artificiosamte salir vn canale d'acqua viua, ò morta, in cima d'ogn'alta torre*. Translated by Bernardino Baldi. Ferrara: per Vittorio Baldini Stampator ducale, 1589.

Hero of Alexandria. *Hero Alexandrinus Spiritali di Herone Alessandrino ridotti in lingua volgare da Alessandro Giorgi da Vrbino (Versi da G.B. Fatio)*. Translated by Alessandro Giorgi. Urbino: Appresso B. e S. Ragusij fratelli, 1592.

Hero of Alexandria. *Heronis Alexandrini Spiritalium liber*. Edited by Federico Commandino. Paris: Apud Ægidium Gorbinum, 1583.

Hero of Alexandria. *Heronis mechanici Liber de machinis bellicis necnon Liber de geodaesia*. Edited by Francesco Barozzi and Francesco de Franceschi. Venetiis: Apud Franciscum Franciscium Senensem, 1572.

Herrera, Gabriel Alonso de. *Libro di Agricoltura utilissimo, tratto da diuersi auttori ... Dalla Spagnuola nell'Italiana lingua traportato (per Mambrino da Fabriano)*. Translated by Mambrino Roseo. Venetia: Michel Tramezzino, 1557.

Hill, D.R. "Engineering." Edited by Rushdī Rāshid and Régis Morelon. *Encyclopedia of the History of Arabic Science*. London; New York: Routledge, 1996.

Hillard, Denise, and Emmanuel Poulle. *Oronce Fine et l'horloge planétaire de la Bibliothèque Sainte-Geneviève*. Genève: Librairie Droz, 1971.

Hirst, Michael. "Bandinelli, Baccio (Bartolomeo)." *Dizionario biografico degli Italiani*. Roma: Istituto della Enciclopedia italiana, 1963.

Holt, Richard. "Medieval England's Water-Related Technologies." In *Working with Water in Medieval Europe: Technology and Resource-Use*, edited by Paolo Squatriti, 51-100. Leiden: Brill, 2000.

Huizinga, Johan, and Eugenio Garin. *L'autunno del Medioevo*. Milano: Rizzoli, 1995.

Hurtado de Toledo, Luis. "Memorial de Algunas Cosas Notables Que Tiene La Imperial Ciudad de Toledo." In *Relaciones Histórico-Geográfico-Estadísticas de Los Pueblos de de España Hechas Por Iniciativa de Felipe II*, edited by Carmelo Viñas Mey and Ramón Paz. Madrid: Reino de Toledo, 1951.

Ibn al-'Awwām, Yaḥyá ibn Muḥammad. *Le livre de l'agriculture d'Ibn-al-Awam (kitab-al-felahah)* ... Edited by Jean Jacques Clément-Mullet. Paris: A. Franck, 1864.

Ilardi, Vincent. "Eyeglasses and Concave Lenses in Fifteenth-Century Florence and Milan: New Documents." *Renaissance Quarterly*, 1976, 341-60.

Iovino, Andrea, Alessandro Vezzosi, Maria Rascaglia, and Rosanna De Simine. *Acqua, continuum vitae: ...il divenire Mediterraneo nel racconto dell'arte e della scienza*. Salerno: Artecnica production, 2000.

Jacopetti, Nicola Ircas. "Il censimento annonario cremonese nel 1576." *Bollettino Storico Cremonese* XXII (1964 1961): 121-48.

Janson, Horst W. "Bardi, Donato, detto Donatello." *Dizionario biografico degli Italiani*. Roma: Istituto della Enciclopedia italiana, 1964.

Joachim, Harold. "About a Landscape by Altdorfer." *The Art Institute of Chicago Quarterly* 49, no. 3 (1955): 51.

Johnson, Francis R. "Astronomical Text-Books in the Sixteenth Century." In *Science, Medicine, and History : Essays on the Evolution of Scientific Thought and Medical Practice Written in Honour of Charles Singer*, edited by E. Ashworth Underwood, 1:285-302. Oxford: Oxford University Press, 1953.

"Karl V imperatore: L'imperatore Carlo V conferma a Giacomo Trecchi e ai sui fratelli i privilegi loro concessi da Francesco Sforza, duca di Milano," n.d. Ms. Gov. 273.

Keating, Jessica. "The Machinations of German Court Culture, Early Modern Automata." Doctoral dissertation, Northwest University, 2010.

Keller, Alex G. "Zilsel, the Artisans, and the Idea of Progress in the Renaissance." *Journal of the History of Ideas* 11, no. 2 (1950): 235-40.

Keller, Alex G. "A Byzantine Admirer of 'Western' Progress: Cardinal Bessarion." *Cambridge Historical Journal* 11, no. 3 (1955): 343-48.

Keller, Alex G. "A Renaissance Humanist Looks At 'new' inventions: The article 'Horologium' in Giovanni Tortelli's 'De Orthographia.'" *Technology and Culture* 11, no. 3 (1970): 345-65.

King, Henry C., and John R. Millburn. *Geared to the Stars: The Evolution of Planetariums, Orreries, and Astronomical Clocks*. Toronto: University of Toronto Press, 1978.

King, Margaret L. *The Renaissance in Europe*. London: Laurence King Publishing, 2003.

King, Margaret L. "The School of Infancy: The Emergence of Mother as Teacher in the Early Modern Times." In *The Renaissance in the Streets, Schools, and Studies: Essays in Honour of Paul F. Grendler*, edited by Konrad Eisenbichler and Nicholas Terpstra, 41-86. Toronto: Centre for Reformation and Renaissance Studies, 2008.

Klemm, Friedrich. *Technik: eine Geschichte ihrer Probleme*. Freiburg: Alber, 1954.

Knobloch, Eberhard. "Les ingénieurs de la Renaissance et leurs manuscrits et traités illustrés." In *Engineering and engineers: Proceedings of the xxth International Congress of History of Science, (Liège, 20-26 July 1997)*, edited by Michael Claran Duffy, 60:23-65. De Diversis Artibus, XVII. Turnhout, 2002.

Kohl, Benjamin G. "Lovati, Lovato." *Dizionario biografico degli Italiani*. Roma: Istituto della Enciclopedia italiana, 2007.

Kohl, Benjamin G., and Alison Andrews Smith, eds. "The Ordinances of Justice of Florence (1295)." In *Major Problems in the History of the Italian Renaissance*. Lexington, Mass.: D.C. Heath and Co., 1995.

König, Albert. "Giulio Aleni SJ (1582-1649) and the Introduction of Western Water Supply Methods in 17th Century China." In *Proceedings of WWAC2016*. Coimbra, 2016.

Korey, Michael. *The Geometry of Power- the Power of Geometry: Mathematical Instruments and Princely Mechanical Devices from around 1600 in the Mathematisch-Physikalischer Salon*. Dresden: Staatliche Kunstsammlungen Dresden, 2007.

Kristeller, Paul Oskar. "Humanism and Scholasticism in the Italian Renaissance." In *Humanism and Scholasticism in the Italian Renaissance*, edited by Benjamin G. Kohl and Alison Andrews Smith, 285-96. Major Problems in European History Series. Lexington, Mass.: D.C. Heath and Co., 1995.

Kuhn, Thomas S. *The Structure of Scientific Revolutions*. Chicago: University of Chicago Press, 1970.

Lafuente, Modesto. *Historia general de España,*. Vol. 13. Madrid: Establecimiento Tipográfico de Mellado, 1869.

Lamberini, Daniela. *Il principe difeso: vita e opere di Bernardo Puccini*. Firenze: La Giuntina, 1990.

Lamberini, Daniela. "Inventori di macchine e privilegi cinque-seicenteschi dall'Archivio Fiorentino delle Riformagioni." *Journal de la Renaissance*, 2005, 177-91.

Lamo, Alessandro. *Discorso di Alessandro Lamo intorno alla scoltura, e pittura, doue ragiona della vita, ed opere in molti luoghi, ed a diuersi principi, e personaggi fatte dall'eccellentissimo, e nobile pittore cremonese*. Cremona: Nella Stamperia del Ricchini, 1774.

Lamo, Alessandro. *Discorso di Alessandro Lamo intorno alla scoltvra, et pittvra doue ragiona della vita & opere in molti luoghi & à diuersi prencipi & personaggi fatte dall'eccell. & nobile M. Bernardino Campi, pittore cremonese.* Cremona: Appresso Christoforo Draconi, 1584.

Lamo, Alessandro. *Sogno non meno piacevole, che morale.* Cremona: Appresso Cristoforo Draconi, 1572.

Lancetti, Vincenzo. *Della vita e degli scritti di Marco Girolamo Vida Cremonese.* Milano: G. Crespi, 1831.

Landes, David S. *Revolution in Time: Clocks and the Making of the Modern World.* Cambridge, Mass.: Belknap Press of Harvard University Press, 1983.

Landes, David S. *The Unbound Prometheus: Technological Change and Industrial Development in Western Europe from 1750 to the Present.* London: Cambridge U.P., 1969.

Lane, Frederic Chapin. *Venice, a Maritime Republic.* Baltimore: Johns Hopkins University Press, 1973.

Lázaro, Antonio. *Memorias de un hombre de palo.* Madrid: Santillana, 2009.

Leino, Marika, and Charles Burnett. "Myth and Astronomy in the Frescoes at Sant'Abbondio in Cremona." *Journal of the Warburg and Courtauld Institutes,* 2004, 273-88.

Lenner, Antonio. "La scuola di Urbino: gli orologi rinascimentali italiani, dai Barocci ai camerini." In *La misura del tempo: l'antico splendore dell'orologeria italiana dal XV al XVIII secolo,* edited by Giuseppe Brusa, 220-27. Trento: Castello del Buonconsiglio. Monumenti e collezioni provinciali, 2005.

Leonardo, da Vinci. *Leonardo on Painting: An Anthology of Writings.* Edited by Martin Kemp and Margaret Walker. New Haven: Yale University Press, 1989.

Leoni, Valeria, and Matteo Morandi, eds. *I professionisti a Cremona: eventi e figure di una storia centenaria.* Cremona: Libreria del Convegno, 2011.

Leopold, John H., and Clare Vincent. "An Extravagant Jewel: The George Watch." *Metropolitan Museum Journal,* 2000, 137-49.

Leydi, Silvio. "'Al fì, chi vol de tut cora a Milan': arti suntuarie milanesi del Cinquecento." In *Arcimboldo: artista milanese tra Leonardo e Caravaggio/ Palazzo Reale,* edited by Sylvia Ferino-Padgen, 51-63. Milano: Skira, 2011.

Leydi, Silvio. "Un cremonese del Cinquecento 'aspectu informis sed ingenio clarus': qualche precisazione per Giannello Torriani a Milano (con una nota sui suoi ritratti)." *Bollettino Storico Cremonese,* Nuova Serie, 1997, no. 4 (1998): 133-38.

Lhermite, Jehan. *El pasatiempos de Jeham Lhermite: memorias de un gentilhombre flamenco en la corte de Felipe II y Felipe III.* Edited by Jesús Sáenz de Miera. Translated by José Luis Checa Cremades. Aranjuez: Doce Calles, 2005.

Libro de las leyes, privilegios, y provisiones reales del Honrado Concejo de la Mesta general, y cabaña Real destos Reynos: confirmados, y mandados guardar por Su Magestad. Madrid: en casa de Iuan de la Cuesta, 1609.

Liceti, Fortunio. *De anulis antiquis, librum singularem in quo diligenter explicantur eorum nomina multa, prim[a]eua origo, materia multiplex, figurae complures, causa efficiens, fines, vsusve plurimi, diffenrentiae ... & contumulatio cum cadauere priscis temporibus.* Vtini: Typis Nicolai Schiratti, 1645.

Litta, Pompeo. *Celebri famiglie italiane.* Milano: P.E. Giusti, 1819.

Liva, Giovanni. "Il Collegio degli ingegneri architetti e agrimensori di Milano." In *Il Collegio degli Ingegneri e Architetti di Milano*, edited by Giorgio Bigatti and Maria Canella, 9-26. Milano: Franco Angeli, 2008.

Llaguno y Amirola, Eugenio. *Noticias de los arquitectos y arquitectura de España desde su restauración, por –, ilustradas y acrecentadas con notas, adiciones y documentos por Juan Agustín Ceán Bermúdez.* Vol. II. Madrid: Imprenta Real, 1829.

Lomazzo, Giovanni Paolo. *Idea del tempio della pittura.* Milano: per Paolo Gottardo Pontio, 1590.

Lomazzo, Giovanni Paolo. *Trattato dell'arte della pittvra, scoltvra, et architettvra di Gio. Paolo Lomazzo milanese pittore.* Milano: Per Paolo Gottardo Pontio ..., a instantia di Pietro Tini, 1584.

Long, Pamela O. *Artisan/Practitioners and the Rise of the New Sciences, 1400-1600.* Corvallis, OR: Oregon State University Press, 2011.

Long, Pamela O. "Hydraulic Engineering and the Study of Antiquity: Rome, 1557-70." *Renaissance Quarterly*, 2008, 1098-1138.

Long, Pamela O. *Openness, Secrecy, Authorship, Technical Arts and the Culture of Knowledge from Antiquity to the Renaissance.* Baltimore: Johns Hopkins University Press, 2001.

Long, Pamela O. "Trading Zones: Arenas of Exchange during the Late-Medieval/Early Modern Transition to the New Empirical Sciences." *History of Technology* 31 (2012): 5-25.

López Terrada, María Luz, José Pardo Tomás, and John Slater, eds. *Medical Cultures of the Early Modern Spanish Empire.* New Hispanisms : Cultural and Literary Studies. Farnham, Surrey: Ashgate, 2014.

Losano, Mario G. *Storie di automi: dalla Grecia classica alla Belle Epoque.* Torino: Einaudi, 1990.

Losito, Maria. *Pirro Ligorio e il casino di Paolo IV in Vaticano: l"essempio" delle "cose passate".* Roma: Palombi, 2000.

Lubkin, Gregory. *A Renaissance Court Milan under Galeazzo Maria Sforza.* Berkeley: University of California Press, 1994.

Luzio, Alessandro, ed. *L' Archivio Gonzaga di Mantova: la corrispondenza familiare, amministrativa e diplomatica dei Gonzaga.* Vol. II. Mantova: Mondadori, 1993.

Maffioli, Cesare S. "Hydraulics in the Late Renaissance 1550-1625: Mathematicians' Involvement in Hydraulic Engineering and the Mathematical Architects." In *Engineering and Engineers: Proceedings of the xxth International Congress of History of Science, (Liège, 20-26 July 1997)*, edited by Michael Claran Duffy, 60:67-75. De Diversis Artibus, XVII. Tumhout: Brepols, 2002.

Magistretti, Ludovico. "Geni della scienza e straordinari progressi nella misura del tempo: l'eredità di Galileo." In *La misura del tempo: l'antico splendore dell'orologeria italiana dal XV al XVIII secolo*, edited by Giuseppe Brusa, 189-200. Trento: Castello del Buonconsiglio. Monumenti e collezioni provinciali, 2005.

Magro, Baltasar. *El Círculo de Juanelo*. Madrid: Brand Editorial, 2000.

Mainardi, Giuseppe. "Due biblioteche private cremonesi del secolo XV." *Italia medievale e umanistica* 2 (1959): 449-51.

Maiolo, Simeone. *Dies caniculares seu colloquia tria, & viginti ...* Romae: ex officina Ioan Angeli Ruffinelli. Typis Aloysij Zannetti, 1597.

Malaguzzi Valeri, Francesco. "I Parolari da Reggio e una medaglia di Pastorino da Siena." *Archivio storico dell'Arte*, no. 1 (1892): 36-37.

Manetti, Antonio. *Vita di Filippo Brunelleschi*. Edited by Carla Chiara Perrone. Minima 34. Roma: Salerno Editrice, 1992.

Manfredi, Antonio. "Gli umanisti e le biblioteche tra l'Italia e l'Europa." In *Il Rinascimento italiano e l'Europa*, edited by Annalisa Belloni and Riccardo Drusi, 2, Umanesimo ed educazione:267-84. Treviso: A. Colla, 2007.

Marcobruni, Paolo Emilio. *Raccolta di lettere di diuersi principi, & altri signori: che contengono negotij et complimenti in molte graui & importantissime occorrenze*. In Venetia: Appresso Pietro Dusinelli, 1595.

Marías, Fernando. "Entre modernos y el antiguo romano Vitruvio: lectores y escritores de arquitectura en la España del siglo XVI." In *Teoría y literatura artística en España*, edited by Nuria Rodríguez Ortega and Miguel Taín Guzmán, 199-233. Madrid: Real Academia de Bellas Artes de San Fernando, 2015.

Marinelli, Sergio. "The Author of the Codex Huygens." *Journal of the Warburg and Courtauld Institutes* 44 (1981): 214-20.

Marini, Lino. "Lo Stato estense." In *Storia d'Italia*, edited by Giuseppe Galasso, 17, I Ducati padani, Trento e Trieste:3-59. Torino: UTET, 1979.

Marr, Alexander. *Between Raphael and Galileo: Mutio Oddi and the Mathematical Culture of Late Renaissance Italy*. Chicago: The University of Chicago Press, 2011.

Marr, Alexander. "'Gentille Curiosité': Wonder-Working and the Culture of Automata in the Late Renaissance." In *Curiosity and Wonder from the Renaissance to the Enlightenment*, edited by Robert John Weston Evans and Alexander Marr, 149-70. Aldershot,: Ashgate, 2006.

Marr, Alexander. "Understanding Automata in the Late Renaissance." *Le Journal de La Renaissance* 2 (2004): 205-22.

Martellozzo Forin, Elda. *La bottega dei fratelli Mazzoleni, orologiai in Padova, 1569 : la sorprendente attività dell'artigianato padovano nella età di Galileo svelata da inedita documentazione archivistica*. Quaderni dell'artigianato padovano. Saonara Pd: Il prato, 2005, 2005.

Martin, John Jeffries. "The Myth of Renaissance Individualism." In *A Companion to the Worlds of the Renaissance*, edited by Guido Ruggiero, 209-24. Malden, MA: Blackwell Publishers, 2002.

Martín Acosta, María Emelina. "Las remesas de Indias y la política imperial." In *Dinero, moneda y crédito en la monarquía hispánica : actas del Simposio internacional Dinero, moneda y credito. De la Monarquía hispánica a la integración monetaria europea : Madrid, 4-7 de mayo de 1999*, edited by Antonio M. Bernal, 405-24. Madrid: Marcial Pons : Fundación ICO, 2000.

Martín Gamero, Antonio. *Aguas potables de Toledo*. Edited by Gabriel Mora del Pozo. Clásicos toledanos, v. Toledo: Instituto Provincial de Investigaciones y Estudios Toledanos, 1997.

Martínez Millán, José, ed. *La corte de Carlos V*. Madrid: Sociedad Estatal para la Conmemoración de los Centenarios de Felipe II y Carlos V, 2000.

Martínez Millán, José, and Santiago Fernández Conti. *La monarquía de Felipe II: la casa del rey*. Vol. II. Madrid: Fundación Mapfre Tavera, 2005.

Marubbi, Mario. "Le 'Storie del testamento Nuovo': cronaca di un cantiere." In *La cattedrale di Cremona Alessandro Tomei. Saggi di Francesco Gandolfo ... Fotografie di Pietro Diotti.*, edited by Alessandro Tomei and Francesco Gandolfo, 84-161. Cinisello Balsamo (Milano): Silvana, 2001.

Marzari, Giacomo. *La historia di Vicenza*. Vicenza: G. Greco, 1604.

Matthes, Dietrich. "From Methaphysics to Astrophysics: Clocks to Represent the Cosmos." In *Janello Torriani: A Renaissance Genius*, edited by Cristiano Zanetti, 127-30. Cremona: Comune di Cremona, 2016.

Maurice, Klaus, and Otto Mayr, eds. *The Clockwork Universe: German Clocks and Automata, 1550-1650*. New York: Smithsonian Institution, 1980.

McLachlan, James. "Experimenting in the History of Science." *Isis* 89, no. 1 (March 1998): 90-92.

Memorie e documenti per la storia dell'Università di Pavia e degli uomini più illustri che v'insegnarono. Vol. 1, Serie dei rettori e priofessori con annotazioni. Pavia: Bizzoni, 1877.

Micheli, Gianni. "L'assimilazione della scienza greca." In *Scienza e tecnica nella cultura e nella società dal Rinascimento a oggi*, 3:199-257. Storia d'Italia. Annali. Torino: Einaudi, 1980.

Micheli, Giuseppe. *I fatti di Cola di Rienzo*. Roma: Sovera Edizioni, 2002.

Mokyr, Joel. *The Gifts of Athena: Historical Origins of the Knowledge Economy*. Princeton, [N.J.]: Princeton University Press, 2002.

Molà, Luca. "Privilegi per l'introduzione di nuove arti e brevetti." In *Il Rinascimento italiano e l'Europa*, edited by Philippe Braunstein and Luca Molà, 3, Produzione e tecniche:533-72. Treviso: Angelo Colla Editore, 2007.

Molà, Luca. "States and Crafts: Relocating Technical Skills in Renaissance Italy." In *The Material Renaissance*, edited by Michelle O'Malley and Evelyn Welch, 133-53. Manchester: Manchester University Press, 2007.

Molho, Anthony. "Cosimo de' Medici: 'Pater patriae' or 'Padrino?'" *Stanford Italian Review* I/I (1979).

Molinari, Franco, and Daniele Montanari. "Rapporti con i vescovi italiani." In *San Carlo e il suo tempo: atti del Convegno Internazionale nel IV centenario della morte (Milano, 21-26 maggio 1984)*, edited by Danilo Zardin, 303-44. Studi e fonti su San Carlo Borromeo 2. Roma: Edizioni di storia e letteratura, 1986.

Montaigne, Michel Eyquem de. *Viaggio in Italia*. Edited by Giovanni Greco and Ettore Camesasca. Milano: Rizzoli, 2008.

Montañés, Luis. "Los Relojes del Emperador." In *Relojes olvidados: sumario de relojeria historica española ...*, 20. Madrid: Artes Gráf. Faure, 1961.

Montemayor, Julian. *Tolède entre fortune et déclin (1530-1640)*. Limoges: PULIM, 1996.

Montuono, Giuseppe M. "L'approvvigionamento idrico della città di Napoli: l'acquedotto del Serino e il Formale Reale in un manoscritto della Biblioteca Nazionale di Madrid." In *Atti del 2° Convegno nazionale di Storia dell'Ingegneria (Napoli, 7-9 aprile 2008)*, edited by Salvatore D'Agostino, 1029-50. Napoli: Cuzzolin, 2008.

Morales, Ambrosio de. *Las antiguedades de las ciudades de España que van nombradas en la Coronica, con la aueriguacion de sus sitios, y nõbres antiguos*. En Alcala de Henares: en casa de Iuan Iñiguez de Lequeríca, 1575.

Moran, Bruce T. "German Prince-Practitioners: Aspects in the Development of Courtly Science, Technology, and Procedures in the Renaissance." *Technology and Culture* 22 (1981): 253-74.

Moran, Bruce T. "Princes, Machines and the Valuation of Precision in the 16th Century." *Sudhoffs Archiv* 61 (1977): 209-28.

Moreno Nieto, Luis, and Ángel Moreno Santiago. *Juanelo y su artificio: antología*. Toledo: D.B. ediciones, 2006.

Morigi, Paolo, and Girolamo Borsieri. *La Nobilità di Milano, descritta dal R.P.F. Paolo Morigi... aggiunte si il supplimento del Girolamo Borsieri...*. Milano: appresso G.B. Bidelli, 1619.

Morigia, Paolo. *La nobilta di Milano, diuisa in sei libri*. Milano: Nella stampa del quon. Pacifico Pontio, 1595.

Morozzo della Rocca, Raimondo. "Sulle orme di Polo." *L'Italia che scrive* XXXVII, no. 10, Numero speciale dedicato a Marco Polo (1954): 121-22.

Morpurgo, Enrico, ed. *Dizionario degli orologiai italiani: 1300-1800*. Roma: La clessidra, 1950.

Morpurgo, Enrico, ed. *Dizionario degli orologiai italiani: 1300-1800*. Milano: Nicola De Toma, 1974.

Morpurgo, Enrico. *L'origine dell'orologio tascabile*. Roma: La clessidra, 1954.

Morpurgo, Enrico. *L'orologeria italiana dalle origini al Quattrocento*. Roma: La Clessidra, 1986.

Morpurgo, Enrico. "L'orologeria italiana: l'orologio di Bologna e il cardinale Bessarione." *La Clessidra* anno 30, no. 12 (1974).

Muratori, Ludovico Antonio, ed. *Antiquitates Italicae medii aevi*. Mediolani: ex typographia Societatis Palatinae, 1742.

Musoni, Johannes. *Apollo Italicus, nuper in lucem restitutus. His etiam Emblemata accedunt, VIII. Ad Jacobum Albensem Juris consultiss. Ode I*. Ticini: Ex typis Francisci Moscheni, 1551.

Muto, Giovanni, and Rossana Sacchi, eds. "Tra le carte degli Anguissola." In *La città di Sofonisba: vita urbana a Cremona tra XVI e XVII secolo ; mostra documentaria ; (Centro culturale "Città di Cremona" Santa Maria della Pietà, 17 settembre-11 dicembre 1994*. Milano: Leonardo, 1994.

Najemy, John M. *A History of Florence 1200-1575*. Malden, MA: Blackwell, 2006.

Nash, Susie. "Claus Sluter's 'Well of Moses' for the Chartreuse de Champmol Reconsidered: Part I." Edited by Benedict Nicolson. *The Burlington Magazine*, 2005, 798-809.

Navarro Brotóns, Victor. "El Colegio Imperial de Madrid: el colegio de San Telmo de Sevilla." In *Historia de la ciencia y de la técnica en la Corona de Castilla*, edited by José María Lopez Piñero, 53-72. III, Siglos XVI y XVII. Valladolid: Junta de Castilla y León, Consejería de Educación y Cultura, 2002.

Navarro Brotóns, Víctor, and William Eamon. *Mas allá de la Leyenda Negra : España y la revolución científica = Beyond the Black Legend : Spain and the scientific revolution*. Valencia: Instituto de Historia de la Ciencia y Documentación López Piñero : Universitat de Valéncia : C.S.I.C., 2007.

Necipoğlu, Gülru. "From Byzantine Constantinople to Ottoman Kostantiniyye: Creation of a Cosmopolitan Capital and Visual Culture under Sultan Mehmed II." In *From Byzantion to İstanbul: 8000 Years of a Capital*, 262-77. Istanbul: Sabanci University Sakip Sabanci Museum, 2010.

Needham, Joseph. *Science and Civilisation in China*. Vol. 3, Mathematics and the sciences of the heavens and the earth. Cambridge: Cambridge university press, 1959.

Needham, Joseph. *The Grand Titration : Science and Society in East and West*. London: Routledge, 1969.

Neudörffer, Johann. *Des Johann Neudörfer, Schreib- und Rechenmeisters zu Nürnberg, Nachrichten von Künstlern und Werkleuten daselbst aus dem Jahre 1547*. Edited by Andreas Gulden and Georg Wolfgang Karl Lochner. Osnabrück: Zeller, 1970.

North, John David. "Cultura e storia economica: il mercato dell'arte europeo." *Studi storici. Rivista trimestrale dell'Istituto Gramsci* anno 50, no. Luglio-Settembre 2009 (2009).

North, John David. *God's Clockmaker Richard of Wallingford and the Invention of Time*. London; New York: Hambledon Continuum, 2006.

Novati, Francesco. "Due matematici cremonesi del secolo XV: frà Leonardo Antonii e Leonardo Mainardi." *Archivio Storico Lombardo*, IV-anno XXXII, 7 (1905): 218-25.

Östmann, Günther, Anthony Turner, and Giorgio Gregato. "The Origins and Diffusion of Watches in the Renaissance: Germany, France, and Italy." In *Janello Torriani: A Renaissance Genius*, edited by Cristiano Zanetti, 141-52. Cremona: Comune di Cremona, 2016.

Ortega Valcárcel, José. "El microcosmo humanizado." In *Historia de la ciencia y de la técnica en la Corona de Castilla*, edited by Luis García Ballester, 277-444. 1, Edad Media. Valladolid: Junta de Castilla y León, Consejería de Educación y Cultura, 2002.

Padoa Schioppa, Antonio. *Giurisdizione e statuti delle arti nella dottrina del diritto comune*. [1964]. Studia et documenta histriae et iuris 30. Milano: Saggi di storia del diritto commerciale, 1992.

Panciroli, Guido. *Rerum memorabilium iam olim deperditarum, & contrà recens atque ingeniose inventarum libri duo*. Ambergae: Typis Fosterianis, 1599.

Panciroli, Guido. *The History of Many Memorable Things Lost, Which Were in Use among the Ancients: And an Account of Many Excellent Things Found, Now in Use among the Moderns, Both Natural and Artificial*. Translated by Henricus Salmuth. John Nicholson ..., and sold, 1715.

Paoletti, John T., and Gary M. Radke. *Art in Renaissance Italy*. London: Laurence King Publishing, 2011.

Pappus of Alexandria, *Mathematical Collection*, trans. D. Jackson, vol. Book 8, 1970 <http://archimedes.mpiwg-berlin.mpg.de/cgi-bin/toc/toc.cgi?page=2;dir=pappu_coll8_095_en_1970;step=textonly>.

Paravicini Bagliani, Agostino. "Campano da Novara." *Dizionario biografico degli italiani*. Roma: Istituto della Enciclopedia italiana, 1974.

Parker, Geoffrey. *The Military Revolution: Military Innovation and the Rise of the West, 1500-1800*. 2nd ed. Cambridge: Cambridge University Press, 2003.

Parmiggiani, Paolo. "Ranieri." *Dizionario Biografico Degli Italiani*. Roma: Istituto della Enciclopedia italiana, 2016.

Parro, Sisto Ramon. *Toledo en la mano*. Toledo: S. Lopez Fando, 1857.

Pedraza Ruiz, Esperanza. "Corregidores toledanos." *Toletum* 8 (1977): 153-75.

Pérez Pastor, Cristóbal. *Bibliografía madrileña, ó, Descripción de las obras impresas en Madrid*. Madrid: Tip. de los Huérfanos, 1891.

Perkinson, Stephen. "Engin and Artifice: Describing Creative Agency at the Court of France, Ca. 1400." *Gesta / International Center of Medieval Art*, 2002, 51-67.

Perla, Antonio, and Fernando Checa Cremades. "Una visita al monasterio de San Jerónimo de Yuste." In *El monasterio de Yuste*, 15-82. Madrid: Fundación Caja, 2007.

Petrucci, Franca. "Francesco Grassi (Grasso, Crassi, Crasso)." *Dizionario biografico degli Italiani*. Roma: Istituto della Enciclopedia italiana, 2000.

Peurbach, Georg von. *Theoricae novae planetarum*. Norimbergae: Johannes Müller Regiomontanus, 1472.

Piccolomini, Alessandro. *De la institutione di tutta la vita de l'homo nato nobile e in citta libera: libri x*. Venetijs: Apud Hieronymum Scotum, 1545.

Pietrasanta, Silvestro. *De symbolis heroicis, libri IX*. Antuerpiæ: Ex officina Plantiniana Balthasaris Moreti, 1634.

Pinessi, Orietta. *Sofonisba Anguissola: un "pittore" alla corte di Filippo II*. Milano: Selene, 1998.

Piovan, Francesca. "Fondulo, Girolamo." *Dizionario biografico degli Italiani*. Roma: Istituto della Enciclopedia italiana, 1997.

Pissavino, Paolo C. "Le forme della conservazione politica: ragion di stato e utopia." In *Le filosofie del Rinascimento*, edited by Cesare Vasoli. Milano: Bruno Mondadori, 2002.

Plinius Secundus, Gaius. *Storia delle arti antiche*. Edited by Silvio Ferri and Maurizio Harrari. Milano: BUR, 2007.

Poli, Valeria. *Architetti, ingegneri, periti agrimensori: le professioni tecniche a Piacenza tra XIII e XIX secolo*. Piacenza: Banca di Piacenza, 2002.

Politi, Giorgio. *Aristocrazia e potere politico nella Cremona di Filippo II*. Milano: SugarCo, 1976.

Politi, Giorgio. "Ultimi anni d'attività di Gianfrancesco Amidani, mercante-banchiere cremonese (1569-1579)." *Archivio Storico Lombardo: Giornale della società storica lombarda* XI, no. I (dic 1984): 44-91.

Polo, Marco. *Il Milione di Marco Polo*. Edited by Giovanni Battista Baldelli Boni. Firenze: Da'Torchi di Giuseppe Pagani, 1827.

Pompeo Farancovi, Ornella. "La riforma dell'astrologia." In *Le Scienze*, edited by Antonio Clericuzio and Germana Ernst. Il Rinascimento italiano e l'Europa 5. Costabissara (Vicenza): Angelo Colla Editore, 2008.

Ponz, Antonio. *Viage de España: en que se da noticia de las cosas mas apreciables, y dignas de saberse, que hay en ella*. Madrid: Por D. Joachin Ibarra ..., 1776.

Popplow, Marcus. "Hydraulic Engines in Renaissance Privileges for Inventions and "theatres of Machines"." In *Arte e scienza delle acque nel Rinascimento*, edited by Andrea Fiocca, Daniela Lamberini, and Cesare Maffioli, 73-84. Venezia: Marsilio, 2003.

Porres Martín-Cleto, Julio. "El Artificio de Juanelo en 1639." *Anales Toledanos*, 1982, 175-86.

Portuondo, María M. "Lunar Eclipses, Longitude and the New World." *Journal for the History of Astronomy* 40 (2009).

Portuondo, María M. *Secret Science: Spanish Cosmography and the New World*. Chicago: The University of Chicago Press, 2009.

Poulle, Emmanuel. "La produzione di strumenti scientifici." In *Produzione e tecniche*, edited by Philippe Braunstein and Luca Molà, Vol. 3. Il Rinascimento italiano e l'Europa. Costabissara Vicenza: Fondazione Cassamarca : Angelo Colla Editore, 2007.

Poulle, Emmanuel. "L'equatoire de Guillaume Gilliszoon de Wissekerke." *Physis* 3, no. 3 (1961): 223-51.

Poulle, Emmanuel. *Les instruments de la théorie des planètes selon Ptolémée: équatoires et horlogerie planétaire du XIIIe au XVIe siècle.* Genève; Paris: Droz ; H. Champion, 1980.

Poulle, Emmanuel. *Science et astrologie au XVI siècle: Oronce Fine et son horloge planétaire.* Paris: Bibliotheque Sainte-Geneviève, 1971.

Predari, Francesco. *Bibliografia enciclopedica milanese...* Milano: Tipografia M. Carrara, 1857.

Price, Derek John de Solla. "An Ancient Greek Computer." *Scientific American* 201, no. June (1959): 60-67.

Price, Derek John de Solla. "Automata and the Origins of Mechanism and Mechanistic Philosophy." *Technology and Culture* 5, no. 1 (1964): 9-23.

Ptolemeu, Claudi. *Il planisferio di Tolomeo.* Edited by Rocco Sinisgalli and Salvatore Vastola. Firenze: Cadmo, 1992.

Puppi, Lionello. "Capobianco, Giorgio." *Dizionario biografico degli italiani.* Roma: Istituto della Enciclopedia italiana, 1975.

Raby, Julian. "A Sultan of Paradox: Mehmed the Conqueror as a Patron of the Arts." *The Oxford Art Journal,* 1982, 3-8.

Ramírez Lozano, José Antonio. *El relojero de Yuste: los últimos días de Carlos V.* La Coruña: Ediciones del Viento, 2015.

Ramus, Petrus. *Petri Rami Scholarum mathematicarum, libri unus et triginta.* Basileae: Per Eusebium Episcopium, & Nicolai fratris haeredes, 1569.

Renn, Jürgen, and Matteo Valleriani. "Galileo and the Challenge of the Arsenal." *Preprint of the Max Planck Institute for the History of Science* 179 (2001): 1-32.

Reti, Ladislao. *El artificio de Juanelo en Toledo: su historia y su tecnica.* Toledo: Diputación Provincial de Toledo, 1967.

Revel, Jacques. *Giochi di scala: la microstoria alla prova dell'esperienza.* Roma: Viella, 2006.

Rill, Gerhard. "Biglia, Melchiorre." *Dizionario biografico degli Italiani.* Roma: Istituto della Enciclopedia italiana, 1968.

Rizzo, M. "Ottima gente da guerra: Cremonesi al servizio della strategia imperiale." In *Storia di Cremona : l'età degli Asburgo di Spagna (1535-1707),* edited by Giorgio Politi, 126-45. Bergamo: Bolis ed., 2006.

Robolotti, Francesco. "Dei medici cremonesi." In *Effemeridi delle scienze mediche, compilate da Giovambattista Fantonetti,* edited by Giambattista Fantonetti, IX:34-127. Milano: Paolo-Andrea Molina, 1839.

Rodriguez Salgado, Mia J. "Terracotta and Iron: Mantuan Politics (ca. 1450-1550)." In *La Corte di Mantova nell'età di Andrea Mantegna, 1450-1550: atti del convegno : Londra, 6-8 marzo 1992, Mantova, 28 marzo 1992 = The court of the Gonzaga in the age of*

Mantegna, 1450-1550, edited by Cesare Mozzarelli, Robert Oresko, and Leandro Ventura, 15-57. Roma: Bulzoni, 1997.

Rojas Rodríguez-Malo, Juan Manuel, and Alejandro Vicente Navarro. "The Contribution of Archaeology to Understanding Janello's Device." In *Janello Torriani, a Renaissance Genius*, edited by Cristiano Zanetti, 173-74. Cremona: Comune di Cremona, 2016.

Rojas Villandrando, Agustín de. *El viaje entretenido*. Edited by Jacques Revel. [1604]. Espasa-Calpe, 1977.

Rose, Paul Lawrence, and Stillman Drake. "The Pseudo-aristotelian 'Questions of Mechanics' in Renaissance Culture." *Studies in the Renaissance* 18 (1971): 65-104.

Rossi, Paolo. *I filosofi e le macchine, 1400-1700*. [1962]. Milano: Feltrinelli, 2007.

Rossi, Paolo. "La nuova Scienza e il simbolo di Prometeo." In *I filosofi e le macchine, 1400-1700*, Seconda edizione., 177-88. Milano: Feltrinelli, 2007.

Rovetta, Alessandro, and Maria Luisa Gatti Peter, eds. *Cesare Cesariano e il classicismo di primo Cinquecento*. Milano: Vita e pensiero, 1996.

Rustin, Michael. "Reflections on the Biographical Turn in Social Science." In *The Turn to Biographical Methods in Social Science: Comparative Issues and Examples*, edited by Prue Chamberlayne, Joanna Bornat, and Tom Wengraf. London; New York: Routledge, 2000.

Rutkin, H. Darrell. "L'astrologia da Alberto Magno a Giovanni Pico della Mirandola," 2008.

Sacco, Bernardo. *Bernardi Sacci Patritii Papiensis De Italicarvm Rervm Varietate Et Elegantia Libri X ; In Qvibvs Mvlta Scitv Digna Recensentur, De Populorum vetustate, dominio, & mutatione ; Item de Prouinciarum proprietate, & Ro. Ecclesiae amplificatione ...* Pavia: [s.n.], 1565.

Sacco, Bernardo. *De Italicarum rerum varietate et elegantia libri X*. Paiae: Hieronymus Bartholus, 1566.

Sacco, Bernardo. *De Italicarum rerum varietate et elegantia libri X*. Ticini: apud Hieronymum Bartolum, 1587.

Sáenz de Miera, Jesús. "The Emperor's retreat from public life." In *Carolus*, edited by Fernando Checa Cremades. Toledo: Museo de Santa Cruz, 2001.

Sánchez, Antonio. "La voz de los artesanos en el Renacimiento científico: cosmógrafos y cartógrafos en el preludio de la 'nueva filosofía natural.'" *Arbor* 186 (2010): 449-60.

Sánchez Cantón, Francisco Javier, ed. *Inventarios reales: bienes muebles que pertenecieron a Felipe II*. Madrid: Real Academia de la Historia, 1959.

Sánchez Mayendía, José Cristóbal. "El artificio de Juanelo en la literatura española." *Cuadernos Hispanoamericanos* n° 103 (1958): 73-93.

Santa Cruz, Alonso de, Ricardo Beltrán y Rózpide, Antonio Blázquez y Delgado-Aguilera, and Francisco de Laiglesia y Auset, eds. *Crónica del emperador Carlos V*. Madrid: Real Academia de la Historia, 1920.

Santoro, Caterina. *Collegi professionali e corporazioni d'arti e mestieri della vecchia Milano: catalogo della mostra*. Milano: Edizioni dell'Ente manifestazioni milanesi – Archivio storico civico di Milano, 1955.

Saracino, Maria Teresa. *Il Torrazzo e il suo restauro*. Cremona: Banco popolare di Cremona, 1979.

Sawday, Jonathan. *Engines of the Imagination: Renaissance Culture and the Rise of the Machine*. London; New York: Routledge, 2007.

Schama, Simon. *Landscape and Memory*. New York: A.A. Knopf, 1995.

Schloss, Martin F. "Grünewald and the Chicago Portrait." *The Art Journal*, 1963, 10-15.

Schmitt, Jean-Claude. *L'invenzione del compleanno*. Roma; Bari: Editori Laterza, 2012.

Schulz, Knut. "La migrazione di tecnici, artigiani e artisti." In *Il Rinascimento italiano e l'Europa*, edited by Philippe Braunstein and Luca Molà, 3, Produzione e tecniche:89-114. Costabissara Vicenza: Angelo Colla, 2007.

Sella, Domenico, and Carlo Capra. *Il Ducato di Milano dal 1535 al 1796*. Storia d'Italia 11. Torino: UTET, 1984.

Shank, Michael H. "L'astronomia nel Quattrocento tra corti e università." In *Le scienze*, edited by Antonio Clericuzio and Germana Ernst, 3-20. Il Rinacimento Italiano e L'Europa 5. Treviso: Angelo Colla Editore, 2008.

Shapin, Steven. "Invisible Technicians: Masters, Servants, and the Making of Experimental Knowledge." In *A Social History of Truth: Civility and Science in Seventeenth-Century England*, 355-407. Chicago: University of Chicago Press, 1994.

Shapin, Steven. *The Scientific Revolution*. Chicago, IL: University of Chicago Press, 1996.

Shapin, Steven, and Simon Schaffer. *Leviathan and the Air-Pump : Hobbes, Boyle, and the Experimental Life*. Princeton, N.J.: Princeton University Press, 1985.

Signorini, Rodolfo, ed. *Fortuna dell'astrologia a Mantova: arte, letteratura, carte d'archivio*. Mantova: Sometti, 2007.

Sigüenza, José de. *Tercera parte de la Historia de la Orden de San Geronimo Doctor de la Iglesia*. Madrid: En la Imprenta Real, 1605.

Simoni, Antonio. "L'orologio pubblico di Bologna del 1451 e la sua sfera." *Culta Bononia*. 5 (1973): 3-19.

Siraisi, Nancy G. *The Clock and the Mirror: Girolamo Cardano and Renaissance Medicine*. Princeton, NJ: Princeton University Press, 1997.

Sitoni, Giovanni Francesco. *Giovanni Francesco Sitoni: ingeniero renacentista al servicio de la Corona de Espanã : con su códice inédito, Trattato delle virtù et proprietà dell'éacque, en su idioma original y traducido al castellano*. Edited by José A García-Diego and Alex Keller. [Madrid]: Fundación Juanelo Turriano : Editorial Castalia, 1990.

Skaarup, Bjørn Okholm. *Anatomy and Anatomists in Early Modern Spain*. Farnham [u.a.: Ashgate, 2015.

Skinner, Quentin. *Visions of Politics*. Vol. 2, Renaissance virtues. Cambridge: Cambridge University Press, 2002.

Smith, Adam. *Wealth of Nations*. Raleigh, NC: Hayes Barton Press, 2005.

Smith, Jeffrey Chipps. "Nuremberg and the Topographies of Expectation." *Journal of the Northern Renaissance*, November 5, 2009 <http://www.northern renaissance.org/nuremberg-and-the-topographies-of-expectation/>.

Smith, Pamela H. *The Body of the Artisan: Art and Experience in the Scientific Revolution*. Chicago: University of Chicago Press, 2004.

Sommi Picenardi, Guido. *Luigi Dovara, gentiluomo cremonese, agente Mediceo alla corte di Filippo II*. Firenze: Tipografia Galileiana, 1911.

Sommi Picenardi, Guido. "Tentativo fatto dai Francesi per impadronirsi del castello di Cremona nel 1537." *Miscellanea di Storia Italiana* XXIV (1885).

Squatriti, Paolo, ed. *Working with Water in Medieval Europe: Technology and Resource-Use*. Leiden: Brill, 2000.

Squatriti, Paolo, and Roberta Magnusson. "The Technologies of Water in Medieval Italy." In *Working with Water in Medieval Europe: Technology and Resource-Use*, edited by Paolo Squatriti, 244-51. Leiden: Brill, 2000.

Stirling Maxwell, William. *The Cloister Life of the Emperor Charles the V*. Second edition. Boston; New York: Crosby, Nichols & company; C.S. Francis & Co., 1853.

Stirling Maxwell, William. *The Cloister Life of the Emperor Charles V*. 4th ed. London: J.C. Nimmo, 1891.

Strada, Famiano. *De bello Belgico decas prima*. Romae: Scheus, 1648.

Strano, Giorgio, ed. *European Collections of Scientific Instruments, 1550-1750*. Boston: Brill, 2009.

Strano, Giorgio, ed. "The in-Existent Instruments." In *Musa Musaei: Studies on Scientific Instruments and Collections in Honour of Mara Miniati*, edited by Marco Beretta, Paolo Galluzzi, and Carlo Triarico. Firenze: L.S. Olschki-Istituto e museo di storia della scienza, 2003.

Strumia, Alberto. *Introduzione alla filosofia delle scienze*. Bologna: Edizioni Studio domenicano, 1992.

Syson, Luke. "Holes and Loops: The Display and Collection of Medals in Renaissance Italy." *Journal of Design History Journal of Design History* 15, no. 4 (2002): 229-44.

Taddei, Mario, Edoardo Zanon, and Castello sforzesco, eds. *Leonardo, l'acqua e il Rinascimento*. Milano: Federico Motta, 2004.

Tallini, Gennaro. "Giovanni Tarcagnota: bibliografia." *Cinquecento plurale. Bibliografie*, 2012 <http://www.nuovorinascimento.org/cinquecento/tarcagnota.pdf>.

Tamalio, Raffaele. *Francesco Gonzaga di Guastalla, cardinale alla corte romana di Pio IV: nel carteggio privato con Mantova (1560-1565)*. Guastalla: Biblioteca Maldotti, 2004.

Tamalio, Raffaele. "Tra Parigi e Madrid. Strategie famigliari gonzaghesche al principio del Cinquecento." In *La Corte di Mantova nell'età di Andrea Mantegna, 1450-1550: atti del convegno : Londra, 6-8 marzo 1992, Mantova, 28 marzo 1992 = The court of the Gonzaga in the age of Mantegna, 1450-1550*, edited by Cesare Mozzarelli, Robert Oresko, and Leandro Ventura. Roma: Bulzoni, 1997.

Tanton, James Stuart. *Encyclopedia of Mathematics*. New York: Facts on File, 2005.

Tarcagnota, Giovanni. *La città di Napoli dopo la rivoluzione urbanistica di Pedro di Toledo*. Edited by Franco Strazzullo. Roma: Gabriele e Mariateresa Benincasa, 1988.

Tartaglia, Niccolò. *Quesiti et inuentioni diuerse de Nicolo Tartaglia: di nouo restampati con una gionta al sesto libro, nella quale si mostra duoi modi di redur una città inespugnabile. La diuisione et continentia di tutta l'opra nel seguente foglio si trouara notata*. [1546]. In Venetia: Appresso de l'auttore : per Nicolo de Bascarini : ad instantia et requisitione, et a proprie spese de Nicolo Tartaglia autore, 1554.

Tartaglia, Niccolò. *Euclide Megarense acutissimo philosopho solo introduttore delle scientie mathematice*. In Venetia: appresso gli heredi di Troian Nauo alla libraria dal Lione, 1586.

TeBrake, William H. "Hydraulic Engineering in the Netherlands during the Middle Ages." In *Working with Water in Medieval Europe: Technology and Resource-Use*, edited by Paolo Squatriti, 101-27. Leiden: Brill, 2000.

Tesoros de la Real Academia de la Historia. Madrid: Real Academia de la Historia, 2001.

Terrall, Mary. "Biography as Cultural History of Science." *Isis* 27 (2006): 306-13.

The Holy Bible: Containing the Old and New Testaments with the Apocryphal/ Deuterocanonical Books. New Revised Standard Version. New York: Oxford University Press, 1989.

Thompson, David. "Lo sviluppo dell'orologio meccanico: il contesto europeo." In *La misura del tempo: l'antico splendore dell'orologeria italiana dal XV al XVIII secolo*, edited by Giuseppe Brusa, 111-17. Trento: Castello del Buonconsiglio. Monumenti e collezioni provinciali, 2005.

Thorndike, Lynn. *A History of Magic and Experimental Science*. Vol. VI. New York: The Macmillan Company, 1941.

Tilley, Arthur Augustus. *Humanism under Francis I*. London, 1900.

Tiraboschi, Girolamo. *Storia della letteratura italiana*. Vol. 4. 4 vols. Milano: Per Nicolò Bettoni e comp., 1833.

Tovar Martín, Virginia. "Lo urbano y lo suburbano: la capital y los sitios reales." In *Madrid, Felipe II y las ciudades de la monarquía*, edited by Enrique Martínez Ruiz, II, Capitalismo y economía:199-212. Madrid: Actas, 2000.

Turner, Gerard L'E. "Two Early Renaissance Astrolabes by Falcono of Bergamo." Edited by Marco Beretta, Paolo Galluzzi, and Carlo Triarico. *Musa Musaei : Studies on Scientific Instruments and Collections in Honour of Mara Miniati*, 2003, 53-62.

Turriano, Juanelo. *Breve discurso a su majestad el Rey Católico en torno a la reducción del año y reforma del calendario: con la explicación de los instrumentos inventados para enseñar su uso en la prática*. Edited by José A García-Diego and José María González Aboin. [Spain]: Fundación Juanelo Turriano : Castalia, 1990.

Turriano, Juanelo. *Los veínte y vn libros de los ingenios, y maquínas de Iuanelo: los quales le mando escribir y demostrar al Chatolico Rei D. Felipe Segundo Rey de las Hespãnas y Nuevo Mundo*. Edited by José Antonio García-Diego and Alex Keller. Madrid: Fundación Juanelo Turriano, 1996.

Valverde Sepúlveda, Joaquín. *Juanelo Turriano: el relojero del emperador*. Madrid: Rubiños 1860, 2001.

Vasari, Giorgio. *Le opere di Giorgio Vasari, pittore e architetto aretino*. Vol. Parte 1, Volume 1. Firenze: David Passigli e Soci, 1832.

Vasari, Giorgio. *Le vite de' piú eccellenti architetti, pittori, et scultori italiani: da Cimabue insino a' tempi nostri : nell edizione per i tipi di Lorenzo Torrentino, Firenze 1550*. Edited by Luciano Bellosi and Aldo Rossi. Torino: G. Einaudi, 1986.

Vasari, Giorgio. *Le vite de' più eccellenti pittori, scultori e architettori*. Novara: Ist. Geografico de Agostini, 1967.

Vasari, Giorgio. *The Lives of the Artists*. Edited by Julia Conaway Bondanella and Peter E Bondanella. Oxford: Oxford University Press, 1991.

Vasoli, Cesare. "Bonatti, Guido." *Dizionario biografico degli Italiani*. Roma: Istituto della Enciclopedia italiana, 1969.

Vecchia, Damiana. "Nuove ricerche sulla Biblioteca Capitolare di Cremona (secc. IX-XVI)." Università degli studi di Parma- Facoltà di Lettere e Filosofia, relatore Prof. A. Belloni, 1997.

Venturelli, Paola. "La lavorazione di pietre dure e cristalli." In *Produzione e tecniche*, edited by Philippe Braunstein and Luca Molà, 3:261-82. Il Rinascimento italiano e l'Europa. Costabissara Vicenza: Fondazione Cassamarca : Angelo Colla Editore, 2007.

Vergerio, Pietro Paolo. "De Ingenuis Moribus et Liberalibus Adolescentiae Studiis." In *Vittorino Da Feltre and Other Humanist Educators: Essays and Versions : An Introduction to the History of Classical Education*, edited by William Harrison Woodward. Cambridge: Cambridge University Press, 1897.

Vergerio, Pietro Paolo. "Pier Paolo Vergerio Defines Liberal Learning." In *Vittorino Da Feltre and Others Humanist Educators*, edited by William Harrison Woodward, 102-9. Cambridge: Cambridge University Press, 1897.

Vérin, Hélène. *La gloire des ingénieurs: l'intelligence technique du XVIe au XVIIIe siècle*. Paris: Albin Michel, 1993.

Verri, Pietro. *Storia di Milano*. Vol. II. Milano: Soc. Tipografica de' Classici Italiani, 1835.

Vicente Maroto, María Isabel. "Juan de Herrera: un hombre de ciencia." In *Actas del Simposio Juan de Herrera y su influencia (Camargo : 14/17 de julio, 1992)*, edited by Miguel Angel Aramburu-Zabala and Javier Gómez Martínez, 79-89. Santander: Fundación Obra Pía Juan de Herrera, 1993.

Vicente Maroto, María Isabel, and Mariano Esteban Piñeiro. *Aspectos de la ciencia aplicada en la España del Siglo de Oro*. [España]: Junta de Castilla y León, Consejería de Cultura y Bienestar Social, 1991.

Vida, Marco Girolamo. *Cremonensium Orationes III adversus Papienses in Controversia Principatus*. Cremonæ: Giovanni Muzio e Bernardino Locheta, 1550.

Viganò, Marino. "Parente et alievo del già messer Janello." In *Leonardo Turriano, inge-niero del rey*, edited by Alicia Cámara Muñoz, Rafael Moreira, and Marino Viganò, 203-27. Madrid: Fundación Juanelo Turriano, 2010.

Vigo, Giovanni. "Il volto economico della città." In *Storia di Cremona*, edited by Giorgio Politi, 4, L'età degli Asburgo di Spagna: 1535-1707:350-415. Bergamo: Bolis, 2006.

Viola, Pietro. *Petri Violae,... de Veteri novaque Romanorum temporum ratione libellus.* Venetiis: impressum apud N. de Bascarinis, 1546.

Virgil. *The Works of Virgil.* Edited by Malcolm Campbell. New-York: Printed for E. Duyckinck, 1803.

Vitruvius Pollio, Marco. *I Dieci libri dell'Architettura di M. Vitruvio, tradotti et commentati da Monsig. Daniel Barbaro eletto Patriarca d'Aquileia, da lui riueduti & ampliati; & hora in piu commoda forma ridotti.* Venetia: appresso Francesco de'Franceschi Senese et Giovanni Chrieger Alemano compagni, 1567.

Vitruvius Pollio, Marco. *The Ten Books on Architecture.* Translated by Morris H. Morgan. De Architectura. Cambridge, etc.,etc.: Harvard university press, 1919.

Viviani, Vincenzio. "Racconto istorico della vita del sig. Galileo Galilei, nobil fiorentino." In *Fasti consolari dell'Accademia Fiorentina*, edited by Salvino Salvini, 397-431. Firenze: nella stamperia di S.A.R., per Gio. Gaetano Tartini, e Santi Franchi, 1717.

Wahlen, Auguste. *Nouveau dictionnaire de la conversation; ou, Répertoire universel... sur le plan du Conversation's lexicon ... Par une Société de Littérateurs, de Savants et d' Artistes ...* Bruxelles: Librairie-Historique-Artistique, 1842.

Walker, Christopher, and Elena Joli. *L'astronomia prima del telescopio.* Bari: Dedalo, 1997.

Warnke, Martin. *Bau und Überbau: Soziologie der mittelalterlichen Architektur nach den Schriftquellen.* Frankfurt am Main: Syndikat, 1976.

Warnke, Martin. *Der Hofkünstler: zur Vorgeschichte des modernen Künstlers.* Köln: DuMont, 1985.

Weissman, Ronald F.E. "Taking Patronage Seriously: Mediterranean Values and Renaissance Society." In *Patronage, Art, and Society in Renaissance Italy*, edited by Francis W. Kent and Patricia Simons, 25-45. Canberra: Huanities Research Centre/ Clarendon Press, 1987.

Westfall, Carroll W. "Painting and the Liberal Arts : Alberti's View." *Journal of the History of Ideas* 30, no. 4 (1969): 487-506.

White, Lynn Townsend. "Medical Astrologers and Late Medieval Technology." *Viator*, 1975, 295-308.

Williamson, George Charles. *Stories of an Expert.* London: H. Jenkins Ltd., 1925.

Wittkower, Rudolf, and Margot Wittkower. *Nati sotto Saturno: la figura dell'artista dall'an-tichità alla Rivoluzione francese.* Torino: Einaudi, 1996.

Wixom, William D. "A Missal for a King: A First Exhibition ; an Introduction to the Gotha Missal and a Catalogue to the Exhibition Held at the Cleveland Museum of Art, August 8 through September 15, 1963." *The Bulletin of the Cleveland Museum of Art / Cleveland Museum of Art.* 50, no. 7 (1963).

Woodbury, Robert S. *History of the Gear-Cutting Machine: A Historical Study in Geometry and Machines*. Cambridge, Mass.: Technology Press, Massachusetts Institute of Technology, 1958.

Woodward, William Harrison. *La pedagogia del Rinascimento, 1400-1600*. Firenze: Vallecchi, 1923.

Yun Casalilla, Bartolomé. "Economical Cycles and Structural Changes." In *Handbook of European History, 1400-1600: Late Middle Ages, Renaissance, and Reformation*, edited by Thomas A. Brady, Heiko Augustinus Oberman, and James D. Tracy, 1:377-411. Leiden: Brill, 1994.

Yun Casalilla, Bartolomé. "Misurazioni e decisioni: la storia economica dell'Europa preindustriale oggi." *Studi storici. Rivista trimestrale dell'Istituto Gramsci* 50, no. 3 (2009): 581-605.

Zaist, Giambattista. *Notizie istoriche de' pittori, scultori, ed architetti cremonesi*. 2 vols. Cremona: Nella Stamperia di Pietro Ricchini, 1774.

Zambelli, Paola. *White Magic, Black Magic in the European Renaissance*. Leiden; Boston: Brill, 2007.

Zamboni, Silla. "Campi, Bernardino." *Dizionario biografico degli Italiani*. Roma: Istituto della Enciclopedia italiana, 1974.

Zamponi, Stefano. "'Exemplaria', manoscritti con indicazioni di pecia e liste di tassazione di opere giuridiche." In *La Production du livre universitaire au Moyen Âge: exemplar et pecia : actes du symposium de Grottaferrata, mai 1983*, edited by Louis-Jacques Bataillon, Bertrand Georges Guyot, and Richard H. Rouse, 125-32. Paris: Éditions du CNRS, 1988.

Zamponi, Stefano, ed. "Manoscritti con indicazioni di pecia nell'Archivio Capitolare di Pistoia." In *Università e società nei secoli XII-XVI: nono Convegno internazionale : Pistoia, 20-25 settembre 1979*, 447-84. Pistoia: Centro italiano di studi di storia e d'arte, 1982.

Zanetti, Cristiano, ed. *Janello Torriani, a Renaissance Genius*. Cremona: Comune di Cremona, 2016.

Zanetti, Cristiano, ed. *Janello Torriani, genio del Rinascimento*. Cremona: Comune di Cremona, 2016.

Zanetti, Cristiano, ed. "Atti della conferenza Janello Torriani, genio del Rinascimento." *Voce di Hora*, forthcoming 2017.

Zanetti, Cristiano. *Juanelo Turriano, de Cremona a la corte: formación y red social de un ingenio del Renacimiento*. Colección Juanelo Turriano de historia de la ingenieria. Madrid: Fundación Juanelo Turriano, 2015.

Zanetti, Cristiano. "The Microcosm: Technological Innovation and Transfer of Mechanical Knowledge in Sixteenth-Century Habsburg Empire." *History of Technology* 32, no. III (2014): 35-65.

Zanetti, Cristiano. "Videmus nunc per speculum in aenigmate : ¿y si además miramos con una lupa? La biografía en la Historia de la Ciencia y de la Tecnología." In *La historia biográfica en Europa: nuevas perspectivas*, edited by Isabel Burdiel and Roy Foster, 119-44. Historia global 7. Zaragoza: Institución Fernando el Católico, 2015.

Zangheri, Luigi. *Pratolino: il giardino delle meraviglie*. Seconda edizione con aggiunta di disegni e tavole. Firenze: Gonnelli, 1987.

Zanker, Paul. *The Power of Images in the Age of Augustus*. Ann Arbor: University of Michigan Press, 2002.

Zanoni, Felice. "Un brevetto pontificio d'invenzioni del 500: Janello Torriano e un documento dell'Archivio segreto vaticano." *Bollettino Storico Cremonese* X, no. settembre-dicembre (1940): 145-53.

Zapata de Chaves, Luis. *Obra completa de Luis Zapata de Chaves (1526-1595)*. Badajoz: Institución Cultural "Pedro de Valencia," 1979.

Zenocarus Snouckaert, Gulielmus. *De republica, vita ... imperatoris, Caesaris, Augusti, Quinti, Caroli ...* Gandavi: excudebat Gislenus Manilius tipographus, 1559.

Zilsel, Edgar, and Joseph Needham. *The Social Origins of Modern Science*. Edited by Diederick Raven, Wolfgang Krohn, and R. S Cohen. 1942 as The Sociological Roots of Science. Dordrecht; Boston: Kluwer Academic Publishers, 2000.

Zinner, Ernst. *Deustche und niederländische astronomische Instrumente des 11.-18. Jahrhunderts, von Ernst Zinner*. München: C.H. Beck, 1956.

Zofío Llorente, Juan Carlos. "Trabajo y socialización: los aprendices en Madrid durante la segunda mitad del siglo XVI." In *Madrid, Felipe II y las ciudades de la monarquía*, edited by Enrique Martínez Ruiz, II, Capitalismo y economía:521-35. Madrid: Actas, 2000.

Zuccolin, Gabriella. "Le figure sanitarie: secoli XIV-XVII." In *I professionisti a Cremona: eventi e figure di una storia centenaria*, edited by Valeria Leoni and Matteo Morandi, 96-97. Cremona: Libreria del Convegno, 2011.

Zuffi, Stefano. *European art of the fifteenth century*. Los Angeles: J. Paul Getty Museum, 2005.

Zulueta, Julian de. "The cause of death of Emperor Charles V." *Parassitologia* 49, no. 1/2 (2007): 107-9.

Index

Printed in the United States
By Bookmasters